DISCOVERING COMPUTER SCIENCE

Interdisciplinary Problems, Principles, and Python Programming

CHAPMAN & HALL/CRC
TEXTBOOKS IN COMPUTING

Series Editors

John Impagliazzo
Professor Emeritus, Hofstra University

Andrew McGettrick
Department of Computer
and Information Sciences
University of Strathclyde

Aims and Scope

This series covers traditional areas of computing, as well as related technical areas, such as software engineering, artificial intelligence, computer engineering, information systems, and information technology. The series will accommodate textbooks for undergraduate and graduate students, generally adhering to worldwide curriculum standards from professional societies. The editors wish to encourage new and imaginative ideas and proposals, and are keen to help and encourage new authors. The editors welcome proposals that: provide groundbreaking and imaginative perspectives on aspects of computing; present topics in a new and exciting context; open up opportunities for emerging areas, such as multi-media, security, and mobile systems; capture new developments and applications in emerging fields of computing; and address topics that provide support for computing, such as mathematics, statistics, life and physical sciences, and business.

Published Titles

Paul Anderson, Web 2.0 and Beyond: Principles and Technologies

Henrik Bærbak Christensen, Flexible, Reliable Software: Using Patterns and Agile Development

John S. Conery, Explorations in Computing: An Introduction to Computer Science

John S. Conery, Explorations in Computing: An Introduction to Computer Science and Python Programming

Jessen Havill, Discovering Computer Science: Interdisciplinary Problems, Principles, and Python Programming

Ted Herman, A Functional Start to Computing with Python

Pascal Hitzler, Markus Krötzsch, and Sebastian Rudolph, Foundations of Semantic Web Technologies

Mark J. Johnson, A Concise Introduction to Data Structures using Java

Mark J. Johnson, A Concise Introduction to Programming in Python

Lisa C. Kaczmarczyk, Computers and Society: Computing for Good

Mark C. Lewis, Introduction to the Art of Programming Using Scala

Bill Manaris and Andrew R. Brown, Making Music with Computers: Creative Programming in Python

Uvais Qidwai and C.H. Chen, Digital Image Processing: An Algorithmic Approach with MATLAB®

David D. Riley and Kenny A. Hunt, Computational Thinking for the Modern Problem Solver

Henry M. Walker, The Tao of Computing, Second Edition

CHAPMAN & HALL/CRC
TEXTBOOKS IN COMPUTING

DISCOVERING COMPUTER SCIENCE

Interdisciplinary Problems, Principles, and Python Programming

Jessen Havill

Denison University
Granville, Ohio, USA

CRC Press
Taylor & Francis Group
Boca Raton London New York

CRC Press is an imprint of the
Taylor & Francis Group, an **informa** business

A CHAPMAN & HALL BOOK

CRC Press
Taylor & Francis Group
6000 Broken Sound Parkway NW, Suite 300
Boca Raton, FL 33487-2742

© 2016 by Taylor & Francis Group, LLC
CRC Press is an imprint of Taylor & Francis Group, an Informa business

No claim to original U.S. Government works

Printed on acid-free paper
Version Date: 20150318

International Standard Book Number-13: 978-1-4822-5414-3 (Pack - Book and Ebook)

This book contains information obtained from authentic and highly regarded sources. Reasonable efforts have been made to publish reliable data and information, but the author and publisher cannot assume responsibility for the validity of all materials or the consequences of their use. The authors and publishers have attempted to trace the copyright holders of all material reproduced in this publication and apologize to copyright holders if permission to publish in this form has not been obtained. If any copyright material has not been acknowledged please write and let us know so we may rectify in any future reprint.

Except as permitted under U.S. Copyright Law, no part of this book may be reprinted, reproduced, transmitted, or utilized in any form by any electronic, mechanical, or other means, now known or hereafter invented, including photocopying, microfilming, and recording, or in any information storage or retrieval system, without written permission from the publishers.

For permission to photocopy or use material electronically from this work, please access www.copyright.com (http://www.copyright.com/) or contact the Copyright Clearance Center, Inc. (CCC), 222 Rosewood Drive, Danvers, MA 01923, 978-750-8400. CCC is a not-for-profit organization that provides licenses and registration for a variety of users. For organizations that have been granted a photocopy license by the CCC, a separate system of payment has been arranged.

Trademark Notice: Product or corporate names may be trademarks or registered trademarks, and are used only for identification and explanation without intent to infringe.

Library of Congress Cataloging-in-Publication Data

Havill, Jessen.
   Discovering computer science : interdisciplinary problems, principles, and Python programming / author, Jessen Havill.
      pages cm -- (Chapman & Hall/CRC textbooks in computing ; 15)
   Includes bibliographical references and index.
   ISBN 978-1-4822-5414-3 (alk. paper)
  1. Computer science--Textbooks. 2. Python (Computer program language)--Textbooks. I. Title.

QA76.H3735 2015
005.13'3--dc23
                                      2015004805

Visit the Taylor & Francis Web site at
http://www.taylorandfrancis.com

and the CRC Press Web site at
http://www.crcpress.com

Contents

Preface   xv
Acknowledgments   xxiii
About the author   xxv

CHAPTER 1 ▪ What is computation?   1
   1.1 PROBLEMS AND ABSTRACTION   2
   1.2 ALGORITHMS AND PROGRAMS   4
   1.3 EFFICIENT ALGORITHMS   11
        Organizing a phone tree   11
        A smoothing algorithm   13
        A better smoothing algorithm 17
   1.4 COMPUTERS ARE DUMB   20
        Inside a computer   20
        Machine language   21
        Everything is bits   22
        The universal machine   26
   1.5 SUMMARY   29
   1.6 FURTHER DISCOVERY   30

CHAPTER 2 ▪ Elementary computations   31
   2.1 WELCOME TO THE CIRCUS   31
   2.2 ARITHMETIC   32
        Finite precision   34
        Division   34
        Order of operations   35
        Complex numbers   37
   2.3 WHAT'S IN A NAME?   38
   2.4 USING FUNCTIONS   45

		Built-in functions	45
		Strings	47
		Modules	51
*2.5		BINARY ARITHMETIC	54
		Finite precision	55
		Negative integers	56
		Designing an adder	57
		Implementing an adder	58
2.6		SUMMARY	62
2.7		FURTHER DISCOVERY	62

CHAPTER 3 ▪ Visualizing abstraction — 65

3.1	DATA ABSTRACTION	66
3.2	VISUALIZATION WITH TURTLES	70
	Drawing with iteration	72
3.3	FUNCTIONAL ABSTRACTION	76
	Function parameters	78
	Let's plant a garden	84
3.4	PROGRAMMING IN STYLE	89
	Program structure	89
	Documentation	91
	Descriptive names and magic numbers	95
3.5	A RETURN TO FUNCTIONS	97
	Return vs. print	100
3.6	SCOPE AND NAMESPACES	103
	Local namespaces	104
	The global namespace	107
3.7	SUMMARY	111
3.8	FURTHER DISCOVERY	112

CHAPTER 4 ▪ Growth and decay — 113

4.1	DISCRETE MODELS	114
	Managing a fishing pond	114
	Measuring network value	121
	Organizing a concert	124
4.2	VISUALIZING POPULATION CHANGES	136
4.3	CONDITIONAL ITERATION	140

*4.4	CONTINUOUS MODELS		145
	Difference equations		145
	Radiocarbon dating		148
	Tradeoffs between accuracy and time		150
	Propagation of errors		152
	Simulating an epidemic		153
*4.5	NUMERICAL ANALYSIS		159
	The harmonic series		159
	Approximating π		162
	Approximating square roots		164
4.6	SUMMING UP		167
4.7	FURTHER DISCOVERY		171
4.8	PROJECTS		171
	Project 4.1	Parasitic relationships	171
	Project 4.2	Financial calculators	173
	*Project 4.3	Market penetration	177
	*Project 4.4	Wolves and moose	180

CHAPTER 5 ■ Forks in the road — 185

5.1	RANDOM WALKS	185
	A random walk in Monte Carlo	192
	Histograms	195
*5.2	PSEUDORANDOM NUMBER GENERATORS	200
	Implementation	201
	Testing randomness	203
*5.3	SIMULATING PROBABILITY DISTRIBUTIONS	205
	The central limit theorem	206
5.4	BACK TO BOOLEANS	209
	Short circuit evaluation	212
	Complex expressions	214
	*Using truth tables	216
	Many happy returns	218
5.5	A GUESSING GAME	224
5.6	SUMMARY	233
5.7	FURTHER DISCOVERY	234
5.8	PROJECTS	234

| | | Project 5.1 The magic of polling | 234 |
| | | Project 5.2 Escape! | 237 |

CHAPTER 6 ▪ Text, documents, and DNA — 241

	6.1	COUNTING WORDS	242
	6.2	TEXT DOCUMENTS	250
		Reading from text files	250
		Writing to text files	253
		Reading from the web	254
	6.3	ENCODING STRINGS	259
		Indexing and slicing	259
		Creating modified strings	261
		Encoding characters	263
	6.4	LINEAR-TIME ALGORITHMS	270
		Asymptotic time complexity	274
	6.5	ANALYZING TEXT	279
		Counting and searching	279
		A concordance	284
	6.6	COMPARING TEXTS	289
*6.7		GENOMICS	297
		A genomics primer	297
		Basic DNA analysis	301
		Transforming sequences	302
		Comparing sequences	304
		Reading sequence files	306
	6.8	SUMMARY	312
	6.9	FURTHER DISCOVERY	313
	6.10	PROJECTS	313
		Project 6.1 Polarized politics	313
		*Project 6.2 Finding genes	316

CHAPTER 7 ▪ Designing programs — 321

	7.1	HOW TO SOLVE IT	322
		Understand the problem	323
		Design an algorithm	324
		Implement your algorithm as a program	327
		Analyze your program for clarity, correctness, and efficiency	330

*7.2	DESIGN BY CONTRACT		331
	Preconditions and postconditions		331
	Checking parameters		332
	Assertions		334
*7.3	TESTING		340
	Unit testing		340
	Regression testing		342
	Designing unit tests		343
	Testing floating point values		347
7.4	SUMMARY		350
7.5	FURTHER DISCOVERY		350

CHAPTER 8 ■ Data analysis   351

8.1	SUMMARIZING DATA	351
8.2	CREATING AND MODIFYING LISTS	360
	List accumulators, redux	360
	Lists are mutable	361
	Tuples	365
	List operators and methods	366
	*List comprehensions	368
8.3	FREQUENCIES, MODES, AND HISTOGRAMS	373
	Tallying values	373
	Dictionaries	374
8.4	READING TABULAR DATA	384
*8.5	DESIGNING EFFICIENT ALGORITHMS	390
	A first algorithm	391
	A more elegant algorithm	399
	A more efficient algorithm	400
*8.6	LINEAR REGRESSION	403
*8.7	DATA CLUSTERING	409
	Defining similarity	410
	A k-means clustering example	411
	Implementing k-means clustering	414
	Locating bicycle safety programs	416
8.8	SUMMARY	421
8.9	FURTHER DISCOVERY	421

8.10	PROJECTS	422
	Project 8.1 Climate change	422
	Project 8.2 Does education influence unemployment?	425
	Project 8.3 Maximizing profit	427
	Project 8.4 Admissions	428
	*Project 8.5 Preparing for a 100-year flood	430
	Project 8.6 Voting methods	435
	Project 8.7 Heuristics for traveling salespeople	438

CHAPTER 9 ▪ Flatland 443

9.1	TWO-DIMENSIONAL DATA	443
9.2	THE GAME OF LIFE	449
	Creating a grid	451
	Initial configurations	452
	Surveying the neighborhood	453
	Performing one pass	454
	Updating the grid	457
9.3	DIGITAL IMAGES	461
	Colors	461
	Image filters	463
	Transforming images	467
9.4	SUMMARY	471
9.5	FURTHER DISCOVERY	471
9.6	PROJECTS	471
	Project 9.1 Modeling segregation	471
	Project 9.2 Modeling ferromagnetism	473
	Project 9.3 Growing dendrites	474

CHAPTER 10 ▪ Self-similarity and recursion 477

10.1	FRACTALS	477
	A fractal tree	479
	A fractal snowflake	481
10.2	RECURSION AND ITERATION	488
	Solving a problem recursively	491
	Palindromes	492
	Guessing passwords	495
10.3	THE MYTHICAL TOWER OF HANOI	500

		*Is the end of the world nigh?	502
10.4	RECURSIVE LINEAR SEARCH		503
		Efficiency of recursive linear search	505
10.5	DIVIDE AND CONQUER		508
		Buy low, sell high	508
		Navigating a maze	512
*10.6	LINDENMAYER SYSTEMS		518
		Formal grammars	518
		Implementing L-systems	522
10.7	SUMMARY		525
10.8	FURTHER DISCOVERY		526
10.9	PROJECTS		526
		*Project 10.1 Lindenmayer's beautiful plants	526
		Project 10.2 Gerrymandering	531
		Project 10.3 Percolation	536

CHAPTER 11 ■ Organizing data — 541

11.1	BINARY SEARCH		542
		Efficiency of iterative binary search	546
		A spelling checker	548
		Recursive binary search	549
		Efficiency of recursive binary search	550
11.2	SELECTION SORT		553
		Implementing selection sort	553
		Efficiency of selection sort	557
		Querying data	558
11.3	INSERTION SORT		563
		Implementing insertion sort	564
		Efficiency of insertion sort	566
11.4	EFFICIENT SORTING		570
		Internal vs. external sorting	574
		Efficiency of merge sort	574
*11.5	TRACTABLE AND INTRACTABLE ALGORITHMS		577
		Hard problems	579
11.6	SUMMARY		580
11.7	FURTHER DISCOVERY		581

11.8	PROJECTS	581
	Project 11.1 Creating a searchable database	581
	Project 11.2 Binary search trees	581

CHAPTER 12 ▪ Networks 587

12.1	MODELING WITH GRAPHS	588
	Making friends	590
12.2	SHORTEST PATHS	594
	Finding the actual paths	598
12.3	IT'S A SMALL WORLD...	601
	Clustering coefficients	603
	Scale-free networks	605
12.4	RANDOM GRAPHS	608
12.5	SUMMARY	611
12.6	FURTHER DISCOVERY	611
12.7	PROJECTS	612
	Project 12.1 Diffusion of ideas and influence	612
	Project 12.2 Slowing an epidemic	614
	Project 12.3 The Oracle of Bacon	616

CHAPTER 13 ▪ Abstract data types 621

13.1	DESIGNING CLASSES	622
	Implementing a class	625
	Documenting a class	632
13.2	OPERATORS AND SPECIAL METHODS	637
	String representations	637
	Arithmetic	638
	Comparison	640
	Indexing	642
13.3	MODULES	645
	Namespaces, redux	646
13.4	A FLOCKING SIMULATION	648
	The World ADT	649
	The Boid ADT	655
13.5	A STACK ADT	665
13.6	A DICTIONARY ADT	671
	Hash tables	672

		Implementing a hash table	673
		Implementing indexing	676
		ADTs vs. data structures	678
13.7	SUMMARY		682
13.8	FURTHER DISCOVERY		682
13.9	PROJECTS		683
		Project 13.1 Tracking GPS coordinates	683
		Project 13.2 Economic mobility	687
		Project 13.3 Slime mold aggregation	690
		Project 13.4 Boids in space	692

APPENDIX A ▪ Installing Python — 697

A.1	AN INTEGRATED DISTRIBUTION	697
A.2	MANUAL INSTALLATION	697

APPENDIX B ▪ Python library reference — 701

B.1	MATH MODULE	701
B.2	TURTLE METHODS	702
B.3	SCREEN METHODS	703
B.4	MATPLOTLIB.PYPLOT MODULE	704
B.5	RANDOM MODULE	704
B.6	STRING METHODS	705
B.7	LIST METHODS	706
B.8	IMAGE MODULE	706
B.9	SPECIAL METHODS	707

Bibliography 709

Index 713

Preface

IN my view, an introductory computer science course should strive to accomplish three things. First, it should demonstrate to students how computing has become a powerful mode of inquiry, and a vehicle of discovery, in a wide variety of disciplines. This orientation is also inviting to students of the natural and social sciences, who increasingly benefit from an introduction to computational thinking, beyond the limited "black box" recipes often found in manuals. Second, the course should engage students in computational problem solving, and lead them to discover the power of abstraction, efficiency, and data organization in the design of their solutions. Third, the course should teach students how to implement their solutions as computer programs. In learning how to program, students more deeply learn the core principles, and experience the thrill of seeing their solutions come to life.

Unlike most introductory computer science textbooks, which are organized around programming language constructs, I deliberately lead with interdisciplinary problems and techniques. This orientation is more interesting to a more diverse audience, and more accurately reflects the role of programming in problem solving and discovery. A computational discovery does not, of course, originate in a programming language feature in search of an application. Rather, it starts with a compelling problem which is modeled and solved algorithmically, by leveraging abstraction and prior experience with similar problems. Only then is the solution implemented as a program.

Like most introductory computer science textbooks, I introduce programming skills in an incremental fashion, and include many opportunities for students to practice them. The topics in this book are arranged to ease students into computational thinking, and encourage them to incrementally build on prior knowledge. Each chapter focuses on a general class of problems that is tackled by new algorithmic techniques and programming language features. My hope is that students will leave the course, not only with strong programming skills, but with a set of problem solving strategies and simulation techniques that they can apply in their future work, whether or not they take another computer science course.

I use Python to introduce computer programming for two reasons. First, Python's intuitive syntax allows students to focus on interesting problems and powerful principles, without unnecessary distractions. Learning how to think algorithmically is hard enough without also having to struggle with a non-intuitive syntax. Second, the expressiveness of Python (in particular, low-overhead lists and dictionaries) expands tremendously the range of accessible problems in the introductory course. Teaching with Python over the last ten years has been a revelation; introductory computer science has become fun again.

Web resources

The text, exercises, and projects often refer to files on the book's accompanying web site, which can be found at

http://discoverCS.denison.edu .

This web site also includes pointers for further exploration, links to additional documentation, and errata.

To students

Learning how to solve computational problems and implement them as computer programs requires daily practice. Like an athlete, you will get out of shape and fall behind quickly if you skip it. There are no shortcuts. Your instructor is there to help, but he or she cannot do the work for you.

With this in mind, it is important that you type in and try the examples throughout the text, and then go beyond them. Be curious! There are numbered "Reflection" questions throughout the book that ask you to stop and think about, or apply, something that you just read. Often, the question is answered in the book immediately thereafter, so that you can check your understanding, but peeking ahead will rob you of an important opportunity.

There are many opportunities to delve into topics more deeply. Boxes scattered throughout the text briefly introduce related, but more technical, topics. For the most part, these are not strictly required to understand what comes next, but I encourage you to read them anyway. In the "Further discovery" section of each chapter, you can find additional pointers to explore chapter topics in more depth.

At the end of most sections are several programming exercises that ask you to further apply concepts from that section. Often, the exercises assume that you have already worked through all of the examples in that section. All later chapters conclude with a selection of more involved interdisciplinary projects that you may be asked by your instructor to tackle.

The book assumes no prior knowledge of computer science. However, it does assume a modest comfort with high school algebra and mathematical functions. Occasionally, trigonometry is mentioned, as is the idea of convergence to a limit, but these are not crucial to an understanding of the main topics in this book.

To instructors

This book may be appropriate for a traditional CS1 course for majors, a CS0 course for non-majors (at a slower pace and omitting more material), or an introductory computing course for students in the natural and/or social sciences.

As suggested above, I emphasize computer science principles and the role of abstraction, both functional and data, throughout the book. I motivate functions as implementations of functional abstractions, and point out that strings, lists, and dictionaries are all abstract data types that allow us to solve more interesting problems than would otherwise be possible. I introduce the idea of time complexity

Preface ■ xvii

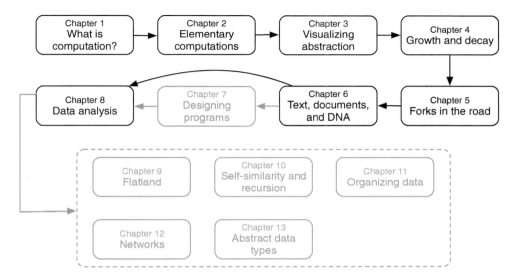

Figure 1 An overview of chapter dependencies.

intuitively, without formal definitions, in the first chapter and return to it several times as more sophisticated algorithms are developed. The book uses a spiral approach for many topics, returning to them repeatedly in increasingly complex contexts. Where appropriate, I also weave into the book topics that are traditionally left for later computer science courses. A few of these are presented in boxes that may be covered at your discretion. None of these topics is introduced rigorously, as they would be in a data structures course. Rather, I introduce them informally and intuitively to give students a sense of the problems and techniques used in computer science. I hope that the tables below will help you navigate the book, and see where particular topics are covered.

This book contains over 600 end-of-section exercises and over 300 in-text reflection questions that may be assigned as homework or discussed in class. At the end of most chapters is a selection of projects (about 30 in all) that students may work on independently or in pairs over a longer time frame. I believe that projects like these are crucial for students to develop both problem solving skills and an appreciation for the many fascinating applications of computer science.

Because this book is intended for a student who may take additional courses in computer science and learn other programming languages, I intentionally omit some features of Python that are not commonly found elsewhere (e.g., simultaneous swap, chained comparisons, **enumerate** in **for** loops). You may, of course, supplement with these additional syntactical features.

There is more in this book than can be covered in a single semester, giving an instructor the opportunity to tailor the content to his or her particular situation and interests. Generally speaking, as illustrated in Figure 1, Chapters 1–6 and 8 form the core of the book, and should be covered sequentially. The remaining chapters can be covered, partially or entirely, at your discretion, although I would expect that most instructors will cover at least parts of Chapters 7, 10, 11, and 13. Chapter 7 contains additional material on program design, including design by contract, assertions and unit testing that may be skipped without consequences for later chapters. Chapters 9–13 are, more or less, independent of each other. Sections marked with an asterisk are optional, in the sense that they are not assumed for future sections in that chapter. When projects depend on optional sections, they are also marked with an asterisk, and the dependency is stated at the beginning of the project.

Chapter outlines

The following tables provide brief overviews of each chapter. Each table's three columns, reflecting the three parts of the book's subtitle, provide three lenses through which to view the chapter. The first column lists a selection of representative *problems* that are used to motivate the material. The second column lists computer science *principles* that are introduced in that chapter. Finally, the third column lists Python *programming* topics that are either introduced or reinforced in that chapter to implement the principles and/or solve the problems.

Chaper 1. What is computation?

Sample problems	Principles	Programming
• digital music • search engines • GPS devices • smoothing data • phone trees	• problems, input/output • abstraction • algorithms and programs • computer architecture • binary representations • time complexity • Turing machine	—

Chapter 2. Elementary computations

Sample problems	Principles	Programming
• wind chill • geometry • compounding interest • Mad Libs	• finite precision • names as references • using functional abstractions • binary addition	• `int` and `float` numeric types • arithmetic and the `math` module • variable names and assignment • calling built-in functions • using strings, + and * operators • `print` and `input`

Chapter 3. Visualizing abstraction

Sample problems	Principles	Programming
• visualizing an archaeological dig • random walks • ideal gas • groundwater flow • demand functions	• using abstract data types • creating functional abstractions • basic functional decomposition	• using classes and objects • `turtle` module • basic `for` loops • writing functions • namespaces • docstrings and comments

Chapter 4. Growth and decay

Sample problems	Principles	Programming

• network value
• demand and profit
• loans and investing
• bacterial growth
• radiocarbon dating
• diffusion models
  – SIR, SIS, Bass
• competition models
  – Nicholson-Bailey
  – Lotka-Volterra
  – indirect
	• accumulators
	• list accumulators
	• difference equations
	• approximating continuous models
	• accuracy vs. time
	• error propagation
	• numerical approximation
	• classes of growth
		• `for` loops
		• format strings
		• `range`
		• `matplotlib`
		• appending to lists
		• `while` loops

Chapter 5. Forks in the road

Sample problems	Principles	Programming
• random walks • guessing games • polling and sampling • particle escape	• Monte Carlo simulation • pseudorandom number generators • simulating probabilities • flag variables • using uniform and normal distributions • DeMorgan's laws	• `random` module • `if/elif/else` • comparison operators • Boolean operators • `matplotlib` histograms • `while` loops

Chapter 6. Text, documents, and DNA

Sample problems	Principles	Programming
• word count • textual analysis • parsing XML • checksums • concordances • detecting plagiarism • congressional votes • genomics	• ASCII, Unicode • linear-time algorithms • asymptotic time complexity • linear search • dot plots • string accumulators	• `str` class and methods • iterating over strings • indexing and slices • iterating over indices • reading and writing text files • nested loops

Chapter 7. Designing programs

Sample problems	Principles	Programming
• word frequency analysis	• problem solving • top-down design • pre and postconditions • assertions • unit testing	• `assert` statement • conditional execution of `main` • writing modules

Chapter 8. Data analysis

Sample problems	Principles	Programming
• 100-year floods	• histograms	• `list` class
• traveling salesman	• hash tables	• iterating over lists
• Mohs scale	• tabular data files	• indexing and slicing
• meteorite sites	• efficient algorithms	• list operators and methods
• zebra migration	• linear regression	• lists in memory; mutability
• tumor diagnosis	• k-means clustering	• list parameters
• education levels	• heuristics	• tuples
• supply and demand		• list comprehensions
• voting methods		• dictionaries

Chapter 9. Flatland

Sample problems	Principles	Programming
• earthquake data	• 2-D data	• 2-D data in list of lists
• Game of Life	• cellular automata	• nested loops
• image filters	• digital images	• 2-D data in a dictionary
• racial segregation	• color models	
• ferromagnetism		
• dendrites		

Chapter 10. Self-similarity and recursion

Sample problems	Principles	Programming
• fractals	• self-similarity	• writing recursive functions
• cracking passwords	• recursion	
• Tower of Hanoi	• linear search	
• maximizing profit	• recurrence relations	
• path through a maze	• divide and conquer	
• Lindenmayer system	• depth-first search	
• electoral districting	• grammars	
• percolation		

Chapter 11. Organizing data

Sample problems	Principles	Programming
• spell check	• binary search	• nested loops
• querying data sets	• recurrence relations	• writing recursive functions
	• basic sorting algorithms	
	• quadratic-time algorithms	
	• parallel lists	
	• merge sort	
	• intractability	
	• P=NP (intuition)	
	• Moore's law	
	• binary search trees	

Chapter 12. Networks

Sample problems	Principles	Programming
• Facebook, Twitter, web graphs • diffusion of ideas • epidemics • Oracle of Bacon	• graphs • adjacency list • adjacency matrix • breadth-first search • distance and shortest paths • depth-first search • small-world networks • scale-free networks • clustering coefficient • uniform random graphs	• dictionaries

Chapter 13. Abstract data types

Sample problems	Principles	Programming
• data sets • genomic sequences • rational numbers • flocking behavior • slime mold aggregation	• abstract data types • data structures • stacks • hash tables • agent-based simulation • swarm intelligence	• writing classes • special methods • overriding operators • modules

Software assumptions

To follow along in this book and complete the exercises, you will need to have installed Python 3.4 (or later) on your computer, and have access to IDLE or another programming environment. The book also assumes that you have installed the `matplotlib` and `numpy` modules. Please refer to Appendix A for more information.

Errata

While I (and my students) have ferreted out many errors, readers will inevitably find more. You can find an up-to-date list of errata on the book web site. If you find an error in the text or have another suggestion, please let me know at `havill@denison.edu`.

Acknowledgments

I was extraordinarily naïve when I embarked on this project two years ago. "How hard can it be to put these ideas into print?" Well, much harder than I thought, as it turns out. I owe debts of gratitude to many who saw me through to the end.

First and foremost, my family not only tolerated me during this period, but offered extraordinary support and encouragement. Thank you Beth, for your patience and strength, and all of the time you have made available to me to work on the book. I am grateful to my in-laws, Roger and Nancy Vincent, who offered me their place in Wyoming for a month-long retreat in the final stretch. And, to my four children, Nick, Amelia, Caroline, and Lillian, I promise to make up for lost time.

My colleagues Matt Kretchmar, Ashwin Lall, and David White used drafts in their classes, and provided invaluable feedback. They have been fantastic sounding boards, and have graciously provided many ideas for exercises and projects. Students in Denison University's CS 111 and 112 classes caught many typos, especially Gabe Schenker, Christopher Castillo, Christine Schmittgen, Alivia Tacheny, Emily Lamm, and Ryan Liedke. Dana Myers read much of the book and offered an abundance of detailed suggestions. Joan Krone also read early chapters and offered constructive feedback. I am grateful to Todd Feil for his support, and his frank advice after reading the almost-final manuscript.

I have benefitted tremendously from many conversations about computational science, geology, and life with my friend and colleague David Goodwin. Project 8.1 is based on an assignment that he has used in his classes. I have also learned a great deal from collaborations with Jeff Thompson. Jeff also advised me on Section 6.7 and Project 6.2. Frank Hassebrock enthusiastically introduced me to theories of problem solving in cognitive psychology. And Dee Ghiloni, the renowned cat herder, has supported me and my work in more ways than I can count.

I am indebted to the following reviewers, who read early chapters and offered expert critiques: Terry Andres (University of Manitoba), John Impagliazzo (Qatar University), Daniel Kaplan (Macalester College), Nathaniel Kell (Duke University), Andrew McGettrick (University of Strathclyde); Hamid Mokhtarzadeh (University of Minnesota), George Novacky (University of Pittsburgh), and J. F. Nystrom (Ferris State University).

I could not have completed this book without the Robert C. Good Fellowship awarded to me by Denison University.

Finally, thank you to Randi Cohen, for believing in this project, and for her advice and patience throughout.

About the author

Jessen Havill is a Professor of Computer Science and the Benjamin Barney Chair of Mathematics at Denison University, where he has been on the faculty since 1998. Dr. Havill teaches courses across the computer science curriculum, as well as an interdisciplinary elective in computational biology. He was awarded the college's highest teaching honor, the Charles A. Brickman Teaching Excellence Award, in 2013.

Dr. Havill is also an active researcher, with a primary interest in the development and analysis of online algorithms. In addition, he has collaborated with colleagues in biology and geosciences to develop computational tools to support research and teaching in those fields. Dr. Havill earned his bachelor's degree from Bucknell University and his Ph.D. in computer science from The College of William and Mary.

CHAPTER 1

What is computation?

> We need to do away with the myth that computer science is about computers. Computer science is no more about computers than astronomy is about telescopes, biology is about microscopes or chemistry is about beakers and test tubes. Science is not about tools, it is about how we use them and what we find out when we do.
>
> Michael R. Fellows and Ian Parberry
> *Computing Research News (1993)*

COMPUTERS are the most powerful tools ever invented, but not because of their versatility and speed, per se. Computers are powerful because they empower *us* to innovate and make unprecedented discoveries.

A computer is a machine that carries out a *computation*, a sequence of simple steps that transforms some initial information, an *input*, into some desired result, the *output*. Computer scientists harness the power of computers to solve complex problems by designing solutions that can be expressed as computations. The output of a computation might be a more efficient route for a spacecraft, a more effective protocol to control an epidemic, or a secret message hidden in a digital photograph.

Computer science has always been interdisciplinary, as computational problems arise in virtually every domain imaginable. Social scientists use computational models to better understand social networks, epidemics, population dynamics, markets, and auctions. Scholars working in the digital humanities use computational tools to curate and analyze classic literature. Artists are increasingly incorporating digital technologies into their compositions and performances. Computational scientists work in areas related to climate prediction, genomics, particle physics, neuroscience, and drug discovery.

In this book, we will explore the fundamental problem solving techniques of computer science, and discover how they can be used to model and solve a variety of interdisciplinary problems. In this first chapter, we will provide an orientation and lay out the context in which to place the rest of the book. We will further develop all of these ideas throughout, so don't worry if they are not all crystal clear at first.

2 ■ What is computation?

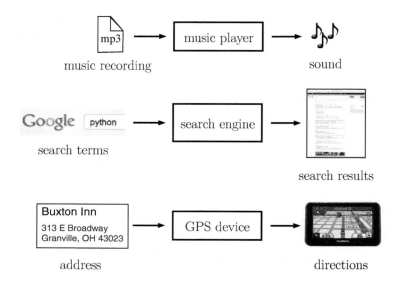

Figure 1.1 Some examples of computational problems.

1.1 PROBLEMS AND ABSTRACTION

Every useful computation solves a *problem* of some sort. A problem is fundamentally defined by the relationship between its input and its output, as illustrated in Figure 1.1. For each problem, we have an input entering on the left and a corresponding output exiting on the right. In between, a computation transforms the input into a correct output. When you listen to a song, the music player performs a computation to convert the digital music file (input) into a sound pattern (output). When you submit a web search request (input), your computer, and many others across the Internet, perform computations to get you results (outputs). And when you use GPS navigation, a device computes directions (output) based on your current position, your destination, and its stored maps (inputs).

Inputs and outputs are probably also familiar to you from high school algebra. When you were given an expression like

$$y = 18x + 31$$

or

$$f(x) = 18x + 31,$$

you may have thought about the variable x as a representation of the input and y, or $f(x)$, as a representation of the output. In this example, when the input is $x = 2$, the output is $y = 67$, or $f(x) = 67$. The arithmetic that turns x into y is a very simple (and boring) example of a computation.

Reflection 1.1 *What kinds of problems are you interested in? What are their inputs and outputs? Are the inputs and outputs, as you have defined them, sufficient to define the problem completely?*

To use the technologies illustrated in Figure 1.1 you do not need to understand how the underlying computation transforms the input to the output; we can think of the computation as a "black box" and still use the technology effectively. We call this idea *functional abstraction*, a very important concept that we often take for granted. Put simply,

> *A functional abstraction describes how to use a tool or technology without necessarily providing any knowledge about how it works.*

We exist in a world of abstractions; we could not function without them. We even think about our own bodies in terms of abstractions. Move your fingers. Did you need to understand how your brain triggered your nervous and musculoskeletal systems to make that happen? As far as most of us are concerned, a car is also an abstraction. To drive a car, do you need to know how turning the steering wheel turns the car or pushing the accelerator makes it go faster? We understand what should happen when we do these things, but not necessarily how they happen. Without abstractions, we would be paralyzed by an avalanche of minutiae.

Reflection 1.2 *Imagine that it was necessary to understand how a GPS device works in order to use it. Or a music player. Or a computer. How would this affect your ability to use these technologies?*

New technologies and automation have introduced new functional abstractions into everyday life. Our food supply is a compelling example of this. Only a few hundred years ago, our ancestors knew exactly where their food came from. Inputs of hard work and suitable weather produced outputs of grain and livestock to sustain a family. In modern times, we input money and get packaged food; the origins of our food have become much more abstract.

Reflection 1.3 *Think about a common functional abstraction that you use regularly, such as your phone or a credit card. How has this functional abstraction changed over time? Can you think of instances in which better functional abstractions have enhanced our ability to use a technology?*

We also use layers of functional abstractions to work more efficiently. For example, suppose you are the president of a large organization (like a university) that is composed of six divisions and 5,000 employees. Because you cannot effectively manage every detail of such a large organization, you assign a vice president to oversee each of the divisions. You expect each VP to keep you informed about the general activity and performance of that division, but insulate you from the day-to-day details. In this arrangement, each division becomes a functional abstraction to you: you know what each division does, but not necessarily how it does it. Benefitting from these abstractions, you are now free to focus your resources on more important organization-level activity. Each VP may utilize a similar arrangement within his or her division. Indeed, organizations are often subdivided many times until the number of employees in a unit is small enough to be overseen by a single manager.

Computers are similarly built from many layers of functional abstractions. When you use a computer, you are presented with a "desktop" abstraction on which you can store files and use applications (i.e., programs) to do work. That there appear to be many applications executing simultaneously is also an abstraction. In reality, some applications may be executing in parallel while others are being executed one at a time, but alternating so quickly that they just appear to be executing in parallel. This interface and the basic management of all of the computer's resources (e.g., hard drives, network, memory, security) is handled by the *operating system*. An operating system is a complicated beast that is often mistakenly described as "the program that is always running in the background on a computer." In reality, the core of an operating system provides several layers of functional abstractions that allow us to use computers more efficiently.

Computer scientists invent computational processes (e.g., search engines, GPS software, and operating systems), that are then packaged as functional abstractions for others to use. But, as we will see, they also harness abstraction to correctly and efficiently solve real-world problems. These problems are usually complicated enough that they must be decomposed into smaller problems that human beings can understand and solve. Once solved, each of these smaller pieces becomes a functional abstraction that can be used in the solution to the original problem.

The ability to think in abstract terms also enables us to see similarities in problems from different domains. For example, by isolating the fundamental operations of sexual reproduction (i.e., mutation and recombination) and natural selection (i.e., survival of the fittest), the process of evolution can be thought of as a randomized computation. From this insight was borne a technique called *evolutionary computation* that has been used to successfully solve thousands of problems. Similarly, a technique known as *simulated annealing* applies insights gained from the process of slow-cooling metals to effectively find solutions to very hard problems. Other examples include techniques based on the behaviors of crowds and insect swarms, the operations of cellular membranes, and how networks of neurons make decisions.

1.2 ALGORITHMS AND PROGRAMS

Every useful computation follows a sequence of steps to transform an input into a correct output. This sequence of steps is called an *algorithm*. To illustrate, let's look at a very simple problem: computing the volume of a sphere. In this case, the input is the radius r of the sphere and the output is the volume of the sphere with radius r. To compute the output from the input, we simply use the well-known formula $V = (4/3)\pi r^3$. This can be visualized as follows.

radius r ⟶ $(4/3)\pi r^3$ ⟶ volume of sphere with radius r

In the box is a representation of the algorithm: multiply 4/3 by π by the radius cubed, and output the result. However, the formula is not quite the same as an algorithm because there are several alternative sequences of steps that one could use to carry out this formula. For example, each of these algorithms computes the volume of a sphere.

1. Divide 4 by 3. 2. Multiply the previous result by π. 3. Repeat the following three times: multiply the previous result by r.	or	1. Compute $r \times r \times r$. 2. Multiply the previous result by π. 3. Multiply the previous result by 4. 4. Divide the previous result by 3.

Both of these algorithms use the same formula, but they execute the steps in different ways. This distinction may seem trivial to us but, depending on the level of abstraction, we may need to be this explicit when "talking to" a computer.

Reflection 1.4 *Write yet another algorithm that follows the volume formula.*

To execute an algorithm on a computer as a computation, we need to express the algorithm in a language that the computer can "understand." These computer languages are called *programming languages*, and an implementation of an algorithm in a programming language is called a *program*. Partial or whole programs are often called *source code*, or just *code*, which is why computer programming is also known as *coding*. Packaged programs, like the ones you see on your computer desktop, are also called *applications* or *apps*, and are collectively called *software* (as opposed to *hardware*, which refers to physical computer equipment).

There are many different programming languages in use, each with its own strengths and weaknesses. In this book, we will use a programming language called *Python*. To give you a sense of what is to come, here is a Python program that implements the sphere-volume algorithm:

```
def sphereVolume(radius):
    volume = (4 / 3) * 3.14159 * (radius ** 3)
    return volume

result = sphereVolume(10)
print(result)
```

Each line in a program is called a *statement*. The first statement in this program, beginning with **def**, defines a new *function*. Like an algorithm, a function contains a sequence of steps that transforms an input into an output. In this case, the function is named **sphereVolume**, and it takes a single input named **radius**, in the parentheses following the function name. The second line, which is indented to indicate that it is part of the **sphereVolume** function, uses the volume formula to compute the volume of the sphere, and then assigns this value to a variable named **volume**. The third line indicates that the value assigned to **volume** should be "returned" as the function's output. These first three lines only define what the **sphereVolume** function *should* do; the fourth line actually invokes the **sphereVolume** function with input **radius**

6 ■ What is computation?

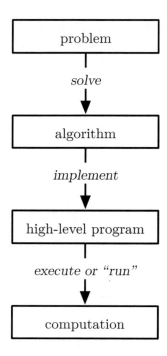

Figure 1.2 The layers of functional abstraction in a computation.

equal to 10, and assigns the result (41.88787) to the variable named `result`. The fifth line, you guessed it, prints the value assigned to `result`.

If you have never seen a computer program before, this probably looks like "Greek" to you. But, at some point in your life, so did $(4/3)\pi r^3$. (What is this symbol π? How can you do arithmetic with a letter? What is that small 3 doing up there? There is no "plus" or "times" symbol; is this even math?) The same can be said for notations like H_2O or $19°\,C$. But now that you are comfortable with these representations, you can appreciate that they are more convenient and less ambiguous than representations like "multiply 4/3 by π by the radius cubed" and "two atoms of hydrogen and one atom of oxygen." You should think of a programming language as simply an extension of the familiar arithmetic notation. But programming languages enable us to express algorithms for problems that reach well beyond simple arithmetic.

The process that we have described thus far is illustrated in Figure 1.2. We start with a problem having well-defined input and output. We then *solve* the problem by designing an algorithm. Next, we *implement* the algorithm by expressing it as a program in a programming language. Programming languages like Python are often called *high-level* because they use familiar words like "return" and "print," and enable a level of abstraction that is familiar to human problem solvers. (As we will see in Section 1.4, computers do not really "understand" high-level language programs.) Finally, *executing* the program on a computer initiates the computation that gives us our results. (Executing a program is also called "running" it.) As we will see soon, this picture is hiding some details, but it is sufficient for now.

Let's look at another, slightly more complicated problem. Suppose, as part of an ongoing climate study, we are tracking daily maximum water temperatures recorded by a drifting buoy in the Atlantic Ocean. We would like to compute the average (or mean) of these temperatures over one year. Suppose our list of high temperature readings (in degrees Celsius) starts like this:

```
18.9, 18.9, 19.0, 19.2, 19.3, 19.3, 19.2, 19.1, 19.4, 19.3, ...
```

Reflection 1.5 *What are the input and output for this problem? Think carefully about all the information you need in the input. Express your output in terms of the input.*

The input to the mean temperature problem obviously includes the list of temperatures. We also need to know how many temperatures we have, since we need to divide by that number. The output is the mean temperature of the list of temperatures.

Reflection 1.6 *What algorithm can we use to find the mean daily temperature? Think in terms of the steps we followed in the algorithms to compute the volume of a sphere.*

Of course, we know that we need to add up all the temperatures and divide by the number of days. But even a direction as simple as "add up all the temperatures" is too complex for a computer to execute directly. As we will see in Section 1.4, computers are, at their core, only able to execute instructions on par with simple arithmetic on two numbers at a time. Any complexity or "intelligence" that we attribute to computers is actually attributable to human beings harnessing this simple palette of instructions in creative ways. Indeed, this example raises two necessary characteristics of computer algorithms: their steps must be both *unambiguous* and *executable* by a computer. In other words, the steps that a computer algorithm follows must correlate to things a computer can actually do, without inviting creative interpretation by a human being.

> *An algorithm is a sequence of unambiguous, executable statements that solves a problem by transforming an input into a correct output.*

These two requirements are not really unique to computer algorithms. For example, we hope that new surgical techniques are unambiguously presented and reference actual anatomy and real surgical tools. Likewise, when an architect designs a building, she must take into account the materials available and be precise about their placement. And when an author writes a novel, he must write to his audience, using appropriate language and culturally familiar references.

To get a handle on how to translate "add up all the temperatures" into something a computer can understand, let's think about how we would do this without a calculator or computer: add two numbers at a time to a running sum. First, we initialize our running sum to 0. Then we add the first temperature value to the running sum: 0 + 18.9 = 18.9. Then we add the second temperature value to this running sum: 18.9 + 18.9 = 37.8. Then we add the third temperature: 37.8 + 19.0 = 56.8. And we continue like this until we have added all the temperatures. Suppose our final sum is 8,696.8 and we have summed over 365 days (for a non-leap year). Then we can compute our final average by dividing the sum by the number of days: $8,696.8/365 \approx 23.8$. A step-by-step sequence of instructions for this algorithm might look like the following:

Algorithm MEAN TEMPERATURE

Input: a list of 365 daily temperatures

1. Initialize the running sum to 0.
2. Add the first temperature to the running sum, and assign the result to be the new running sum.
3. Add the second temperature to the running sum, and assign the result to be the new running sum.
4. Add the third temperature to the running sum, and assign the result to be the new running sum.
   $\vdots$
366. Add the 365th temperature to the running sum, and assign the result to be the new running sum.
367. Divide the running sum by the number of days, 365, and assign the result to be the mean temperature.

Output: the mean temperature

Since this 367-step algorithm is pretty cumbersome to write, and steps 2–366 are essentially the same, we can shorten the description of the algorithm by substituting steps 2–366 with

For each temperature t in our list, repeat the following:

add t to the running sum, and assign the result to be the new running sum.

In this shortened representation, which is called a *loop*, t stands in for each temperature. First, t is assigned the first temperature in the list, 18.9, which the indented statement adds to the running sum. Then t is assigned the second temperature, 18.9, which the indented statement next adds to the running sum. Then t is assigned the third temperature, 19.0. And so on. With this substitution (in red), our algorithm becomes:

> **Algorithm** MEAN TEMPERATURE 2
>
> **Input:** a list of 365 daily temperatures
>
> 1. Initialize the running sum to 0.
> 2. For each temperature t in our list, repeat the following:
>   add t to the running sum, and assign the result to be the new running sum.
> 3. Divide the running sum by the number of days, 365, and assign the result to be the mean temperature.
>
> **Output:** the mean temperature

We can visualize the execution of this loop more explicitly by "unrolling" it into its equivalent sequence of statements. The statement indented under step 2 is executed once for each different value of t in the list. Each time, the running sum is updated. So, if our list of temperatures is the same as before, this loop is equivalent to the following sequence of statements:

(a) Add $\underbrace{18.9}_{t}$ to $\underbrace{0}_{\text{the running sum}}$, and assign $\underbrace{18.9}_{\text{the result}}$ to be the new running sum.

(b) Add $\underbrace{18.9}_{t}$ to $\underbrace{18.9}_{\text{the running sum}}$, and assign $\underbrace{37.8}_{\text{the result}}$ to be the new running sum.

(c) Add $\underbrace{19.0}_{t}$ to $\underbrace{37.8}_{\text{the running sum}}$, and assign $\underbrace{56.8}_{\text{the result}}$ to be the new running sum.

⋮

Another important benefit of the loop representation is that it renders the algorithm representation independent of the length of the list of temperatures. In MEAN TEMPERATURE 1, we had to know up front how many temperatures there were because we needed one statement for each temperature. However, in MEAN TEMPERATURE 2, there is no mention of 365, except in the final statement; the loop simply repeats as long as there are temperatures remaining. Therefore, we can generalize MEAN TEMPERATURE 2 to handle any number of temperatures. Actually, there is nothing in the algorithm that depends on these values being temperatures at all, so we should also generalize it to handle any list of numbers. If we let n denote the length of the list, then our generalized algorithm looks like the following:

> **Algorithm** MEAN
>
> **Input:** a list of n numbers
>
> 1. Initialize the running sum to 0.
> 2. For each number t in our list, repeat the following:
>   Add t to the running sum, and assign the result to be the new running sum.
> 3. Divide the running sum by n, and assign the result to be the mean.
>
> **Output:** the mean

As you can see, there are often different, but equivalent, ways to represent an algorithm. Sometimes the one you choose depends on personal taste and the programming language in which you will eventually express it. Just to give you a sense of what is to come, we can express the MEAN algorithm in Python like this:

```
def mean(values):
    n = len(values)
    sum = 0
    for number in values:
        sum = sum + number
    mean = sum / n
    return mean
```

Try to match the statements in MEAN to the statements in the program. By doing so, you can get a sense of what each part of the program must do. We will flesh this out in more detail later, of course; in a few chapters, programs like this will be second nature. As we pointed out earlier, once you are comfortable with this notation, you will likely find it much *less* cumbersome and *more* clear than writing everything in full sentences, with the inherent ambiguities that tend to emerge in the English language.

Exercises

1.2.1. Describe three examples from your everyday life in which an abstraction is beneficial. Explain the benefits of each abstraction versus what life would be like without it.

1.2.2. Write an algorithm to find the minimum value in a list of numbers. Like the MEAN algorithm, you may only examine one number at a time. But instead of remembering the current running sum, remember the current minimum value.

1.2.3. You are organizing a birthday party where cookies will be served. On the invite list are some younger children and some older children. The invite list is given in shorthand as a list of letters, y for a young child and o for an older child. Each older child will eat three cookies, while each younger child will eat two cookies.

Write an algorithm that traces through the list, and prints how many cookies are needed. For example, if the input were y, y, o, y, o then the output should be 12 cookies.

1.2.4. Write an algorithm for each player to follow in a simple card game like Blackjack or Go Fish. Assume that the cards are dealt in some random order.

1.2.5. Write an algorithm to sort a stack of any 5 cards by value in ascending order. In each step, your algorithm may compare or swap the positions of any two cards.

1.2.6. Write an algorithm to walk between two nearby locations, assuming the only legal instructions are "Take s steps forward," and "Turn d degrees to the left," where s and d are positive integers.

1.3   EFFICIENT ALGORITHMS

For any problem, there may be many possible algorithms that correctly solve it. However, not every algorithm is equally good. We will illustrate this point with two problems: organizing a phone tree and "smoothing" data from a "noisy" source.

Organizing a phone tree

Imagine that you have been asked to organize an old-fashioned "phone tree" for your organization to personally alert everyone with a phone call in the event of an emergency. You have to make the first call, but after that, you can delegate others to make calls as well. Who makes which calls and the order in which they are made constitutes an algorithm. For example, suppose there are only eight people in the organization, and they are named Amelia, Beth, Caroline, Dave, Ernie, Flo, Gretchen, and Homer. Then here is a simple algorithm:

Algorithm ALPHABETICAL PHONE TREE

1. Amelia calls Beth.
2. Beth calls Caroline.
3. Caroline calls Dave.
   ⋮
7. Gretchen calls Homer.

(For simplicity, we will assume that everyone answers the phone right away and every phone call takes the same amount of time.) Is this the best algorithm for solving the problem? What does "best" even mean?

Reflection 1.7 *For the "phone tree" problem, what characteristics would make one algorithm better than another?*

As the question suggests, deciding whether one algorithm is better than another

12 ■ What is computation?

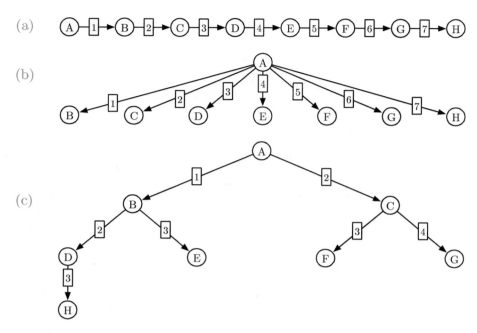

Figure 1.3 Illustrations of three possible phone tree algorithms. Each name is represented by its first letter, calls are represented by arrows, and the numbers indicate the order in which the calls are made.

depends on your criteria. For the "phone tree" problem, your primary criterion may be to make as few calls as possible yourself, delegating most of the work to others. The Alphabetical Phone Tree algorithm, graphically depicted in Figure 1.3(a), is one way to satisfy this criterion. Alternatively, you may feel guilty about making others do any work, so you decide to make all the calls yourself; this is depicted in Figure 1.3(b). However, in the interest of community safety, we should really organize the phone tree so that the last person is notified of the emergency as soon as possible. Both of the previous two algorithms fail miserably in this regard. In fact, they both take the longest possible time! A better algorithm would have you call two people, then each of those people call two people, and so on. This is depicted in Figure 1.3(c). Notice that this algorithm notifies everyone in only four steps because more than one phone call happens concurrently; A and B are both making calls during the second step; and B, C, and D are all making calls in the third step.

If all of the calls were utilizing a shared resource (such as a wireless network), we might need to balance the time with the number of simultaneous calls. This may not seem like an issue with only eight people in our phone tree, but it would become a significant issue with many more people. For example, applying the algorithm depicted in Figure 1.3(c) to thousands of people would result in thousands of simultaneous calls.

Let's now consider how these concerns might apply to algorithms more generally.

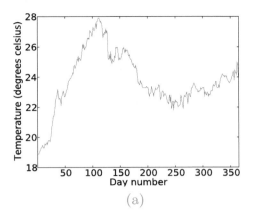 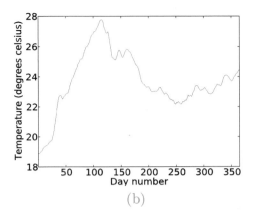

Figure 1.4 Plot of (a) a year's worth of daily high temperature readings and (b) the temperature readings smoothed over a five-day window.

Reflection 1.8 *What general characteristics might make one algorithm for a particular problem better than another algorithm for the same problem?*

As was the case in the phone tree problem, the most important hallmark of a good algorithm is speed; given the choice, we almost always want the algorithm that requires the least amount of time to finish. (Would you rather wait five minutes for a web search or half of a second?) But there are other attributes as well that can distinguish one algorithm from another. For example, we saw in Section 1.2 how a long, tedious algorithm can be represented more compactly using a loop; the more compact version is easier to write down and translate into a program. The compact version also requires less space in a computer's memory. Because the amount of storage space in a computer is limited, we want to create algorithms that use as little of this resource as possible to store both their instructions and their data. Efficient resource usage may also apply to network capacity, as in the phone tree problem on a wireless network. In addition, just as writers and designers strive to create elegant works, computer scientists pride themselves on writing elegant algorithms that are easy to understand and do not contain extraneous steps. And some advanced algorithms are considered to be especially important because they introduce new techniques that can be applied to solve other hard problems.

A smoothing algorithm

To more formally illustrate how we can evaluate the time required by an algorithm, let's revisit the sequence of temperature readings from the previous section. Often, when we are dealing with large data sets (much longer than this), anomalies can arise due to errors in the sensing equipment, human fallibility, or in the network used to send results to a lab or another collection point. We can mask these erroneous measurements by "smoothing" the data, replacing each value with the mean of the values in a "window" of values containing it. This technique is also useful for

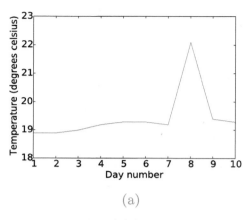
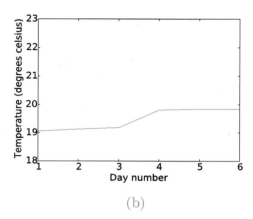

Figure 1.5 Plot of (a) the ten high temperature readings and (b) the temperature readings smoothed over a five-day window.

extracting general patterns in data by eliminating distracting "bumpy" areas. For example, Figure 1.4 shows a year's worth of raw temperature data from a floating ocean buoy, next to the same data smoothed over a five day window.

Let's design an algorithm for this problem. To get a better sense of the problem and the technique we will use, we will begin by looking at a small example consisting of the first ten temperature readings from the previous section, with an anomalous reading inserted. Specifically, we will replace the 19.1° C reading in the eighth position with an erroneous temperature of 22.1° C:

18.9, 18.9, 19.0, 19.2, 19.3, 19.3, 19.2, 22.1, 19.4, 19.3

The plot of this data in Figure 1.5(a) illustrates this erroneous "bump."

Now let's smooth the data by averaging over windows of size five. For each value in the original list, its window will include itself and the four values that come before it. Our algorithm will need to compute the mean of each of these windows, and then add each of these means to a new smoothed list. The last four values do not have four more values after them, so our smoothed list will contain four fewer values than our original list. The first window looks like this:

$$\underbrace{18.9, 18.9, 19.0, 19.2, 19.3}_{\text{mean} = 95.3 \,/\, 5 = 19.06}\ 19.3, 19.2, 22.1, 19.4, 19.3$$

To find the mean temperature for the window, we sum the five values and divide by 5. The result, 19.06, will represent this window in the smoothed list. The remaining five windows are computed in the same way:

$$18.9, \underbrace{18.9, 19.0, 19.2, 19.3, 19.3}_{\text{mean} = 95.7 \,/\, 5 = 19.14},\ 19.2, 22.1, 19.4, 19.3$$

18.9, 18.9, |19.0, 19.2, 19.3, 19.3, 19.2,| 22.1, 19.4, 19.3

$$\text{mean} = 96.0 \,/\, 5 = 19.20$$

18.9, 18.9, 19.0, |19.2, 19.3, 19.3, 19.2, 22.1,| 19.4, 19.3

$$\text{mean} = 99.1 \,/\, 5 = 19.82$$

18.9, 18.9, 19.0, 19.2, |19.3, 19.3, 19.2, 22.1, 19.4,| 19.3

$$\text{mean} = 99.3 \,/\, 5 = 19.86$$

18.9, 18.9, 19.0, 19.2, 19.3, |19.3, 19.2, 22.1, 19.4, 19.3|

$$\text{mean} = 99.3 \,/\, 5 = 19.86$$

In the end, our list of smoothed temperature readings is:

$$19.06,\ 19.14,\ 19.20,\ 19.82,\ 19.86,\ 19.86$$

We can see from the plot of this smoothed list in Figure 1.5(b) that the "bump" has indeed been smoothed somewhat.

We can express our smoothing algorithm using notation similar to our previous algorithms. Notice that, to find each of the window means in our smoothing algorithm, we call upon our MEAN algorithm from Section 1.2.

Algorithm SMOOTH

Input: a list of n numbers and a window size w

1. Create an empty list of mean values.
2. For each position d, from 1 to $n - w + 1$, repeat the following:
    (a) invoke the MEAN algorithm on the list of numbers between positions d and $d + w - 1$;
    (b) append the result to the list of mean values.

Output: the list of mean values

Step 1 creates an empty list to hold the window means. We will append the mean for each window to the end of this list after we compute it. Step 2 uses a loop to compute the means for all of the windows. This loop is similar to the loop in the MEAN algorithm: the variable d takes on the values between 1 and $n - w + 1$, like t took on each value in the list in the MEAN algorithm. First, d is assigned the value 1, and the mean of the window between positions 1 and $1 + w - 1 = w$ is added to the list of mean values. Then d is assigned the value 2, and the mean of the window between positions 2 and $2 + w - 1 = w + 1$ is added to the list of mean values. And so on, until d takes on the value of $n - w + 1$, and the mean of the window between positions $n - w + 1$ and $(n - w + 1) + w - 1 = n$ is added to the list of mean values.

Reflection 1.9 *Carry out each step of the* SMOOTH *algorithm with w = 5 and the ten-element list in our example:*

> 18.9, 18.9, 19.0, 19.2, 19.3, 19.3, 19.2, 22.1, 19.4, 19.3

Compare each step with what we worked out above to convince yourself that the algorithm is correct.

How long does it take?

To determine how much time this algorithm requires to smooth the values in a list, we could implement it as a program and execute it on a computer. However, this approach presents some problems. First, which list(s) do we use for our timing experiments? Second, which value(s) of w do we use? Third, which computer(s) do we use? Will these specific choices give us a complete picture of our algorithm's efficiency? Will they allow us to predict the time required to smooth other lists on other computers? Will these predictions still be valid ten years from now?

A better way to predict the amount of time required by an algorithm is to count the number of *elementary steps* that are required. An elementary step is one that always requires the same amount of time, regardless of the input. By counting elementary steps, we can estimate how long an algorithm will take on any computer, relative to another algorithm for the same problem. For example, if algorithm A requires 100 elementary steps while algorithm B requires 300 elementary steps, we know that algorithm A will always be about three times faster than algorithm B, regardless of which computer we run them both on.

Let's look at which steps in the SMOOTH algorithm qualify as elementary steps. Step 1 qualifies because creating an empty list makes no reference to the input at all. But step 2 does not qualify because its duration depends on the values of n and w, which are both part of the input. Similarly, step 2(a) does not qualify as an elementary step because its duration depends on the time required by the MEAN algorithm, which depends on the value of w. Finding the mean temperature of the window between positions 1 and w will take longer when w is 10 than it will when w is 3. However, each addition and division operation required to compute a mean does qualify as an elementary step because the time required for an arithmetic operation is independent of the operands (up to a point). Step 2(b) also qualifies as an elementary step because, intuitively, adding something to the end of a list should not depend on how many items are in the list (as long as we have direct access to the end of the list).

Reflection 1.10 *Based on this analysis, what are the elementary steps in this algorithm? How many times are they each executed?*

The only elementary steps we have identified are creating a list, appending to a list, and arithmetic operations. Let's start by counting the number of arithmetic operations that are required. If each window has size five, then we perform five addition operations and one division operation each time we invoke the MEAN algorithm, for a total of six arithmetic operations per window. Therefore, the total number of arithmetic operations is six times the number of windows. In general, the window size is denoted w and there are $n-w+1$ windows. So the algorithm performs w additions and one division per window, for a total of $(w+1) \cdot (n-w+1)$ arithmetic operations. For example, smoothing a list of 365 temperatures with a window size of five days will require $(w+1) \cdot (n-w+1) = 6 \cdot (365-5+1) = 2{,}166$ arithmetic operations. In addition to the arithmetic operations, we create the list once and append to the list once for every window, a total of $n-w+1$ times. Therefore, the total number of elementary steps is

$$\underbrace{(w+1) \cdot (n-w+1)}_{\text{arithmetic}} + \underbrace{1}_{\substack{\text{create} \\ \text{list}}} + \underbrace{n-w+1}_{\text{appends}}.$$

This expression can be simplified a bit to $(w+2) \cdot (n-w+1) + 1$.

This number of elementary steps is called an algorithm's *time complexity*. An algorithm's time complexity is always measured with respect to the size of the input. In this case, the size of the input depends on n, the length of the list, and w, the window size. As we will discuss later, we could simplify this unpleasant time complexity expression by just taking into account the terms that matter when the input gets large. But, for now, we will leave it as is.

Reflection 1.11 *Can you think of a way to solve the smoothing problem with fewer arithmetic operations?*

A better smoothing algorithm

We can design a more efficient smoothing algorithm, one requiring fewer elementary steps, by exploiting the following simple observation. While finding the sums of neighboring windows in the SMOOTH algorithm, we unnecessarily performed some addition operations more than once. For example, we added the fourth and fifth temperature readings four different times, once in each of the first four windows. We can eliminate this extra work by taking advantage of the relationship between the sums of two contiguous windows. For example, consider the first window:

$$\underbrace{\boxed{18.9,\ 18.9,\ 19.0,\ 19.2,\ 19.3}}_{\text{sum} = 95.3}\ 19.3,\ 19.2,\ 22.1,\ 19.4,\ 19.3$$

The sum for the second window must be almost the same as the first window, since they have four numbers in common. The only difference in the second window is that it loses the first 18.9 and gains 19.3. So once we have the sum for the first window (95.3), we can get the sum of the second window with only two additional arithmetic operations: $95.3 - 18.9 + 19.3 = 95.7$.

$$18.9,\ \underbrace{\boxed{18.9,\ 19.0,\ 19.2,\ 19.3,\ 19.3,}}_{\text{sum} = 95.7}\ 19.2,\ 22.1,\ 19.4,\ 19.3$$

18 ■ What is computation?

We can apply this process to every subsequent window as well. At the end of the algorithm, once we have the sums for all of the windows, we can simply divide each by its window length to get the final list of means. Written in the same manner as the previous algorithms, our new smoothing algorithm looks like this:

Algorithm SMOOTH 2

Input: a list of n numbers and a window size w

1. Create an empty list of window sums.
2. Compute the sum of the first w numbers and append the result to the list of window sums.
3. For each position d, from 2 to $n - w + 1$, repeat the following:
   (a) subtract the number in position $d - 1$ from the previous window sum and then add the number in position $d + w - 1$;
   (b) append the result to the list of window sums.
4. For each position d, from 1 to $n - w + 1$, repeat the following:
   (a) divide the dth window sum by w.

Output: the list of mean values

Step 2 computes the sum for the first window and adds it to a list of window sums. Step 3 then computes the sum for each subsequent window by subtracting the number that is lost from the previous window and adding the number that is gained. Finally, step 4 divides all of the window sums by w to get the list of mean values.

Reflection 1.12 *As with the previous algorithm, carry out each step of the* SMOOTH *2 algorithm with $w = 5$ and the ten-element list in our example:*

$$18.9,\ 18.9,\ 19.0,\ 19.2,\ 19.3,\ 19.3,\ 19.2,\ 22.1,\ 19.4,\ 19.3$$

Does the SMOOTH 2 algorithm actually require fewer elementary steps than our first attempt? Let's look at each step individually. As before, step 1 counts as one elementary step. Step 2 requires $w - 1$ addition operations and one append, for a total of w elementary steps. Step 3 performs an addition, a subtraction, and an append for every window but the first, for a total of $3(n - w)$ elementary steps. Finally, step 4 performs one division operation per window, for a total of $n - w + 1$ arithmetic operations. This gives us a total of

$$\underbrace{1}_{\substack{\text{create}\\\text{list}}} + \underbrace{w}_{\substack{\text{first}\\\text{window}}} + \underbrace{3(n - w)}_{\substack{\text{other}\\\text{windows}}} + \underbrace{n - w + 1}_{\text{divisions}}$$

elementary steps. Combining all of these terms, we find that the time complexity of SMOOTH 2 is $4n - 3w + 2$. Therefore, our old algorithm requires

$$\frac{(w+2) \cdot (n-w+1) + 1}{4n - 3w + 2}$$

times as many operations as the new one. It is hard to tell from this fraction, but our new algorithm is about $w/4$ times faster than our old one. To see this, suppose our list contains $n = 10,000$ temperature readings. The following table shows the value of the fraction for increasing window sizes w.

w	Speedup
5	1.7
10	3.0
20	5.5
100	25.4
500	123.9
1,000	243.7

When w is small, our new algorithm does not make much difference, but the difference becomes quite pronounced when w gets larger. In real applications of smoothing on extremely large data sets containing billions or trillions of items, window sizes can be as high as $w = 100,000$. For example, smoothing is commonly used to visualize statistics over DNA sequences that are billions of units long. So our refined algorithm can have a marked impact! Indeed, as we will examine further in Section 6.4, it is often the case that a faster algorithm can reduce the time required for a computation significantly more than faster hardware.

Exercises

1.3.1. The phone tree algorithm depicted in Figure 1.3(c) comes very close to making all of the phone calls in three steps. Is it possible to actually achieve three steps? How?

1.3.2. Suppose the phone tree algorithm illustrated in Figure 1.3(c) was being executed with a large number of people. We saw one call made during the first time step, two calls during the second time step, and three calls during the third time step.

(a) How many concurrent calls would be made during time steps 4, 5, and 6?

(b) In general, can you provide a formula (or algorithm) to determine the number of calls made during any time step t?

1.3.3. Describe the most important criterion for evaluating how good an algorithm is. Then add at least one more criterion and describe how it would be applied to algorithms for some everyday problem.

1.3.4. What is the time complexity of the MEAN algorithm on Page 10, in terms of n (the size of the input list)?

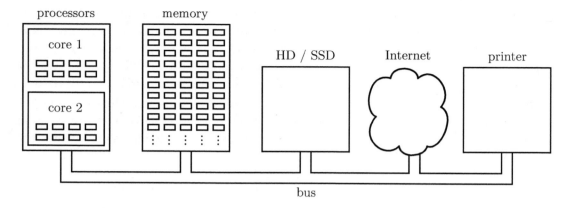

Figure 1.6 A simplified schematic diagram of a computer.

1.4 COMPUTERS ARE DUMB

By executing clever algorithms very fast, computers can sometimes appear to exhibit primitive intelligence. In a historic example, in 1997, the IBM Deep Blue computer defeated reigning world champion Garry Kasparov in a six-game match of chess. More recently, in 2011, IBM's Watson computer beat two champions in the television game show Jeopardy.

The intelligence that we sometimes attribute to computers, however, is actually human intelligence that was originally expressed as an algorithm, and then as a program. Even the statements in a high-level programming language are themselves abstract conveniences built upon a much more rudimentary set of instructions, called a *machine language*, that a computer can execute natively.

Every statement in a Python program must be translated into a sequence of equivalent machine language instructions before it can be executed by a computer. This process is handled automatically, and somewhat transparently, by either a program called an *interpreter* or a program called a *compiler*. An interpreter translates one line of a high-level program into machine language, executes it, then translates the next line and executes it, etc. A compiler instead translates a high-level language program all at once into machine language. Then the compiled machine language program can be executed from start to finish without additional translation. This tends to make compiled programs faster than interpreted ones. However, interpreted languages allow us to more closely interact with a program during its execution. As we will see in Chapter 2, Python is an interpreted programming language.

Inside a computer

The types of instructions that constitute a machine language are based on the internal design of a modern computer. As illustrated in Figure 1.6, a computer essentially consists of one or more *processors* connected to a *memory*. A computer's memory, often called *RAM* (short for *random access memory*), is conceptually organized as a long sequence of *cells*, each of which can contain one unit of information. Each cell is labeled with a unique *memory address* that allows the processor to reference it specifically. So a computer's memory is like a huge sequence of equal-sized post office boxes, each of which can hold exactly one letter. Each P.O. box number is analogous to a memory address and a letter is analogous to one unit of information. The information in each cell can represent either one instruction or one unit of data.[1] So a computer's memory stores both programs and the data on which the programs work.

A processor, often called a *CPU* (short for *central processing unit*) or a *core*, contains both the machinery for executing instructions and a small set of memory locations called *registers* that temporarily hold data values needed by the current instruction. If a computer contains more than one core, as most modern computers do, then it is able to execute more than one instruction at a time. These instructions may be from different programs or from the same program. This means that our definition of an algorithm as a *sequence* of instructions on Page 7 is not strictly correct. In fact, an algorithm (or a program) may consist of several semi-independent sequences of steps called *threads* that cooperatively solve a problem.

As illustrated in Figure 1.6, the processors and memory are connected by a communication channel called a *bus*. When a processor needs a value in memory, it transmits the request over the bus, and then the memory returns the requested value the same way. The bus also connects the processors and memory to several other components that either improve the machine's efficiency or its convenience, like the Internet and *secondary storage* devices like hard drives, solid state drives, and flash memory. As you probably know, the contents of a computer's memory are lost when the computer loses power, so we use secondary storage to preserve data for longer periods of time. We interact with these devices through a "file system" abstraction that makes it appear as if our hard drives are really filing cabinets. When you execute a program or open a file, it is first copied from secondary storage into memory where the processor can access it.

Machine language

The machine language instructions that a processor can execute are very simple. For example, consider something as simple as computing $x = 2 + 5$. In a program, this statement adds 2 and 5, and stores the result in a memory location referred to by the variable x. But even this is likely to be too complex for one machine language instruction. The machine language equivalent likely, depending on the computer, consists of four instructions. The first instruction loads the value 2 into a register in the processor. The second instruction loads the value 5 into another register in the processor. The third instruction adds the values in these two registers, and stores

---

[1] In reality, each instruction or unit of data usually occupies multiple contiguous cells.

> Box 1.1: High performance computing
>
> Although today's computers are extremely fast, there are some problems that are so big that additional power is necessary. These include weather forecasting, molecular modeling and drug design, aerodynamics, and deciphering encrypted data. For these problems, scientists use *supercomputers*. A supercomputer, or *high performance computer*, is made of up to tens of thousands of *nodes*, connected by a very fast data network that shuttles information between nodes. A node, essentially a standalone computer, can contain multiple processors, each with multiple cores. The fastest supercomputers today have millions of cores and a million gigabytes of memory.
>
> To realize the power of these computers, programmers need to supply their cores with a constant stream of data and instructions. The results computed by the individual cores are then aggregated into an overall solution. The algorithm design and programming techniques, known as *parallel programming*, are beyond the scope of this book, but we provide additional resources at the end of the chapter if you would like to learn more.

the result in a third register. Finally, the fourth instruction stores the value in the third register into the memory cell with address x.

From the moment a computer is turned on, its processors are operating in a continuous loop called the *fetch and execute cycle* (or *machine cycle*). In each cycle, the processor fetches one machine language instruction from memory and executes it. This cycle repeats until the computer is turned off or loses power. In a nutshell, this is all a computer does. The rate at which a computer performs the fetch and execute cycle is loosely determined by the rate at which an internal clock "ticks" (the processor's *clock rate*). The ticks of this clock keep the machine's operations synchronized. Modern personal computers have clocks that tick a few billion times each second; a 3 gigahertz (GHz) processor ticks 3 billion times per second ("giga" means "billion" and a "hertz" is one tick per second).

So computers, at their most basic level, really are quite dumb; the processor blindly follows the fetch and execute cycle, dutifully executing whatever sequence of simple instructions we give it. The frustrating errors that we yell at computers about are, in fact, human errors. The great thing about computers is not that they are smart, but that they follow our instructions so very quickly; they can accomplish an incredible amount of work in a very short amount of time.

Everything is bits

Our discussion so far has glossed over a very important consideration: in what form does a computer store programs and data? In addition to machine language instructions, we need to store numbers, documents, maps, images, sounds, presentations, spreadsheets, and more. Using a different storage medium for each type of information would be insanely complicated and inefficient. Instead, we need a simple storage format that can accommodate any type of data. The answer is bits. A *bit* is the simplest possible unit of information, capable of taking on one of only two values: 0 or 1 (or equivalently, off/on, no/yes, or false/true). This simplicity makes both storing information (i.e., memory) and computing with it (e.g., processors) relatively simple and efficient.

Bits are switches

Storing the value of a bit is absurdly simple: a bit is equivalent to an on/off switch, and the value of the bit is equivalent to the state of the switch: off = 0 and on = 1. A computer's memory is essentially composed of billions of microscopic switches, organized into memory cells. Each memory cell contains 8 switches, so each cell can store 8 bits, called a *byte*. We represent a byte simply as a sequence of 8 bits, such as 01101001. A computer with 8 gigabytes (GB) of memory contains about 8 billion memory cells, each storing one byte. (Similarly, a kilobyte (KB) is about one thousand bytes, a megabyte (MB) is about one million bytes, and a terabyte (TB) is about one trillion bytes.)

Bits can store anything

A second advantage of storing information with bits is that, as the simplest possible unit of information, it serves as a "lowest common denominator": all other information can be converted to bits in fairly straightforward manners. Numbers, words, images, music, and machine language instructions are all encoded as sequences of bits before they are stored in memory. Information stored as bits is also said to be in *binary notation*. For example, consider the following sequence of 16 bits (or two bytes).

$$0100010001010101$$

This bit sequence can represent each of the following, depending upon how it is interpreted:

(a) the integer value 17,493;

(b) the decimal value 2.166015625;

(c) the two characters "DU";

(d) the Intel machine language instruction `inc x` (`inc` is short for "increment"); or

(e) the following 4 × 4 black and white image, called a *bitmap* (0 represents a white square and 1 represents a black square).

For now, let's just look more closely at (a). Integers are represented in a computer using the *binary number system*, which is understood best by analogy to the decimal number system. In decimal, each position in a number has a value that is a power of ten: from right to left, the positional values are $10^0 = 1$, $10^1 = 10$, $10^2 = 100$, etc. The value of a decimal number comes from adding the digits, each first multiplied by the value of its position. So the decimal number 1,831 represents the value

$$1 \times 10^3 + 8 \times 10^2 + 3 \times 10^1 + 1 \times 10^0 = 1000 + 800 + 30 + 1.$$

The binary number system is no different, except that each position represents a power of two, and there are only two digits instead of ten. From right to left, the binary number system positions have values $2^0 = 1$, $2^1 = 2$, $2^2 = 4$, $2^3 = 8$, etc. So, for example, the binary number 110011 represents the value

$$1 \times 2^5 + 1 \times 2^4 + 0 \times 2^3 + 0 \times 2^2 + 1 \times 2^1 + 1 \times 2^0 = 32 + 16 + 2 + 1 = 51$$

in decimal. The 16 bit number above, 0100010001010101, is equivalent to

$$2^{14} + 2^{10} + 2^6 + 2^4 + 2^2 + 2^0 = 16,384 + 1,024 + 64 + 16 + 4 + 1 = 17,493$$

in decimal.

Reflection 1.13 *To check your understanding, show why the binary number* 1001000 *is equivalent to the decimal number 72.*

This idea can be extended to numbers with a fractional part as well. In decimal, the positions to the right of the decimal point have values that are negative powers of 10: the tenths place has value $10^{-1} = 0.1$, the hundredths place has value $10^{-2} = 0.01$, etc. So the decimal number 18.31 represents the value

$$1 \times 10^1 + 8 \times 10^0 + 3 \times 10^{-1} + 1 \times 10^{-2} = 10 + 8 + 0.3 + 0.01.$$

Similarly, in binary, the positions to the right of the "binary point" have values that are negative powers of 2. For example, the binary number 11.0011 represents the value

$$1 \times 2^1 + 1 \times 2^0 + 0 \times 2^{-1} + 0 \times 2^{-2} + 1 \times 2^{-3} + 1 \times 2^{-4} = 2 + 1 + 0 + 0 + \frac{1}{8} + \frac{1}{16} = 3\frac{3}{16}$$

in decimal. This is not, however, how we derived (b) above. Numbers with fractional components are stored in a computer using a different format that allows for a much greater range of numbers to be represented. We will revisit this in Section 2.2.

Reflection 1.14 *To check your understanding, show why the binary number* 1001.101 *is equivalent to the decimal number 9 5/8.*

Binary computation is simple

A third advantage of binary is that computation on binary values is exceedingly easy. In fact, there are only three fundamental operators, called **and**, **or**, and **not**. These are known as the *Boolean operators*, after 19th century mathematician George Boole, who is credited with inventing modern mathematical logic, now called *Boolean logic* or *Boolean algebra*.

To illustrate the Boolean operators, let the variables a and b each represent a bit with a value of 0 or 1. Then a **and** b is equal to 1 only if both a and b are equal to 1; otherwise a **and** b is equal to 0.[2] This is conveniently represented by the following *truth table*:

a	b	a **and** b
0	0	0
0	1	0
1	0	0
1	1	1

Each row of the truth table represents a different combination of the values of a and b. These combinations are shown in the first two columns. The last column of the truth table contains the corresponding values of a **and** b. We see that a **and** b is 0 in all cases, except where both a and b are 1. If we let 1 represent "true" and 0 represent "false," this conveniently matches our own intuitive meaning of "and." The statement "the barn is red *and* white" is true only if the barn both has red on it and has white on it.

Second, a **or** b is equal to 1 if at least one of a or b is equal to 1; otherwise a **or** b is equal to 0. This is represented by the following truth table:

a	b	a **or** b
0	0	0
0	1	1
1	0	1
1	1	1

Notice that a **or** b is 1 even if both a and b are 1. This is different from our normal understanding of "or." If we say that "the barn is red or white," we usually mean it is either red or white, not both. But the Boolean operator can mean both are true. (There is another Boolean operator called "exclusive or" that does mean "either/or," but we won't get into that here.)

Finally, the **not** operator simply inverts a bit, changing a 0 to 1, or a 1 to 0. So, **not** a is equal to 1 if a is equal to 0, or 0 if a is equal to 1. The truth table for the **not** operator is simple:

---

[2]In formal Boolean algebra, a **and** b is usually represented $a \wedge b$, a **or** b is usually represented $a \vee b$, and **not** a is usually represented as $\neg a$.

a	**not** a
0	1
1	0

With these basic operators, we can build more complicated expressions. For example, suppose we wanted to find the value of the expression **not** a **and** b. (Note that the **not** operator applies only to the variable immediately after it, in this case a, not the entire expression a **and** b. For **not** to apply to the expression, we would need parentheses: **not** (a **and** b).) We can evaluate **not** a **and** b by building a truth table for it. We start by listing all of the combinations of values for a and b, and then creating a column for **not** a, since we need that value in the final expression:

a	b	**not** a
0	0	1
0	1	1
1	0	0
1	1	0

Then, we create a column for **not** a **and** b by **and**ing the third column with the second.

a	b	**not** a	**not** a **and** b
0	0	1	0
0	1	1	1
1	0	0	0
1	1	0	0

So we find that **not** a **and** b is 1 only when a is 0 and b is 1. Or, equivalently, **not** a **and** b is *true* only when a is *false* and b is *true*. (Think about that for a moment; it makes sense, doesn't it?)

The universal machine

The previous advantages of computing in binary—simple switch implementations, simple Boolean operators, and the ability to encode anything—would be useless if we could not actually compute in binary everything we want to compute. In other words, we need binary computation (i.e., Boolean algebra) to be *universal*: we need to be able to compute any computable problem by using only the three basic Boolean operators. But is such a thing really possible? And what do we mean by *any computable problem*? Can we really perform any computation at all—a web browser, a chess-playing program, Mars rover software—just by converting it to binary and using the three elementary Boolean operators to get the answer?

Fortunately, the answer is yes, when we combine logic circuits with a sufficiently large memory and a simple controller called a *finite state machine* (FSM) to conduct traffic. A finite state machine consists of a finite set of *states*, along with *transitions* between states. A state represents the current value of some object or the degree

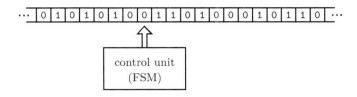

Figure 1.7 A schematic representation of a Turing machine.

of progress made toward some goal. For example, a simple elevator, with states representing floors, is a finite state machine, as illustrated below.

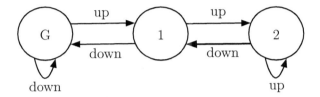

States are represented by circles and transitions are arrows between circles. In this elevator, there are only up and down buttons (no ability to choose your destination floor when you enter). The label on each transition represents the button press event that causes the transition to occur. For example, when we are on the ground floor and the down button is pressed, we stay on the ground floor. But when the up button is pressed, we transition to the first floor. Many other simple systems can also be represented by finite state machines, such as vending machines, subway doors, traffic lights, and tool booths. The implementation of a finite state machine in a computer coordinates the fetch and execute cycle, as well as various intermediate steps involved in executing machine language instructions.

This question of whether a computational system is universal has its roots in the very origins of computer science itself. In 1936, Alan Turing proposed an abstract computational model, now called a *Turing machine*, that he proved could compute any problem considered to be mathematically computable. As illustrated in Figure 1.7, a Turing machine consists of a control unit containing a finite state machine that can read from and write to an infinitely long tape. The tape contains a sequence of "cells," each of which can contain a single character. The tape is initially inscribed with some sequence of input characters, and a pointer attached to the control unit is positioned at the beginning of the input. In each step, the Turing machine reads the character in the current cell. Then, based on what it reads and the current state, the finite state machine decides whether to write a character in the cell and whether to move its pointer one cell to the left or right. Not unlike the fetch and execute cycle in a modern computer, the Turing machine repeats this simple process as long as needed, until a designated final state is reached. The output is the final sequence of characters on the tape.

The Turing machine still stands as our modern definition of computability. The

28 ■ What is computation?

Church-Turing thesis states that a problem is computable if and only if it can be computed by a Turing machine. Any mechanism that can be shown to be equivalent in computational power to a Turing machine is considered to be computationally universal, or *Turing complete*. A modern computer, based on Boolean logic and a finite state machine, falls into this category.

Exercises

1.4.1. Show how to convert the binary number 1101 to decimal.

1.4.2. Show how to convert the binary number 1111101000 to decimal.

1.4.3. Show how to convert the binary number 11.0011 to decimal.

1.4.4. Show how to convert the binary number 11.110001 to decimal.

1.4.5. Show how to convert the decimal number 22 to binary.

1.4.6. Show how to convert the decimal number 222 to binary.

1.4.7. See how closely you can represent the decimal number 0.1 in binary using six places to the right of the binary point. What is the actual value of your approximation?

1.4.8. Consider the following 6 × 6 black and white image. Describe two plausible ways to represent this image as a sequence of bits.

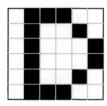

1.4.9. Design a truth table for the Boolean expression **not** (*a* **and** *b*).

1.4.10. Design a truth table for the Boolean expression **not** *a* **or not** *b*. Compare the result to the truth table for **not** (*a* **and** *b*). What do you notice? The relationship between these two Boolean expressions is one of *De Morgan's laws*.

1.4.11. Design a truth table for the Boolean expression **not** *a* **and not** *b*. Compare the result to the truth table for **not** (*a* **or** *b*). What do you notice? The relationship between these two Boolean expressions is the other of *De Morgan's laws*.

1.4.12. Design a finite state machine that represents a highway toll booth controlling a single gate.

1.4.13. Design a finite state machine that represents a vending machine. Assume that the machine only takes quarters and vends only one kind of drink, for 75 cents. First, think about what the states should be. Then design the transitions between states.

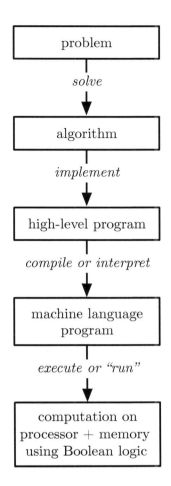

Figure 1.8 An expanded (from Figure 1.2) illustration of the layers of functional abstraction in a computer.

1.5 SUMMARY

In the first chapter, we developed a top-to-bottom view of computation at many layers of abstraction. As illustrated in Figure 1.8, we start with a well-defined problem and solve it by designing an algorithm. This step is usually the most difficult by far. Professional algorithm designers rely on experience, creativity, and their knowledge of a battery of general design techniques to craft efficient, correct algorithms. In this book, we will only scratch the surface of this challenging field. Next, an algorithm can be programmed in a high-level programming language like Python, which is translated into machine language before it can be executed on a computer. The instructions in this machine language program may either be executed directly on a processor or request that the operating system do something on their behalf (e.g., saving a file or allocating more memory). Each instruction in the resulting computation is implemented using Boolean logic, controlled by a finite state machine.

Try to keep this big picture in mind as you work through this book, especially the role of abstraction in effective problem solving. The problems we will tackle are modest, but they should give you an appreciation for the big questions and techniques used by computer scientists, and some of the fascinating interdisciplinary opportunities that await you.

1.6   FURTHER DISCOVERY

The epigraph at the beginning of this chapter is from an article by computer scientists Michael Fellows and Ian Parberry [13]. A similar quote is often attributed to the late Dutch computer scientist Edsger Dijkstra.

There are several excellent books available that give overviews of computer science and how computers work. In particular, we recommend *The Pattern on the Stone* by Danny Hillis [18], *Algorithmics* by David Harel [17], *Digitized* by Peter Bentley [5], and *CODE: The Hidden Language of Computer Hardware and Software* by Charles Petzold [40].

There really are drifting buoys in the world's oceans that are constantly taking temperature readings to monitor climate change. For example, see

http://www.coriolis.eu.org .

A list of the fastest supercomputers in the world is maintained at

http://top500.org .

You can learn more about IBM's Deep Blue supercomputer at

http://www-03.ibm.com/ibm/history/ibm100/us/en/icons/deepblue/

and IBM's Watson supercomputer at

http://www.ibm.com/smarterplanet/us/en/ibmwatson/ .

To learn more about high performance computing in general, we recommend looking at the website of the National Center for Supercomputing Applications (NCSA) at the University of Illinois, one of the first supercomputer centers funded by the National Science Foundation, at

http://www.ncsa.illinois.edu/about/faq .

There are several good books available on the life of Alan Turing, the father of computer science. The definitive biography, *Alan Turing: The Enigma*, was written by Andrew Hodges [20].

CHAPTER 2

Elementary computations

> It has often been said that a person does not really understand something until after teaching it to someone else. Actually a person does not really understand something until after teaching it to a computer, i.e., expressing it as an algorithm.
>
> Donald E. Knuth
> *American Scientist (1973)*

NUMBERS are some of the simplest and most familiar objects on which we can compute. Numbers are also quite versatile; they can represent quantities, measurements, identifiers, and are a means of ordering items. In this chapter, we will introduce some of the fundamentals of carrying out computations with numbers in the Python programming language, and point out the ways in which numbers in a computer sometimes behave differently from numbers in mathematics. In later chapters, we will build on these ideas to design algorithms that solve more sophisticated problems requiring objects like text, lists, tables, and networks.

2.1 WELCOME TO THE CIRCUS

Writing programs (or "programming") is a hands-on activity that allows us to test ideas, and harness the results of our algorithms, in a tangible and satisfying way. Solving problems and writing programs should be fun and creative. Guido van Rossum, the inventor of Python, "being in a slightly irreverent mood," understood this when he named the programming language after the British comedy series "Monty Python's Flying Circus!"

Because writing a program requires us to translate our ideas into a concise form that a computer can "understand," it also necessarily improves our understanding of the problem itself. The quote above by Donald Knuth, one of the founders of modern computer science, expresses this perfectly. In a similar vein, the only way to really learn how to design algorithms and write programs is to actually do it!

To get started, launch the application called IDLE that comes with every Python distribution (or another similar programming environment recommended by your instructor). You should see a window appear with something like this at the top:

```
Python 3.4.2 (v3.4.2:ab2c023a9432, Oct  5 2014, 20:42:22)
[GCC 4.2.1 (Apple Inc. build 5666) (dot 3)] on darwin
Type "copyright", "credits" or "license()" for more information.
>>>
```

The program executing in this window is known as a *Python shell*. The first line tells you which version of Python you are using (in this case, 3.4.2). The examples in this book are based on Python version 3, so make sure that your version number starts with a 3. If you need to install a newer version, you can find one at `python.org`. The

```
>>>
```

on the fourth line in the IDLE window is called the *prompt* because it is prompting you to type in a Python statement. To start, type in `print('Hello world!')` at the prompt and hit return.

```
>>> print('Hello world!')
Hello world!
```

Congratulations, you have just written the traditional first program! This one-statement program simply prints the characters in single quotes, called a *character string* or just *string*, on the screen. Now you are ready to go!

As you read this book, we highly recommend that you do so in front of a computer. Python statements that are preceded by a prompt, like those above, are intended to be entered into the Python shell by the reader. Try out each example for yourself. Then go beyond the examples, and experiment with your own ideas. Instead of just wondering, "What if I did this?", type it in and see what happens! Similarly, when you read a "Reflection" question, pause for a moment and think about the answer or perform an experiment to find the answer.

2.2 ARITHMETIC

We have to start somewhere, so it may as well be in elementary school, where you learned your first arithmetic computations. Python, of course, can do arithmetic too. At the prompt, type `8 * 9` and hit return.

```
>>> 8 * 9
72
>>>
```

The spaces around the `*` multiplication operator are optional and ignored. In general, Python does not care if you include spaces in expressions, but you always want to make your programs readable to others, and spaces often help.

> **Box 2.1: Floating point notation**
>
> Python floating point numbers are usually stored in 64 bits of memory, in a format known as IEEE 754 double-precision floating point. One bit is allocated to the sign of the number, 11 bits are allocated to the exponent, and 52 bits are allocated to the *mantissa*, the fractional part to the right of the point. After a number is converted to binary (as in Section 1.4), the binary point and exponent are adjusted so that there is a single 1 to the left of the binary point. Then the exponent is stored in its 11 bit slot and as much of the mantissa that can fit is stored in the 52 bit slot. If the mantissa is longer than 52 bits, it is simply truncated and information is lost. This is exactly what happened when we computed 2.0 ** 100. Since space is limited in computers' memory, tradeoffs between accuracy and space are common, and it is important to understand these limitations to avoid errors. In Chapter 4, we will see a more common situation in which this becomes quite important.

Notice that the shell responded to our input with the result of the computation, and then gave us a new prompt. The shell will continue this prompt–compute–respond cycle until we quit (by typing `quit()`). Recall from Section 1.4 that Python is an interpreted language; the shell is displaying the interpreter at work. Each time we enter a statement at the prompt, the Python interpreter transparently converts that statement to machine language, executes the machine language, and then prints the result.

Now let's try something that most calculators cannot do:

```
>>> 2 ** 100
1267650600228229401496703205376
```

The `**` symbol is the exponentiation operator; so this is computing 2^{100}, a 31-digit number. Python can handle even longer integer values; try computing one googol (10^{100}) and 2^{500}. Indeed, Python's integer values have *unlimited precision*, meaning that they can be arbitrarily long.

Next, try a slight variation on the previous statement:

```
>>> 2.0 ** 100
1.2676506002282294e+30
```

This result looks different because adding a decimal point to the 2 caused Python to treat the value as a *floating point number* instead of an integer. A floating point number is any number with a decimal point. Floating point derives its name from the manner in which the numbers are stored, a format in which the decimal point is allowed to "float," as in scientific notation. For example, $2.345 \times 10^4 = 23.45 \times 10^3 = 0.02345 \times 10^7$. To learn a bit more about floating point notation, see Box 2.1. Whenever a floating point number is involved in a computation, the result of the computation is also a floating point number. Very large floating point numbers, like `1.2676506002282294e+30`, are printed in scientific notation (the `e` stands for "exponent"). This number represents

$$1.2676506002282294 \times 10^{30} = 1267650600228229400000000000000.$$

It is often convenient to enter numbers in scientific notation also. For example, knowing that the earth is about 4.54 billion years old, we can compute the approximate number of days in the age of the earth with

```
>>> 4.54e9 * 365.25
1658235000000.0
```

Similarly, if the birth rate in the United States is 13.42 per 1,000 people, then the number of babies born in 2012, when the population was estimated to be 313.9 million, was approximately

```
>>> 13.42e-3 * 313.9e6
4212538.0
```

Finite precision

In normal arithmetic, 2^{100} and 2.0^{100} are, of course, the same number. However, on the previous page, the results of 2 ** 100 and 2.0 ** 100 were different. The first answer was correct, while the second was off by almost 1.5 trillion! The problem is that floating point numbers are stored differently from integers, and have *finite precision*, meaning that the range of numbers that can be represented is limited. Limited precision may mean, as in this case, that we get approximate results, or it may mean that a value is too large or too small to even be approximated well. For example, try:

```
>>> 10.0 ** 500
OverflowError: (34, 'Result too large')
```

An *overflow error* means that the computer did not have enough space to represent the correct value. A similar fatal error will occur if we try to do something illegal, like divide by zero:

```
>>> 10.0 / 0
ZeroDivisionError: division by zero
```

Division

Python provides two different kinds of division: so-called "true division" and "floor division." The true division operator is the slash (/) character and the floor division operator is two slashes (//). True division gives you what you would probably expect. For example, 14 / 3 will give 4.6666666666666667. On the other hand, floor division rounds the quotient down to the nearest integer. For example, 14 // 3 will give the answer 4. (Rounding down to the nearest integer is called "taking the floor" of a number in mathematics, hence the name of the operator.)

```
>>> 14 / 3
4.6666666666666667
>>> 14 // 3
4
```

When both integers are positive, you can think of floor division as the "long division" you learned in elementary school: floor division gives the whole quotient without the remainder. In this example, dividing 14 by 3 gives a quotient of 4 and a remainder of 2 because 14 is equal to $4 \cdot 3 + 2$. The operator to get the remainder is called the "modulo" operator. In mathematics, this operator is denoted mod, e.g., 14 mod 3 = 2; in Python it is denoted %. For example:

```
>>> 14 % 3
2
```

To see why the // and % operators might be useful, think about how you would determine whether an integer is odd or even. An integer is even if it is evenly divisible by 2; i.e., when you divide it by 2, the remainder is 0. So, to decide whether an integer is even, we can "mod" the number by 2 and check the answer.

```
>>> 14 % 2
0
>>> 15 % 2
1
```

Reflection 2.1 *The floor division and modulo operators also work with negative numbers. Try some examples, and try to infer what is happening. What are the rules governing the results?*

Consider dividing some integer value x by another integer value y. The floor division and modulo operators always obey the following rules:

1. x % y has the same sign as y, and
2. (x // y) * y + (x % y) is equal to x.

Confirm these observations yourself by computing

- -14 // 3 and -14 % 3
- 14 // -3 and 14 % -3
- -14 // -3 and -14 % -3

Order of operations

Now let's try something just slightly more interesting: computing the area of a circle with radius 10. (Recall the formula $A = \pi r^2$.)

```
>>> 3.14159 * 10 ** 2
314.159
```

36 ■ Elementary computations

	Operators	Description
1.	**	exponentiation (power)
2.	+, -	unary positive and negative, e.g., -(4 * 9)
3.	*, /, //, %	multiplication and division
4.	+, -	addition and subtraction

Table 2.1 Arithmetic operator precedence, highest to lowest. Operators with the same precedence are evaluated left to right.

Notice that Python correctly performs exponentiation before multiplication, as in standard arithmetic. These operator precedence rules, all of which conform to standard arithmetic, are summarized in Table 2.1. As you might expect, you can override the default precedence with parentheses. For example, suppose we wanted to find the average salary at a retail company employing 100 salespeople with salaries of $18,000 per year and one CEO with a salary of $18 million:

```
>>> ((100 * 18000) + 18e6) / 101
196039.60396039605
```

In this case, we needed parentheses around `(100 * 18000) + 18e6` to compute the total salaries before dividing by the number of employees. However, we also included parentheses around `100 * 18000` for clarity.

We should also point out that the answers in the previous two examples were floating point numbers, even though some of the numbers in the original expression were integers. When Python performs arithmetic with an integer and a floating point number, it first converts the integer to floating point. For example, in the last expression, `(100 * 18000)` evaluates to the integer value `1800000`. Then `1800000` is added to the floating point number `18e6`. Since these two operands are of different types, `1800000` is converted to the floating point equivalent `1800000.0` before adding. Then the result is the floating point value `19800000.0`. Finally, `19800000.0` is divided by the integer `101`. Again, `101` is converted to floating point first, so the actual division operation is `19800000.0 / 101.0`, giving the final answer. This process is also depicted below.

$$((100 * 18000) + 18e6) / 101$$
$$\underbrace{}_{1800000}$$
$$\underbrace{}_{19800000.0}$$
$$\underbrace{}_{196039.60396039605}$$

In most cases, this works exactly as we would expect, and we do not need to think about this automatic conversion taking place.

Complex numbers

Although we will not use them in this book, it is worth pointing out that Python can also handle complex numbers. A complex number has both a real part and an imaginary part involving the imaginary unit i, which has the property that $i^2 = -1$. In Python, a complex number like $3.2 + 2i$ is represented as `3.2 + 2j`.(The letter j is used instead of i because in some fields, such as electrical engineering, i has another well-established meaning that could lead to ambiguities.) Most of the normal arithmetic operators work on complex numbers as well. For example,

```
>>> (5 + 4j) + (-4 + -3.1j)
(1+0.8999999999999999j)
>>> (23 + 6j) / (1 + 2j)
(7-8j)
>>> (1 + 2j) ** 2
(-3+4j)
>>> 1j ** 2
(-1+0j)
```

The last example illustrates the definition of i: $i^2 = -1$.

Exercises

Use the Python interpreter to answer the following questions. Where appropriate, provide both the answer and the Python expression you used to get it.

2.2.1. The Library of Congress stores its holdings on 838 miles of shelves. Assuming an average book is one inch thick, how many books would this hold? (There are 5,280 feet in one mile.)

2.2.2. If I gave you a nickel and promised to double the amount you have every hour for the next 24, how much money would you have at the end? What if I only increased the amount by 50% each hour, how much would you have? Use exponentiation to compute these quantities.

2.2.3. The Library of Congress stores its holdings on 838 miles of shelves. How many round trips is this between Granville, Ohio and Columbus, Ohio?

2.2.4. The earth is estimated to be 4.54 billion years old. The oldest known fossils of anatomically modern humans are about 200,000 years old. What fraction of the earth's existence have humans been around? Use Python's scientific notation to compute this.

2.2.5. If you counted at an average pace of one number per second, how many years would it take you to count to 4.54 billion? Use Python's scientific notation to compute this.

2.2.6. Suppose the internal clock in a modern computer can "count" about 2.8 billion ticks per second. How long would it take such a computer to tick 4.54 billion times?

2.2.7. A hard drive in a computer can hold about a terabyte (2^{40} bytes) of information. An average song requires about 8 megabytes (8×2^{20} bytes). How many songs can the hard drive hold?

38 ■ Elementary computations

2.2.8. What is the value of each of the following Python expressions? Make sure you understand why in each case.

(a) 15 * 3 - 2
(b) 15 - 3 * 2
(c) 15 * 3 // 2
(d) 15 * 3 / 2
(e) 15 * 3 % 2
(f) 15 * 3 / 2e0

2.3 WHAT'S IN A NAME?

Both in algorithms and in programs, a name associated with a value is called a *variable*. For example, in computing the volume of a sphere (using the formula $V = (4/3)\pi r^3$), we might use a variable to refer to the sphere's radius and another variable to refer to the (approximate) value of π:

```
>>> pi = 3.14159
>>> radius = 10
>>> (4 / 3) * pi * (radius ** 3)
4188.786666666666
```

The equal sign (=) is called the *assignment operator* because it is used to assign values to names. In this example, we assigned the value **3.14159** to the name **pi** and we assigned the value **10** to the name **radius**. It is convenient to think of variable names as "Sticky notes"[1] attached to locations in the computer's memory.

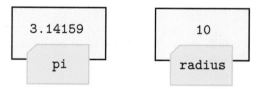

Recall from Section 1.4 that numbers are stored in cells in a computer's memory. These cells are analogous to post office boxes; the cell's address is like the number on the post office box and the value stored in the cell is like the box's content. In the picture above, each rectangle containing a number represents a memory cell. A variable name is a reference to a particular cell, analogous to a Sticky note with a name on it on a post office box. In this picture, the name **pi** refers to the cell containing the value **3.14159** and the name **radius** refers to the cell containing the value **10**. We do not actually know the addresses of each of these memory cells, but there is no reason for us to because we will always refer to them by a variable name.

As with a sticky note, a value can be easily reassigned to a different name at a later time. For example, if we now assign a different value to the name **radius**, the **radius** Sticky note in the picture moves to a different value.

---

[1] "Sticky note" is a registered trademark of the BIC Corporation.

and	continue	except	global	lambda	pass	while
as	def	False	if	None	raise	with
assert	del	finally	import	nonlocal	return	yield
break	elif	for	in	not	True	
class	else	from	is	or	try	

Table 2.2 The 33 Python keywords.

```
>>> radius = 17
```

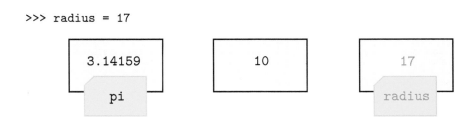

After the reassignment, a reference to a different memory cell containing the value 17 is assigned `radius`. The value 10 may briefly remain in memory without a name attached to it, but since we can no longer access that value without a reference to it, the Python "garbage collection" mechanism will soon free up the memory that it occupies.

Naming values serves three important purposes. First, assigning oft-used values to descriptive names can make our algorithms much easier to understand. For example, if we were writing an algorithm to model competition between two species, such as wolves and moose (as in Project 4.4), using names for the birth rates of the two species, like `birthRateMoose` and `birthRateWolves`, rather than unlabeled values, like 0.5 and 0.005, makes the algorithm much easier to read. Second, as we will see in Section 3.5, naming inputs will allow us to generalize algorithms so that, instead of being tied to one particular input, they work for a variety of possible inputs. Third, names will serve as labels for computed values that we wish to use later, obviating the need to compute them again at that time.

Notice that we are using descriptive names, not just single letters, as is the convention in algebra. As noted above, using descriptive names is always important when you are writing programs because it makes your work more accessible to others. In the "real world," programming is almost always a collaborative endeavor, so it is important for others to be able to understand your intentions. Sometimes assigning an intermediate computation to a descriptive name, even if not required, can improve readability.

Names in Python can be any sequence of characters drawn from letters, digits, and the underscore (_) character, but they may not start with a digit. You also cannot use any of Python's *keywords*, shown in Table 2.2. Keywords are elements of the Python language that have predefined meanings. We saw a few of these already in the snippets of Python code in Chapter 1 (`def`, `for`, `in`, and `return`), and we will encounter most of the others as we progress through this book.

Let's try breaking some of these naming rules to see what happens.

```
>>> 49ers = 50
SyntaxError: invalid token
```

A *syntax error* indicates a violation of the *syntax*, or *grammar*, of the Python language. It is normal for novice programmers to encounter many, many syntax errors; it is part of the learning process. Most of the time, it will be immediately obvious what you did wrong, you will fix the mistake, and move on. Other times, you will need to look harder to discover the problem but, with practice, these instances too will become easier to diagnose. In this case, "invalid token" means that we have mistakenly used a name that starts with a digit. *Token* is a technical term that refers to any group of characters that the interpreter considers meaningful; in this case, it just refers to the name we chose.

Next, try this one.

```
>>> my-age = 21
SyntaxError: can't assign to operator
```

This syntax error is referring to the dash/hyphen/minus sign symbol (-) that we have in our name. Python interprets the symbol as the minus operator, which is why it is not allowed in names. Instead, we can use the underscore (_) character (i.e., **my_age**) or vary the capitalization (i.e., **myAge**) to distinguish the two words in the name.

In addition to assigning constant values to names, we can assign names to the results of whole computations.

```
>>> volume = (4 / 3) * pi * (radius ** 3)
```

When you assign values to names in the Python shell, the shell does not offer any feedback. But you can view the value assigned to a variable by simply typing its name.

```
>>> volume
20579.50889333333
```

So we see that we have assigned the value **20579.50889333333** to the name **volume**, as depicted below. (The values assigned to **pi** and **radius** are unchanged.)

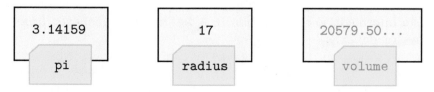

Alternatively, you can display a value using **print**, followed by the value you wish to see in parentheses. Soon, when we begin to write programs outside of the shell, **print** will be the only way to display values.

```
>>> print(volume)
20579.50889333333
```

Now let's change the value of `radius` again:

```
>>> radius = 20
```

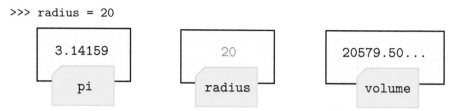

Since the value of `volume` was based on the value of `radius`, it makes sense to check whether the value of `volume` has also changed. Try it:

```
>>> volume
20579.50889333333
```

What's going on here? While the value assigned to `radius` has changed, the value assigned to `volume` has not. This example demonstrates that assignment is a one-time event; Python does not "remember" how the value of `volume` was computed. Put another way, an assignment is *not* creating an equivalence between a name and a computation. Rather, it performs the computation on the righthand side of the assignment operator only when the assignment happens, and then assigns the result to the name on the lefthand side. That value remains assigned to the name until it is explicitly assigned some other value or it ceases to exist. To compute a new value for `volume` based on the new value of `radius`, we would need to perform the volume computation again.

```
>>> volume = (4 / 3) * pi * (radius ** 3)
>>> volume
33510.29333333333
```

Now the value assigned to `volume` has changed, due to the explicit assignment statement above.

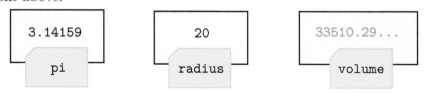

Let's try another example. The formula for computing North American wind chill temperatures, in degrees Celsius, is

$$W = 13.12 + 0.6215\,t + (0.3965\,t - 11.37)\,v^{0.16}$$

where t is the ambient temperature in degrees Celsius and v is the wind speed in km/h.[2] To compute the wind chill for a temperature of $-3°$ C and wind speed of 13 km/h, we will first assign the temperature and wind speed to two variables:

---

[2]Technically, wind chill is only defined at or below 10° C and for wind speeds above 4.8 km/h.

```
>>> temp = -3
>>> wind = 13
```

Then we type in the formula in Python, assigning the result to the name `chill`.

```
>>> chill = 13.12 + 0.6215 * temp + (0.3965 * temp - 11.37) * wind**0.16
>>> chill
-7.676796032159553
```

Notice once again that changing `temp` does not change `chill`:

```
>>> temp = 4.0
>>> chill
-7.676796032159553
```

To recompute the value of `chill` with the new temperature, we need to re-enter the entire formula above, which will now use the new value assigned to `temp`.

```
>>> chill = 13.12 + 0.6215 * temp + (0.3965 * temp - 11.37) * wind**0.16
>>> chill
0.8575160333891443
```

This process is, of course, very tedious and error-prone. Fortunately, there is a much better way to define such computations, as we will see in Section 3.5.

It is very important to realize that, despite the use of the equals sign, assignment is not the same as equality. In other words, when we execute an assignment statement like `temp = 4.0`, we are not asserting that `temp` and `4.0` are equivalent. Instead, we are assigning a value to a name in two steps:

1. Evaluate the expression on the righthand side of the assignment operator.

2. Assign the resulting value to the name on the lefthand side of the assignment operator.

For example, consider the following assignment statements:

```
>>> radius = 20
>>> radius = radius + 1
```

If the equals sign denoted equality, then the second statement would not make any sense! However, if we interpret the statement using the two-step process, it is perfectly reasonable. First, we evaluate the expression on the righthand side, ignoring the lefthand side entirely. Since, at this moment, the value 20 is assigned to `radius`, the righthand side evaluates to 21. Second, we assign the value 21 to the name `radius`. So this statement has added 1 to the current value of `radius`, an operation we refer to as an *increment*. To verify that this is what happened, check the value of radius:

```
>>> radius
21
```

What if we had not assigned a value to `radius` before we tried to increment it? To find out, we have to use a name that has not yet been assigned anything.

```
>>> trythis = trythis + 1
NameError: name 'trythis' is not defined
```

This *name error* occurred because, when the Python interpreter tried to evaluate the righthand side of the assignment, it found that `trythis` was not assigned a value, i.e., it was not defined. So we need to make sure that we first define any variable that we try to increment. This sounds obvious but, in the context of some larger programs later on, it might be easy to forget.

Let's look at one more example. Suppose we want to increment the number of minutes displayed on a digital clock. To initialize the minutes count and increment the value, we essentially copy what we did above.

```
>>> minutes = 0
>>> minutes = minutes + 1
>>> minutes
1
>>> minutes = minutes + 1
>>> minutes
2
```

But incrementing is not going to work properly when `minutes` reaches `59`, and we need the value of `minutes` to wrap around to 0 again. The solution lies with the modulo operator: if we mod the incremented value by 60 each time, we will get exactly the behavior we want. We can see what happens if we reset `minutes` to 58 and increment a few times:

```
>>> minutes = 58
>>> minutes = (minutes + 1) % 60
>>> minutes
59
>>> minutes = (minutes + 1) % 60
>>> minutes
0
>>> minutes = (minutes + 1) % 60
>>> minutes
1
```

When `minutes` is between 0 and 58, `(minutes + 1) % 60` gives the same result as `minutes + 1` because `minutes + 1` is less than 60. But when `minutes` is 59, `(minutes + 1) % 60` equals `60 % 60`, which is 0. This example illustrates why arithmetic using the modulo operator, formally called *modular arithmetic*, is often also called "clock arithmetic."

44 ■ Elementary computations

Exercises

Use the Python interpreter to answer the following questions. Where appropriate, provide both the answer and the Python expression you used to get it.

2.3.1. Every cell in the human body contains about 6 billion base pairs of DNA (3 billion in each set of 23 chromosomes). The distance between each base pair is about 3.4 angstroms (3.4×10^{-10} meters). Uncoiled and stretched, how long is the DNA in a single human cell? There are about 50 trillion cells in the human body. If you stretched out all of the DNA in the human body end to end, how long would it be? How many round trips to the sun is this? The distance to the sun is about 149,598,000 kilometers. Write Python statements to compute the answer to each of these three questions. Assign variables to hold each of the values so that you can reuse them to answer subsequent questions.

2.3.2. Set a variable named **radius** to have the value 10. Using the formula for the area of a circle ($A = \pi r^2$), assign to a new variable named **area** the area of a circle with radius equal to your variable **radius**. (The number 10 should not appear in the formula.)

2.3.3. Now change the value of **radius** to 15. What is the value of **area** now? Why?

2.3.4. Suppose we wanted to swap the values of two variables named x and y. Why doesn't the following work? Show a method that does work.

```
x = y
y = x
```

2.3.5. What are the values of x and y at the end of the following? Make sure you understand why this is the case.

```
x = 12.0
y = 2 * x
x = 6
```

2.3.6. What is the value of x at the end of the following? Make sure you understand why this is the case.

```
x = 0
x = x + 1
x = x + 1
x = x + 1
```

2.3.7. What are the values of x and y at the end of the following? Make sure you understand why this is the case.

```
x = 12.0
y = 6
y = y * x
```

2.3.8. Given a variable x that refers to an integer value, show how to extract the individual digits in the ones, tens and hundreds places. (Use the // and % operators.)

2.4 USING FUNCTIONS

A programming language like Python provides a rich set of functional abstractions that enable us to solve a wide variety of interesting problems. Variable names and arithmetic operators like ** are simple examples of this. Variable names allow us to easily refer to computation results without having to remember their exact values (e.g., volume in place of 33510.29333333333) and operators like ** allow us to perform more complex operations without explicitly specifying how (e.g., power notation in place of a more explicit algorithm that repeatedly multiplies).

In addition to the standard arithmetic operators, Python offers several more complex mathematical functions, each of which is itself another functional abstraction. A *function* takes one or more inputs, called *arguments*, and produces an output from them. You have probably seen mathematical functions like $f(x) = x^2 + 3$ before. In this case, f is the name of the function, x is the single argument, and $f(x)$ represents the output. The output is computed from the input (i.e., argument) by a simple algorithm represented on the righthand side of the equals sign. In this case, the algorithm is "square the input and add 3." If we substitute x with a specific input value, then the algorithm produces a specific output value. For example, if x is 4, then the output is $f(4) = 4^2 + 3 = 19$, as illustrated below.

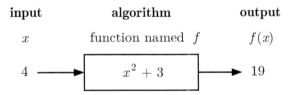

Does this look familiar? Like the example problems in Figure 1.1 (at the beginning of Chapter 1), a mathematical function has an input, an output, and an algorithm that computes the output from the input. So a mathematical function is another type of functional abstraction.

Built-in functions

As we suggested above, in a programming language, functional abstractions are also implemented as functions. In Chapter 3, we will learn how to define our own functions to implement a variety of functional abstractions, most of which are much richer than simple mathematical functions. In this section, we will set the stage by looking at how to use some of Python's built-in functions. Perhaps the simplest of these is the Python function abs, which outputs the absolute value (i.e., magnitude) of its argument.

```
>>> result2 = abs(-5.0)
>>> result2
5.0
```

On the righthand side of the assignment statement above is a *function call*, also called a *function invocation*. In this case, the abs function is being called with argument -5.0. The function call executes the (hidden) algorithm associated with the **abs** function, and **abs(-5.0)** evaluates to the output of the function, in this case, 5.0. The output of the function is also called the function's *return value*. Equivalently, we say that the function *returns* its output value. The return value of the function call, in this case 5.0, is then assigned to the variable name **result2**.

Two other helpful functions are **float** and **int**. The **float** function converts its argument to a floating point number, if it is not already.

```
>>> float(3)
3.0
>>> float(2.718)
2.718
```

The **int** function converts its argument to an integer by truncating it, i.e., removing the fractional part to the right of the decimal point. For example,

```
>>> int(2.718)
2
>>> int(-1.618)
-1
```

Recall the wind chill computation from Section 2.3:

```
>>> temp = -3
>>> wind = 13
>>> chill = 13.12 + 0.6215 * temp + (0.3965 * temp - 11.37) * wind**0.16
>>> chill
-7.676796032159553
```

Normally, we would want to truncate or round the final wind chill temperature to an integer. (It would be surprising to hear the local television meteorologist tell us, "the expected wind chill will be −7.676796032159553° C today.") We can easily compute the truncated wind chill with

```
>>> truncChill = int(chill)
>>> truncChill
-7
```

There is also a **round** function that we can use to round the temperature instead.

```
>>> roundChill = round(chill)
>>> roundChill
-8
```

Alternatively, if we do not need to retain the original value of **chill**, we can simply assign the modified value to the name **chill**:

```
>>> chill = round(chill)
>>> chill
-8
```

Function arguments can be more complex than just single constants and variables; they can be anything that evaluates to a value. For example, if we want to convert the wind chill above to Fahrenheit and then round the result, we can do so like this:

```
>>> round(chill * 9/5 + 32)
18
```

The expression in parentheses is evaluated first, and then the result of this expression is used as the argument to the **round** function.

Not all functions output something. For example, the **print** function, which simply prints its arguments to the screen, does not. For example, try this:

```
>>> x = print(chill)
-8
>>> print(x)
None
```

The variable **x** was assigned whatever the **print** function returned, which is different from what it printed. When we print **x**, we see that it was assigned something called **None**. **None** is a Python keyword that essentially represents "nothing." Any function that does not define a return value itself returns **None** by default. We will see this again shortly when we learn how to define our own functions.

Strings

As in the first "Hello world!" program, we often use the **print** function to print text:

```
>>> print('I do not like green eggs and ham.')
I do not like green eggs and ham.
```

The sentence in quotes is called a *string*. A string is simply text, a sequence of characters, enclosed in either single (') or double quotes ("). The **print** function prints the characters inside the quotation marks to the screen. The **print** function can actually take several arguments, separated by commas:

```
>>> print('The wind chill is', chill, 'degrees Celsius.')
The wind chill is -8 degrees Celsius.
```

The first and last arguments are strings, and the second is the variable we defined above. You will notice that a space is inserted between arguments, and there are no quotes around the variable name **chill**.

Reflection 2.2 *Why do you think quotation marks are necessary around strings? What error do you think would result from typing the following instead? (Try it.)*

```
>>> print(The expected wind chill is, chill, degrees Celsius.)
```

48 ■ Elementary computations

The quotation marks are necessary because otherwise Python has no way to distinguish text from a variable or function name. Without the quotation marks, the Python interpreter will try to make sense of each argument inside the parentheses, assuming that each word is a variable or function name, or a reserved word in the Python language. Since this sequence of words does not follow the syntax of the language, and most of these names are not defined, the interpreter will print an error.

String values can also be assigned to variables and manipulated with the + and * operators, but the operators have different meanings. The + operator *concatenates* two strings into one longer string. For example,

```
>>> first = 'Monty'
>>> last = 'Python'
>>> name = first + last
>>> print(name)
MontyPython
```

To insert a space between the first and last names, we can use another + operator:

```
>>> name = first + ' ' + last
>>> print(name)
Monty Python
```

Alternatively, if we just wanted to print the full name, we could have bypassed the `name` variable altogether.

```
>>> print(first + ' ' + last)
Monty Python
```

Reflection 2.3 *Why do we not want quotes around* `first` *and* `last` *in the statement above? What would happen if we did use quotes?*

Placing quotes around `first` and `last` would cause Python to interpret them literally as strings, rather than as variables:

```
>>> print('first' + ' ' + 'last')
first last
```

Applied to strings, the * operator becomes the *repetition operator*, which repeats a string some number of times. The operand on the left side of the repetition operator is a string and the operand on the right side is an integer that indicates the number of times to repeat the string.

```
>>> last * 4
'PythonPythonPythonPython'
>>> print(first + ' ' * 10 + last)
Monty          Python
```

We can also interactively query for string input in our programs with the `input` function. The `input` function takes a string prompt as an argument and returns a string value that is typed in response. For example, the following statement prompts for your name and prints a greeting.

```
>>> name = input('What is your name? ')
What is your name? George
>>> print('Howdy,', name, '!')
Howdy, George !
```

The call to the `input` function above prints the string `'What is your name? '` and then waits. After you type something (above, we typed `George`, shown in red) and hit Return, the text that we typed is returned by the `input` function and assigned to the variable called `name`. The value of `name` is then used in the `print` function.

Reflection 2.4 *How can we use the + operator to avoid the inserted space before the exclamation point above?*

To avoid the space, we can construct the string that we want to print using the concatenation operator instead of allowing the `print` function to insert spaces between the arguments:

```
>>> print('Howdy, ' + name + '!')
Howdy, George!
```

If we want to input a numerical value, we need to convert the string returned by the `input` function to an integer or float, using the `int` or `float` function, before we can use it as a number. For example, the following statements prompt for your age, and then print the number of days you have been alive.

```
>>> text = input('How old are you? ')
How old are you? 18
>>> age = int(text)
>>> print('You have been alive for', age * 365, 'days!')
You have been alive for 6570 days!
```

In the response to the prompt above, we typed 18 (shown in red). Then the `input` function assigned what we typed to the variable `text` as a string, in this case `'18'` (notice the quotes). Then, using the `int` function, the string is converted to the integer value 18 (no quotes) and assigned to the variable `age`. Now that `age` is a numerical value, it can be used in the arithmetic expression `age * 365`.

Reflection 2.5 *Type the statements above again, omitting* `age = int(text)`:

```
>>> text = input('How old are you? ')
How old are you? 18
>>> print('You have been alive for', text * 365, 'days!')
```

What happened? Why?

50 ■ Elementary computations

Because the value of `text` is a string rather than an integer, the `*` operator was interpreted as the repetition operator instead!

Sometimes we want to perform conversions in the other direction, from a numerical value to a string. For example, if we printed the number of days in the following form instead, it would be nice to remove the space between the number of days and the period.

```
>>> print('The number of days in your life is', age * 365, '.')
The number of days in your life is 6570 .
```

But the following will not work because we cannot concatenate the numerical value `age * 365` with the string `'.'`.

```
>>> print('The number of days in your life is', age * 365 + '.')
TypeError: unsupported operand type(s) for +: 'int' and 'str'
```

To make this work, we need to first convert `age * 365` to a string with the `str` function:

```
>>> print('The number of days in your life is', str(age * 365) + '.')
The number of days in your life is 6570.
```

Assume the value 18 is assigned to `age`, then the `str` function converts the integer value 6570 to the string '6570' before concatenating it to '.'.

For a final, slightly more involved, example here is how we could prompt for a temperature and wind speed with which to compute the current wind chill.

```
>>> text = input('Temperature: ')
Temperature: 2.3
>>> temp = float(text)
>>> text = input('Wind speed: ')
Wind speed: 17.5
>>> wind = float(text)
>>> chill = 13.12 + 0.6215 * temp + (0.3965 * temp - 11.37) * wind**0.16
>>> chill = round(chill)
>>> print('The wind chill is', chill, 'degrees Celsius.')
The wind chill is -2 degrees Celsius.
```

Reflection 2.6 *Why do we use* `float` *above instead of* `int`*? Replace one of the calls to* `float` *with a call to* `int`*, and then enter a floating point value (like 2.3). What happens and why?*

```
>>> text = input('Temperature: ')
Temperature: 2.3
>>> temp = int(text)
```

Modules

In Python, there are many more mathematical functions available in a *module* named `math`. A module is an existing Python program that contains predefined values and functions that you can use. To access the contents of a module, we use the `import` keyword. To access the `math` module, type:

```
>>> import math
```

After a module has been imported, we can access the functions in the module by preceding the name of the function we want with the name of the module, separated by a period (`.`). For example, to take the square root of 5, we can use the square root function `math.sqrt`:

```
>>> math.sqrt(5)
2.23606797749979
```

Other commonly used functions from the `math` module are listed in Appendix B.1. The `math` module also contains two commonly used constants: `pi` and `e`. Our volume computation earlier would have been more accurately computed with:

```
>>> radius = 20
>>> volume = (4 / 3) * math.pi * (radius ** 3)
>>> volume
33510.32163829113
```

Notice that, since `pi` and `e` are variable names, not functions, there are no parentheses after their names.

Function calls can also be used in longer expressions, and as arguments of other functions. In this case, it is useful to think about a function call as equivalent to the value that it returns. For example, we can use the `math.sqrt` function in the computation of the volume of a tetrahedron with edge length $h = 7.5$, using the formula $V = h^3/(6\sqrt{2})$.

```
>>> h = 7.5
>>> volume = h ** 3 / (6 * math.sqrt(2))
>>> volume
49.71844555217912
```

In the parentheses, the value of `math.sqrt(2)` is computed first, and then multiplied by 6. Finally, `h ** 3` is divided by this result, and the answer is assigned to `volume`. If we want the rounded volume, we can use the entire volume computation as the argument to the `round` function:

```
>>> volume = round(h ** 3 / (6 * math.sqrt(2)))
>>> volume
50
```

We illustrate below the complete sequence of events in this evaluation:

$$\text{round}(\underbrace{\underbrace{\underbrace{\underbrace{h\ **\ 3}_{421.875}/\underbrace{(6\ *\ \underbrace{\text{math.sqrt}(2)}_{1.4142...})}_{8.4852...}}_{49.7184...}}_{50})$$

Now suppose we wanted to find the cosine of a 52° angle. We can use the `math.cos` function to compute the cosine, but the Python trigonometric functions expect their arguments to be in radians instead of degrees. (360 degrees is equivalent to 2π radians.) Fortunately, the `math` module provides a function named `radians` that converts degrees to radians. So we can find the cosine of a 52° angle like this:

```
>>> math.cos(math.radians(52))
0.6156614753256583
```

The function call `math.radians(52)` is evaluated first, giving the equivalent of 52° in radians, and this result is used as the argument to the `math.cos` function:

$$\underbrace{\text{math.cos}(\underbrace{\text{math.radians}(52)}_{0.9075...})}_{0.6156...}$$

Finally, we note that, if you need to compute with complex numbers, you will want to use the `cmath` module instead of `math`. The names of the functions in the two modules are largely the same, but the versions in the `cmath` module understand complex numbers. For example, attempting to find the square root of a negative number using `math.sqrt` will result in an error:

```
>>> math.sqrt(-1)
ValueError: math domain error
```

This *value error* indicates that −1 is outside the domain of values for the `math.sqrt` function. On the other hand, calling the `cmath.sqrt` function to find $\sqrt{-1}$ returns the value of i:

```
>>> import cmath
>>> cmath.sqrt(-1)
1j
```

Exercises

Use the Python interpreter to answer the following questions. Where appropriate, provide both the answer and the Python expression you used to get it.

2.4.1. Try taking the absolute value of both −8 and −8.0. What do you notice?

2.4.2. Find the area of a circle with radius 10. Use the value of π from the `math` module.

2.4.3. The geometric mean of two numbers is the square root of their product. Find the geometric mean of 18 and 31.

2.4.4. Suppose you have P dollars in a savings account that will pay interest rate r, compounded n times per year. After t years, you will have

$$P\left(1 + \frac{r}{n}\right)^{nt}$$

dollars in your account. If the interest were compounded continuously (i.e., with n approaching infinity), you would instead have

$$Pe^{rt}$$

dollars after t years, where e is Euler's number, the base of the natural logarithm. Suppose you have P = $10,000 in an account paying 1% interest (r = 0.01), compounding monthly. How much money will you have after t = 10 years? How much more money would you have after 10 years if the interest were compounded continuously instead? (Use the `math.exp` function.)

2.4.5. Show how you can use the `int` function to truncate any positive floating point number x to two places to the right of the decimal point. In other words, you want to truncate a number like `3.1415926` to `3.14`. Your expression should work with any value of x.

2.4.6. Show how you can use the `int` function to find the fractional part of any positive floating point number. For example, if the value 3.14 is assigned to x, you want to output 0.14. Your expression should work with any value of x.

2.4.7. The well-known quadratic formula, shown below, gives solutions to the quadratic equation $ax^2 + bx + c = 0$.

$$x = \frac{-b \pm \sqrt{b^2 - 4ac}}{2a}$$

Show how you can use this formula to compute the two solutions to the equation $3x^2 + 4x - 5 = 0$.

2.4.8. Suppose we have two points (x1, y1) and (x2, y2). The distance between them is equal to

$$\sqrt{(x1 - x2)^2 + (y1 - y2)^2}.$$

Show how to compute this in Python. Make sure you test your answer with real values.

54 ■ Elementary computations

2.4.9. A parallelepiped is a three-dimensional box in which the six sides are parallelograms. The volume of a parallelepiped is

$$V = abc\sqrt{1 + 2\cos(x)\cos(y)\cos(z) - \cos(x)^2 - \cos(y)^2 - \cos(z)^2}$$

where a, b, and c are the edge lengths, and x, y, and z are the angles between the edges, in radians. Show how to compute this in Python. Make sure you test your answer with real values.

2.4.10. Repeat the previous exercise, but now assume that the angles are given to you in degrees.

2.4.11. Repeat Exercise 2.4.4, but prompt for each of the four values first using the `input` function, and then print the result.

2.4.12. Repeat Exercise 2.4.7, but prompt for each of the values of a, b, and c first using the `input` function, and then print the results.

2.4.13. The following program implements a Mad Lib.

```
adj1 = input('Adjective: ')
noun1 = input('Noun: ')
noun2 = input('Noun: ')
adj2 = input('Adjective: ')
noun3 = input('Noun: ')

print('How to Throw a Party')
print()
print('If you are looking for a/an', adj1, 'way to')
print('celebrate your love of', noun1 + ', how about a')
print(noun2 + '-themed costume party?  Start by')
print('sending invitations encoded in', adj2, 'format')
print('giving directions to the location of your', noun3 + '.')
```

Write your own Mad Lib program, requiring at least five parts of speech to insert. (You can download the program above from the book web site to get you started.)

*2.5 BINARY ARITHMETIC

Recall from Section 1.4 that numbers are stored in a computer's memory in binary. Therefore, computers must also perform arithmetic in binary. In this section, we will briefly explore how binary addition works.

Although it may be unfamiliar, binary addition is actually much easier than decimal addition since there are only three basic binary addition facts: 0 + 0 = 0, 0 + 1 = 1 + 0 = 1, and 1 + 1 = 10. (Note that this is 10 in binary, which has the value 2 in decimal.) With these basic facts, we can add any two arbitrarily long binary numbers using the same right-to-left algorithm that we all learned in elementary school.

For example, let's add the binary numbers `1110` and `0111`. Starting on the right, we add the last column: 0 + 1 = 1.

```
      1 1 1 0
  +   0 1 1 1
  ─────────────
              1
```

In the next column, we have 1 + 1 = 10. Since the answer contains more than one bit, we carry the 1.

```
        1
      1 1 1 0
  +   0 1 1 1
  ─────────────
            0 1
```

In the next column, we have 1 + 1 = 10 again, but with a carry bit as well. Adding in the carry, we have 10 + 1 = 11 (or 2 + 1 = 3 in decimal). So the answer for the column is 1, with a carry of 1.

```
      1 1
      1 1 1 0
  +   0 1 1 1
  ─────────────
          1 0 1
```

Finally, in the leftmost column, with the carry, we have 1 + 0 + 1 = 10. We write the 0 and carry the 1, and we are done.

```
    1 1 1
      1 1 1 0
  +   0 1 1 1
  ─────────────
    1 0 1 0 1
```

We can easily check our work by converting everything to decimal. The top number in decimal is 8 + 4 + 2 + 0 = 14 and the bottom number in decimal is 0 + 4 + 2 + 1 = 7. Our answer in decimal is 16 + 4 + 1 = 21. Sure enough, 14 + 7 = 21.

Finite precision

Although Python integers can store arbitrarily large values, this is not true at the machine language level. In other words, Python integers are another abstraction built atop the native capabilities of the computer. At the machine language level (and in most other programming langauges), every integer is stored in a fixed amount of memory, usually four bytes (32 bits). This is another example of *finite precision*, and it means that sometimes the result of an arithmetic operation is too large to fit.

We can illustrate this by revisiting the previous problem, but assuming that we only have four bits in which to store each integer. When we add the four-bit integers 1110 and 0111, we arrived at a sum, 10101, that requires five bits to be represented. When a computer encounters this situation, it simply discards the leftmost bit. In our example, this would result in an incorrect answer of 0101, which is 5 in decimal. Fortunately, there are ways to detect when this happens, which we leave to you to discover as an exercise.

Negative integers

We assumed above that the integers we were adding were positive, or, in programming terminology, *unsigned*. But of course computers must also be able to handle arithmetic with *signed* integers, both positive and negative.

Everything, even a negative sign, must be stored in a computer in binary. One option for representing negative integers is to simply reserve one bit in a number to represent the sign, say 0 for positive and 1 for negative. For example, if we store every number with eight bits and reserve the first (leftmost) bit for the sign, then 00110011 would represent 51 and 10110011 would represent −51. This approach is known as *sign and magnitude* notation. The problem with this approach is that the computer then has to detect whether a number is negative and handle it specially when doing arithmetic.

For example, suppose we wanted to add −51 and 102 in sign and magnitude notation. In this notation, −51 is 10110011 and 102 is 01100110. First, we notice that 10110011 is negative because it has 1 as its leftmost bit and 01100110 is positive because it has 0 as its leftmost bit. So we need to *subtract* positive 51 from 102:

```
           0  10 10    0 10 10
     0   1̸  1̸  0̸  0  1̸  1̸  0̸     ← 102
  −  0   0  1  1  0  0  1  1     ←  51
     ─────────────────────────
     0   0  1  1  0  0  1  1     ←  51
```

Borrowing in binary works the same way as in decimal, except that we borrow a 2 (10 in binary) instead of a 10. Finally, we leave the sign of the result as positive because the largest operand was positive.

To avoid these complications, computers use a clever representation called *two's complement notation*. Integers stored in two's complement notation can be added directly, regardless of their sign. The leftmost bit is also the sign bit in two's complement notation, and positive numbers are stored in the normal way, with leading zeros if necessary to fill out the number of bits allocated to an integer. To convert a positive number to its negative equivalent, we *invert every bit to the left of the rightmost* 1. For example, since 51 is represented in eight bits as 00110011, −51 is represented as 11001101. Since 4 is represented in eight bits as 00000100, −4 is represented as 11111100.

To illustrate how addition works in two's complement notation, let's once again add −51 and 102:

```
        1  1        1  1
        1  1  0  0  1  1  0  1     ←  −51
   +    0  1  1  0  0  1  1  0     ←  102
      ─────────────────────────
    1̸  0  0  1  1  0  0  1  1      ←   51
```

As a final step in the addition algorithm, we always disregard an extra carry bit. So, indeed, in two's complement, −51 + 102 = 51.

Note that it is still possible to get an incorrect answer in two's complement if the answer does not fit in the given number of bits. Some exercises below prompt you to investigate this further.

Designing an adder

Let's look at how to an *adder* that takes in two single bit inputs and outputs a two bit answer. We will name the rightmost bit in the answer the "sum" and the leftmost bit the "carry." So we want our abstract adder to look this:

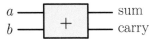

The two single bit inputs enter on the left side, and the two outputs exit on the right side. Our goal is to replace the inside of this "black box" with an actual logic circuit that computes the two outputs from the two inputs.

The first step is to design a truth table that represents what the values of sum and carry should be for all of the possible input values:

a	b	carry	sum
0	0	0	0
0	1	0	1
1	0	0	1
1	1	1	0

Notice that the value of carry is 0, except for when a and b are both 1, i.e., when we are computing 1 + 1. Also, notice that, listed in this order (carry, sum), the two output bits can also be interpreted as a two bit sum: 0 + 0 = 00, 0 + 1 = 1, 1 + 0 = 1, and 1 + 1 = 10. (As in decimal, a leading 0 contributes nothing to the value of a number.)

Next, we need to create an equivalent Boolean expression for each of the two outputs in this truth table. We will start with the sum column. To convert this column to a Boolean expression, we look at the rows in which the output is 1. In this case, these are the second and third rows. The second row says that we want sum to be 1 when a is 0 *and* b is 1. The *and* in this sentence is important; in order for an **and** expression to be 1, both inputs must be 1. But, in this case, a is 0 so we need to flip it with **not** a. The b input is already 1, so we can leave it alone. Putting these two halves together, we have **not** a **and** b. Now the third row says that we want sum to be 1 when a is 1 *and* b is 0. Similarly, we can convert this to the Boolean expression a **and not** b.

a	b	carry	sum	
0	0	0	0	
0	1	0	1	← **not** a **and** b
1	0	0	1	← a **and not** b
1	1	1	0	

58 ■ Elementary computations

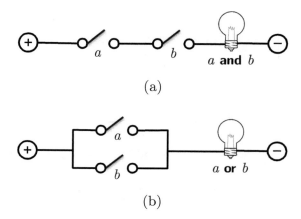

Figure 2.1 Simple electrical implementations of an (a) **and** and (b) **or** gate.

Finally, let's combine these two expressions into one expression for the sum column: taken together, these two rows are saying that sum is 1 if a is 0 and b is 1, *or* if a is 1 and b is 0. In other words, we need at least one of these two cases to be 1 for the sum column to be 1. This is just equivalent to (**not** a **and** b) **or** (a **and not** b). So this is the final Boolean expression for the sum column.

Now look at the carry column. The only row in which the carry bit is 1 says that we want carry to be 1 if a is 1 and b is 1. In other words, this is simply a **and** b. In fact, if you look at the entire carry column, you will notice that this column is the same as in the truth table for a **and** b. So, to compute the carry, we just compute a **and** b.

a	b	carry	sum
0	0	0	0
0	1	0	1
1	0	0	1
1	1	1	0
		↑	↑
		a **and** b	(**not** a **and** b) **or** (a **and not** b)

Implementing an adder

To implement our adder, we need physical devices that implement each of the binary operators. Figure 2.1(a) shows a simple electrical implementation of an **and** operator. Imagine that electricity is trying to flow from the positive terminal on the left to the negative terminal on the right and, if successful, light up the bulb. The binary inputs, a and b, are each implemented with a simple switch. When the switch is open, it represents a 0, and when the switch is closed, it represents a 1. The light bulb represents the output (off = 0 and on = 1). Notice that the bulb will only light up if both of the switches are closed (i.e., both of the inputs are 1). An **or** operator can be implemented in a similar way, represented in Figure 2.1(b). In

Figure 2.2 Schematic representations of logic gates.

this case, the bulb will light up if at least one of the switches is closed (i.e., if at least one of the inputs is 1).

Physical implementations of binary operators are called *logic gates*. It is interesting to note that, although modern gates are implemented electronically, they can be implemented in other ways as well. Enterprising inventors have implemented hydraulic and pneumatic gates, mechanical gates out of building blocks and sticks, optical gates, and recently, gates made from molecules of DNA.

Logic gates have standard, implementation-independent schematic representations, shown in Figure 2.2. Using these symbols, it is a straightforward matter to compose gates to create a *logic circuit* that is equivalent to any Boolean expression. For example, the expression **not** a **and** b would look like the following:

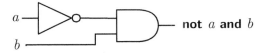

Both inputs a and b enter on the left. Input a enters a **not** gate before the **and** gate, so the top input to the **and** gate is **not** a and the bottom input is simply b. The single output of the circuit on the right leaves the **and** gate with value **not** a **and** b. In this way, logic circuits can be built to an arbitrary level of complexity to perform useful functions.

The circuit for the carry output of our adder is simply an **and** gate:

The circuit for the sum output is a bit more complicated:

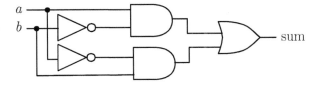

By convention, the solid black circles represent connections between "wires"; if there is no solid black circle at a crossing, this means that one wire is "floating" above the other and they do not touch. In this case, by virtue of the connections, the a input is flowing into both the top **and** gate and the bottom **not** gate, while the b input is flowing into both the top **not** gate and the bottom **and** gate. The top **and**

gate outputs the value of a **and not** b and the bottom **and** gate outputs the value of **not** a **and** b. The **or** gate then outputs the result of **or**ing these two values.

Finally, we can combine these two circuits into one grand adder circuit with two inputs and two outputs, to replace the "black box" adder we began with. The shaded box represents the "black box" that we are replacing.

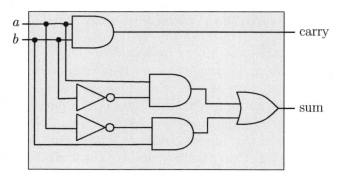

Notice that the values of both a and b are each now flowing into three different gates initially, and the two outputs are conceptually being computed in parallel. For example, suppose a is 0 and b is 1. The figure below shows how this information flows through the adder to arrive at the final output values.

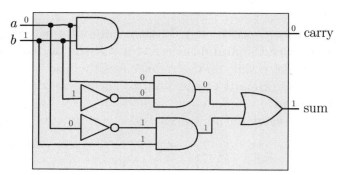

In this way, the adder computes $0 + 1 = 1$, with a carry of 0.

Exercises

2.5.1. Show how to add the unsigned binary numbers 001001 and 001101.

2.5.2. Show how to add the unsigned binary numbers 0001010 and 0101101.

2.5.3. Show how to add the unsigned binary numbers 1001 and 1101, assuming that all integers must be stored in four bits. Convert the binary values to decimal to determine if you arrived at the correct answer.

2.5.4. Show how to add the unsigned binary numbers 001010 and 101101, assuming that all integers must be stored in six bits. Convert the binary values to decimal to determine if you arrived at the correct answer.

2.5.5. Suppose you have a computer that stores unsigned integers in a fixed number of bits. If you have the computer add two unsigned integers, how can you tell if the answer is correct (without having access to the correct answer from some other source)? (Refer back to the unsigned addition example in the text.)

2.5.6. Show how to add the two's complement binary numbers 0101 and 1101, assuming that all integers must be stored in four bits. Convert the binary values to decimal to determine if you arrived at the correct answer.

2.5.7. What is the largest positive integer that can be represented in four bits in two's complement notation? What is the smallest negative number? (Think especially carefully about the second question.)

2.5.8. Show how to add the two's complement binary numbers 1001 and 1101, assuming that all integers must be stored in four bits. Convert the binary values to decimal to determine if you arrived at the correct answer.

2.5.9. Show how to add the two's complement binary numbers 001010 and 101101, assuming that all integers must be stored in six bits. Convert the binary values to decimal to determine if you arrived at the correct answer.

2.5.10. Suppose you have a computer that stores two's complement integers in a fixed number of bits. If you have the computer add two two's complement integers, how can you tell if the answer is correct (without having access to the correct answer from some other source)?

2.5.11. Subtraction can be implemented by adding the first operand to the two's complement of the second operand. Using this algorithm, show how to subtract the two's complement binary number 0101 from 1101. Convert the binary values to decimal to determine if you arrived at the correct answer.

2.5.12. Show how to subtract the two's complement binary number 0011 from 0110. Convert the binary values to decimal to determine if you arrived at the correct answer.

2.5.13. Copy the completed adder circuit, and show, as we did above, how the two outputs (carry and sum) obtain their final values when the input a is 1 and the input b is 0.

2.5.14. Convert the Boolean expression **not** (a **and** b) to a logic circuit.

2.5.15. Convert the Boolean expression **not** a **and not** b to a logic circuit.

2.5.16. The single Boolean operator **nand** (short for "not and") can replace all three traditional Boolean operators. The truth table and logic gate symbol for **nand** are shown below.

a	b	a **nand** b
0	0	1
0	1	1
1	0	1
1	1	0

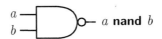

Show how you can create three logic circuits, using only **nand** gates, each of which is equivalent to one of the **and**, **or**, and **not** gates. (Hints: you can use constant inputs, e.g., inputs that are always 0 or 1, or have both inputs of a single gate be the same.)

2.6 SUMMARY

To learn how to write algorithms and programs, you have to dive right in! We started our introduction to programming in Python by performing simple computations on numbers. There are two numerical types in Python: *integers* and *floating point numbers*. Floating point numbers have a decimal point and sometimes behave differently than we expect due to their *finite precision*. Aside from these rounding errors, most arithmetic operators in Python behave as we would expect, except that there are two kinds of division: so-called *true division* and *floor division*.

We use descriptive variable names in our programs for two fundamental reasons. First, they make our code easier to understand. Second, they can "remember" our results for later so that they do not have to be recomputed. The *assignment statement* is used to assign values to variable names. Despite its use of the equals sign, using an assignment statement is *not* the same thing as saying that two things are equal. Instead, the assignment statement evaluates the expression on the righthand side of the equals sign first, and then assigns the result to the variable on the lefthand side.

Functional abstractions in Python are implemented as *functions*. Functions take arguments as input and then *return* a value as output. When used in an expression, it is useful to think of a function call as equivalent to the value that it returns. We used some of Python's built-in functions, and the mathematical functions provided by the `math` module. In addition to numerical values, Python lets us print, input, and manipulate *string* values, which are sequences of characters enclosed in quotation marks.

Because everything in a computer is represented in *binary notation*, all of these operators and functions are really computed in binary, and then expressed to us in more convenient formats. As an example, we looked at how positive and negative integers are stored in a computer, and how two binary integers can be added together just by using the basic Boolean operators of **and**, **or**, and **not**.

In the next chapter, we will begin to write our own functions and incorporate them into longer programs. We will also explore Python's "turtle graphics" module, which will allow us to draw pictures and visualize our data.

2.7 FURTHER DISCOVERY

The epigraph at the beginning of this chapter is from the great Donald Knuth [26]. When he was still in graduate school in the 1960's Dr. Knuth began his life's work, a multi-volume set of books titled, *The Art of Computer Programming* [24]. In 2011, he published the first part of Volume 4, and has plans to write seven volumes total. Although incomplete, this work was cited at the end of 1999 in *American Scientist*'s list of "100 or so Books that Shaped a Century of Science" [35]. Dr. Knuth also invented the typesetting program TEX, which was used to write this book. He is the recipient of many international awards, including the Turing Award, named after Alan Turing, which is considered to be the "Nobel Prize of computer science."

Guido van Rossum is a Dutch computer programmer who invented the Python programming language. The assertion that Van Rossum was in a "slightly irreverent mood" when he named Python after the British comedy show is from the Foreword to *Programming Python* [29] by Mark Lutz. IDLE is an acronym for "Integrated DeveLopment Environment," but is also considered to be a tribute to Eric Idle, one of the founders of Monty Python.

The "Hello world!" program is the traditional first program that everyone learns when starting out. See

http://en.wikipedia.org/wiki/Hello_world_program

for an interesting history.

As you continue to program in Python, it might be helpful to add the following web site containing Python documentation to your "favorites" list:

https://docs.python.org/3/index.html

There is also a list of links on the book web site, and references to commonly used classes and function in Appendix B.

CHAPTER 3

Visualizing abstraction

We have seen that computer programming is an art, because it applies accumulated knowledge to the world, because it requires skill and ingenuity, and especially because it produces objects of beauty. A programmer who subconsciously views himself as an artist will enjoy what he does and will do it better.

Donald E. Knuth
Turing Award Lecture (1974)

We may say most aptly that the Analytical Engine weaves algebraical patterns just as the Jacquard-loom weaves flowers and leaves.

Ada Lovelace
Notes (1843)

S UPPOSE we have a list of 1,000 two-dimensional points representing the sites of ancient artifacts discovered during an excavation, and we would like to know whether there is any pattern to our discoveries. By looking at a list of raw data like the following, would you be able to draw any conclusions?

Site 1:	(5.346, 3.537)
Site 2:	(5.249, 8.418)
Site 3:	(7.177, 1.937)
⋮	⋮
Site 1000:	(1.933, 8.547)

Probably not. However, simply plotting the points instantly provides insight:

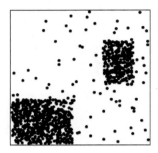

A picture really is worth a thousand words, especially when we are faced with a slew of data. To visualize data like this, we will often turn to *turtle graphics*. To draw in turtle graphics, we create an abstraction called a "turtle" in a window and move it with directional commands. As a turtle moves, its "tail" leaves behind a trail, as shown in Figure 3.1. If we lift a turtle's tail up, it can move without leaving a trace. In this chapter, in the course of learning about turtle graphics, we will also explore how abstractions can be created, used, and combined to solve problems.

3.1 DATA ABSTRACTION

The description of a turtle in turtle graphics is an example of an *abstract data type* (ADT). An abstract data type is an implementation-independent description of some category of things; it supplies the information we need to *use* the thing without specifying how it actually works. An ADT is composed of two parts:

(a) the types of information, called *attributes*, that we need to maintain about the things, and

(b) the *operations* that we are allowed to use to access or modify that information.

For example, a turtle abstract data type specifies that the turtle must maintain the following attributes about its current state.

	Turtle ADT Attributes
Name	Description
position	the turtle's current (x, y) position
heading	the turtle's current heading (in degrees)
color	the turtle's current drawing color
width	the turtle's current pen width
tail position	whether the turtle's tail is up or down

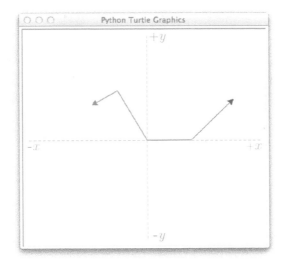

Figure 3.1 A turtle graphics window containing two turtles. The blue turtle moved forward, turned left 45°, and then moved forward again. The red turtle turned left 120°, moved forward, turned left again 90°, and then moved forward again.

The **Turtle** ADT also describes the following operations that change or report on the turtle's state. If the operation requires an argument as input, that is listed in the second column.

		Turtle ADT Operations
Name	Argument	Description
create	—	create a new drawing turtle, positioned at $(0,0)$ and facing east
forward	distance	move the turtle forward in its current direction
backward	distance	move the turtle opposite to its current direction
right	angle	turn the turtle clockwise
left	angle	turn the turtle counterclockwise
setheading	angle	set the turtle's heading
goto	(x, y)	move the turtle to the given position
up, down	—	put the turtle's tail up or down
pensize	width	set the turtle's pen width
pencolor	color	set the turtle's pen color
xcor, ycor	—	return the turtle's x or y coordinate
heading	—	return the turtle's heading

In Python, an abstract data type is implemented with a *class*. In a class, attributes are maintained in a set of *instance variables* and the operations that can access these attributes are special functions called *methods*. The Python class that implements the `Turtle` ADT is named `Turtle`. The `Turtle` class contains instance variables to store the five pieces of information above (plus some others). But we will not actually see any of these instance variables directly. Instead, we will indirectly access and/or modify their values by calling `Turtle` methods like `left`, `right`, `forward`, and `pencolor`. More `Turtle` methods are listed in Appendix B.2.

The `Turtle` class is defined inside a module named `turtle` (notice the different capitalization). So, to access the `Turtle` class, we first need to import this module:

```
>>> import turtle
```

To confirm the existence of the `Turtle` class, try this:

```
>>> turtle.Turtle
<class 'turtle.Turtle'>
```

Just as a blueprint describes the structure of a house, but is not actually a house, the `Turtle` class, or more abstractly the `Turtle` ADT, describes the structure (i.e., attributes and methods) of a drawing turtle, but is not actually a drawing turtle. Actual turtles in turtle graphics, like those pictured in Figure 3.1, are called turtle *objects*. When we create a new turtle object belonging to the `Turtle` class, the turtle object is endowed with its own *independent* values of orientation, position, color, and so on, as described in the class definition. For this reason, there can be more than one turtle object, as illustrated in Figure 3.1.

The distinction between a class and an object can also be loosely described by analogy to animal taxonomy. A species, like a class, describes a category of animals sharing the same general (morphological and/or genetic) characteristics. An actual living organism is an instance of a species, like an object is an instance of a class. For example, the species of Galápagos giant tortoise (*Chelonoidis nigra*) is analogous to a class, while Lonesome George, the famous Galápagos giant tortoise who died in 2012, is analogous to an object of that class. Super Diego, another famous Galápagos giant tortoise, is a member of the same species but, like another object of the same class, is a distinct individual with its own unique attributes.

Reflection 3.1 *Can you think of another analogy for a class and its associated objects?*

Virtually any consumer product can be thought of an object belonging to a class of products. For example, a pair of headphones is an object belonging to the class of all headphones with that particular make and model. The ADT or class specification is analogous to the user manual since the user manual tells you how to use the product without necessarily giving any information about how it works or is made. A course assignment is also analogous to an ADT because it describes the requirements for

> **Box 3.1: Hexadecimal notation**
>
> Hexadecimal is base 16 notation. Just as the positional values in decimal are powers of 10 and the positional values in binary are powers of 2, the positional values in hexadecimal are powers of 16. Because of this, hexadecimal needs 16 digits, which are 0 through 9, followed by the letters `a` through `f`. The letter `a` has the value 10, `b` has the value 11, ..., `f` has the value 15. For example, the hexadecimal number `51ed` has the decimal value
>
> $$5 \cdot 16^3 + 1 \cdot 16^2 + 14 \cdot 16^1 + 13 \cdot 16^0 = 5 \cdot 4096 + 1 \cdot 256 + 14 \cdot 16 + 13 \cdot 1 = 20{,}973.$$
>
> Hexadecimal is used a convenient shorthand for binary. Because any 4 binary digits can represent the values 0 through 15, they can be conveniently replaced by a single hexadecimal digit. So the hexadecimal number `100522f10` is equivalent to `000100000000010100100010111100010000` in binary, as shown below:
>
> ```
>    1    0    0    5    2    2    f    1    0
> 0001 0000 0000 0101 0010 0010 1111 0001 0000
> ```
>
> Instead of displaying this 36 bit binary number, it is more convenient to display the 9 digit hexadecimal equivalent.

the assignment. When a student completes the assignment, she is creating an object that (hopefully) adheres to those requirements.

To create a turtle object in Python, we call a function with the class' name, preceded by the name of the module in which the class resides.

```
>>> george = turtle.Turtle()
```

The empty parentheses indicate that we are calling a function with no arguments. The `Turtle()` function returns a reference to a `Turtle` object, which is then assigned to the name `george`. You should also notice that a window appears on your screen with a little arrow-shaped "turtle" in the center. The center of the window has coordinates $(0,0)$ and is called the *origin*. In Figure 3.1, the axes are superimposed on the window in light gray to orient you to the coordinate system. We can confirm that `george` is a `Turtle` object by printing the object's value.

```
>>> george
<turtle.Turtle object at 0x100522f10>
```

The odd-looking "`0x100522f10`" is the address in memory where this `Turtle` object resides. The address is displayed in *hexadecimal*, or base 16, notation. The `0x` at the front is a prefix that indicates hexadecimal; the actual hexadecimal memory address is `100522f10`. See Box 3.1 for more about how hexadecimal works.

To call a method belonging to an object, we precede the name of the method with the name of the object, separated by a period. For example, to ask `george` to move forward 200 units, we write

70 ■ Visualizing abstraction

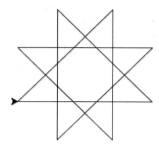

Figure 3.2 A simple geometric "flower" drawn with turtle graphics.

```
>>> george.forward(200)
```

Since the origin has coordinates $(0,0)$ and **george** is initially pointing east (toward positive x values), **george** has moved to position $(200, 0)$; the **forward** method silently changed **george**'s hidden position attribute to reflect this, which you can confirm by calling **george**'s **position** method.

```
>>> george.position()
(200.00,0.00)
```

Notice that we did not change the object's position attribute directly. Indeed, we do not even know the name of that attribute because the class definition remains hidden to us. This is by design. By interacting with objects only through their methods, and not tinkering directly with their attributes, we maintain a clear separation between the ADT specification and the underlying implementation. This allows for the possibility that the underlying implementation may change, to make it more efficient, for example, without affecting programs that use it. We will discuss these issues in much greater detail in Chapter 13.

Exercises

   3.1.1. Explain the difference between an abstract data type (ADT) and a class.

   3.1.2. Give another analogy for the difference between a class and an object. Explain.

   3.1.3. Why do we use methods to change the state of a **Turtle** object instead of directly changing the values of its attributes?

3.2 VISUALIZATION WITH TURTLES

Let's continue by drawing something resembling a geometric flower bloom, as in Figure 3.2. We have just drawn the lower horizontal line segment in the figure. Before we draw the next segment, we need to ask **george** to turn left 135 degrees.

```
>>> george.left(135)
```

With this method call, we have changed **george**'s hidden orientation attribute, which we can confirm by calling the **heading** method.

```
>>> george.heading()
135.0
```

To finish the drawing, we just have to repeat the previous `forward` and `left` calls seven more times! (Hint: see IDLE help for how retrieve previous statements.)

```
>>> george.forward(200)
>>> george.left(135)
>>> george.forward(200)
>>> george.left(135)
>>> george.forward(200)
>>> george.left(135)
>>> george.forward(200)
>>> george.left(135)
>>> george.forward(200)
>>> george.left(135)
>>> george.forward(200)
>>> george.left(135)
>>> george.forward(200)
>>> george.left(135)
```

That was tedious. But before we look at how to avoid similar tedium in the future, we are going to transition out of the Python shell. This will allow us to save our programs so that we can easily modify them or fix mistakes, and then re-execute them without retyping everything. In IDLE, we can create a new, empty program file by choosing **New Window** from the **File** menu.[1] In the new window, retype (or copy and paste) the work we have done so far, shown below, plus four additional lines highlighted in red. (If you copy and paste, be sure to remove the prompt symbols before each line.)

```
import turtle

george = turtle.Turtle()
george.hideturtle()
george.speed(6)

george.forward(200)
george.left(135)
george.forward(200)
george.left(135)
george.forward(200)
george.left(135)
george.forward(200)
george.left(135)
george.forward(200)
george.left(135)
george.forward(200)
```

---

[1] If you are using a different text editor, the steps are probably very similar.

```
george.left(135)
george.forward(200)
george.left(135)
george.forward(200)
george.left(135)

screen = george.getscreen()
screen.exitonclick()
```

The two statements after the first assignment statement hide **george** and speed up the drawing a bit. (The argument to the **speed** method is a number from 0 to 10, with 1 being slow, 10 being fast, and 0 being fastest.) The second to last statement assigns to the variable **screen** an object of the class **Screen**, which represents the drawing area in which **george** lives. The last statement calls the **Screen** method **exitonclick** which will close the window when we click on it.

When you are done typing, save your file by selecting **Save As...** from the **File** menu. The file name of a Python program must always end with the extension **.py**, for example, **george.py**. To execute your new program in IDLE, select **Run Module** from the **Run** menu. If IDLE prompts you to save your file again, just click **OK**. The program should draw the "flower" shape again. When it is done, click on the turtle graphics window to dismiss it.

Drawing with iteration

You may recall from Chapter 1 that, when we were faced with a repetitive computation (e.g., computing the sum of 365 numbers), we rewrote it in a more compact form called a *loop*. Similarly, we would like to replace our eight identical pairs of calls to the **forward** and **left** methods with a loop that repeats one pair eight times. To do so, replace the 16 drawing statements with the following **for** loop:

```
for segment in range(8):
    george.forward(200)
    george.left(135)
```

A *loop* is a structure that repeats a computation some number of times. Repetitive computation, such as that created by a loop, is called *iteration*. The first line of the **for** loop must end with a colon (:), and the statements that are to be repeated by the loop must be indented underneath. The indented statements are collectively known as the *body* of the loop. This **for** loop repeats the two indented statements eight times, indicated by the **8** in parentheses. After the loop is done, Python executes the next non-indented statement (in this case, there is none).

Reflection 3.2 *What happens if you forget the colon?*

It is easy to forget the colon. If you do forget it, you will be notified by a syntax error, like the following, that points to the end of the line containing the **for** keyword.

```
for segment in range(8)
                      ^
SyntaxError: invalid syntax
```

In the `for` loop syntax, `for` and `in` are Python keywords, and `segment` is called the *index variable*. The name of the index variable can be anything we want, but it should be descriptive of its role in the program. In this case, we chose the name `segment` to represent one line segment in the shape, which is what the body of the loop draws. `range(8)` represents the range of integers between 0 and 7. So the `for` loop is essentially saying "execute the body of the loop once for each integer in the range between 0 and 7." Since there are 8 integers in this range, the body of the loop is executed 8 times.

Reflection 3.3 *Try changing* `segment` *to some other name. Did changing the name change the behavior of the program?*

Let's look a little more closely at what is happening in this loop. Before each iteration of the loop, the next value in the range between 0 and 7 is assigned to the index variable, `segment`. Then the statements in the body of the loop are executed. So, quite literally, this loop is equivalent to executing the following 24 statements:

```
segment = 0
george.forward(200)    } iteration 1
george.left(135)
segment = 1
george.forward(200)    } iteration 2
george.left(135)
segment = 2
george.forward(200)    } iteration 3
george.left(135)
           ⋮
segment = 6
george.forward(200)    } iteration 7
george.left(135)
segment = 7
george.forward(200)    } iteration 8
george.left(135)
```

Reflection 3.4 *Try different values between* 1 *and* 10 *in place of* 8 *in* `range(8)`. *Can you see the connection between the value and the picture?*

Now that we have a basic flower shape, let's add some color. To set the color that the turtle draws in, we use the `pencolor` method. Insert

```
george.pencolor('red')
```

> **Box 3.2: Defining colors**
>
> The most common way to specify an arbitrary color is to specify its red, green, and blue (RGB) components individually. Each of these components is often described by an integer between 0 and 255. (These are the values that can be represented by 8 bits. Together then, a color is specified by 24 bits. If you have heard a reference to "24-bit color," now you know its origin.) Alternatively, each component can be described by a real number between 0 and 1.0. In Python turtle graphics, call `screen.colormode(255)` or `screen.colormode(1.0)` to chose the desired representation.
>
> A higher value for a particular component represents a brighter color. So, at the extremes, $(0, 0, 0)$ represents black, and $(255, 255, 255)$ and $(1.0, 1.0, 1.0)$ both represent white. Other common colors include $(255, 255, 0)$ for yellow, $(127, 0, 127)$ for purple, and $(153, 102, 51)$ for brown. So, assuming `george` is a `Turtle` object and `screen` has been assigned `george`'s `Screen` object,
>
> ```
> screen.colormode(255)
> george.pencolor((127, 0, 127))
> ```
>
> would make `george` purple. So would
>
> ```
> screen.colormode(1.0)
> george.pencolor((0.5, 0, 0.5))
> ```

Figure 3.3 A simple geometric "flower," outlined in red and filled in yellow.

before the `for` loop, and run your program again. A color can be specified in one of two ways. First, common colors can be specified with strings such as `'red'`, `'blue'`, and `'yellow'`. Notice that, as we saw in the previous chapter, a string must be enclosed in quotes to distinguish it from a variable or function name. A color can also be defined by explicitly specifying its red, green, and blue (RGB) components, as explained in Box 3.2.

Finally, we will specify a color with which to fill the "flower" shape. The fill color is set by the `fillcolor` method. The statements that draw the area to be filled must be contained between calls to the `begin_fill` and `end_fill` methods. To color our flower yellow, precede the `for` loop with

```
george.fillcolor('yellow')
george.begin_fill()
```

and follow the `for` loop with

```
george.end_fill()
```

Be sure to *not* indent the call to `george.end_fill()` in the body of the `for` loop since we want that statement to execute just once after the loop is finished. Your flower should now look like Figure 3.3, and the complete flower bloom program should look like the following:

```
import turtle

george = turtle.Turtle()
george.hideturtle()
george.speed(6)

george.pencolor('red')
george.fillcolor('yellow')
george.begin_fill()
for segment in range(8):
    george.forward(200)
    george.left(135)
george.end_fill()

screen = george.getscreen()
screen.exitonclick()
```

Reflection 3.5 *Can you figure out why the shape was filled this way?*

Exercises

Write a short program to answer each of the following questions. Submit each as a separate python program file with a `.py` extension (e.g., `picture.py`).

3.2.1. Create two turtles and use them to draw the picture in Figure 3.1.

3.2.2. Write a modified version of the flower bloom program that draws a flower with 18 sides, using an angle of 100°.

3.2.3. Write a program that uses a `for` loop to draw a square with side length 200.

3.2.4. Write a program that uses a `for` loop to draw a rectangle with length 200 and width 100.

3.2.5. Draw an interesting picture using turtle graphics. Consult Appendices B.2 and B.3 for a list of methods. You might want to draw your picture on graph paper first.

3.3 FUNCTIONAL ABSTRACTION

What if we wanted to draw a garden of flowers, each with a different color and size? We can draw one flower bloom with the code in the previous section. To draw each additional flower, we could copy our flower bloom code, paste it below the original code, and then change some method arguments to alter the size and color. For example, the following program draws two flowers. The first three statements in red move `george` to a new location so as not to draw over the first flower. The remaining code in red is a copy of the code above it with three changes.

```
import turtle

george = turtle.Turtle()
george.hideturtle()
george.speed(6)

george.pencolor('red')
george.fillcolor('yellow')
george.begin_fill()
for segment in range(8):
    george.forward(200)
    george.left(135)
george.end_fill()

george.up()
george.goto(200, 200)
george.down()

george.pencolor('yellow')
george.fillcolor('purple')
george.begin_fill()
for segment in range(8):
    george.forward(150)
    george.left(135)
george.end_fill()

screen = george.getscreen()
screen.exitonclick()
```

Reflection 3.6 *What is the purpose of the statements* `george.up()` *and* `george.down()`? *How is the second flower different from the first?*

However, this strategy is a *very bad idea*. First, it is very time-consuming and error-prone; when you repeatedly copy and paste, it is very easy to make mistakes, or forget to make appropriate changes. Second, it makes your code unnecessarily long and hard to read. Doing this a few times can quickly lead to dozens or hundreds of lines of dense code. Third, it is difficult to make changes. For example, what if you copied enough code to draw twenty flowers, and then decided that you wanted to give all of them six petals instead of eight?

Instead, we can *create a new function* to draw a flower. Then we can repeatedly call this function with different arguments to draw different flowers. To create a function in Python, we use the `def` keyword, followed by the function name and, for now, empty parentheses (we will come back to those shortly). As with a `for` loop, the `def` line must end with a colon (`:`).

```
def bloom():
```

The body of the function is then indented relative to the `def` line. The body of our new function will consist of the flower bloom code. Insert this new function after the `import` statement:

```
import turtle

def bloom():
    george.pencolor('red')
    george.fillcolor('yellow')
    george.begin_fill()
    for segment in range(8):
        george.forward(200)
        george.left(135)
    george.end_fill()

george = turtle.Turtle()
george.hideturtle()
george.speed(6)

bloom()

screen = george.getscreen()
screen.exitonclick()
```

The `def` construct *only defines* the new function; it *does not execute* it. We need to *call* the function for it to execute. As we saw earlier, a function call consists of the function name, followed by a list of arguments. Since this function does not have any arguments (yet), and does not return a value, we can call it with

```
bloom()
```

inserted, at the outermost indentation level, where the flower bloom code used to be (as shown above).

Reflection 3.7 *Try running the program with and without the* `bloom()` *function call to see what happens.*

Before continuing, let's take a moment to review what the program is doing. As illustrated below, execution begins at the top, labeled "start," so the program first imports the `turtle` module.

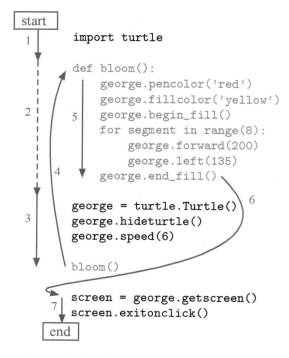

Next, the `bloom` function is defined, but not executed, as signified by the dashed line above marked with a 2. The next three statements define a new `Turtle` object named `george`, hide the turtle and speed it up a bit. Next, the call to `bloom()` causes execution to jump up to the beginning of the function, signified by the arrow marked with a 4. The statements in the function then draw the flower (step 5). When the function is complete, execution continues with the statement after the function call, signified with arrow marked with a 6, and continues to the end of the program.

Function parameters

The function named `bloom` that we have written is not as useful as it could be because it always draws the same flower: yellow with segment length 200. To make the function more general, we want to be able to pass in arguments for values like the fill color and the segment length. We do this by adding one or more *parameters* to the function definition. A parameter is a variable that is assigned the value of an argument when the function is called. (Parameters and arguments are also called *formal parameters* and *actual parameters*, respectively.) In the new version below, we have defined two formal parameters (in red) to represent the fill color and the segment length. We have also replaced the old constants `'yellow'` and 200 with the names of these new parameters.

```
def bloom(fcolor, length):
    george.pencolor('red')
    george.fillcolor(fcolor)
    george.begin_fill()
    for segment in range(8):
        george.forward(length)
        george.left(135)
    george.end_fill()
```

Now, to replicate the old behavior, we would call the function with

```
bloom('yellow', 200)
```

When this function is called, the value of the first argument `'yellow'` is assigned to the first parameter `fcolor` and the value of the second argument 200 is assigned to the second parameter `length`. Then the body of the function executes. Whenever `fcolor` is referenced, it is replaced with `'yellow'`, and whenever `length` is referenced, it is replaced with 200.

Reflection 3.8 *After making these changes, run the program again. Then try running it a few more times with different arguments passed into the* `bloom` *function call. For example, try* `bloom('orange', 50)` *and* `bloom('purple', 350)`. *What happens if you switch the order of the arguments in one these function calls?*

Notice that we have just created a functional abstraction! When we want to draw a flower bloom, we pass two arguments as input to the function, and the function draws it, without us having to understand how it was drawn.

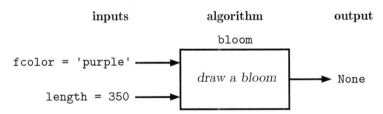

(Of course, because we wrote the function, we *do* understand how it was drawn. But to call it, we do not need to understand how it works.)

We are going to make one more change to this function before moving on, motivated by the following question.

Reflection 3.9 *Look at the variable name* `george` *that is used inside the* `bloom` *function. Where is it defined?*

When the `bloom` function executes, the Python interpreter encounters the variable name `george` in the first line, but `george` has not been defined in that function. Realizing this, Python looks for the name `george` outside the function. This behavior

80 ■ Visualizing abstraction

is called a *scoping rule*. The *scope* of a variable name is the part of the program where the name is defined, and hence can be used.

The scope of a variable name that is defined inside a function, such as `segment` in the `bloom` function, is limited to that function. Such a variable is called a *local variable*. If we tried to refer to `segment` outside of the the `bloom` function, we would get an error. We will look at local variables in more detail in Section 3.6.

A variable name that is defined at the outermost indentation level can be accessed from anywhere in the program, and is called a *global variable*. In our program, `george` and `screen` are global variable names. It is generally a bad idea to have any global variables at all in a program, a topic that we will further discuss in Sections 3.4 and 3.5. But even aside from that issue, we should be concerned that our function is tied to one specific turtle named `george` that is defined outside our function. It would be much better to make the turtle a parameter to the function, so that we can call it with any turtle we want. Replacing `george` with a parameter named `tortoise` gives us the following modified function:

```
def bloom(tortoise, fcolor, length):
    tortoise.pencolor('red')
    tortoise.fillcolor(fcolor)
    tortoise.begin_fill()
    for segment in range(8):
        tortoise.forward(length)
        tortoise.left(135)
    tortoise.end_fill()
```

We also need to update the function call by passing `george` as the first argument, to be assigned to the first parameter, `tortoise`.

```
bloom(george, 'yellow', 200)
```

To finish our flower, let's create another function that draws a stem. Our stem-drawing function will take two parameters: `tortoise`, which is the name of the turtle object, and `length`, the length of the stem. For convenience, we will assume that the stem length is the same as the length of a segment in the associated flower. Include this function in your program immediately after where the `bloom` function is defined.

```
def stem(tortoise, length):
    tortoise.pencolor('green')
    tortoise.pensize(length / 20)
    tortoise.up()
    tortoise.forward(length / 2)
    tortoise.down()
    tortoise.right(90)
    tortoise.forward(length)
```

Figure 3.4 A simple geometric "flower" with a stem.

Since the `bloom` function nicely returns the turtle to the origin, pointing east, we will assume that `tortoise` is in this state when `stem` is called. We start the function by setting the pen color to green, and thickening the turtle's tail by calling the method `pensize`. Notice that the pen size is based on the parameter `length`, so that it scales properly with different size flowers. Next, we need to move halfway across the flower and turn south to start drawing the stem. So that we do not draw over the existing flower, we put the turtle's tail up with the `up` method before we move, and return it to its resting position again with `down` when we are done. Finally, we turn right and move the turtle forward to draw a thick green stem. To draw a stem for our yellow flower, we can insert the function call

```
stem(george, 200)
```

after the call to the `bloom` function. When you run your program, the flower should look like Figure 3.4.

We now have two functions — two functional abstractions, in fact — that draw a flower bloom and an associated stem. We can now focus on the larger problem of creating a virtual garden without thinking further about how the flower-drawing functions work.

Reflection 3.10 *Do you see an opportunity in the program for yet another functional abstraction?*

What do the following two statements together accomplish?

```
bloom(george, 'yellow', 200)
stem(george, 200)
```

Right, they draw a flower! So we can replace these two lines with another function (another functional abstraction) that draws a flower:

Visualizing abstraction

```
import turtle

def bloom(tortoise, fcolor, length):
    tortoise.pencolor('red')
    tortoise.fillcolor(fcolor)
    tortoise.begin_fill()
    for segment in range(8):
        tortoise.forward(length)
        tortoise.left(135)
    tortoise.end_fill()

def stem(tortoise, length):
    tortoise.pencolor('green')
    tortoise.pensize(length / 20)
    tortoise.up()
    tortoise.forward(length / 2)
    tortoise.down()
    tortoise.right(90)
    tortoise.forward(length)

def flower(tortoise, fcolor, length):
    bloom(tortoise, fcolor, length)
    stem(tortoise, length)

george = turtle.Turtle()
george.hideturtle()
george.speed(6)
flower(george, 'yellow', 200)
screen = george.getscreen()
screen.exitonclick()
```

Figure 3.5 The flower program before the flower planting code.

```
import turtle
import random

def bloom(tortoise, fcolor, length):
    tortoise.pencolor('red')
    tortoise.fillcolor(fcolor)
    tortoise.begin_fill()
    for segment in range(8):
        tortoise.forward(length)
        tortoise.left(135)
    tortoise.end_fill()

def stem(tortoise, length):
    tortoise.pencolor('green')
    tortoise.pensize(length / 20)
    tortoise.up()
    tortoise.forward(length / 2)
    tortoise.down()
    tortoise.right(90)
    tortoise.forward(length)

def flower(tortoise, fcolor, length):
    bloom(tortoise, fcolor, length)
    stem(tortoise, length)

def growFlower(x, y):
    span = random.randrange(20, 200)
    fill = random.choice(['yellow',
            'pink', 'red', 'purple'])
    tortoise = turtle.Turtle()
    tortoise.hideturtle()
    tortoise.speed(6)
    tortoise.up()
    tortoise.goto(x, y)
    tortoise.setheading(0)
    tortoise.pensize(1)
    tortoise.down()
    flower(tortoise, fill, span)

george = turtle.Turtle()
george.hideturtle()
screen = george.getscreen()
screen.onclick(growFlower)
screen.mainloop()
```

Figure 3.6 The final flower program with the flower planting code.

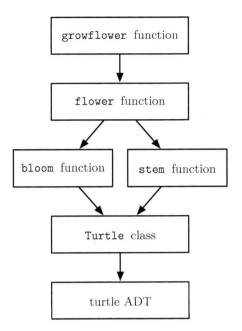

Figure 3.7 Layers of abstraction in the flower-drawing program.

Figure 3.8 A "garden" of random flowers.

```
def flower(tortoise, fcolor, length):
    bloom(tortoise, fcolor, length)
    stem(tortoise, length)
```

Because the `bloom` and `stem` functions together require a turtle, a fill color and a length, and we want to be able to customize our flower in these three ways, these are the parameters to our `flower` function. We pass all three of these parameters straight through to the `bloom` function, and then we pass two of them to the `stem` function. Finally, we can call our new `flower` function with

```
flower(george, 'yellow', 200)
```

to accomplish the same thing as the two statements above. The program is shown in Figure 3.5. Make sure you have all of this down before moving on.

In these three sections, we have learned how to draw with turtle graphics, but more importantly, we learned how to identify and create abstractions. The `Turtle` class implements a turtle abstract data type. Using this class, we have implemented two functional abstractions with Python functions, `bloom` and `stem`. We used these, in turn, to build another functional abstraction, `flower`. Creating a hierarchy of functional abstractions is what allows us to solve large problems. This hierarchy of abstractions, with one additional layer that will develop next, is depicted in Figure 3.7.

Let's plant a garden

We are now ready to plant a virtual garden of flowers. Just for fun, we will draw a flower (using the `flower` function) with a random size and color wherever we click in the drawing window. The code to do this uses some Python functions and methods that we have not yet seen. For reference, all of the turtle graphics methods are listed in Appendices B.2 and B.3.

To generate this random behavior, we will utilize the `random` module, which we will discuss much more extensively in Chapter 5. For now, we need just two functions. First, the `random.randrange` function returns a randomly chosen integer between its two arguments. To generate a random integer between 20 and 199 for the flower's size, we will use

```
span = random.randrange(20, 200)
```

Second, the `random.choice` function returns a randomly chosen element from a list. As we will see later, a Python list is contained in square brackets ([ ]). To generate a random color for the flower, we will call `random.choice` with a list of color strings:

```
fill = random.choice(['yellow', 'pink', 'red', 'purple'])
```

After this function call, one of the strings in the list will be assigned to `fill`.

The following function incorporates these two function calls to draw a random flower at coordinates (x, y). Insert it into your program after the definition of the `flower` function, and insert `import random` at the top. (See Figure 3.6.) Make sure you understand how this function works before you continue.

```
def growFlower(x, y):
    span = random.randrange(20, 200)
    fill = random.choice(['yellow', 'pink', 'red', 'purple'])
    tortoise = turtle.Turtle()
    tortoise.hideturtle()
    tortoise.speed(6)
    tortoise.up()
    tortoise.goto(x, y)
    tortoise.setheading(0)
    tortoise.pensize(1)
    tortoise.down()
    flower(tortoise, fill, span)
```

Notice that this function creates a new `Turtle` object each time it is called (one per flower).

Next, we want to make our program respond to a mouse click by calling the `growFlower` function with arguments equal to the coordinates of the click. This is accomplished by the `onclick` method of the `Screen` class. In particular, the following statement tells the Python interpreter that the function passed as an argument (in this case, the `growFlower` function) should be called every time we click in the window.

    `screen.onclick(growFlower)`

The `onclick` function passes the (x, y) coordinates of each mouse click as parameters to the `growFlower` function. (The function that we pass into `onclick` can only take these two parameters, which is why we did not also include a `Turtle` object as a parameter the `growFlower` function above.) Insert the statement above after the call to `george.getscreen()`. Finally, replace `screen.exitonclick()` with

    `screen.mainloop()`

as the last line of the program. The `mainloop` method repeatedly checks for mouse click events and calls the appropriate function (which we indicated should be `growFlower`) when click events happen. This final program is shown in Figure 3.6, and a possible flower garden is shown in Figure 3.8.

Exercises

Write a short program to answer each of the following questions. Submit each as a separate python program file with a .py extension (e.g., `picture.py`).

3.3.1. Modify the `flower` function so that it allows different numbers of petals. The revised function will need to take an additional parameter:

    `flower(tortoise, fcolor, length, petals)`

Note that the original function has the turtle travel a total of $8 \cdot 135 = 1080$ degrees. When you generalize the number of petals, make sure that the total number of degrees is still a multiple of 360.

3.3.2. Modify the `flower` function so that it creates a daffodil-like double bloom like the one below. The revised function will need two fill color parameters:

    `flower(tortoise, fcolor1, fcolor2, length)`

It might help to know that the distance between any two opposite points of a bloom is about 1.08 times the segment length.

3.3.3. Write a program that draws the word "CODE," as shown below. Use the `circle` method to draw the arcs of the "C" and "D." The `circle` method takes two arguments: the radius of the circle and the extent of the circle in degrees. For example, `george.circle(100, 180)` would draw half of a circle with radius 100. Making the extent negative draws the arc in the reverse direction. In addition, use the `up` method to move the turtle between letters without drawing, and the `down` method to resume drawing.

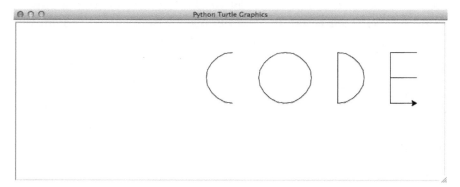

3.3.4. Rewrite your program from Exercise 3.3.3 so that each letter is drawn by its own function. Then use your functions to draw "DECODE." (Call your "D" and "E" functions each twice.)

3.3.5. Write a function

```
drawSquare(tortoise, width)
```

that uses the turtle named `tortoise` to draw a square with the given `width`. This function generalizes the code you wrote for Exercise 3.2.3 so that it can draw a square with any width. Use a `for` loop.

3.3.6. Write a function

```
drawRectangle(tortoise, length, width)
```

that uses the turtle named `tortoise` to draw a rectangle with the given `length` and `width`. This function generalizes the code you wrote for Exercise 3.2.4 so that it can draw a rectangle of any size. Use a `for` loop.

3.3.7. Write a function

```
drawPolygon(tortoise, sideLength, numSides)
```

that uses the turtle named `tortoise` to draw a regular polygon with the given number of sides and side length. This function is a generalization of your `drawSquare` function from Exercise 3.3.5. Use the value of `numSides` in your `for` loop and create a new variable for the turn angle.

3.3.8. Write a function

```
drawCircle(tortoise, radius)
```

that calls your `drawPolygon` function from Exercise 3.3.7 to approximate a circle with the given radius.

3.3.9. Write a function

> `horizontalCircles(tortoise)`

that draws ten non-overlapping circles, each with radius 50, that run horizontally across the graphics window. Use a `for` loop.

3.3.10. Write a function

> `diagonalCircles(tortoise)`

that draws ten non-overlapping circles, each with radius 50, that run diagonally, from the top left to the bottom right, of the graphics window. Use a `for` loop.

3.3.11. Write a function

> `drawRow(tortoise)`

that draws one row of an 8 × 8 red/black checkerboard. Use a `for` loop and the `drawSquare` function you wrote in Exercise 3.3.5.

3.3.12. Write a function

> `drawRow(tortoise, color1, color2)`

that draws one row of an 8 × 8 checkerboard in which the colors of the squares alternate between `color1` and `color2`. The parameters `color1` and `color2` are both strings representing colors. For example, calling `drawRow(george, 'red', 'black')` should draw a row that alternates between red and black.

3.3.13. Write a function

> `checkerBoard(tortoise)`

that draws an 8 × 8 red/black checkerboard, using a `for` loop and the function you wrote in Exercise 3.3.12.

3.3.14. Interesting flower-like shapes can also be drawn by repeatedly drawing polygons that are rotated some number of degrees each time. Write a new function

> `polyFlower(tortoise, sideLength, numSides, numPolygons)`

that calls the `drawPolygon` function from Exercise 3.3.7 to draw an interesting flower design. The function will repeatedly call `drawPolygon` a number of times equal to the parameter `numPolygons`, rotating the turtle each time to make a flower pattern. You will need to figure out the rotation angle based on the number of polygons drawn.

For example, the following was drawn by calling `drawFlower(george, 40, 12, 7)`.

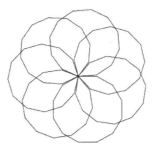

3.3.15. A random walk simulates a particle, or person, randomly moving in a two-dimensional space. At each step, the particle turns in a random direction and walks a fixed distance (e.g., 10 units) in the current direction. If this step is repeated many times, it can simulate Brownian motion or animal foraging behavior. Write a function

```
randomWalk(steps)
```

that draws a random walk for the given number of steps. To randomly choose an angle in each step, use

```
angle = random.randrange(360)
```

You will need to import the `random` module to use this function. (We will talk more about random numbers and random walks in Chapter 5.) One particular 1000-step random walk is shown below.

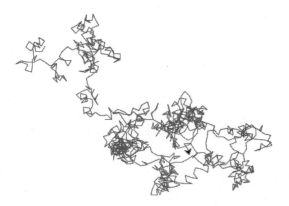

The following are additional exercises that ask you to write functions that do not involve turtle graphics. Be sure to test each one by calling it with appropriate arguments.

3.3.16. Write a function

```
basketball(fieldGoals, threePointers)
```

that prints your team's basketball score if the numbers of two pointers and three pointers are given in the parameters `fieldgoal` and `threePointers`.

3.3.17. Write a function

```
age(birthYear)
```

that prints a person's age when given his or her birth year as a parameter. You can assume that this function only works this year and that the person has not had his or her birthday yet this year.

3.3.18. Write a function

```
cheer(teamName)
```

that takes as a parameter a team name and prints "Go" followed by the team name. For example, if the function is called as `cheer('Buzzards')`, it should print the string `'Go Buzzards'` to the screen.

3.3.19. Write a function

    `sum(number1, number2)`

that prints the sum of `number1` and `number2` to the screen.

3.3.20. Write a function

    `printTwice(word)`

that prints its parameter twice on two separate lines.

3.3.21. Write a function

    `printMyName()`

that uses a `for` loop to print your name 100 times.

3.4 PROGRAMMING IN STYLE

Programming style and writing style share many of the same concerns. When we write an essay, we want the reader to clearly understand our thesis and the arguments that support it. We want it to be clear and concise, and have a logical flow from beginning to end. Similarly, when we write a program, we want to help collaborators understand our program's goal, how it accomplishes that goal, and how it flows from beginning to end. Even if you are the only one to ever read your code, good style will pay dividends both while you are working through the solution, and in the future when you try to reacquaint yourself with your work. We can accomplish these goals by organizing our programs neatly and logically, using descriptive variable and function names, writing code that accomplishes its goal in a non-obfuscated manner, and documenting our intentions within the program.

Program structure

Let's return to the program that we wrote in the previous section (Figure 3.5), and reorganize it a bit to reflect better programming habits. As shown in Figure 3.9, every program should begin with documentation that identifies the program's author and its purpose. This type of documentation is called a *docstring*; we will look more closely at docstrings and other types of documentation in the next section.

    We follow this with our `import` statements. Putting these at the top of our program both makes our program neater and ensures that the imported modules are available anywhere later on.

    Next, we define all of our functions. Because programs are read by the interpreter from top to bottom, we should always define our functions above where we call them. For example, in our flower program, we had to place the definition of the `flower` function above the line where we called it. If we had tried to call the `flower` function at the top of the program, before it was defined, we would have generated an error message.

    At the end of the flower-drawing program in Figure 3.5, there are six statements

90 ■ Visualizing abstraction

program docstring
```
"""
Purpose: Draw a flower
Author: Ima Student
Date: September 15, 2020
CS 111, Fall 2020
"""
```

import statements
```
import turtle
```

function definitions
```
def bloom(tortoise, fcolor, length):
    """Draws a geometric flower bloom.

    Parameters:
        tortoise: a Turtle object with which to draw the bloom.
        fcolor: a color string to use to fill the bloom.
        length: the length of each segment of the bloom.

    Return value:
        None
    """
    tortoise.pencolor('red')       # set tortoise's pen color to red
    tortoise.fillcolor(fcolor)     # and fill color to fcolor
    tortoise.begin_fill()
    for segment in range(8):       # draw a filled 8-sided
        tortoise.forward(length)   #   geometric flower bloom
        tortoise.left(135)
    tortoise.end_fill()

# other functions omitted ...
```

main function
```
def main():
    """Draws a yellow flower with segment length 200, and
        waits for a mouse click to exit.
    """
    george = turtle.Turtle()
    george.hideturtle()
    george.speed(6)
    flower(george, 'yellow', 200)
    screen = george.getscreen()
    screen.exitonclick()
```

main function call
```
main()
```

Figure 3.9 An overview of a program's structure.

at the outermost indentation level. Recall that the first and fifth of these statements define a global variable name that is visible and potentially modifiable anywhere in the program. When the value assigned to a global variable is modified in a function, it is called a *side effect*. In large programs, where the values of global variables can be potentially modified in countless different places, errors in their use become nearly impossible to find. For this reason, we should get into the habit of never using them, unless there is a *very* good reason, and these are pretty hard to come by. See Box 3.3 for more information on how global names are handled in Python.

To prevent the use of global variables, and to make programs more readable, we will move statements at the global level of our programs into a function named `main`, and then call `main` as the last statement in the program, as shown at the end of the program in Figure 3.9. With this change, the call to the `main` function is where the

> **Box 3.3: Globals in Python**
>
> The Python interpreter handles global names inside functions differently, depending on whether the name's value is being read or the name is being assigned a value. When the Python interpreter encounters a name that needs to be evaluated (e.g., on the righthand side of an assignment statement), it first looks to see if this name is defined inside the scope of this function. If it is, the name in the local scope is used. Otherwise, the interpreter successively looks at outer scopes until the name is found. If it reaches the global scope and the name is still not found, we see a "name error."
>
> On the other hand, if we assign a value to a name, that name is always considered to be local, unless we have stated otherwise by using a `global` statement. For example, consider the following program:
>
> ```
> spam = 13
>
> def func1():
>     spam = 100
>
> def func2():
>     global spam
>     spam = 200
>
> func1()
> print(spam)
> func2()
> print(spam)
> ```
>
> The first `print` will display 13 because the assignment statement that is executed in `func1` defines a new local variable; it does not modify the global variable with the same name. But the second `print` will display 200 because the `global` statement in `func2` indicates that `spam` should refer to the global variable with that name, causing the subsequent assignment statement to change the value assigned to the global variable. This convention prevents accidental side effects because it forces the programmer to explicitly decide to modify a global variable. In any case, using `global` is strongly discouraged.

action begins in this program. (Remember that the function definitions above only define functions; they do not execute them.) The **main** function eventually calls our **flower** function, which then calls the **bloom** and **stem** functions. Getting used to this style of programming has an additional benefit: it is very similar to the style of other common programming languages (e.g., C, C++, Java) so, if you go on to use one of these in the future, it should seem relatively familiar.

Documentation

Python program documentation comes in two flavors: *docstrings* and *comments*. A docstring is meant to articulate everything that someone needs to know to use a program or module, or to call a function. Comments, on the other hand, are used to document individual program statements or groups of statements. In other words, a docstring explains *what* a program or function does, while comments explain *how* the code works; a docstring describes an abstraction while comments describe what happens inside the black box. The Python interpreter ignores both docstrings and comments while it is executing a program; both are intended for human eyes only.

Docstrings

A docstring is enclosed in a matching pair of triple double quotes (""""), and may occupy several lines. We use a docstring at the beginning of every program to identify the program's author and its purpose, as shown at the top of Figure 3.9.[2] We also use a docstring to document each function that we write, to ensure that the reader understands what it does. A function docstring should articulate everything that someone needs to know to call the function: the overall purpose of the function, and descriptions of the function's parameters and return value.

The beginning of a function's docstring is indented on the line immediately following the `def` statement. Programmers prefer a variety of different styles for docstrings; we will use one that closely resembles the style in Google's official Python style guide. The following illustrate this docstring style, applied to the three functions in the program from Figure 3.5. (The bodies of the functions are omitted.)

```
def bloom(tortoise, fcolor, length):
    """Draws a geometric flower bloom.

    Parameters:
        tortoise: a Turtle object with which to draw the bloom
        fcolor: a color string to use to fill the bloom
        length: the length of each segment of the bloom

    Return value: None
    """

def stem(tortoise, length):
    """Draws a flower stem.

    Parameters:
        tortoise: a Turtle object, located at the bloom starting position
        length: the length of the stem and each segment of the bloom

    Return value: None
    """
```

---

[2]Your instructor may require a different format, so be sure to ask.

```
def flower(tortoise, fcolor, length):
    """Draws a flower.

    Parameters:
        tortoise: a Turtle object with which to draw the flower
        fcolor: a color string to use to fill the bloom
        length: the length of each segment of the bloom

    Return value: None
    """
```

In the first line of the docstring, we succinctly explain what the function does. This is followed by a parameter section that lists each parameter with its intended purpose and the class to which it should belong. If there are any assumptions made about the value of the parameter, these should be stated also. For example, the turtle parameter of the `stem` function is assumed to start at the origin of the bloom. Finally, we describe the return value of the function. We did not have these functions return anything, so they return `None`. We will look at how to write functions that return values in Section 3.5.

Another advantage of writing docstrings is that Python can automatically produce documentation from them, in response to calling the `help` function. For example, try this short example in the Python shell:

```
>>> def spam(x):
        """Pointlessly returns x.

        Parameters:
            x: some value of no consequence

        Return value:
            x: same value of no consequence
        """

        return x

>>> help(spam)
Help on function spam in module __main__:

spam(x)
    Pointlessly returns x.

    Parameters:
        x: some value of no consequence

    Return value:
        x: same value of no consequence
```

You can also use `help` with modules and built-in functions. For example, try this:

```
>>> import turtle
>>> help(turtle)
>>> help(turtle.color)
```

In the documentation that appears, hit the space bar to scroll forward a page and the Q key to exit.

Comments

A comment is anything between a hash symbol (#) and the end of the line. As with docstrings, the Python interpreter ignores comments. Comments should generally be neatly lined up to the right of your code. However, there are times when a longer comment is needed to explain a complicated section of code. In this case, you might want to precede that section with a comment on one or more lines by itself.

There is a fine line between under-commenting and over-commenting. As a general rule, you want to supply high-level descriptions of what your code intends to do. You do *not* want to literally repeat what each individual line does, as this is not at all helpful to someone reading your code. Remember, anyone who reads your program is going to also be a programmer, so you don't need to explain what each Python statement does. Doing so tends to clutter your code and make it *harder* to read! As Albert Einstein purportedly said,

> *If you can't explain it simply, you don't understand it well enough.*

Here are examples of good comments for the body of the `bloom` function.

```
tortoise.pencolor('red')        # set tortoise's pen color to red
tortoise.fillcolor(fcolor)      # and fill color to fcolor
tortoise.begin_fill()
for segment in range(8):        # draw a filled 8-sided
    tortoise.forward(length)    #   geometric flower bloom
    tortoise.left(135)
tortoise.end_fill()
```

Notice that the five lines that draw the bloom are commented together, just to note the programmer's intention. In contrast, the following comments illustrate what *not* to do. The following comments are both hard to read and uninformative.

```
tortoise.pencolor('red')      # set tortoise's pen color to red
tortoise.fillcolor(fcolor)    # set tortoise's fill color to fcolor
tortoise.begin_fill()         # begin to fill a shape
for segment in range(8):      # for segment = 0, 1, 2, ..., 7
    tortoise.forward(length)  # move tortoise forward length
    tortoise.left(135)        # turn tortoise left 135 degrees
tortoise.end_fill()           # stop filling the shape
```

We leave the task of commenting the other functions in this program as an exercise.

Descriptive names and magic numbers

As we discussed in Section 2.3, using descriptive variable names is a very important step in making your program's intentions clear. The variable names in the flower program are already in good shape, let's look at another example. Consider the following statements.

```
x = 462
y = (3.95 - 1.85) * x - 140
```

Without any context, it is impossible to infer what this is supposed to represent. However, if we rename the two variables, as follows, the meaning becomes clearer.

```
cupsSold = 462
profit = (3.95 - 1.85) * cupsSold - 140
```

Now it is clear that this code is computing the profit generated from selling cups of something. But the meaning of the numbers is still a mystery. These are examples of "magic numbers," so-called in programming parlance because they seem to appear out of nowhere. There are at least two reasons to avoid "magic numbers." First, they make your code less readable and obscure its meaning. Second, they make it more difficult and error-prone to change your code, especially if you use the same value multiple times. By assigning these numbers to descriptive variable names, the code becomes even clearer.

```
cupsSold = 462
pricePerCup = 3.95
costPerCup = 1.85
fixedCost = 140
profit = (pricePerCup - costPerCup) * cupsSold - fixedCost
```

What we have now is what we call "self-documenting code." Since we have named all of our variables and values with descriptive names, just reading the code is enough to deduce its intention. These same rules, of course, apply to function names and parameters. By naming our functions with descriptive names, we make their purposes clearer and we contribute to the readability of the functions from which we call them. This practice will continue to be demonstrated in the coming chapters.

In this book, we use a naming convention that is sometimes called "camelCase," in which the first letter is in lower case and then the first letters of subsequent words are capitalized. But other programmers prefer different styles. For example, some programmers prefer "snake_case," in which an underscore character is placed between words (`cupsSold` would be `cups_sold`). Unless you are working in an environment with a specific mandated style, the choice is yours, as long as it results in self-documenting code.

Exercises

3.4.1. Incorporate all the changes we discussed in this section into your flower-drawing program, and finish commenting the bodies of the remaining functions.

3.4.2. Write a function that implements your Mad Lib from from Exercise 2.4.13, and then write a complete program (with `main` function) that calls it. Your Mad Lib function should take the words needed to fill in the blanks as parameters. Your `main` function should get these values with calls to the `input` function, and then pass them to your function. Include docstrings and comments in your program. For example, here is a function version of the example in Exercise 2.4.13 (without docstrings or comments).

```
def party(adj1, noun1, noun2, adj2, noun3):
    print('How to Throw a Party')
    print()
    print('If you are looking for a/an', adj1, 'way to')
    print('celebrate your love of', noun1 + ', how about a')
    print(noun2 + '-themed costume party?  Start by')
    print('sending invitations encoded in', adj2, 'format')
    print('giving directions to the location of your', noun3 + '.')

def main():
    firstAdj = input('Adjective: ')
    firstNoun = input('Noun: ')
    secondNoun = input('Noun: ')
    secondAdj = input('Adjective: ')
    thirdNoun = input('Noun: ')
    party(firstAdj, firstNoun, secondNoun, secondAdj, thirdNoun)

main()
```

3.4.3. Study the following program, and then reorganize it with a `main` function that calls one or more other functions. Your `main` function should be very short: create a turtle, call your functions, and then wait for a mouse click. Document your program with appropriate docstrings and comments.

```
import turtle
george = turtle.Turtle()
george.setposition(0, 100)
george.pencolor('red')
george.fillcolor('red')
george.begin_fill()
george.circle(-100, 180)
george.right(90)
george.forward(200)
george.end_fill()
george.up()
george.right(90)
george.forward(25)
george.right(90)
george.forward(50)
```

```
george.left(90)
george.down()
george.pencolor('white')
george.fillcolor('white')
george.begin_fill()
george.circle(-50, 180)
george.right(90)
george.forward(100)
george.end_fill()
screen = george.getscreen()
screen.exitonclick()
```

3.5 A RETURN TO FUNCTIONS

We previously described a function as a computation that takes one or more inputs, called *parameters*, and produces an output, called a *return value*. But, up until now, the functions that we have written have not had return values. In this section, we will remedy that.

Let's return to the simple mathematical function $f(x) = x^2 + 3$. We understand that this function takes an input x and evaluates to the output $x^2 + 3$. So, for example, $f(4) = 4^2 + 3 = 19$; if 4 is the input, then 19 is the output. The algorithm that transforms the input into the output is, of course, "add 3 to the square of x."

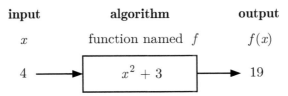

The Python equivalent of our mathematical function f is the following.

```
def f(x):
    return x ** 2 + 3
```

The `return` statement defines the output of the function. Remember that the function definition by itself does not compute anything. We must call the function for it to be executed. For example, in a program, we might call our function `f` like this:

```
def main():
    y = f(4)
    print(y)

main()
```

When the assignment statement in `main` is executed, the righthand side calls the function `f` with the argument 4. So the function `f` is executed next with 4 assigned to the parameter `x`. It is convenient to visualize this as follows, with the value 4 replacing `x` in the function:

```
def f(x):
    return x ** 2 + 3
```
with annotations showing x replaced by 4, and 4 ** 2 + 3 = 19.

Next, the value 4 ** 2 + 3 = 19 is computed and returned by the function. This return value becomes the value associated with the function call `f(4)`, and is therefore the value assigned to `y` in `main`. Since `f(4)` evaluates to `19`, we could alternatively replace the two lines in `main` with

```
def main():
    print(f(4))
```

Similarly, we can incorporate calls to functions with return values into longer expressions, such as

```
y = (3 * f(4) + 2) ** 2
```

In this example, `f(4)` evaluates to 19, and then the rest of the expression is evaluated:

$$y = (3 * \underbrace{f(4)}_{19} + 2) ** 2$$

with f(4) = 19, 3*19+2 = 59, and 59**2 = 3481.

Recall that, in Section 2.3, we computed quantities like volume and wind chill on the righthand side of an assignment statement.

```
volume = (4 / 3) * pi * (radius ** 3)
```

Then, to recompute the volume with a different value of `radius`, or wind chill with a different value of `wind`, we had to re-type the same formula again. This was obviously tedious and error-prone, analogous to building a new microwave oven from scratch every time we wanted to pop a bag of popcorn. What we were missing then were functions. After creating a function for each of these computations, computing the results with a different input simply amounts to calling the function again. For example, the volume of a sphere can be computed with the following function:

```
import math

def volumeSphere(radius):
    """Computes the volume of a sphere.

    Parameter:
        radius: radius of the sphere

    Return value: volume of a sphere with the given radius
    """

    return (4 / 3) * math.pi * (radius ** 3)
```

```
def main():
    radius1 = float(input('Radius of first sphere: '))
    radius2 = float(input('Radius of second sphere: '))
    volume1 = volumeSphere(radius1)
    volume2 = volumeSphere(radius2)
    print('The volumes of the spheres are', volume1, 'and', volume2)

main()
```

The `return` statement in the function assigns its return value to be the volume of the sphere with the given `radius`. Now, as shown in the `main` function above, computing the volumes of two different spheres just involves calling the function twice. Similarly, suppose that inside an empty sphere with radius 20, we had a solid sphere with radius 10. To compute the empty space remaining inside the larger sphere, we can compute

$$\text{emptyVolume} = \underbrace{\underbrace{\text{volumeSphere}(20)}_{33510.3216...} - \underbrace{\text{volumeSphere}(10)}_{4188.7902...}}_{29321.5314...}$$

In addition to defining a function's return value, the `return` statement *also* causes the function to end and return this value back to the function call. So the `return` statement actually does two things. A `return` statement

1. defines the function's return value, *and*

2. causes the function to end.

Therefore, if we add statements to a function after the return statement, they will never be executed. For example, the red call to the print function in the following program is useless.

```
def volumeSphere(radius):
    """Computes the volume of a sphere.

    Parameter:
        radius: radius of the sphere

    Return value:
        volume of a sphere with the given radius
    """

    return (4 / 3) * math.pi * (radius ** 3)

    print('This will never, ever be printed.')
```

Return vs. print

A common beginner's mistake is to end a function with a call to `print` instead of a `return` statement. For example, suppose we replaced the `return` with `print` in the `volumeSphere` function:

```
def volumeSphere(radius):
    """Computes the volume of a sphere.

    Parameter:
        radius: radius of the sphere

    Return value:
        volume of a sphere with the given radius
    """

    print((4 / 3) * math.pi * (radius ** 3))        # WRONG
```

This will print the desired answer but, because we did not provide a return value, the function's return value defaults to `None`. So if we try to compute

```
emptyVolume = volumeSphere(20) - volumeSphere(10)
```
$$\underbrace{\text{volumeSphere(20)}}_{\text{None}} - \underbrace{\text{volumeSphere(10)}}_{\text{None}}$$
$$\underbrace{\phantom{\text{volumeSphere(20)} - \text{volumeSphere(10)}}}_{\text{None - None ???}}$$

we get an error because we are trying to assign `emptyVolume = None - None`, which is nonsensical.

Exercises

*Write a function for each of the following problems. In each case, make sure the function **returns** the specified value (instead of just printing it).*

3.5.1. Write a function

```
sum(number1, number2)
```

that returns the sum of `number1` and `number2`. Also write a complete program (with a `main` function) that gets these two values using the `input` function, passes them to your `sum` function, and then prints the value returned by the `sum` function.

3.5.2. Write a function

```
power(base, exponent)
```

that returns the value $\text{base}^{\text{exponent}}$. Also write a complete program (with a `main` function) that gets these two values using the `input` function, passes them to your `power` function, and then prints the returned value of $\text{base}^{\text{exponent}}$.

3.5.3. Write a function

```
football(touchdowns, fieldGoals, safeties)
```

that returns your team's football score if the number of touchdowns (7 points), field goals (3 points), and safeties (2 points) are passed as parameters. Then write a complete program (with `main` function) that gets these three values using the `input` function, passes them to your `football` function, and then prints the score.

3.5.4. Exercise 2.4.8 asked how to compute the distance between two points. Now write a function

```
distance(x1, y1, x2, y2)
```

that returns the distance between points (x1,y1) and (x2,y2).

3.5.5. The ideal gas law states that $PV = nRT$ where

- P = pressure in atmospheres (atm)
- V = volume in liters (L)
- n = number of moles (mol) of gas
- R = gas constant = 0.08 L atm / mol K
- T = absolute temperature of the gas in Kelvin (K)

Write a function

```
moles(V, P, T)
```

that returns the number of moles of an ideal gas in V liters contained at pressure P and T degrees Celsius. (Be sure to convert Celsius to Kelvin in your function.) Also write a complete program (with a `main` function) that gets these three values using the `input` function, passes them to your `moles` function, and then prints the number of moles of ideal gas.

3.5.6. Suppose we have two containers of an ideal gas. The first contains 10 L of gas at 1.5 atm and 20 degrees Celsius. The second contains 25 L of gas at 2 atm and 30 degrees Celsius. Show how to use two calls to your function in the previous exercise to compute the total number of moles of ideal gas in the two containers.

Now replace the `return` statement in your `moles` function with a call to `print` instead. (So your function does not contain a `return` statement.) Can you still compute the total number of moles in the same way? If so, show how. If not, explain why not.

3.5.7. Most of the world is highly dependent upon groundwater for survival. Therefore, it is important to be able to monitor groundwater flow to understand potential contamination threats. Darcy's law states that the flow of a liquid (e.g., water) through a porous medium (e.g., sand, gravel) depends upon the capacity of the medium to carry the flow and the gradient of the flow:

$$Q = K \frac{dh}{dl}$$

where

- K is the hydraulic conductivity of the medium, the rate at which the liquid can move through it, measured in area/time
- dh/dl is the hydraulic gradient

- dh is the drop in elevation (negative for flow down)
- dl is the horizontal distance of the flow

Write a function

```
darcy(K, dh, dl)
```

that computes the flow with the given parameters.

Use your function to compute the amount of groundwater flow inside a hill with hydraulic conductivity of 130 m²/day, and a 50 m drop in elevation over a distance of 1 km.

3.5.8. A person's Body Mass Index (BMI) is calculated by the following formula:

$$\text{BMI} = \frac{w}{h^2} \cdot 703$$

where w is the person's weight in pounds and h is the person's height in inches. Write a function

```
bmi(weight, height)
```

that uses this formula to return the corresponding BMI.

3.5.9. When you (or your parents) rip songs from a CD, the digital file is created by sampling the sound at some rate. Common rates are 128 kbps (128×2^{10} bits per second), 192 kbps, and 256 kbps. Write a function

```
songs(capacity, bitrate)
```

that returns the number of 4-minute songs someone can fit on his or her iPod. The function's two parameters are the capacity of the iPod in gigabytes (GB) and the sampling rate in kbps. A gigabyte is 2^{30} bytes and a byte contains 8 bits. Also write a complete program (with a `main` function) that gets these two values using the `input` function, passes them to your `songs` function, and then prints the number of songs.

3.5.10. The speed of a computer is often (simplistically) expressed in gigahertz (GHz), the number of billions of times the computer's internal clock "ticks" per second. For example, a 2 GHz computer has a clock that "ticks" 2 billion times per second. Suppose that a single computer instruction requires 3 "ticks" to execute. Write a function

```
time(instructions, gigahertz)
```

that returns the time in seconds required to execute the given number of `instructions` on a computer with clock rate `gigahertz`. For example, `time(1e9, 3)` should return 1 (second).

3.5.11. Exercise 2.3.4 asked how to swap the values in two variables. Can we write a function to swap the values of two parameters? In other words, can we write a function

```
swap(a, b)
```

and call it like

```
x = 10
y = 1
swap(x, y)
```

so that after the function returns, x has the value 1 and y has the value 10? (The function should not return anything.) If so, write it. If not, explain why not.

3.5.12. Given an integer course grade from 0 to 99, we convert it to the equivalent grade point according to the following scale: 90–99: 4, 80–89: 3, 70–79: 2, 60–69: 1, < 60: 0. Write a function

```
gradePoint(score)
```

that returns the grade point (i.e., GPA) equivalent to the given score. (Hint: use floor division and the built-in max function which returns the maximum of two or more numbers.)

3.5.13. The function time.time() (in the time module) returns the current time in seconds since January 1, 1970. Write a function

```
year()
```

that uses this function to return the current year as an integer value.

3.6 SCOPE AND NAMESPACES

In this section, we will look more closely at using variable names inside functions. As an example, let's consider our chill computation from Section 2.3, implemented as a function that is called from a main function.

```
def windChill(temp, wind):
    """Gives the North American metric wind chill equivalent
       for given temperature and wind speed.

    Parameters:
        temp: temperature in degrees Celsius
        wind: wind speed at 10m in km/h

    Return value:
        equivalent wind chill in degrees Celsius, rounded to
        the nearest integer
    """

    chill = 13.12 + 0.6215*temp + (0.3965*temp - 11.37) * wind**0.16
    return round(chill)

def main():
    chilly = windChill(-3, 13)
    print('The wind chill is', chilly)

main()
```

Notice that we have introduced a variable inside the `windChill` function named `chill` to break up the computation a bit. We assign `chill` the result of the wind chill computation, using the parameters `temp` and `wind`, and then return this value rounded to the nearest integer. Because we created the name `chill` inside the function `windChill`, its *scope* is local to the function. In other words, if we tried to refer to `chill` anywhere in the program outside of the function `windChill` (e.g., in the `main` function), we would get the following error:

```
NameError: name 'chill' is not defined
```

Because `chill` has a local scope, it is called a *local variable*.

Local namespaces

To better understand how the scoping rules of local variable and parameter names work, let's look more closely at how these names are managed in Python. To make this a little more interesting, let's modify our wind chill program a bit more, as highlighted in red:

```
def windChill(temp, wind):
    """ (docstring omitted) """

    chill = 13.12 + 0.6215*temp + (0.3965*temp - 11.37) * wind**0.16
    temp = round(chill)
    return temp

def main():
    t = -3
    w = 13
    chilly = windChill(t, w)
    print('The wind chill is', chilly)

main()
```

In this program, just after we call the `windChill` function, but just before the values of the arguments `t` and `w` are assigned to the parameters `temp` and `wind`, we can visualize the situation like this:

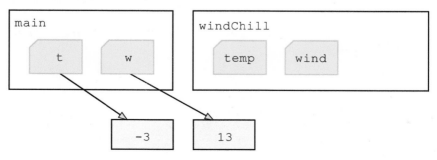

The box around t and w represents the scope of the main function, and the box around temp and wind represents the scope of the windChill function. In each case, the scope defines what names have been defined, or have meaning, in that function. In the picture, we are using arrows instead of affixing the "Sticky notes" directly to the values to make clear that the names, not the values, reside in their respective scopes. The names are references to the memory cells in which their values reside.

The scope corresponding to a function in Python is managed with a *namespace*. A namespace of a function is simply a list of names that are defined in that function, together with references to their values. We can view the namespace of a particular function by calling the locals function from within it. For example, insert the following statement into the main function, just before the call to windChill:

```
print('Namespace of main before calling windChill:', locals())
```

When we run the program, we will see

```
Namespace of main before calling windChill: {'t': -3, 'w': 13}
The wind chill is -8
```

This is showing us that, at that point in the program, the local namespace in the main function consists of two names: t, which is assigned the value -3, and w, which is assigned the value 13 (just as we visualized above). The curly braces ({ }) around the namespace representation indicate that the namespace is a *dictionary*, another abstract data type in Python. We will explore dictionaries in more detail in Chapter 8.

Returning to the program, when windChill is called and the parameters are assigned the values of their respective arguments, we are implicitly assigning temp = t and wind = w, so the picture changes to this:

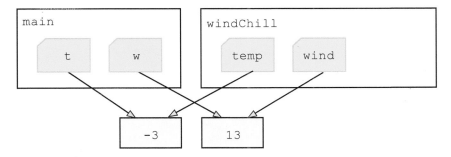

We can see this in the program by inserting another call to locals as the first line of the windChill function:

```
print('Local namespace at the start of windChill:', locals())
```

Now when we run the program, we will see

```
Namespace of main before calling windChill: {'w': 13, 't': -3}
Local namespace at the start of windChill: {'temp': -3, 'wind': 13}
The wind chill is -8
```

This is showing us that, at the beginning of the `windChill` function, the only visible names are `temp` and `wind`, which have been assigned the values of `t` and `w`, respectively. Notice, however, that `t` and `w` do not exist inside `windChill`, and there is no direct connection between `t` and `temp`, or between `w` and `wind`; rather they are only indirectly connected through the value to which they are both assigned.

Next, the first statement in the function, the assignment statement involving the local variable `chill` inserts the new name `chill` into the namespace of the `windChill` function and assigns it the result of the wind chill computation.

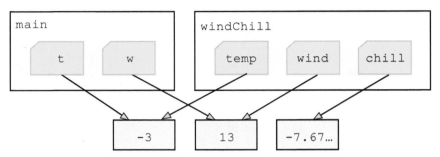

Finally, the function reassigns the rounded wind chill value to the local parameter `temp` before returning it:

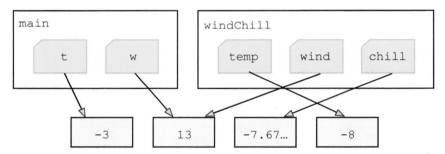

To see this sequence of events in the context of the program, insert another call to `locals` in `windChill`, just before the `return` statement:

```
print('Local namespace at the end of windChill:', locals())
```

Now when we run the program, we will see

```
Namespace of main before calling windChill: {'t': -3, 'w': 13}
Local namespace at the start of windChill: {'wind': 13, 'temp': -3}
Local namespace at the end of windChill: {'wind': 13,
   'chill': -7.676796032159553, 'temp': -8}
The wind chill is -8
```

After the `windChill` function returns −8, the namespace of `windChill`, and all of the local names in that namespace, cease to exist, leaving `t` and `w` untouched in the `main` namespace. However, a new name, `chilly`, is created in the `main` namespace and assigned the return value of the `windChill` function:

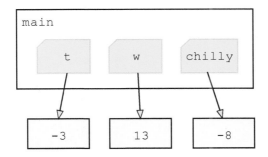

To confirm this, add one more call to `locals` at the end of `main`:

```
print('Local namespace in main after calling windChill:', locals())
```

Then the final program shown in Figure 3.10 will print:

```
Namespace of main before calling windChill: {'t': -3, 'w': 13}
Local namespace at the start of windChill: {'wind': 13, 'temp': -3}
Local namespace at the end of windChill: {'wind': 13,
    'chill': -7.676796032159553, 'temp': -8}
The wind chill is -8
Local namespace in main after calling windChill: {'t': -3, 'w': 13,
    'chilly': -8}
```

One important take-away from this example is that when we are dealing with parameters that are numbers, a parameter and its corresponding argument are not directly connected; if you change the value assigned to the parameter, as we did to the parameter `temp`, it does not change the corresponding argument, in this case, `t`.

The global namespace

Recall that, in Section 3.3, we briefly discussed global variable names, and why we should never use them. The namespace in which global variable names reside is called the *global namespace*. We can view the contents of the global namespace by calling the `globals` function. For example, add the following call to `globals` to the end of `main` in our program above:

```
print('Global namespace:', globals())
```

The result will be something like the following (some names are not shown):

```
Global namespace: {'main': <function main at 0x10065e290>,
            'windChill': <function windChill at 0x10065e440>,
            '__name__': '__main__',
            '__builtins__': <module 'builtins' (built-in)>, ... }
```

Notice that the only global names that we created refer to our two functions, `main` and `windChill`. We can think of each of these names as referring to the functions' respective namespaces, as illustrated below (references for some names are omitted):

108 ■ Visualizing abstraction

```
def windChill(temp, wind):
    """ (docstring omitted) """

    print('Local namespace at the start of windChill:', locals())
    chill = 13.12 + 0.6215*temp + (0.3965*temp - 11.37) * wind**0.16
    temp = round(chill)
    print('Local namespace at the end of windChill:', locals())
    return temp

def main():
    t = -3
    w = 13
    print('Namespace of main before calling windChill:', locals())
    chilly = windChill(t, w)
    print('The wind chill is', chilly)
    print('Local namespace in main after calling windChill:', locals())

main()
```

Figure 3.10 The complete wind chill program, with calls to the `locals` function.

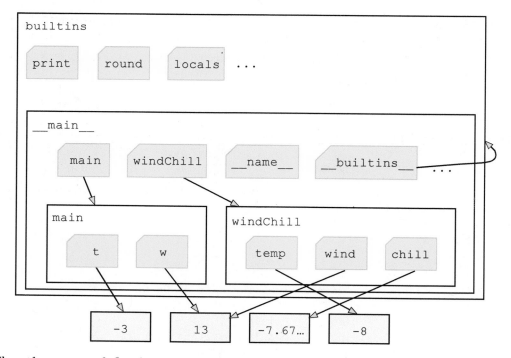

The other names defined in the global namespace are standard names defined in every Python program. The name `__name__` refers to the name of the current module, which, in this case, is '`__main__`' (not to be confused with the namespace of the **main** function) because we are in the file that we initially executed in the Python interpreter. The name `__builtins__` refers to an implicitly imported module that contains all of Python's built-in functions.

As the illustration suggests, we can think of these namespaces as being nested inside each other because names that are not located in a local namespace are sought in enclosing namespaces. For example, when we are in the `main` function and call the function `print`, the Python interpreter first looks in the local namespace for this function name. Not finding it there, it looks in the next outermost namespace, `__main__`. Again, not finding it there, it looks in the `builtins` namespace.

Each module that we import also defines its own namespace. For example, when we import the `math` module with `import math`, a new namespace is created within `builtins`, at the same nesting level as the `__main__` namespace, as illustrated below.

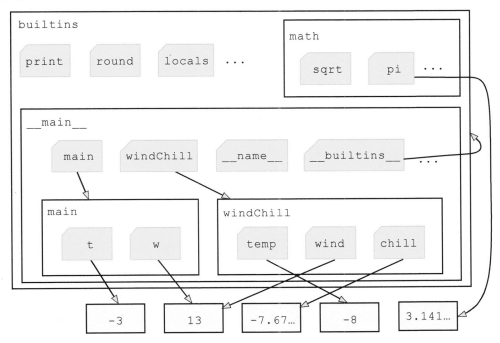

When we preface each of the function names in the `math` module with `math` (e.g., `math.sqrt(7)`), we are telling the Python interpreter to look in the `math` namespace for the function.

Maintaining a mental model like this should help you manage the names that you use in your programs, especially as they become longer.

Exercises

3.6.1. Show how to use the `locals` function to print all of the local variable names in the `distance` function from Exercise 3.5.4. Use your modified function to compute the distance between the points $(3, 7.25)$ and $(9.5, 1)$. What does the local namespace look like while the function is executing?

3.6.2. Consider the following program:

```
import turtle

def drawStar(tortoise, length):
    for segment in range(5):
        tortoise.forward(length)
        tortoise.left(144)

def main():
    george = turtle.Turtle()
    sideLength = 200
    drawStar(george, sideLength)
    screen = george.getscreen()
    screen.exitonclick()

main()
```

Sketch a picture like that on Page 106 depicting the namespaces in this program just before returning from the **drawStar** function. Here is a picture to get you started:

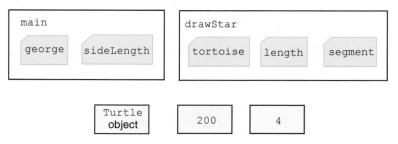

3.6.3. In economics, a demand function gives the price a consumer is willing to pay for an item, given that a particular quantity of that item is available. For example, suppose that in a coffee bean market the demand function is given by

$$D(Q) = 45 - \frac{2.3Q}{1000},$$

where Q is the quantity of available coffee, measured in kilograms, and the returned price is for 1 kg. So, for example, if there are 5000 kg of coffee beans available, the price will be 45 − (2.3)(5000)/1000 = 33.50 dollars for 1 kg. The following program computes this value.

```
    def demand(quantity):
        quantity = quantity / 1000
        return 45 - 2.3 * quantity

    def main():
        coffee = 5000
        price = demand(coffee)
        print(price)

main()
```

Sketch a picture like that on Page 106 depicting the namespaces in this program just before returning from the `demand` function and also just before returning from the `main` function.

3.7 SUMMARY

In this chapter, we made progress toward writing more sophisticated programs. The key to successfully solving larger problems is to break the problem into smaller, more manageable pieces, and then treat each of these pieces as an abstract "black box" that you can use to solve the larger problem. There are two types of "black boxes," those that represent things (i.e., data, information) and those that represent actions. A "black box" representing a thing is described by an *abstract data type* (ADT), which contains both hidden data and a set of functions that we can call to access or modify that data. In Python, an ADT is implemented with a *class*, and instances of a class are called *objects*. The class, such as `Turtle`, to which an object belongs specifies what (hidden) data the object has and what *methods* can be called to access or modify that data. Remember that a class is the "blueprint" for a category of objects, but is not actually an object. We "built" new `Turtle` objects by calling a function with the class' name:

```
george = turtle.Turtle()
```

Once the object is created, we can do things with it by calling its methods, like `george.forward(100)`, without worrying about how it actually works.

A "black box" that performs an action is called a *functional abstraction*. We implement functional abstractions in Python with functions. Earlier in the chapter, we designed functions to draw things in turtle graphics, gradually making them more general (and hence more useful) by adding parameters. We also started using `for` loops to create more interesting iterative algorithms. Later in the chapter, we also looked at how we can add return values to functions, and how to properly think about all of the names that we use in our programs. By breaking our programs up into functions, like breaking up a complex organization into divisions, we can more effectively focus on how to solve the problem at hand.

This increasing complexity becomes much easier to manage if we follow a few guidelines for structuring and documenting our programs. We laid these out in Section 3.4, and encourage you to stick to them as we forge ahead.

In the next chapter, we will design iterative algorithms to model quantities that change over time, like the sizes of dynamic populations. These techniques will also serve as the foundation for many other algorithms in later chapters.

3.8 FURTHER DISCOVERY

The chapter's first epigraph is once again from Donald Knuth, specifically his address after receiving the 1974 Turing award [25]. You can read or watch other Turing award lectures at

> `http://amturing.acm.org` .

The second epigraph is from Ada Lovelace, considered by many to be the first computer programmer. She was born Ada Byron in England in 1815, the daughter of the Romantic poet Lord Byron. (However, she never knew her father because he left England soon after she was born.) In marriage, Ada acquired the title "Countess of Lovelace," and is now commonly known simply as Ada Lovelace. She was educated in mathematics by several prominent tutors and worked with Charles Babbage, the inventor of two of the first computers, the Difference Engine and the Analytical Engine. Although the Analytical Engine was never actually built, Ada wrote a set of "Notes" about its design, including what many consider to be the first computer program. (The quote is from Note A, Page 696.) In her "Notes" she also imagined that future computers would be able to perform tasks far more interesting than arithmetic (like make music). Ada Lovelace died in 1852, at the age of 37.

The giant tortoise named Lonesome George was, sadly, the last surviving member of his subspecies, *Chelonoidis nigra abingdonii*. The giant tortoise named Super Diego is a member of a different subspecies, *Chelonoidis nigra hoodensis*.

The commenting style we use in this book is based on Google's official Python style guide, which you can find at

> `http://google-styleguide.googlecode.com/svn/trunk/pyguide.html` .

CHAPTER 4

Growth and decay

> Our population and our use of the finite resources of planet Earth are growing exponentially, along with our technical ability to change the environment for good or ill.
>
> Stephen Hawking
> *TED talk (2008)*

Some of the most fundamental questions asked in the natural and social sciences concern the dynamic sizes of populations and other quantities over time. For example, we may be interested in the size of a plant population being affected by an invasive species, the magnitude of an infection facing a human population, the quantity of a radioactive material in a storage facility, the penetration of a product in the global marketplace, or the evolving characteristics of a dynamic social network. The possibilities are endless.

To study situations like these, scientists develop a simplified *model* that abstracts key characteristics of the actual situation so that it might be more easily understood and explored. In this sense, a model is another example of abstraction. Once we have a model that describes the problem, we can write a *simulation* that shows what happens when the model is applied over time. The power of modeling and simulation lies in their ability to either provide a theoretical framework for past observations or predict future behavior. Scientists often use models in parallel with traditional experiments to compare their observations to a proposed theoretical framework. These parallel scientific processes are illustrated in Figure 4.1. On the left is the computational process. In this case, we use "model" instead of "algorithm" to acknowledge the possibility that the model is mathematical rather than algorithmic. (We will talk more about this process in Section 7.1.) On the right side is the parallel experimental process, guided by the scientific method. The results of the computational and experimental processes can be compared, possibly leading to model adjustments or new experiments to improve the results.

114 ■ Growth and decay

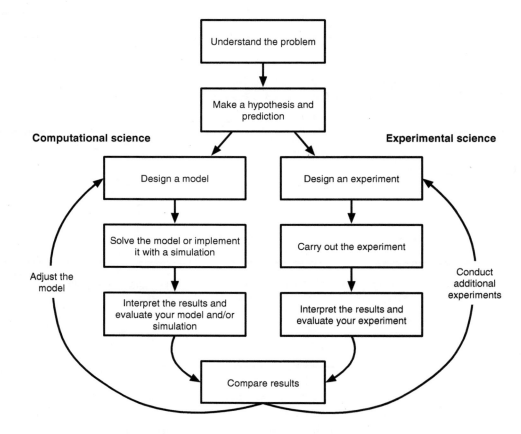

Figure 4.1 Parallel experimental and computational processes.

When we model the dynamic behavior of populations, we will assume that time ticks in discrete steps and, at any particular time step, the current population size is based on the population size at the previous time step. Depending on the problem, a time step may be anywhere from a nanosecond to a century. In general, a new time step may bring population increases, in the form of births and immigration, and population decreases, in the form of deaths and emigration. In this chapter, we will discuss a fundamental algorithmic technique, called an accumulator, that we will use to model dynamic processes like these. Accumulators crop up in all kinds of problems, and lie at the foundation of a variety of different algorithmic techniques. We will continue to see examples throughout the book.

4.1 DISCRETE MODELS

Managing a fishing pond

Suppose we manage a fishing pond that contained a population of 12,000 largemouth bass on January 1 of this year. With no fishing, the bass population is expected to grow at a rate of 8% per year, which incorporates both the birth rate and the death rate of the population. The maximum annual fishing harvest allowed is 1,500 bass. Since this is a popular fishing spot, this harvest is attained every year. Is our maximum annual harvest sustainable? If not, how long until the fish population dies out? Should we reduce the maximum harvest? If so, what should it be reduced to?

We can find the projected population size for any given year by starting with the initial population size, and then computing the population size in each successive year based on the size in the previous year. For example, suppose we wanted to know the projected population size four years from now. We start with the initial population of largemouth bass, assigned to a variable named `population0`:

```
population0 = 12000
```

Then we want to set the population size at the end of the first year to be this initial population size, plus 8% of the initial population size, minus 1,500. In other words, if we let `population1` represent the population size at the end of the first year, then

```
population1 = population0 + 0.08 * population0 - 1500   # 11,460
```

or, equivalently,

```
population1 = 1.08 * population0 - 1500   # 11,460.0
```

The population size at the end of the second year is computed in the same way, based on the value of `population1`:

```
population2 = 1.08 * population1 - 1500   # 10,876.8
```

Continuing,

```
population3 = 1.08 * population2 - 1500   # 10,246.94
population4 = 1.08 * population3 - 1500   # 9,566.6952
```

So this model projects that the bass population in four years will be 9,566 (ignoring the "fractional fish" represented by the digits to the right of the decimal point).

The process we just followed was obviously repetitive (or *iterative*), and is therefore ripe for a **for** loop. Recall that we used the following **for** loop in Section 3.2 to draw our geometric flower with eight petals:

```
for segment in range(8):
    tortoise.forward(length)
    tortoise.left(135)
```

The keywords **for** and **in** are required syntax elements, while the parts in red are up to the programmer. The variable name, in this case **segment**, is called an *index variable*. The part in red following **in** is a list of values assigned to the index variable, one value per iteration. Because **range(8)** represents a list of integers between 0 and 7, this loop is equivalent to the following sequence of statements:

```
segment = 0
tortoise.forward(length)
tortoise.left(135)
segment = 1
tortoise.forward(length)
tortoise.left(135)
segment = 2
tortoise.forward(length)
tortoise.left(135)
   ⋮
segment = 7
tortoise.forward(length)
tortoise.left(135)
```

Because eight different values are assigned to **segment**, the loop executes the two drawing statements in the body of the loop eight times.

In our fish pond problem, to compute the population size at the end of the fourth year, we performed four computations, namely:

```
population0 = 12000
population1 = 1.08 * population0 - 1500     # 11460.0
population2 = 1.08 * population1 - 1500     # 10876.8
population3 = 1.08 * population2 - 1500     # 10246.94
population4 = 1.08 * population3 - 1500     # 9566.6952
```

So we need a **for** loop that will iterate four times:

```
for year in range(4):
    # compute population based on previous population
```

In each iteration of the loop, we want to compute the current year's population size based on the population size in the previous year. In the first iteration, we want to compute the value that we assigned to **population1**. And, in the second, third, and fourth iterations, we want to compute the values that we assigned to **population2**, **population3**, and **population4**.

But how do we generalize these four statements into one statement that we can repeat four times? The trick is to notice that, once each variable is used to compute the next year's population, it is never used again. Therefore, we really have no need for all of these variables. Instead, we can use a single variable **population**, called an *accumulator variable* (or just *accumulator*), that we multiply by 1.08 and subtract 1500 from each year. We initialize **population = 12000**, and then for each successive year we assign

```
population = 1.08 * population - 1500
```

Remember that an assignment statement evaluates the righthand side first. So the value of **population** on the righthand size of the equals sign is the previous year's population, which is used to compute the current year's population that is assigned to **population** of the lefthand side.

```
population = 12000
for year in range(4):
    population = 1.08 * population - 1500
```

This `for` loop is equivalent to the following statements:

```
population = 12000

year = 0
population = 1.08 * population - 1500    # new population = 11460.0
                        ⏟
                      12000
year = 1
population = 1.08 * population - 1500    # new population = 10876.8
                        ⏟
                      11460.0
year = 2
population = 1.08 * population - 1500    # new population = 10246.94
                        ⏟
                      10876.8
year = 3
population = 1.08 * population - 1500    # new population = 9566.6952
                        ⏟
                      10246.94
```

In the first iteration of the `for` loop, 0 is assigned to `year` and `population` is assigned the previous value of `population` (12000) times 1.08 minus 1500, which is 11460.0. Then, in the second iteration, 1 is assigned to `year` and `population` is once again assigned the previous value of `population` (now 11460.0) times 1.08 minus 1500, which is 10876.8. This continues two more times, until the `for` loop ends. The final value of `population` is 9566.6952, as we computed earlier.

Reflection 4.1 *Type in the statements above and add the following statement after the assignment to* `population` *in the body of the* `for` *loop:*

```
print(year + 1, int(population))
```

Run the program. What is printed? Do you see why?

We see in this example that we can use the index variable `year` just like any other variable. Since `year` starts at zero and the first iteration of the loop is computing the population size in year 1, we print `year + 1` instead of `year`.

Reflection 4.2 *How would you change this loop to compute the fish population in five years? Ten years?*

Changing the number of years to compute is now simple. All we have to do is change the value in the `range` to whatever we want: `range(5)`, `range(10)`, etc. If we put this computation in a function, then we can have the parameter be the desired number of years:

118 ■ Growth and decay

```
def pond(years):
    """Simulates a fish population in a fishing pond, and
        prints annual population size.  The population
        grows 8% per year with an annual harvest of 1500.

    Parameter:
        years: number of years to simulate

    Return value: the final population size
    """

    population = 12000
    for year in range(years):
        population = 1.08 * population - 1500
        print(year + 1, int(population))

    return population

def main():
    finalPopulation = pond(10)
    print('The final population is', finalPopulation)

main()
```

Reflection 4.3 *What would happen if* `population = 12000` *was inside the body of the loop instead of before it? What would happen if we omitted the* `population = 12000` *statement altogether?*

The initialization of the accumulator variable before the loop is crucial. If `population` were not initialized before the loop, then an error would occur in the first iteration of the `for` loop because the righthand side of the assignment statement would not make any sense!

Reflection 4.4 *Use the* `pond` *function to answer the original questions: Is this maximum harvest sustainable? If not, how long until the fish population dies out? Should the pond manager reduce the maximum harvest? If so, what should it be reduced to?*

Calling this function with a large enough number of years shows that the fish population drops below zero (which, of course, can't really happen) in year 14:

```
1 11460
2 10876
3 10246
⋮
13 392
14 -1076
⋮
```

This harvesting plan is clearly not sustainable, so the pond manager should reduce it to a sustainable level. In this case, determining the sustainable level is easy: since the population grows at 8% per year and the pond initially contains 12,000 fish, we cannot allow more than $0.08 \cdot 12000 = 960$ fish to be harvested per year without the population declining.

Reflection 4.5 *Generalize the* `pond` *function with two additional parameters: the initial population size and the annual harvest. Using your modified function, compute the number of fish that will be in the pond in 15 years if we change the annual harvest to 800.*

With these modifications, your function might look like this:

```
def pond(years, initialPopulation, harvest):
    """ (docstring omitted) """

    population = initialPopulation
    for year in range(years):
        population = 1.08 * population - harvest
        print(year + 1, int(population))

    return population
```

The value of the `initialPopulation` parameter takes the place of our previous initial population of 12000 and the parameter named `harvest` takes the place of our previous harvest of 1500. To answer the question above, we can replace the call to the `pond` function from `main` with:

```
finalPopulation = pond(15, 12000, 800)
```

The result that is printed is:

```
1 12160
2 12332
3 12519
4 12720
⋮
13 15439
14 15874
15 16344
The final population is 16344.338228396558
```

Reflection 4.6 *How would you call the new version of the* `pond` *function to replicate its original behavior, with an annual harvest of 1500?*

Before moving on, let's look at a helpful Python trick, called a *format string*, that enables us to format our table of annual populations in a more attractive way. To illustrate the use of a format string, consider the following modified version of the previous function.

```
def pond(years, initialPopulation, harvest):
    """ (docstring omitted) """

    population = initialPopulation
    print('Year | Population')
    print('-----|-----------')
    for year in range(years):
        population = 1.08 * population - harvest
        print('{0:^4} | {1:>9.2f}'.format(year + 1, population))

    return population
```

The function begins by printing a table header to label the columns. Then, in the call to the `print` function inside the `for` loop, we utilize a format string to line up the two values in each row. The syntax of a format string is

'<replacement fields>'.format(<values to format>)

(The parts in red above are descriptive and not part of the syntax.) The period between the the string and `format` indicates that `format` is a method of the string class; we will talk more about the string class in Chapter 6. The parameters of the `format` method are the values to be formatted. The format for each value is specified in a *replacement field* enclosed in curly braces (`{}`) in the format string.

In the example in the `for` loop above, the `{0:^4}` replacement field specifies that the first (really the "zero-th"; computer scientists like to start counting at 0) argument to `format`, in this case `year + 1`, should be centered (^) in a field of width 4. The `{1:>9.2f}` replacement field specifies that `population`, as the second argument to `format`, should be right justified (>) in a field of width 9 as a floating point number with two places to the right of the decimal point (.2f). When formatting floating point numbers (specified by the `f`), the number before the decimal point in the replacement field is the minimum width, including the decimal point. The number after the decimal point in the replacement field is the minimum number of digits to the right of the decimal point in the number. (If we wanted to align to the left, we would use <.) Characters in the string that are not in replacement fields (in this case, two spaces with a vertical bar between them) are printed as-is. So, if `year` were assigned the value 11 and `population` were assigned the value 1752.35171, the above statement would print

$$\underbrace{\text{ 12 }}_{\{0:\hat{}4\}} | \underbrace{\text{   1752.35}}_{\{1:>9.2f\}}$$

To fill spaces with something other than a space, we can use a *fill character* immediately after the colon. For example, if we replaced the second replacement field with `{1:*>9.2f}`, the previous statement would print the following instead:

$$\underbrace{\text{ 12 }}_{\{0:\hat{}4\}} | \underbrace{\text{ **1752.35}}_{\{1:*>9.2f\}}$$

Measuring network value

Now let's consider a different problem. Suppose we have created a new online social network (or a new group within an existing social network) that we expect to steadily grow over time. Intuitively, as new members are added, the value of the network to its members grows because new relationships and opportunities become available. The potential value of the network to advertisers also grows as new members are added. But how can we quantify this value?

We will assume that, in our social network, two members can become connected or "linked" by mutual agreement, and that connected members gain access to each other's network profile. The inherent value of the network lies in these connections, or *links*, rather than in the size of its membership. Therefore, we need to figure out how the potential number of links grows as the number of members grows. The picture below visualizes this growth. The circles, called *nodes*, represent members of the social network and lines between nodes represent links between members.

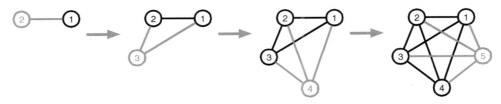

At each step, the red node is added to the network. In each step, the red links represent all of the potential new connections that could result from the addition of the new member.

Reflection 4.7 *What is the maximum number of new connections that could arise when each of nodes 2, 3, 4, and 5 are added? In general, what is the maximum number of new connections that could arise from adding node number n?*

Node 2 adds a maximum of 1 new connection, node 3 adds a maximum of 2 new connections, node 4 adds a maximum of 3 new connections, etc. In general, a maximum of $n-1$ new connections arise from the addition of node number n. This pattern is illustrated in the table below.

node number:	2	3	4	5	...	n
maximum increase in number of links:	1	2	3	4	...	$n-1$

Therefore, as shown in the last row, the maximum number of links in a network with n nodes is the sum of the numbers in the second row:

$$1 + 2 + 3 + \ldots + n - 1.$$

We will use this sum to represent the potential value of the network.

Let's write a function, similar to the previous one, that lists the maximum number of new links, and the maximum total number of links, as new nodes are added to a network. In this case, we will need an accumulator to count the total number of links. Adapting our `pond` function to this new purpose gives us the following:

```
def countLinks(totalNodes):
    """Prints a table with the maximum total number of links
       in networks with 2 through totalNodes nodes.

    Parameter:
        totalNodes: the total number of nodes in the network

    Return value:
        the maximum number of links in a network with totalNodes nodes
    """

    totalLinks = 0
    for node in range(totalNodes):
        newLinks = ???
        totalLinks = totalLinks + newLinks
        print(node, newLinks, totalLinks)

    return totalLinks
```

In this function, we want our accumulator variable to count the total number of links, so we renamed it from `population` to to `totalLinks`, and initialized it to zero. Likewise, we renamed the parameter, which specifies the number of iterations, from `years` to `totalNodes`, and we renamed the index variable of the `for` loop from `year` to `node` because it will now be counting the number of the node that we are adding at each step. In the body of the `for` loop, we add to the accumulator the maximum number of new links added to the network with the current node (we will return to this in a moment) and then print a row containing the node number, the maximum number of new links, and the maximum total number of links in the network at that point.

Before we determine what the value of `newLinks` should be, we have to resolve one issue. In the table above, the node numbers range from 2 to the number of nodes in the network, but in our `for` loop, `node` will range from 0 to `totalNodes - 1`. This turns out to be easily fixed because the `range` function can generate a wider variety of number ranges than we have seen thus far. If we give `range` two arguments instead of one, like `range(start, stop)`, the first argument is interpreted as a starting value and the second argument is interpreted as the stopping value, producing a range of values starting at `start` and going up to, *but not including*, `stop`. For example, `range(-5, 10)` produces the integers $-5, -4, -3, \ldots, 8, 9$.

To see this for yourself, type `range(-5, 10)` into the Python shell (or `print` it in a program).

```
>>> range(-5, 10)
range(-5, 10)
```

Unfortunately, you will get a not-so-useful result, but one that we can fix by converting the range to a *list* of numbers. A list, enclosed in square brackets ([ ]), is another kind of abstract data type that we will make extensive use of in later chapters. To convert the range to a list, we can pass it as an argument to the list function:

```
>>> list(range(-5, 10))
[-5, -4, -3, -2, -1, 0, 1, 2, 3, 4, 5, 6, 7, 8, 9]
```

Reflection 4.8 *What list of numbers does* range(1, 10) *produce? What about* range(10, 1)*? Can you explain why in each case?*

Reflection 4.9 *Back to our program, what do we want our* for *loop to look like?*

For node to start at 2 and finish at totalNodes, we want our for loop to be

```
for node in range(2, totalNodes + 1):
```

Now what should the value of newLinks be in our program? The answer is in the table we constructed above; the maximum number of new links added to the network with node number n is $n - 1$. In our loop, the node number is assigned to the name node, so we need to add node - 1 links in each step:

```
newLinks = node - 1
```

With these substitutions, our function looks like this:

```
def countLinks(totalNodes):
    """ (docstring omitted) """

    totalLinks = 0
    for node in range(2, totalNodes + 1):
        newLinks = node - 1
        totalLinks = totalLinks + newLinks
        print(node, newLinks, totalLinks)

    return totalLinks

def main():
    links = countLinks(10)
    print('The total number of links is', links)

main()
```

As with our previous for loop, we can see more clearly what this loop does by looking at an equivalent sequence of statements. The changing value of node is highlighted in red.

```
        totalLinks = 0

        node = 2
        newLinks = node - 1                 # newLinks is assigned 2 - 1 = 1
        totalLinks = totalLinks + newLinks  # totalLinks is assigned 0 + 1 = 1
        print(node, newLinks, totalLinks)   # prints 2 1 1

        node = 3
        newLinks = node - 1                 # newLinks is assigned 3 - 1 = 2
        totalLinks = totalLinks + newLinks  # totalLinks is assigned 1 + 2 = 3
        print(node, newLinks, totalLinks)   # prints 3 2 3

        node = 4
        newLinks = node - 1                 # newLinks is assigned 4 - 1 = 3
        totalLinks = totalLinks + newLinks  # totalLinks is assigned 3 + 3 = 6
        print(node, newLinks, totalLinks)   # prints 4 3 6

            ⋮

        node = 10
        newLinks = node - 1                 # newLinks is assigned 10 - 1 = 9
        totalLinks = totalLinks + newLinks  # totalLinks is assigned 36 + 9 = 45
        print(node, newLinks, totalLinks)   # prints 10 9 45
```

When we call `countLinks(10)` from the `main` function above, it prints

```
        2 1 1
        3 2 3
        4 3 6
        5 4 10
        6 5 15
        7 6 21
        8 7 28
        9 8 36
        10 9 45
        The total number of links is 45
```

We leave lining up the columns more uniformly using a format string as an exercise.

Reflection 4.10 *What does* `countLinks(100)` *return? What does this represent?*

Organizing a concert

Let's look at one more example. Suppose you are putting on a concert and need to figure out how much to charge per ticket. Your total expenses, for the band and the venue, are $8000. The venue can seat at most 2,000 and you have determined through market research that the number of tickets you are likely to sell is related to a ticket's selling price by the following relationship:

```
        sales = 2500 - 80 * price
```

According to this relationship, if you give the tickets away for free, you will overfill your venue. On the other hand, if you charge too much, you won't sell any tickets at all. You would like to price the tickets somewhere in between, so as to maximize your profit. Your total income from ticket sales will be `sales * price`, so your profit will be this amount minus $8000.

To determine the most profitable ticket price, we can create a table using a `for` loop similar to that in the previous two problems. In this case, we would like to iterate over a range of ticket prices and print the profit resulting from each choice. In the following function, the `for` loop starts with a ticket price of one dollar and adds one to the price in each iteration until it reaches `maxPrice` dollars.

```
def profitTable(maxPrice):
    """Prints a table of profits from a show based on ticket price.

    Parameters:
        maxPrice: maximum price to consider

    Return value: None
    """
    print('Price    Income    Profit')
    print('------  ---------  ---------')
    for price in range(1, maxPrice + 1):
        sales = 2500 - 80 * price
        income = sales * price
        profit = income - 8000
        formatString = '${0:>5.2f}  ${1:>8.2f}  ${2:8.2f}'
        print(formatString.format(price, income, profit))

def main():
    profitTable(25)

main()
```

The number of expected sales in each iteration is computed from the value of the index variable `price`, according to the relationship above. Then we print the price and the resulting income and profit, formatted nicely with a format string. As we did previously, we can look at what happens in each iteration of the loop:

```
price = 1
sales = 2500 - 80 * price       # sales is assigned 2500 - 80 * 1 = 2420
income = sales * price          # income is assigned   2420 * 1 = 2420
profit = income - 8000          # profit is assigned 2420 - 8000 = -5580
print(price, income, profit)    # prints  $ 1.00    $ 2420.00    $-5580.00

price = 2
sales = 2500 - 80 * price       # sales is assigned 2500 - 80 * 2 = 2340
income = sales * price          # income is assigned   2340 * 2 = 4680
profit = income - 8000          # profit is assigned 4680 - 8000 = -3320
print(price, income, profit)    # prints  $ 2.00    $ 4680.00    $-3320.00

price = 3
sales = 2500 - 80 * price       # sales is assigned 2500 - 80 * 3 = 2260
income = sales * price          # income is assigned   2260 * 3 = 6780
profit = income - 8000          # profit is assigned 6780 - 8000 = -1220
print(price, income, profit)    # prints  $ 3.00    $ 6780.00    $-1220.00
    ⋮
```

Reflection 4.11 *Run this program and determine what the most profitable ticket price is.*

The program prints the following table:

Price	Income	Profit
$ 1.00	$ 2420.00	$-5580.00
$ 2.00	$ 4680.00	$-3320.00
$ 3.00	$ 6780.00	$-1220.00
$ 4.00	$ 8720.00	$ 720.00
⋮		
$15.00	$19500.00	$11500.00
$16.00	$19520.00	$11520.00
$17.00	$19380.00	$11380.00
⋮		
$24.00	$13920.00	$ 5920.00
$25.00	$12500.00	$ 4500.00

The profit in the third column increases until it reaches $11,520.00 at a ticket price of $16, then it drops off. So the most profitable ticket price seems to be $16.

Reflection 4.12 *Our program only considered whole dollar ticket prices. How can we modify it to increment the ticket price by fifty cents in each iteration instead?*

The **range** function can only create ranges of integers, so we cannot ask the **range** function to increment by 0.5 instead of 1. But we can achieve our goal by doubling the range of numbers that we iterate over, and then set the price in each iteration to be the value of the index variable divided by two.

Discrete models ■ 127

```
def profitTable(maxPrice):
    """ (docstring omitted) """

    print('Price     Income     Profit')
    print('------    ---------  ---------')
    for price in range(1, 2 * maxPrice + 1):
        realPrice = price / 2
        sales = 2500 - 80 * realPrice
        income = sales * realPrice
        profit = income - 8000
        formatString = '${0:>5.2f}  ${1:>8.2f}  ${2:8.2f}'
        print(formatString.format(realPrice, income, profit))
```

Now when `price` is 1, the "real price" that is used to compute profit is `0.5`. When `price` is 2, the "real price" is `1.0`, etc.

Reflection 4.13 *Does our new function find a more profitable ticket price than $16?*

Our new function prints the following table.

```
Price    Income     Profit
------   ---------  ---------
$ 0.50   $ 1230.00  $-6770.00
$ 1.00   $ 2420.00  $-5580.00
$ 1.50   $ 3570.00  $-4430.00
$ 2.00   $ 4680.00  $-3320.00
  ⋮
$15.50   $19530.00  $11530.00
$16.00   $19520.00  $11520.00
$16.50   $19470.00  $11470.00
  ⋮
$24.50   $13230.00  $ 5230.00
$25.00   $12500.00  $ 4500.00
```

If we look at the ticket prices around $16, we see that $15.50 will actually make $10 more.

Just from looking at the table, the relationship between the ticket price and the profit is not as clear as it would be if we plotted the data instead. For example, does profit rise in a straight line to the maximum and then fall in a straight line? Or is it a more gradual curve? We can answer these questions by drawing a plot with turtle graphics, using the `goto` method to move the turtle from one point to the next.

```
import turtle

def profitPlot(tortoise, maxPrice):
    """ (docstring omitted) """

    for price in range(1, 2 * maxPrice + 1):
        realPrice = price / 2
        sales = 2500 - 80 * realPrice
        income = sales * realPrice
        profit = income - 8000
        tortoise.goto(realPrice, profit)

def main():
    george = turtle.Turtle()
    screen = george.getscreen()
    screen.setworldcoordinates(0, -15000, 25, 15000)
    profitPlot(george, 25)
    screen.exitonclick()

main()
```

Our new `main` function sets up a turtle and then uses the `setworldcoordinates` function to change the coordinate system in the drawing window to fit the points that we are likely to plot. The first two arguments to `setworldcoordinates` set the coordinates of the lower left corner of the window, in this case $(0, -15,000)$. So the minimum visible x value (price) in the window will be zero and the minimum visible y value (profit) will be $-15,000$. The second two arguments set the coordinates in the upper right corner of the window, in this case $(25, 15,000)$. So the maximum visible x value (price) will be 25 and the maximum visible y value (profit) will be $15,000$. In the `for` loop in the `profitPlot` function, since the first value of `realPrice` is `0.5`, the first `goto` is

```
george.goto(0.5, -6770)
```

which draws a line from the origin $(0,0)$ to $(0.5, -6770)$. In the next iteration, the value of `realPrice` is `1.0`, so the loop next executes

```
george.goto(1.0, -5580)
```

which draws a line from the previous position of $(0.5, -6770)$ to $(1.0, -5580)$. The next value of `realPrice` is `1.5`, so the loop then executes

```
george.goto(1.5, -4430)
```

which draws a line from from $(1.0, -5580)$ to $(1.5, -4430)$. And so on, until `realPrice` takes on its final value of 25 and we draw a line from the previous position of $(24.5, 5230)$ to $(25, 4500)$.

Reflection 4.14 *What shape is the plot? Can you see why?*

Reflection 4.15 *When you run this plotting program, you will notice an ugly line from the origin to the first point of the plot. How can you fix this? (We will leave the answer as an exercise.)*

Exercises

*Write a function for each of the following problems. When appropriate, make sure the function **returns** the specified value (rather than just printing it). Be sure to appropriately document your functions with docstrings and comments.*

4.1.1. Write a function

```
print100()
```

that uses a `for` loop to print the integers from 0 to 100.

4.1.2. Write a function

```
triangle1()
```

that uses a `for` loop to print the following:

```
*
**
***
****
*****
******
*******
********
*********
**********
```

4.1.3. Write a function

```
square(letter, width)
```

that prints a square with the given `width` using the string `letter`. For example, `square('Q', 5)` should print:

```
QQQQQ
QQQQQ
QQQQQ
QQQQQ
QQQQQ
```

4.1.4. Write a `for` loop that prints the integers from −50 to 50.

4.1.5. Write a `for` loop that prints the odd integers from 1 to 100. (Hint: use `range(50)`.)

4.1.6. On Page 43, we talked about how to simulate the minutes ticking on a digital clock using modular arithmetic. Write a function

```
clock(ticks)
```

that prints `ticks` times starting from midnight, where the clock ticks once each minute. To simplify matters, the midnight hour can be denoted 0 instead of 12. For example, `clock(100)` should print

```
0:00
0:01
0:02
⋮
0:59
1:00
1:01
⋮
1:38
1:39
```

To line up the colons in the times and force the leading zero in the minutes, use a format string like this:

```
print('{0:>2}:{1:0>2}'.format(hours, minutes))
```

4.1.7. There are actually three forms of the `range` function:

- 1 parameter: `range(stop)`
- 2 parameters: `range(start, stop)`
- 3 parameters: `range(start, stop, skip)`

With three arguments, `range` produces a range of integers starting at the start value and ending at or before `stop - 1`, adding `skip` each time. For example,

```
range(5, 15, 2)
```

produces the range of numbers 5, 7, 9, 11, 13 and

```
range(-5, -15, -2)
```

produces the range of numbers -5, -7, -9, -11, -13. To print these numbers, one per line, we can use a `for` loop:

```
for number in range(-5, -15, -2):
    print(number)
```

(a) Write a `for` loop that prints the even integers from 2 to 100, using the third form of the `range` function.

(b) Write a `for` loop that prints the odd integers from 1 to 100, using the third form of the `range` function.

(c) Write a `for` loop that prints the integers from 1 to 100 in descending order.

(d) Write a `for` loop that prints the values 7, 11, 15, 19.

(e) Write a `for` loop that prints the values 2, 1, 0, −1, −2.

(f) Write a `for` loop that prints the values −7, −11, −15, −19.

4.1.8. Write a function

> `multiples(n)`

that prints all of the multiples of the parameter n between 0 and 100, inclusive. For example, if n were 4, the function should print the values 0, 4, 8, 12, ...

4.1.9. Write a function

> `countdown(n)`

that prints the integers between 0 and n in descending order. For example, if n were 5, the function should print the values 5, 4, 3, 2, 1, 0.

4.1.10. Write a function

> `triangle2()`

that uses a `for` loop to print the following:

```
*****
****
***
**
*
```

4.1.11. Write a function

> `triangle3()`

that uses a `for` loop to print the following:

```
******
 *****
  ****
   ***
    **
     *
```

4.1.12. Write a function

> `diamond()`

that uses `for` loops to print the following:

```
***** *****
 ****   ****
  ***     ***
   **       **
    *         *
    *         *
   **       **
  ***     ***
 ****   ****
***** *****
```

4.1.13. Write a function

    circles(tortoise)

that uses turtle graphics and a `for` loop to draw concentric circles with radii $10, 20, 30, \ldots, 100$. (To draw each circle, you may use the turtle graphics `circle` method or the `drawCircle` function you wrote in Exercise 3.3.8.)

4.1.14. In the `profitPlot` function in the text, fix the problem raised by Reflection 4.15.

4.1.15. Write a function

    plotSine(tortoise, n)

that uses turtle graphics to plot $\sin x$ from $x = 0$ to $x = n$ degrees. Use `setworldcoordinates` to make the x coordinates of the window range from 0 to 1080 and the y coordinates range from -1 t0 1.

4.1.16. Python also allows us to pass function names as parameters. So we can generalize the function in Exercise 4.1.15 to plot any function we want. We write a function

    plot(tortoise, n, f)

where `f` is the name of an arbitrary function that takes a single numerical argument and returns a number. Inside the `for` loop in the `plot` function, we can apply the function `f` to the index variable `x` with

    tortoise.goto(x, f(x))

To call the `plot` function, we need to define one or more functions to pass in as arguments. For example, to plot x^2, we can define

```
def square(x):
    return x * x
```

and then call `plot` with

    plot(george, 20, square)

Or, to plot an elongated $\sin x$, we could define

```
def sin(x):
    return 10 * math.sin(x)
```

and then call `plot` with

    plot(george, 20, sin)

After you create your new version of `plot`, also create at least one new function to pass into `plot` for the parameter `f`. Depending on the functions you pass in, you may need to adjust the window coordinate system with `setworldcoordinates`.

4.1.17. Modify the `profitTable` function so that it considers all ticket prices that are multiples of a quarter.

4.1.18. Generalize the `pond` function so that it also takes the annual growth rate as a parameter.

4.1.19. Generalize the `pond` function further to allow for the pond to be annually restocked with an additional quantity of fish.

4.1.20. Modify the `countLinks` function so that it prints a table like the following:

```
|       |    Links    |
| Nodes | New | Total |
| ----- | --- | ----- |
|   2   |  1  |   1   |
|   3   |  2  |   3   |
|   4   |  3  |   6   |
|   5   |  4  |  10   |
|   6   |  5  |  15   |
|   7   |  6  |  21   |
|   8   |  7  |  28   |
|   9   |  8  |  36   |
|  10   |  9  |  45   |
```

4.1.21. Write a function

> `growth1(totalDays)`

that simulates a population growing by 3 individuals each day. For each day, print the day number and the total population size.

4.1.22. Write a function

> `growth2(totalDays)`

that simulates a population that grows by 3 individuals each day but also shrinks by, on average, 1 individual every 2 days. For each day, print the day number and the total population size.

4.1.23. Write a function

> `growth3(totalDays)`

that simulates a population that increases by 110% every day. Assume that the initial population size is 10. For each day, print the day number and the total population size.

4.1.24. Write a function

> `growth4(totalDays)`

that simulates a population that grows by 2 on the first day, 4 on the second day, 8 on the third day, 16 on the fourth day, etc. Assume that the initial population size is 10. For each day, print the day number and the total population size.

4.1.25. Write a function

> `sum(n)`

that returns the sum of the integers between 1 and `n`, inclusive. For example, `sum(4)` returns $1 + 2 + 3 + 4 = 10$. (Use a `for` loop; if you know a shortcut, don't use it.)

4.1.26. Write a function

    `sumEven(n)`

that returns the sum of the even integers between 2 and `n`, inclusive. For example, `sumEven(6)` returns $2 + 4 + 6 = 12$. (Use a `for` loop.)

4.1.27. Between the ages of three and thirteen, girls grow an average of about six centimeters per year. Write a function

    `growth(finalAge)`

that prints a simple height chart based on this information, with one entry for each age, assuming the average girl is 95 centimeters (37 inches) tall at age three.

4.1.28. Write a function

    `average(low, high)`

that returns the average of the integers between `low` and `high`, inclusive. For example, `average(3, 6)` returns $(3 + 4 + 5 + 6)/4 = 4.5$.

4.1.29. Write a function

    `factorial(n)`

that returns the value of $n! = 1 \times 2 \times 3 \times \cdots \times (n-1) \times n$. (Be careful; how should the accumulator be initialized?)

4.1.30. Write a function

    `power(base, exponent)`

that returns the value of `base` raised to the `exponent` power, without using the `**` operator. Assume that `exponent` is a positive integer.

4.1.31. The geometric mean of n numbers is defined to be the nth root of the product of the numbers. (The nth root is the same as the $1/n$ power.) Write a function

    `geoMean(high)`

that returns the geometric mean of the numbers between 1 and `high`, inclusive.

4.1.32. Write a function

    `sumDigits(number, numDigits)`

that returns the sum of the individual digits in a parameter `number` that has `numDigits` digits. For example, `sumDigits(1234, 4)` should return the value $1 + 2 + 3 + 4 = 10$. (Hint: use a `for` loop and integer division (`//` and `%`).)

4.1.33. Consider the following fun game. Pick any positive integer less than 100 and add the squares of its digits. For example, if you choose 25, the sum of the squares of its digits is $2^2 + 5^2 = 29$. Now make the answer your new number, and repeat the process. For example, if we continue this process starting with 25, we get: 25, 29, 85, 89, 145, 42, etc.

Write a function

    `fun(number, iterations)`

that prints the sequence of numbers generated by this game, starting with the two digit `number`, and continuing for the given number of `iterations`. It will be helpful to know that no number in a sequence will ever have more than three digits.

Execute your function with every integer between 15 and 25, with `iterations` at least 30. What do you notice? Can you classify each of these integers into one of two groups based on the results?

4.1.34. You have $1,000 to invest and need to decide between two savings accounts. The first account pays interest at an annual rate of 1% and is compounded daily, meaning that interest is earned daily at a rate of $(1/365)\%$. The second account pays interest at an annual rate of 1.25% but is compounded monthly. Write a function

    `interest(originalAmount, rate, periods)`

that computes the interest earned in one year on `originalAmount` dollars in an account that pays the given annual interest rate, compounded over the given number of periods. Assume the interest rate is given as a percentage, not a fraction (i.e., 1.25 vs. 0.0125). Use the function to answer the original question.

4.1.35. Suppose you want to start saving a certain amount each month in an investment account that compounds interest monthly. To determine how much money you expect to have in the future, write a function

    `invest(investment, rate, years)`

that returns the income earned by investing `investment` dollars every month in an investment account that pays the given rate of return, compounded monthly (`rate` / 12 % each month).

4.1.36. A mortgage loan is charged some rate of interest every month based on the current balance on the loan. If the annual interest rate of the mortgage is $r\%$, then interest equal to $r/12$ % of the current balance is added to the amount owed each month. Also each month, the borrower is expected to make a payment, which reduces the amount owed.

Write a function

    `mortgage(principal, rate, years, payment)`

that prints a table of mortgage payments and the remaining balance every month of the loan period. The last payment should include any remaining balance. For example, paying $1,000 per month on a $200,000 loan at 4.5% for 30 years should result in the following table:

```
Month   Payment     Balance
1       1000.00     199750.00
2       1000.00     199499.06
3       1000.00     199247.18
⋮
359     1000.00      11111.79
360    11153.46          0.00
```

4.1.37. Suppose a bacteria colony grows at a rate of 10% per hour and that there are initially 100 bacteria in the colony. Write a function

136 ■ Growth and decay

```
bacteria(days)
```

that returns the number of bacteria in the colony after the given number of `days`. How many bacteria are in the colony after one week?

4.1.38. Generalize the function that you wrote for the previous exercise so that it also accepts parameters for the initial population size and the growth rate. How many bacteria are in the same colony after one week if it grows at 15% per hour instead?

4.2 VISUALIZING POPULATION CHANGES

Visualizing changes in population size over time will provide more insight into how population models evolve. We could plot population changes with turtle graphics, as we did in Section 4.1, but instead, we will use a dedicated plotting module called `matplotlib`, so-named because it emulates the plotting capabilities of the technical programming language MATLAB[1]. If you do not already have `matplotlib` installed on your computer, see Appendix A for instructions.

To use `matplotlib`, we first import the module using

```
import matplotlib.pyplot as pyplot
```

`matplotlib.pyplot` is the name of module; "`as pyplot`" allows us to refer to the module in our program with the abbreviation `pyplot` instead of its rather long full name. The basic plotting functions in `matplotlib` take two arguments: a list of x values and an associated list of y values. A list in Python is an object of the `list` class, and is represented as a comma-separated sequence of items enclosed in square brackets, such as

```
[4, 7, 2, 3.1, 12, 2.1]
```

We saw lists briefly in Section 4.1 when we used the `list` function to visualize `range` values; we will use lists much more extensively in Chapter 8. For now, we only need to know how to build a list of population sizes in our `for` loop so that we can plot them. Let's look at how to do this in the fishing pond function from Page 119, reproduced below.

```
def pond(years, initialPopulation, harvest):
    """ (docstring omitted) """

    population = initialPopulation
    print('Year  Population')
    for year in range(years):
        population = 1.08 * population - harvest
        print('{0:^4}  {1:>9.2f}'.format(year + 1, population))

    return population
```

---

[1]MATLAB is a registered trademark of The MathWorks, Inc.

We start by creating an empty list of annual population sizes before the loop:

```
populationList = [ ]
```

As you can see, an empty list is denoted by two square brackets with nothing in between. To add an annual population size to the end of the list, we will use the `append` method of the list class. We will first append the initial population size to the end of the empty list with

```
populationList.append(initialPopulation)
```

If we pass in `12000` for the initial population parameter, this will result in `populationList` becoming the single-element list `[12000]`. Inside the loop, we want to append each value of `population` to the end of the growing list with

```
populationList.append(population)
```

Incorporating this code into our `pond` function, and deleting the calls to `print`, yields:

```
def pond(years, initialPopulation, harvest):
    """Simulates a fish population in a fishing pond, and
       plots annual population size.  The population
       grows 8% per year with an annual harvest.

    Parameters:
        years: number of years to simulate
        initialPopulation: the initial population size
        harvest: the size of the annual harvest

    Return value: the final population size
    """

    population = initialPopulation
    populationList = [ ]
    populationList.append(initialPopulation)
    for year in range(years):
        population = 1.08 * population - harvest
        populationList.append(population)

    return population
```

The table below shows how the `populationList` grows with each iteration by appending the current value of `population` to its end, assuming an initial population of 12,000. When the loop finishes, we have `years + 1` population sizes in our list.

138 ■ Growth and decay

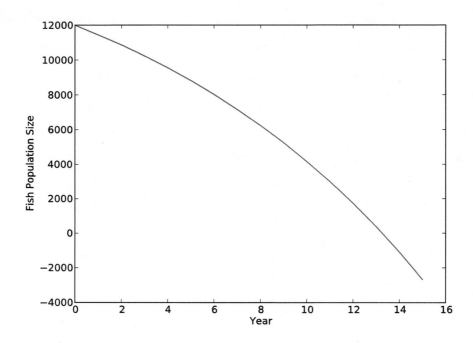

Figure 4.2 Plot of population size in our fishing pond model with **years** = 15.

year	population	populationList
--	12000	[12000]
1	11460.0	[12000, 11460.0]
2	10876.8	[12000, 11460.0, 10876.8]
⋮	⋮	⋮
13	392.53984839	[12000, 11460.0, ..., 392.53984839]
14	-1076.05696374	[12000, 11460.0, ..., 392.53984839, -1076.05696374]

There is a strong similarity between the manner in which we are appending elements to a list and the accumulators that we have been talking about in this chapter. In an accumulator, we accumulate values into a sum by repeatedly adding new values to a running sum. The running sum changes (usually grows) in each iteration of the loop. With the list in the **for** loop above, we are accumulating values in a different way—by repeatedly appending them to the end of a growing list. Therefore, we call this technique a *list accumulator*.

We now want to use this list of population sizes as the list of y values in a **matplotlib** plot. For the x values, we need a list of the corresponding years, which can be obtained with **range(years + 1)**. Once we have both lists, we can create a plot by calling the **plot** function and then display the plot by calling the **show** function:

```
pyplot.plot(range(years + 1), populationList)
pyplot.show()
```

The first argument to the **plot** function is the list of x values and the second parameter is the list of y values. The **matplotlib** module includes many optional ways to customize our plots before we call **show**. Some of the simplest are functions that label the x and y axes:

```
pyplot.xlabel('Year')
pyplot.ylabel('Fish Population Size')
```

Incorporating the plotting code yields the following function, whose output is shown in Figure 4.2.

```
import matplotlib.pyplot as pyplot

def pond(years, initialPopulation, harvest):
    """ (docstring omitted) """

    population = initialPopulation
    populationList = [ ]
    populationList.append(initialPopulation)
    for year in range(years):
        population = 1.08 * population - harvest
        populationList.append(population)

    pyplot.plot(range(years + 1), populationList)
    pyplot.xlabel('Year')
    pyplot.ylabel('Fish Population Size')
    pyplot.show()

    return population
```

For more complex plots, we can alter the scales of the axes, change the color and style of the curves, and label multiple curves on the same plot. See Appendix B.4 for a sample of what is available. Some of the options must be specified as *keyword arguments* of the form **name = value**. For example, to color a curve in a plot red and specify a label for the plot legend, you would call something like this:

```
pyplot.plot(x, y, color = 'red', label = 'Bass population')
pyplot.legend()    # creates a legend from labeled lines
```

Exercises

4.2.1. Modify the **countLinks** function on Page 123 so that it uses **matplotlib** to plot the number of nodes on the x axis and the maximum number of links on the y axis. Your resulting plot should look like the one in Figure 4.3.

4.2.2. Modify the **profitPlot** function on Page 128 so that it uses **matplotlib** to plot ticket price on the x axis and profit on the y axis. (Remove the **tortoise** parameter.) Your resulting plot should look like the one in Figure 4.4. To get the correct prices (in half dollar increments) on the x axis, you will need to create a second list of x values and append the **realPrice** to it in each iteration.

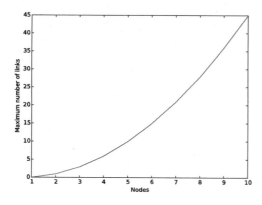
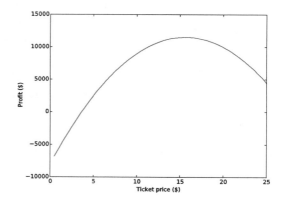

Figure 4.3 Plot for Exercise 4.2.1.   Figure 4.4 Plot for Exercise 4.2.2.

4.2.3. Modify your `growth1` function from Exercise 4.1.21 so that it uses `matplotlib` to plot days on the x axis and the total population on the y axis. Create a plot that shows the growth of the population over 30 days.

4.2.4. Modify your `growth3` function from Exercise 4.1.23 so that it uses `matplotlib` to plot days on the x axis and the total population on the y axis. Create a plot that shows the growth of the population over 30 days.

4.2.5. Modify your `invest` function from Exercise 4.1.35 so that it uses `matplotlib` to plot months on the x axis and your total accumulated investment amount on the y axis. Create a plot that shows the growth of an investment of $50 per month for ten years growing at an annual rate of 8%.

4.3 CONDITIONAL ITERATION

In our fishing pond model, to determine when the population size fell below zero, it was sufficient to simply print the annual population sizes for at least 14 years, and look at the results. However, if it had taken a thousand years for the population size to fall below zero, then looking at the output would be far less convenient. Instead, we would like to have a program tell us the year directly, by ceasing to iterate when `population` drops to zero, and then returning the year it happened. This is a different kind of problem because we no longer know how many iterations are required before the loop starts. In other words, we have no way of knowing what value to pass into `range` in a `for` loop.

Instead, we need a more general kind of loop that will iterate only while some condition is met. Such a loop is generally called a *while loop*. In Python, a while loop looks like this:

```
while <condition>:
    <body>
```

The `<condition>` is replaced with a Boolean expression that evaluates to `True` or

False, and the <body> is replaced with statements constituting the body of the loop. The loop checks the value of the condition before each iteration. If the condition is true, it executes the statements in the body of the loop, and then checks the condition again. If the condition is false, the body of the loop is skipped, and the loop ends.

We will talk more about building Boolean expressions in the next chapter; for now we will only need very simple ones like `population > 0`. This Boolean expression is true if the value of `population` is positive, and false otherwise. Using this Boolean expression in the `while` loop in the following function, we can find the year that the fish population drops to 0.

```
def yearsUntilZero(initialPopulation, harvest):
    """Computes # of years until a fish population reaches zero.
       Population grows 8% per year with an annual harvest.

    Parameters:
        initialPopulation: the initial population size
        harvest: the size of the annual harvest

    Return value: year during which the population reaches zero
    """

    population = initialPopulation
    year = 0
    while population > 0:
        population = 1.08 * population - harvest
        year = year + 1
    return year
```

Let's assume that `initialPopulation` is 12000 and `harvest` is 1500, as in our original `pond` function in Section 4.1. Therefore, before the loop, `population` is 12000 and `year` is 0. Since `population > 0` is true, the loop body executes, causing `population` to become 11460 and `year` to become 1. (You might want to refer back to the annual population sizes on Page 118.) We then go back to the top of the loop to check the condition again. Since `population > 0` is still true, the loop body executes again, causing `population` to become 10876 and `year` to become 2. Iteration continues until `year` reaches 14. In this year, `population` becomes -1076.06. When the condition is checked now, we find that `population > 0` is false, so the loop ends and the function returns 14.

Using `while` loops can be tricky for two reasons. First, a `while` loop may not iterate at all. For example, if the initial value of `population` were zero, the condition in the `while` loop will be false before the first iteration, and the loop will be over before it starts.

Reflection 4.16 *What will be returned by the function in this case?*

A loop that sometimes does not iterate at all is generally not a bad thing, and can even be used to our advantage. In this case, if `population` were initially zero, the function would return zero because the value of `year` would never be incremented in the loop. And this is correct; the population dropped to zero in year zero, before the clock started ticking beyond the initial population size. But it is something that one should always keep in mind when designing algorithms involving `while` loops.

Second, a `while` loop may become an *infinite loop*. For example, suppose `initialPopulation` is 12000 and `harvest` is 800 instead of 1500. In this case, as we saw on Page 119, the population size *increases* every year instead. So the population size will *never* reach zero and the loop condition will *never* be false, so the loop will iterate forever. For this reason, we must always make sure that the body of a `while` loop makes progress toward the loop condition becoming false.

Let's look at one more example. Suppose we have $1000 to invest and we would like to know how long it will take for our money to double in size, growing at 5% per year. To answer this question, we can create a loop like the following that compounds 5% interest each year:

```
amount = 1000
while ???:
    amount = 1.05 * amount
print(amount)
```

Reflection 4.17 *What should be the condition in the `while` loop?*

We want the loop to stop iterating when `amount` reaches 2000. Therefore, we want the loop to continue to iterate *while* `amount < 2000`.

```
amount = 1000
while amount < 2000:
    amount = 1.05 * amount
print(amount)
```

Reflection 4.18 *What is printed by this block of code? What does this result tell us?*

Once the loop is done iterating, the final amount is printed (approximately $2078.93), but this does not answer our question.

Reflection 4.19 *How do figure out how many years it takes for the $1000 to double?*

To answer our question, we need to count the number of times the `while` loop iterates, which is very similar to what we did in the `yearsUntilZero` function. We can introduce another variable that is incremented in each iteration, and print its value after the loop, along with the final value of `amount`:

```
amount = 1000
while amount < 2000:
    amount = 1.05 * amount
    year = year + 1
print(year, amount)
```

Reflection 4.20 *Make these changes and run the code again. Now what is printed?*

Oops, an error message is printed, telling us that the name `year` is undefined.

Reflection 4.21 *How do we fix the error?*

The problem is that we did not initialize the value of `year` before the loop. Therefore, the first time `year = year + 1` was executed, `year` was undefined on the right side of the assignment statement. Adding one statement before the loop fixes the problem:

```
amount = 1000
year = 0
while amount. < 2000:
    amount = 1.05 * amount
    year = year + 1
print(year, amount)
```

Reflection 4.22 *Now what is printed by this block of code? In other words, how many years until the $1000 doubles?*

We will see some more examples of `while` loops later in this chapter, and again in Section 5.5.

Exercises

4.3.1. Suppose you put $1000 into the bank and you get a 3% interest rate compounded annually. How would you use a `while` loop to determine how long will it take for your account to have at least $1200 in it?

4.3.2. Repeat the last question, but this time write a function

       `interest(amount, rate, target)`

that takes the initial amount, the interest rate, and the target amount as parameters. The function should return the number of years it takes to reach the target amount.

4.3.3. Since `while` loops are more general than `for` loops, we can emulate the behavior of a `for` loop with a `while` loop. For example, we can emulate the behavior of the `for` loop

```
for counter in range(10):
    print(counter)
```

with the `while` loop

```
counter = 0
while counter < 10:
    print(counter)
    counter = counter + 1
```

Execute both loops "by hand" to make sure you understand how these loops are equivalent.

(a) What happens if we omit `counter = 0` before the `while` loop? Why does this happen?

(b) What happens if we omit `counter = counter + 1` from the body of the `while` loop? What does the loop print?

(c) Show how to emulate the following `for` loop with a `while` loop:

```
for index in range(3, 12):
    print(index)
```

(d) Show how to emulate the following `for` loop with a `while` loop:

```
for index in range(12, 3, -1):
    print(index)
```

4.3.4. In the `profitTable` function on Page 127, we used a `for` loop to indirectly consider all ticket prices divisible by a half dollar. Rewrite this function so that it instead uses a `while` loop that increments `price` by $0.50 in each iteration.

4.3.5. A zombie can convert two people into zombies everyday. Starting with just one zombie, how long would it take for the entire world population (7 billion people) to become zombies? Write a function

   `zombieApocalypse()`

that returns the answer to this question.

4.3.6. Tribbles increase at the rate of 50% per hour (rounding down if there are an odd number of them). How long would it take 10 tribbles to reach a population size of 1 million? Write a function

   `tribbleApocalypse()`

that returns the answer to this question.

4.3.7. Vampires can each convert v people a day into vampires. However, there is a band of vampire hunters that can kill k vampires per day. If a coven of vampires starts with `vampires` members, how long before a town with a population of `people` becomes a town with no humans left in it? Write a function

   `vampireApocalypse(v, k, vampires, people)`

that returns the answer to this question.

4.3.8. An amoeba can split itself into two once every h hours. How many hours does it take for a single amoeba to become `target` amoebae? Write a function

   `amoebaGrowth(h, target)`

that returns the answer to this question.

*4.4 CONTINUOUS MODELS

If we want to more accurately model the situation in our fishing pond, we need to acknowledge that the size of the fish population does not really change only once a year. Like virtually all natural processes, the change happens continually or, mathematically speaking, *continuously*, over time. To more accurately model continuous natural processes, we need to update our population size more often, using smaller time steps. For example, we could update the size of the fish population every month instead of every year by replacing every annual update of

```
population = population + 0.08 * population - 1500
```

with twelve monthly updates:

```
for month in range(12):
    population = population + (0.08 / 12) * population - (1500 / 12)
```

Since both the growth rate of `0.08` and the harvest of `1500` are based on 1 year, we have divided both of them by 12.

Reflection 4.23 *Is the final value of* `population` *the same in both cases?*

If the initial value of `population` is `12000`, the value of `population` after one annual update is `11460.0` while the final value after 12 monthly updates is `11439.753329049303`. Because the rate is "compounding" monthly, it reduces the population more quickly.

This is exactly how bank loans work. The bank will quote an annual percentage rate (APR) of, say, 6% (or 0.06) but then compound interest monthly at a rate of $6/12\% = 0.5\%$ (or 0.005), which means that the actual annual rate of interest you are being charged, called the annual percentage yield (APY), is actually $(1+0.005)^{12}-1 \approx 0.0617 = 6.17\%$. The APR, which is really defined to be the monthly rate times 12, is sometimes also called the "nominal rate." So we can say that our fish population is increasing at a nominal rate of 8%, but updated every month.

Difference equations

A population model like this is expressed more accurately with a *difference equation*, also known as a *recurrence relation*. If we let $P(t)$ represent the size of the fish population at the end of year t, then the difference equation that defines our original model is

$$P(t) = \underbrace{P(t-1)}_{\text{previous year's population}} + \underbrace{0.08 \cdot P(t-1)}_{\text{proportional increase}} - \underbrace{1500}_{\text{harvest}}$$

or, equivalently,

$$P(t) = 1.08 \cdot P(t-1) - 1500.$$

In other words, the size of the population at the end of year t is 1.08 times the size of the population at the end of the previous year $(t-1)$, minus 1500. The initial population or, more formally, the *initial condition* is $P(0) = 12{,}000$. We can find the projected population size for any given year by starting with $P(0)$, and using the difference equation to compute successive values of $P(t)$. For example, suppose we wanted to know the projected population size four years from now, represented by $P(4)$. We start with the initial condition: $P(0) = 12{,}000$. Then, we apply the difference equation to $t = 1$:

$$P(1) = 1.08 \cdot P(0) - 1500 = 1.08 \cdot 12{,}000 - 1500 = 11{,}460.$$

Now that we have $P(1)$, we can compute $P(2)$:

$$P(2) = 1.08 \cdot P(1) - 1500 = 1.08 \cdot 11{,}460 - 1500 = 10{,}876.8.$$

Continuing,

$$P(3) = 1.08 \cdot P(2) - 1500 = 1.08 \cdot 10{,}876.8 - 1500 = 10{,}246.94$$

and

$$P(4) = 1.08 \cdot P(3) - 1500 = 1.08 \cdot 10{,}246.94 - 1500 = 9{,}566.6952.$$

So this model projects that the bass population in 4 years will be $9{,}566$. This is the same process we followed in our `for` loop in Section 4.1.

To turn this *discrete* model into a *continuous* model, we define a small update interval, which is customarily named Δt (Δ represents "change," so Δt represents "change in time"). If, for example, we want to update the size of our population every month, then we let $\Delta t = 1/12$. Then we express our difference equation as

$$P(t) = P(t - \Delta t) + (0.08 \cdot P(t - \Delta t) - 1500) \cdot \Delta t$$

This difference equation is defining the population size at the end of year t in terms of the population size one Δt fraction of a year ago. For example, if t is 3 and Δt is $1/12$, then $P(t)$ represents the size of the population at the end of year 3 and $P(t - \Delta t)$ represents the size of the population at the end of "year $2\frac{11}{12}$," equivalent to one month earlier. Notice that both the growth rate and the harvest number are scaled by Δt, just as we did in the `for` loop on Page 145.

To implement this model, we need to make some analogous changes to the algorithm from Page 119. First, we need to pass in the value of Δt as a parameter so that we have control over the accuracy of the approximation. Second, we need to modify the number of iterations in our loop to reflect $1/\Delta t$ decay events each year; the number of iterations becomes `years` $\cdot (1/\Delta t) =$ `years`$/\Delta t$. Third, we need to alter how the accumulator is updated in the loop to reflect this new type of difference equation. These changes are reflected below. We use `dt` to represent Δt.

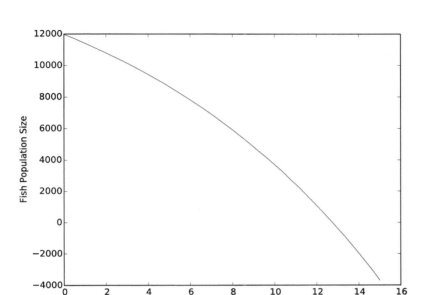

Figure 4.5 The plot produced by calling pond(15, 12000, 1500, 0.01).

```
def pond(years, initialPopulation, harvest, dt):
    """ (docstring omitted) """

    population = initialPopulation
    for step in range(1, int(years / dt) + 1):
        population = population + (0.08 * population - harvest) * dt

    return population
```

Reflection 4.24 *Why do we use* range(1, int(years / dt) + 1) *in the* for *loop instead of* range(int(years / dt))*?*

We start the for loop at one instead of zero because the first iteration of the loop represents the first time step of the simulation. The initial population size assigned to population before the loop represents the population at time zero.

To plot the results of this simulation, we use the same technique that we used in Section 4.2. But we also use a list accumulator to create a list of time values for the x axis because the values of the index variable step no longer represent years. In the following function, the value of step * dt is assigned to the variable t, and then appended to a list named timeList.

```
import matplotlib.pyplot as pyplot

def pond(years, initialPopulation, harvest, dt):
    """Simulates a fish population in a fishing pond, and plots
       annual population size.  The population grows at a nominal
       annual rate of 8% with an annual harvest.

    Parameters:
        years: number of years to simulate
        initialPopulation: the initial population size
        harvest: the size of the annual harvest
        dt: value of "Delta t" in the simulation (fraction of a year)

    Return value: the final population size
    """

    population = initialPopulation
    populationList = [initialPopulation]
    timeList = [0]
    t = 0
    for step in range(1, int(years / dt) + 1):
        population = population + (0.08 * population - harvest) * dt
        populationList.append(population)
        t = step * dt
        timeList.append(t)

    pyplot.plot(timeList, populationList)
    pyplot.xlabel('Year')
    pyplot.ylabel('Fish Population Size')
    pyplot.show()
    return population
```

Figure 4.5 shows a plot produced by calling this function with $\Delta t = 0.01$. Compare this plot to Figure 4.2. To actually come close to approximating a real continuous process, we need to use very small values of Δt. But there are tradeoffs involved in doing so, which we discuss in more detail later in this section.

Radiocarbon dating

When archaeologists wish to know the ages of organic relics, they often turn to *radiocarbon dating*. Both carbon-12 (^{12}C) and carbon-14 (or radiocarbon, ^{14}C) are isotopes of carbon that are present in our atmosphere in a relatively constant proportion. While carbon-12 is a stable isotope, carbon-14 is unstable and decays at a known rate. All living things ingest both isotopes, and die possessing them in the same proportion as the atmosphere. Thereafter, an organism's acquired carbon-14 decays at a known rate, while its carbon-12 remains intact. By examining the current ratio of carbon-12 to carbon-14 in organic remains (up to about 60,000 years old), and comparing this ratio to the known ratio in the atmosphere, we can approximate how long ago the organism died.

The annual decay rate (more correctly, the decay constant[2]) of carbon-14 is about
$$k = -0.000121.$$
Radioactive decay is a continuous process, rather than one that occurs at discrete intervals. Therefore, $Q(t)$, the quantity of carbon-14 present at the beginning of year t, needs to be defined in terms of the value of $Q(t - \Delta t)$, the quantity of carbon-14 a small Δt fraction of a year ago. Therefore, the difference equation modeling the decay of carbon-14 is
$$Q(t) = Q(t - \Delta t) + k \cdot Q(t - \Delta t) \cdot \Delta t\,.$$
Since the decay constant k is based on one year, we scale it down for an interval of length Δt by multiplying it by Δt. We will represent the initial condition with $Q(0) = q$, where q represents the initial quantity of carbon-14.

Although this is a completely different application, we can implement the model the same way we implemented our continuous fish population model.

```
import matplotlib.pyplot as pyplot

def decayC14(originalAmount, years, dt):
    """Approximates the continuous decay of carbon-14.

    Parameters:
        originalAmount: the original quantity of carbon-14 (g)
        years: number of years to simulate
        dt: value of "Delta t" in the simulation (fraction of a year)

    Return value: final quantity of carbon-14 (g)
    """

    k = -0.000121
    amount = originalAmount
    t = 0
    timeList = [0]                  # x values for plot
    amountList = [amount]           # y values for plot

    for step in range(1, int(years/dt) + 1):
        amount = amount + k * amount * dt
        t = step * dt
        timeList.append(t)
        amountList.append(amount)

    pyplot.plot(timeList, amountList)
```

---
[2] The probability of a carbon-14 molecule decaying in a very small fraction Δt of a year is $k\Delta t$.

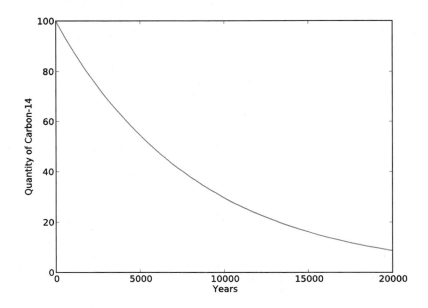

Figure 4.6 Plot of carbon-14 decay generated with `decayC14(100, 20000, 0.01)`.

```
pyplot.xlabel('Years')
pyplot.ylabel('Quantity of carbon-14')
pyplot.show()

return amount
```

Like all of our previous accumulators, this function initializes our accumulator variable, named `amount`, before the loop. Then the accumulator is updated in the body of the loop according to the difference equation above. Figure 4.6 shows an example plot from this function.

Reflection 4.25 *How much of 100 g of carbon-14 remains after 5,000 years of decay? Try various Δt values ranging from 1 down to 0.001. What do you notice?*

Tradeoffs between accuracy and time

Approximations of continuous models are more accurate when the value of Δt is smaller. However, accuracy has a cost. The `decayC14` function has time complexity proportional to $y/\Delta t$, where y is the number of years we are simulating, since this is how many times the `for` loop iterates. So, if we want to improve accuracy by dividing Δt in half, this will directly result in our algorithm requiring twice as much time to run. If we want Δt to be one-tenth of its current value, our algorithm will require ten times as long to run.

> **Box 4.1: Differential equations**
>
> If you have taken a course in calculus, you may recognize each of these continuous models as an approximation of a *differential equation*. A differential equation is an equation that relates a function to the rate of change of that function, called the derivative. For example, the differential equation corresponding to the carbon-14 decay problem is
>
> $$\frac{dQ}{dt} = kQ$$
>
> where Q is the quantity of carbon-14, dQ/dt is the rate of change in the quantity of carbon-14 at time t (i.e., the derivative of Q with respect to t), and k is the decay constant. Solving this differential equation using basic calculus techniques yields
>
> $$Q(t) = Q(0)\, e^{kt}$$
>
> where $Q(0)$ is the original amount of carbon-14. We can use this equation to directly compute how much of 1000 g of carbon-14 would be left after 2,000 years with:
>
> $$Q(2000) = 1000\, e^{2000k} \approx 785.0562.$$
>
> Although simple differential equations like this are easily solved if you know calculus, most realistic differential equations encountered in the sciences are not, making approximate iterative solutions essential. The approximation technique we are using in this chapter, called *Euler's method*, is the most fundamental, and introduces error proportional to Δt. More advanced techniques seek to reduce the approximation error further.

To get a sense of how decreasing values of Δt affect the outcome, let's look at what happens in our `decayC14` function with `originalAmount = 1000` and `years = 2000`. The table below contains these results, with the theoretically derived answer in the last row (see Box 4.1). The error column shows the difference between the result and this value for each value of `dt`. All values are rounded to three significant digits to reflect the approximate nature of the decay constant. We can see from the table that smaller values of `dt` do indeed provide closer approximations.

dt	final amount	error	time
1.0	785.0447	0.0115	500 μs
0.1	785.0550	0.0012	5 ms
0.01	785.0561	0.0001	50 ms
0.001	785.0562	0.0	500 ms
0.0001	785.0562	0.0	5 sec
—	785.0562	—	—

Reflection 4.26 *What is the relationship between the value of* `dt` *and the error? What about between the value of* `dt` *and the execution time?*

Each row in the table represents a computation that took ten times as long as that in the previous row because the value of dt was ten times smaller. But the error is also ten times smaller. Is this tradeoff worthwhile? The answer depends on the situation. Certainly using dt = 0.0001 is not worthwhile because it gives the same answer (to three significant digits) as dt = 0.001, but takes ten times as long.

These types of tradeoffs — quality versus cost — are common in all fields, and computer science is no exception. Building a faster memory requires more expensive technology. Ensuring more accurate data transmission over networks requires more overhead. And finding better approximate solutions to hard problems requires more time.

Propagation of errors

In both the pond and decayC14 functions in this section, we skirted a very subtle error that is endemic to numerical computer algorithms. Recall from Section 2.2 that computers store numbers in binary and with finite precision, resulting in slight errors, especially with very small floating point numbers. But a slight error can become magnified in an iterative computation. This would have been the case in the for loops in the pond and decayC14 functions if we had accumulated the value of t by adding dt in each iteration with

```
t = t + dt
```

instead of getting t by multiplying dt by step. If dt is very small, then there might have been a slight error every time dt was added to t. If the loop iterates for a long time, the value of t will become increasingly inaccurate.

To illustrate the problem, let's mimic what might have happened in decayC14 with the following bit of code.

```
dt = 0.0001
iterations = 1000000

t = 0
for index in range(iterations):
    t = t + dt

correct = dt * iterations       # correct value of t
print(correct, t)
```

The loop accumulates the value 0.0001 one million times, so the correct value for t is $0.0001 \cdot 1{,}000{,}000 = 100$. However, by running the code, we see that the final value of t is actually 100.00000000219612, a small fraction over the correct answer. In some applications, even errors this small may be significant. And it can get even worse with more iterations. Scientific computations can often run for days or weeks, and the number of iterations involved can blow up errors significantly.

Reflection 4.27 *Run the code above with 10 million and 100 million iterations. What do you notice about the error?*

To avoid this kind of error, we instead assigned the product of `dt` and the current iteration number to `t`, `step`:

    t = step * dt

In this way, the value of `t` is computed from only one arithmetic operation instead of many, reducing the potential error.

Simulating an epidemic

Real populations interact with each other and the environment in complex ways. Therefore, to accurately model them requires an interdependent set of difference equations, called *coupled difference equations*. In 1927, Kermack and McKendrick [23] introduced such a model for the spread of infectious disease called the *SIR model*. The "S" stands for the "susceptible" population, those who may still acquire the disease; "I" stands for the "infected" population, and "R" stands for the "recovered" population. In this model, we assume that, once recovered, an individual has built an immunity to the disease and cannot reacquire it. We also assume that the total population size is constant, and no one dies from the disease. Therefore, an individual moves from a "susceptible" state to an "infected" state to a "recovered" state, where she remains, as pictured below.

These assumptions apply very well to common viral infections, like the flu.

A virus like the flu travels through a population more quickly when there are more infected people with whom a susceptible person can come into contact. In other words, a susceptible person is more likely to become infected if there are more infected people. Also, since the total population size does not change, an increase in the number of infected people implies an identical decrease in the number who are susceptible. Similarly, a decrease in the number who are infected implies an identical increase in the number who have recovered. Like most natural processes, the spread of disease is fluid or continuous, so we will need to model these changes over small intervals of Δt, as we did in the radioactive decay model.

We need to design three difference equations that describe how the sizes of the three groups change. We will let $S(t)$ represent the number of susceptible people on day t, $I(t)$ represent the number of infected people on day t, and $R(t)$ represent the number of recovered people on day t.

The recovered population has the most straightforward difference equation. The size of the recovered group only increases; when infected people recover they move from the infected group into the recovered group. The number of people who recover at each step depends on the number of infected people at that time and the *recovery rate* r: the average fraction of people who recover each day.

Reflection 4.28 *What factors might affect the recovery rate in a real outbreak?*

Since we will be dividing each day into intervals of length Δt, we will need to use a scaled recovery rate of $r \cdot \Delta t$ for each interval. So the difference equation describing the size of the recovered group on day t is

$$R(t) = R(t - \Delta t) + \underbrace{r \cdot I(t - \Delta t) \cdot \Delta t}_{\text{newly recovered, from } I}$$

Since no one has yet recovered on day 0, we set the initial condition to be $R(0) = 0$.

Next, we will consider the difference equation for $S(t)$. The size of the susceptible population only decreases by the number of newly infected people. This decrease depends on the number of susceptible people, the number of infected people with whom they can make contact, and the rate at which these potential interactions produce a new infection. The number of possible interactions between susceptible and infected individuals at time $t - \Delta t$ is simply their product: $S(t - \Delta t) \cdot I(t - \Delta t)$. If we let d represent the rate at which these interactions produce an infection, then our difference equation is

$$S(t) = S(t - \Delta t) - \underbrace{d \cdot S(t - \Delta t) \cdot I(t - \Delta t) \cdot \Delta t}_{\text{newly infected, to } I}.$$

Reflection 4.29 *What factors might affect the infection rate in a real outbreak?*

If N is the total size of our population, then the initial condition is $S(0) = N - 1$ because we will start with one infected person, leaving $N - 1$ who are susceptible.

The difference equation for the infected group depends on the number of susceptible people who have become newly infected and the number of infected people who are newly recovered. These numbers are precisely the number leaving the susceptible group and the number entering the recovered group, respectively. We can simply copy those from the equations above.

$$I(t) = I(t - \Delta t) + \underbrace{d \cdot S(t - \Delta t) \cdot I(t - \Delta t) \cdot \Delta t}_{\text{newly infected, from } S} - \underbrace{r \cdot I(t - \Delta t) \cdot \Delta t}_{\text{newly recovered, to } R}.$$

Since we are starting with one infected person, we set $I(0) = 1$.

The program below is a "skeleton" for implementing this model with a recovery rate $r = 0.25$ and an infection rate $d = 0.0004$. These rates imply that the average infection lasts $1/r = 4$ days and there is a 0.04% chance that an encounter between a susceptible person and an infected person will occur and result in a new infection. The program also demonstrates how to plot and label several curves in the same figure, and display a legend. We leave the implementation of the difference equations in the loop as an exercise.

```
import matplotlib.pyplot as pyplot

def SIR(population, days, dt):
    """Simulates the SIR model of infectious disease and
       plots the population sizes over time.

    Parameters:
        population: the population size
        days: number of days to simulate
        dt: the value of "Delta t" in the simulation
            (fraction of a day)

    Return value: None
    """

    susceptible = population - 1   # susceptible count = S(t)
    infected = 1.0                 # infected count = I(t)
    recovered = 0.0                # recovered count = R(t)
    recRate = 0.25                 # recovery rate r
    infRate = 0.0004               # infection rate d
    SList = [susceptible]
    IList = [infected]
    RList = [recovered]
    timeList = [0]

    # Loop using the difference equations goes here.

    pyplot.plot(timeList, SList, label = 'Susceptible')
    pyplot.plot(timeList, IList, label = 'Infected')
    pyplot.plot(timeList, RList, label = 'Recovered')
    pyplot.legend(loc = 'center right')
    pyplot.xlabel('Days')
    pyplot.ylabel('Individuals')
    pyplot.show()
```

Figure 4.7 shows the output of the model with 2,200 individuals (students at a small college, perhaps?) over 30 days. We can see that the infection peaks after about 2 weeks, then decreases steadily. After 30 days, the virus is just about gone, with only about 40 people still infected. We also notice that not everyone is infected after 30 days; about 80 people are still healthy.

Reflection 4.30 *Look at the relationships among the three population sizes over time in Figure 4.7. How do the sizes of the susceptible and recovered populations cause the decline of the affected population after the second week? What other relationships do you notice?*

When we are implementing a coupled model, we need to be careful to compute the change in each population size based on population sizes from the *previous* iteration

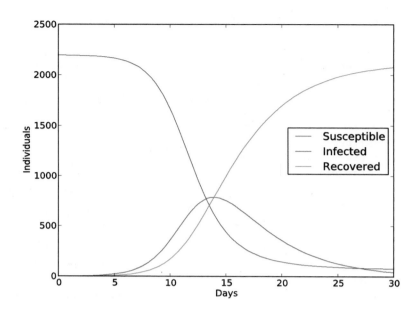

Figure 4.7 Output from the SIR model with 2,200 individuals over 30 days with $\Delta t = 0.01$.

rather than population sizes previously computed in *this* iteration. Therefore, we need to compute all of the changes in population sizes for time t, denoted $\Delta R(t)$, $\Delta S(t)$ and $\Delta I(t)$, based on the previous population sizes, before we update the new population sizes in each iteration. In other words, in each iteration, we first compute

$$\Delta R(t) = r \cdot I(t - \Delta t) \cdot \Delta t,$$
$$\Delta S(t) = d \cdot S(t - \Delta t) \cdot I(t - \Delta t) \cdot \Delta t, \text{ and}$$
$$\Delta I(t) = \Delta S(t) - \Delta R(t).$$

Then, we use these values to update the new population sizes:

$$R(t) = R(t - \Delta t) + \Delta R(t),$$
$$S(t) = S(t - \Delta t) - \Delta S(t), \text{ and}$$
$$I(t) = I(t - \Delta t) + \Delta I(t).$$

Interestingly, the SIR model can also be used to model the diffusion of ideas, fads, or memes in a population. A "contagious" idea starts with a small group of people and can "infect" others over time. As more people adopt the idea, it spreads more quickly because there are more potential interactions between those who have adopted the idea and those who have not. Eventually, people move on to other things and forget about the idea, thereby "recovering."

Exercises

Write a function for each of the following problems.

4.4.1. Suppose we have a pair of rabbits, one male and one female. Female rabbits are mature and can first mate when they are only one month old, and have a gestation period of one month. Suppose that every mature female rabbit mates every month and produces exactly one male and one female offspring one month later. No rabbit ever dies.

(a) Write a difference equation $R(m)$ representing the number of female rabbits at the end of each month. (Note that this is a discrete model; there are no Δt's.) Start with the first month and compute successive months until you see a pattern. Don't forget the initial condition(s). To get you started:
- $R(0) = 1$, the original pair
- $R(1) = 1$, the original pair now one month old
- $R(2) = 2$, the original pair plus a newborn pair
- $R(3) = 3$, the original pair plus a newborn pair plus a one-month-old pair
- $R(4) = 5$, the pairs from the previous generation plus two newborn pairs
- ⋮
- $R(m) = ?$

(b) Write a function

```
rabbits(months)
```

that uses your difference equation from part (a) to compute and return the number of rabbits after the given number of months. Your function should also plot the population growth using matplotlib. (The sequence of numbers generated in this problem is called the *Fibonacci sequence*.)

4.4.2. Consider Exercise 4.1.37, but now assume that a bacteria colony grows continuously with a growth rate of $0.1\Delta t$ per small fraction Δt of an hour.

(a) Write a difference equation for $B(t)$, the size of the bacteria colony, using Δt's to approximately model continuous growth.

(b) Write a function

```
bacteria(population, dt, days)
```

that uses your difference equation from part (a) compute and return the number of bacteria in the colony after the given number of days. The other parameters, population and dt, are the initial size of the colony and the step size (Δt), respectively. Your function should also plot the population growth using matplotlib.

4.4.3. Radioactive elements like carbon-14 are normally described by their *half-life*, the time required for a quantity of the element to decay to half its original quantity. Write a function

```
halflifeC14(originalAmount, dt)
```

that computes the half-life of carbon-14. Your function will need to simulate radioactive decay until the amount is equal to half of `originalAmount`.

The known half-life of carbon-14 is $5,730 \pm 40$ years. How close does this approximation come? Try it with different values of `dt`.

4.4.4. Finding the half-life of carbon-14 is a special case of radiocarbon dating. When an organic artifact is dated with radiocarbon dating, the fraction of extant carbon-14 is found relative to the content in the atmosphere. Let's say that the fraction in an ancient piece of parchment is found to be 70%. Then, to find the age of the artifact, we can use a generalized version of `halflifeC14` that iterates while `amount > originalAmount * 0.7`. Show how to generalize the `halflifeC14` function in the previous exercise by writing a new function

>   `carbonDate(originalAmount, fractionRemaining, dt)`

that returns the age of an artifact with the given fraction of carbon-14 remaining. Use this function to find the approximate age of the parchment.

4.4.5. Complete the implementation of the SIR simulation. Compare your results to Figure 4.7 to check your work.

4.4.6. Run your implementation of the SIR model for longer periods of time than that shown in Figure 4.7. Given enough time, will everyone become infected?

4.4.7. Suppose that enough people have been vaccinated that the infection rate is cut in half. What effect do these vaccinations have?

4.4.8. The SIS model represents an infection that does not result in immunity. In other words, there is no "recovered" population; people who have recovered re-enter the "susceptible" population.

   (a) Write difference equations for this model.

   (b) Copy and then modify the `SIR` function (renaming it `SIS`) so that it implements your difference equations.

   (c) Run your function with the same parameters that we used for the SIR model. What do you notice? Explain the results.

4.4.9. Suppose there are two predator species that compete for the same prey, but do not directly harm each other. We might expect that, if the supply of prey was low and members of one species were more efficient hunters, then this would have a negative impact on the health of the other species. This can be modeled through the following pair of difference equations. $P(t)$ and $Q(t)$ represent the populations of predator species 1 and 2, respectively, at time t.

$$P(t) = P(t - \Delta t) + b_P \cdot P(t - \Delta t) \cdot \Delta t - d_P \cdot P(t - \Delta t) \cdot Q(t - \Delta t) \cdot \Delta t$$
$$Q(t) = Q(t - \Delta t) + b_Q \cdot Q(t - \Delta t) \cdot \Delta t - d_Q \cdot P(t - \Delta t) \cdot Q(t - \Delta t) \cdot \Delta t$$

The initial conditions are $P(0) = p$ and $Q(0) = q$, where p and q represent the initial population sizes of the first and second predator species, respectively. The values b_P and d_P are the birth rate and death rates (or, more formally, proportionality constants) of the first predator species, and the values b_Q and d_Q are the birth rate and death rate for the second predator species. In the first equation, the term $b_P \cdot P(t - \Delta t)$ represents the net number of births per month for the first predator species, and the term $d_P \cdot P(t - \Delta t) \cdot Q(t - \Delta t)$ represents the number of deaths per month. Notice that this term is dependent on the sizes of both populations: $P(t - \Delta t) \cdot Q(t - \Delta t)$ is the number of possible (indirect) competitions between individuals for food and d_P is the rate at which one of these competitions results in a death (from starvation) in the first predator species.

This type of model is known as *indirect, exploitative competition* because the two predator species do not directly compete with each other (i.e., eat each other), but they do compete indirectly by exploiting a common food supply.

Write a function

```
compete(pop1, pop2, birth1, birth2, death1, death2, years, dt)
```

that implements this model for a generalized indirect competition scenario, plotting the sizes of the two populations over time. Run your program using

- $p = 21$ and $q = 26$
- $b_P = 1.0$ and $d_P = 0.2$; $b_Q = 1.02$ and $d_Q = 0.25$
- $dt = 0.001$; 6 years

and explain the results. Here is a "skeleton" of the function to get you started.

```
def compete(pop1, pop2, birth1, birth2, death1, death2, years, dt):
    """ YOUR DOCSTRING HERE """

    pop1List = [pop1]
    pop2List = [pop2]
    timeList = [0]
    for step in range(1, int(years / dt) + 1):
        # YOUR CODE GOES HERE

    pyplot.plot(timeList, pop1List, label = 'Population 1')
    pyplot.plot(timeList, pop2List, label = 'Population 2')
    pyplot.legend()
    pyplot.xlabel('Years')
    pyplot.ylabel('Individuals')
    pyplot.show()
```

*4.5 NUMERICAL ANALYSIS

Accumulator algorithms are also used in the natural and social sciences to approximate the values of common mathematical constants, and to numerically compute values and roots of complicated functions that cannot be solved mathematically. In this section, we will discuss a few relatively simple examples from this field of mathematics and computer science, known as *numerical analysis*.

The harmonic series

Suppose we have an ant that starts walking from one end of a 1 meter long rubber rope. During the first minute, the ant walks 10 cm. At the end of the first minute,

160 ■ Growth and decay

we stretch the rubber rope uniformly by 1 meter, so it is now 2 meters long. During the next minute, the ant walks another 10 cm, and then we stretch the rope again by 1 meter. If we continue this process indefinitely, will the ant ever reach the other end of the rope? If it does, how long will it take?

The answer lies in counting what fraction of the rope the ant traverses in each minute. During the first minute, the ant walks 1/10 of the distance to the end of the rope. After stretching, the ant has *still* traversed 1/10 of the distance because the portion of the rope on which the ant walked was doubled along with the rest of the rope. However, in the second minute, the ant's 10 cm stroll only covers 10/200 = 1/20 of the entire distance. Therefore, after 2 minutes, the ant has covered 1/10 + 1/20 of the rope. During the third minute, the rope is 3 m long, so the ant covers only 10/300 = 1/30 of the distance. This pattern continues, so our problem boils down to whether the following sum ever reaches 1.

$$\frac{1}{10} + \frac{1}{20} + \frac{1}{30} + \cdots = \frac{1}{10}\left(1 + \frac{1}{2} + \frac{1}{3} + \cdots\right)$$

Naturally, we can answer this question using an accumulator. But how do we add these fractional terms? In Exercise 4.1.25, you may have computed $1 + 2 + 3 + \cdots + n$:

```
def sum(n):
    total = 0
    for number in range(1, n + 1):
        total = total + number
    return total
```

In each iteration of this `for` loop, we add the value of `number` to the accumulator variable `total`. Since `number` is assigned the values $1, 2, 3, \ldots, n$, `total` has the sum of these values after the loop. To compute the fraction of the rope traveled by the ant, we can modify this function to add `1 / number` in each iteration instead, and then multiply the result by `1/10`:

```
def ant(n):
    """Simulates the "ant on a rubber rope" problem.  The rope
    is initially 1 m long and the ant walks 10 cm each minute.

    Parameter:
        n: the number of minutes the ant walks

    Return value:
        fraction of the rope traveled by the ant in n minutes
    """
    total = 0
    for number in range(1, n + 1):
        total = total + (1 / number)
    return total * 0.1
```

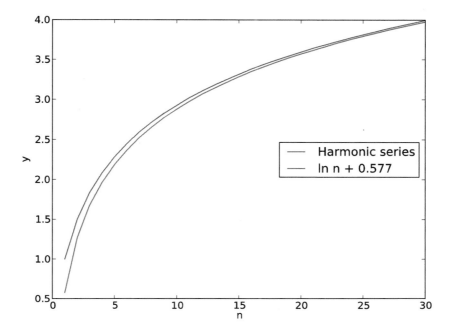

Figure 4.8 Harmonic series approximation of the natural logarithm (ln).

To answer our question with this function, we need to try several values of n to see if we can find a sufficiently high value for which the sum exceeds 1. If we find such a value, then we need to work with smaller values until we find the value of n for which the sum *first* reaches or exceeds 1.

Reflection 4.31 *Using at most 5 calls to the* `ant` *function, find a range of minutes that answers the question.*

For example, `ant(100)` returns about `0.52`, `ant(1000)` returns about `0.75`, `ant(10000)` returns about `0.98`, and `ant(15000)` returns about `1.02`. So the ant will reach the other end of the rope after 10,000–15,000 minutes. As cumbersome as that was, continuing it to find the exact number of minutes required would be far worse.

Reflection 4.32 *How would we write a function to find when the ant first reaches or exceeds the end of the rope? (Hint: this is similar to the carbon-14 half-life problem.) We will leave the answer as an exercise.*

This infinite version of the sum that we computed in our loop,

$$1 + \frac{1}{2} + \frac{1}{3} + \cdots,$$

is called the *harmonic series* and each finite sum

$$H_n = 1 + \frac{1}{2} + \frac{1}{3} + \cdots + \frac{1}{n}$$

is called the nth harmonic number. (Mathematically, the ant reaches the other end of the rope because the harmonic series *diverges*, that is, its partial sums increase forever.) The harmonic series can be used to approximate the natural logarithm (ln) function. For sufficiently large values of n,

$$H_n = 1 + \frac{1}{2} + \frac{1}{3} + \cdots + \frac{1}{n} \approx \ln n + 0.577.$$

This is illustrated in Figure 4.8 for only small values of n. Notice how the approximation improves as n increases.

Reflection 4.33 *Knowing that $H_n \approx \ln n + 0.577$, how can you approximate how long until the ant will reach the end of the rope if it walks only 1 cm each minute?*

To answer this question, we want to find n such that

$$100 = \ln n + 0.577$$

since the ant's first step is not 1/100 of the total distance. This is the same as

$$e^{100 - 0.577} = n.$$

In Python, we can find the answer with `math.exp(100 - 0.577)`, which gives about 1.5×10^{43} minutes, a long time indeed.

Approximating π

The value π is probably the most famous mathematical constant. There have been many infinite series found over the past 500 years that can be used to approximate π. One of the most famous is known as the *Leibniz series*, named after Gottfried Leibniz, the co-inventor of calculus:

$$\pi = 4 \left(1 - \frac{1}{3} + \frac{1}{5} - \frac{1}{7} + \cdots \right)$$

Like the harmonic series approximation of the natural logarithm, the more terms we compute of this series, the closer we get to the true value of π. To compute this sum, we need to identify a pattern in the terms, and relate them to the values of the index variable in a `for` loop. Then we can fill in the red blank line below with an expression that computes the ith term from the value of the index variable `i`.

```
def leibniz(terms):
    """Computes a partial sum of the Leibniz series.

    Parameter:
        terms: the number of terms to add

    Return value:
        the sum of the given number of terms
    """

    sum = 0
    for i in range(terms):
        sum = sum + _____
    pi = sum * 4
    return pi
```

To find the pattern, we can write down the values of the index variable next to the values in the series to identify a relationship:

i	0	1	2	3	4	...
i^{th} term	1	$-\frac{1}{3}$	$\frac{1}{5}$	$-\frac{1}{7}$	$\frac{1}{9}$	...

Ignoring the alternating signs for a moment, we can see that the absolute value of the ith term is

$$\frac{1}{2i+1}.$$

To alternate the signs, we use the fact that -1 raised to an even power is 1, while -1 raised to an odd power is -1. Since the even terms are positive and odd terms are negative, the final expression for the i term is

$$(-1)^i \cdot \frac{1}{2i+1}.$$

Therefore, the red assignment statement in our `leibniz` function should be

```
sum = sum + (-1) ** i / (2 * i + 1)
```

Reflection 4.34 *Call the completed* `leibniz` *function with a series of increasing arguments. What do you notice about how the values converge to π?*

By examining several values of the function, you might notice that they alternate between being greater and less than the actual value of π. Figure 4.9 illustrates this.

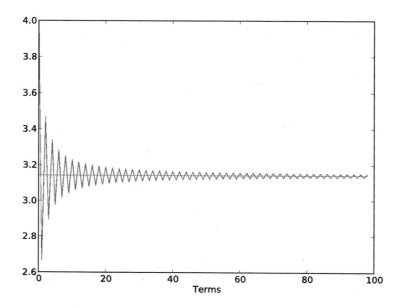

Figure 4.9 The Leibniz series converging to π.

Approximating square roots

The square root function ($\sqrt{n}$) cannot, in general, be computed directly. But it can be approximated very well with many iterations of the following difference equation, known as the *Babylonian method* (or *Newton's method*):

$$X(k) = \begin{cases} 1 & \text{if } k = 0 \\ \frac{1}{2}\left(X(k-1) + \frac{n}{X(k-1)}\right) & \text{otherwise} \end{cases}$$

Here, n is the value whose square root we want. The approximation of $\sqrt{n}$ will be better for larger values of k; $X(20)$ will be closer to the actual square root than $X(10)$.

Similar to our previous examples, we can compute successive values of the difference equation using iteration. In this case, each value is computed from the previous value according to the formula above. If we let the variable name x represent a term in the difference equation, then we can compute the kth term with the following simple function:

```
def sqrt(n, k):
    """Approximates the square root of n with k iterations
       of the Babylonian method.

    Parameters:
        n: the number to take the square root of
        k: number of iterations

    Return value:
        the approximate square root of n
    """

    x = 1.0
    for index in range(k):
        x = 0.5 * (x + n / x)
    return x
```

Reflection 4.35 *Call the function above to approximate $\sqrt{10}$ with various values of k. What value of k is necessary to match the value given by the* `math.sqrt` *function?*

Exercises

4.5.1. Recall Reflection 4.32: How would we write a function to find when the ant first reaches or exceeds the end of the rope? (Hint: this is similar to the carbon-14 half-life problem.)

  (a) Write a function to answer this question.

  (b) How long does it take for the ant to traverse the entire rope?

  (c) If the ant walks 5 cm each minute, how long does it take to reach the other end?

4.5.2. Augment the `ant` function so that it also produces the plot in Figure 4.8.

4.5.3. The value e (Euler's number, the base of the natural logarithm) is equal to the infinite sum

$$e = 1 + \frac{1}{1!} + \frac{1}{2!} + \frac{1}{3!} + \cdots$$

Write a function

    e(n)

that approximates the value of e by computing and returning the value of n terms of this sum. For example, calling `e(4)` should return the value $1+1/1+1/2+1/6 \approx 2.667$. Your function should *call* the `factorial` function you wrote for Exercise 4.1.29 to aid in the computation.

4.5.4. Calling the `factorial` function repeatedly in the function you wrote for the previous problem is very inefficient because many of the same arithmetic operations are being performed repeatedly. Explain where this is happening.

4.5.5. To avoid the problems suggested by the previous exercise, rewrite the function from Exercise 4.5.3 without calling the `factorial` function.

4.5.6. Rather than specifying the number of iterations in advance, numerical algorithms usually iterate until the absolute value of the current term is sufficiently small. At this point, we assume the approximation is "good enough." Rewrite the `leibniz` function so that it iterates while the absolute value of the current term is greater than 10^{-6}.

4.5.7. Similar to the previous exercise, rewrite the `sqrt` function so that it iterates while the absolute value of the difference between the current and previous values of `x` is greater than 10^{-15}.

4.5.8. The following expression, discovered in the 14th century by Indian mathematician Madhava of Sangamagrama, is another way to compute π.

$$\pi = \sqrt{12}\left(1 - \frac{1}{3\cdot 3} + \frac{1}{5\cdot 3^2} - \frac{1}{7\cdot 3^3} + \cdots\right)$$

Write a function

`approxPi(n)`

that computes `n` terms of this expression to approximate π. For example, `approxPi(3)` should return the value

$$\sqrt{12}\left(1 - \frac{1}{3\cdot 3} + \frac{1}{5\cdot 3^2}\right) \approx \sqrt{12}(1 - 0.111 + 0.022) \approx 3.156.$$

To determine the pattern in the sequence of terms, consider this table:

Term number	Term
0	$1/(3^0 \cdot 1)$
1	$1/(3^1 \cdot 3)$
2	$1/(3^2 \cdot 5)$
3	$1/(3^3 \cdot 7)$
$\vdots$	$\vdots$
i	?

What is the term for a general value of i?

4.5.9. The Wallis product, named after 17th century English mathematician John Wallis, is an infinite product that converges to π:

$$\pi = 2\left(\frac{2}{1}\cdot\frac{2}{3}\cdot\frac{4}{3}\cdot\frac{4}{5}\cdot\frac{6}{5}\cdot\frac{6}{7}\cdots\right)$$

Write a function

`wallis(terms)`

that computes the given number of terms in the Wallis product. Hint: Consider the terms in pairs and find an expression for each pair. Then iterate over the number of pairs needed to flesh out the required number of terms. You may assume that `terms` is even.

4.5.10. The Nilakantha series, named after Nilakantha Somayaji, a 15th century Indian mathematician, is another infinite series for π:

$$\pi = 3 + \frac{4}{2 \cdot 3 \cdot 4} - \frac{4}{4 \cdot 5 \cdot 6} + \frac{4}{6 \cdot 7 \cdot 8} - \frac{4}{8 \cdot 9 \cdot 10} + \cdots$$

Write a function

```
nilakantha(terms)
```

that computes the given number of terms in the Nilakantha series.

4.5.11. The following infinite product was discovered by François Viète, a 16th century French mathematician:

$$\pi = 2 \cdot \frac{2}{\sqrt{2}} \cdot \frac{2}{\sqrt{2+\sqrt{2}}} \cdot \frac{2}{\sqrt{2+\sqrt{2+\sqrt{2}}}} \cdot \frac{2}{\sqrt{2+\sqrt{2+\sqrt{2+\sqrt{2}}}}} \cdots$$

Write a function

```
viete(terms)
```

that computes the given number of terms in the Viète's product. (Look at the pattern carefully; it is not as hard as it looks if you base the denominator in each term on the denominator of the previous term.)

4.6 SUMMING UP

Although we have solved a variety of different problems in this chapter, almost all of the functions that we have designed have the same basic format:

```
def accumulator(_____):
    sum = ___                        # initialize the accumulator
    for index in range(___):         # iterate some number of times
        sum = sum + _____          # add something to the accumulator
    return sum                       # return final accumulator value
```

The functions we designed differ primarily in what is added to the accumulator (the red statement) in each iteration of the loop. Let's look at three of these functions in particular: the **pond** function from Page 119, the **countLinks** function from Page 123, and the solution to Exercise 4.1.27 from Page 134, shown below.

```
def growth(finalAge):
    height = 95
    for age in range(4, finalAge + 1):
        height = height + 6
    return height
```

In the **growth** function, a constant value is added to the accumulator in each iteration:

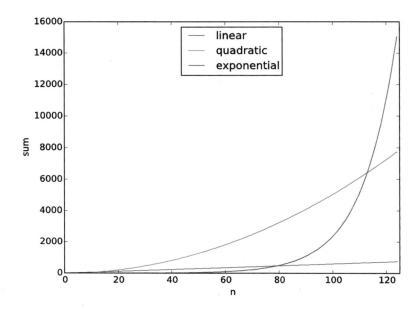

Figure 4.10 An illustration of linear, quadratic, and exponential growth. The curves are generated by accumulators adding 6, the index variable, and 1.08 times the accumulator, respectively, in each iteration.

```
height = height + 6
```

In the `countLinks` function, the value of the index variable, minus one, is added to the accumulator:

```
newLinks = node - 1
totalLinks = totalLinks + newLinks
```

And in the `pond` function, a factor of the accumulator itself is added in each iteration:

```
population = population + 0.08 * population   # ignoring "- 1500"
```

(To simplify things, we will ignore the subtraction present in the original program.)

These three types of accumulators grow in three different ways. Adding a constant value to the accumulator in each iteration, as in the `growth` function, results in a final sum that is equal to the number of iterations times the constant value. In other words, if the initial value is a, the constant added value is c, and the number of iterations is n, the final value of the accumulator is $a + cn$. (In the `growth` function, $a = 95$ and $c = 6$, so the final sum is $95 + 6n$.) As n increases, cn increases by a constant amount. This is called *linear growth*, and is illustrated by the blue line in Figure 4.10.

Adding the value of the index variable to the accumulator, as in the `countLinks` function, results in much faster growth. In the `countLinks` function, the final value

> **Box 4.2: Triangular numbers**
>
> There are a few nice tricks to figure out the value of the sum
>
> $$1 + 2 + 3 + \cdots + (n-2) + (n-1) + n$$
>
> for any positive integer n. The first technique is to add the numbers in the sum from the outside in. Notice that the sum of the first and last numbers is $n+1$. Then, coming in one position from both the left and right, we find that $(n-1) + 2 = n+1$ as well. Next, $(n-2) + 3 = n+1$. This pattern will obviously continue, as we are subtracting 1 from the number on the left and adding 1 to the number on the right. In total, there is one instance of $n+1$ for every two terms in the sum, for a total of $n/2$ instances of $n+1$. Therefore, the sum is
>
> $$1 + 2 + 3 + \cdots + (n-2) + (n-1) + n = \frac{n}{2}(n+1) = \frac{n(n+1)}{2}.$$
>
> For example, $1+2+3+\cdots+8 = (8 \cdot 9)/2 = 36$ and $1+2+3+\cdots+1000 = (1000 \cdot 1001)/2 = 500{,}500$.
>
> The second technique to derive this result is more visual. Depict each number in the sum as a column of circles, as shown on the left below with $n = 8$.
>
>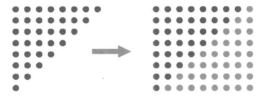
>
> The first column has $n = 8$ circles, the second has $n - 1 = 7$, etc. So the total number of circles in this triangle is equal to the value we are seeking. Now make an exact duplicate of this triangle, and place its mirror image to the right of the original triangle, as shown on the right above. The resulting rectangle has n rows and $n+1$ columns, so there are a total of $n(n+1)$ circles. Since the number of circles in this rectangle is exactly twice the number in the original triangle, the number of circles in the original triangle is $n(n+1)/2$. Based on this representation, numbers like 36 and 500,500 that are sums of this form are called *triangular numbers*.

of the accumulator is

$$1 + 2 + 3 + \cdots + (n-1)$$

which is equal to

$$\frac{1}{2} \cdot n \cdot (n-1) = \frac{n^2 - n}{2}.$$

Box 4.2 explains two clever ways to derive this result. Since this sum is proportional to n^2, we say that it exhibits *quadratic growth*, as shown by the red curve in Figure 4.10. This sum is actually quite handy to know, and it will surface again in Chapter 11.

Finally, adding a factor of the accumulator to itself in each iteration, as in the **pond** function, results in even faster growth. In the **pond** function, if we add `0.08 * population` to `population` in each iteration, the accumulator variable will be equal to the initial value of `population` times 1.08^n at the end of n iterations of the loop. For this reason, we call this *exponential growth*, which is illustrated by the green curve in Figure 4.10. Notice that, as n gets larger, exponential growth quickly outpaces the other two curves, even when the number we are raising n to is small, like 1.08.

So although all accumulator algorithms look more or less alike, the effects of the accumulators can be strikingly different. Understanding the relative rates of these different types of growth is quite important in a variety of fields, not just computer science. For example, mistaking an exponentially growing epidemic for a linearly growing one can be a life or death mistake!

These classes of growth can also be applied to the efficiency of algorithms, as we will see in in later chapters. In this case, n would represent the size of the input and the y-axis would represent the number of elementary steps required by the algorithm to compute the corresponding output.

Exercises

4.6.1. Decide whether each of the following accumulators exhibits linear, quadratic, or exponential growth.

(a)
```
sum = 0
for index in range(n):
    sum = sum + index * 2
```

(b)
```
sum = 10
for index in range(n):
    sum = sum + index / 2
```

(c)
```
sum = 1
for index in range(n):
    sum = sum + sum
```

(d)
```
sum = 0
for index in range(n):
    sum = sum + 1.2 * sum
```

(e)
```
sum = 0
for index in range(n):
    sum = sum + 0.01
```

(f)
```
sum = 10
for index in range(n):
    sum = 1.2 * sum
```

4.6.2. Look at Figure 4.10. For values of n less than about 80, the fast-growing exponential curve is actually below the other two. Explain why.

4.6.3. Write a program to generate Figure 4.10.

4.7 FURTHER DISCOVERY

The epigraph of this chapter is from a TED talk given by Stephen Hawking in 2008. You can watch it yourself at

www.ted.com/talks/stephen_hawking_asks_big_questions_about_the_universe.

If you are interested in learning more about population dynamics models, and computational modeling in general, a great source is *Introduction to Computational Science* [52] by Angela Shiflet and George Shiflet.

4.8 PROJECTS

Project 4.1 Parasitic relationships

For this project, we assume that you have read the material on discrete difference equations on Pages 145–146 of Section 4.4.

A parasite is an organism that lives either on or inside another organism for part of its life. A parasitoid is a parasitic organism that eventually kills its host. A parasitoid insect infects a host insect by laying eggs inside it then, when these eggs later hatch into larvae, they feed on the live host. (Cool, huh?) When the host eventually dies, the parasitoid adults emerge from the host body.

The *Nicholson-Bailey model*, first proposed by Alexander Nicholson and Victor Bailey in 1935 [36], is a pair of difference equations that attempt to simulate the relative population sizes of parasitoids and their hosts. We represent the size of the host population in year t with $H(t)$ and the size of the parasitoid population in year t with $P(t)$. Then the difference equations describing this model are

$$H(t) = r \cdot H(t-1) \cdot e^{-aP(t-1)}$$
$$P(t) = c \cdot H(t-1) \cdot \left(1 - e^{-aP(t-1)}\right)$$

where

- r is the average number of surviving offspring from an uninfected host,
- c is the average number of eggs that hatch inside a single host, and
- a is a scaling factor describing the searching efficiency or search area of the parasitoids (higher is more efficient).

The value $(1 - e^{-aP(t-1)})$ is the probability that a host is infected when there are $P(t-1)$ parasitoids, where e is Euler's number (the base of the natural logarithm). Therefore,

$$H(t-1) \cdot \left(1 - e^{-aP(t-1)}\right)$$

is the number of hosts that are infected during year $t - 1$. Multiplying this by c gives us $P(t)$, the number of new parasitoids hatching in year t. Notice that the probability of infection grows exponentially as the size of the parasitoid population grows. A higher value of a also increases the probability of infection.

Question 4.1.1 *Similarly explain the meaning of the difference equation for $H(t)$. ($e^{-aP(t-1)}$ is the probability that a host is not infected.)*

Part 1: Implement the model

To implement this model, write a function

    NB(hostPop, paraPop, r, c, a, years)

that uses these difference equations to plot both population sizes over time. Your function should plot these values in two different ways (resulting in two different plots). First, plot the host population size on the x-axis and the parasitoid population size on the y-axis. So each point represents the two population sizes in a particular year. Second, plot both population sizes on the y-axis, with time on the x-axis. To show both population sizes on the same plot, call the `pyplot.plot` function for each population list before calling `pyplot.show`. To label each line and include a legend, see the end of Section 4.2.

Question 4.1.2 *Write a `main` function that calls your `NB` function to simulate initial populations of 24 hosts and 12 parasitoids for 35 years. Use values of $r = 2$, $c = 1$, and $a = 0.056$. Describe and interpret the results.*

Question 4.1.3 *Run the simulation again with the same parameters, but this time assign a to be `-math.log(0.5) / paraPop`. (This is $a = -\ln 0.5/12 \approx 0.058$, just slightly above the original value of a.) What do you observe?*

Question 4.1.4 *Run the simulation again with the same parameters, but this time assign $a = 0.06$. What do you observe?*

Question 4.1.5 *Based on these three simulations, what can you say about this model and its sensitivity to the value of a?*

Part 2: Constrained growth

An updated Nicholson-Bailey model incorporates a *carrying capacity* that keeps the host population under control. The carrying capacity of an ecosystem is the maximum number of organisms that the ecosystem can support at any particular time. If the population size exceeds the carrying capacity, there are not enough resources to support the entire population, so some individuals do not survive.

In this revised model, $P(t)$ is the same, but $H(t)$ is modified to be

$$H(t) = H(t-1) \cdot e^{-aP(t-1)} \cdot e^{r(1-H(t-1)/K)}$$

where K is the carrying capacity. In this new difference equation, the average number of surviving host offspring, formerly r, is now represented by

$$e^{r(1-H(t-1)/K)}.$$

Notice that, when the number of hosts $H(t-1)$ equals the carrying capacity K, the exponent equals zero. So the number of surviving host offspring is $e^0 = 1$. In general, as the number of hosts $H(t-1)$ gets closer to the carrying capacity K, the exponent gets smaller and the value of the expression above gets closer to 1. At the other extreme, when $H(t-1)$ is close to 0, the expression is close to e^r. So, overall, the number of surviving offspring varies between 1 and e^r, depending on how close $H(t-1)$ comes to the carrying capacity.

Write a function

```
NB_CC(hostPop, paraPop, r, c, a, K, years)
```

that implements this modified model and generates the same plots as the previous function.

Question 4.1.6 *Call your* `NB_CC` *function to simulate initial populations of 24 hosts and 12 parasitoids for 35 years. Use values of r = 1.5, c = 1, a = 0.056, and K = 40. Describe and interpret the results.*

Question 4.1.7 *Run your simulation with all three values of a that we used in Part 1. How do these results differ from the prior simulation?*

Project 4.2 Financial calculators

In this project, you will write three calculators to help someone (maybe you) plan for your financial future.

Part 1: How long will it take to repay my college loans?

Your first calculator will compare the amount of time it takes to repay student loans with two different monthly payment amounts. Your main program will need to prompt for

- the student loan balance at the time repayment begins
- the nominal annual interest rate (the monthly rate times twelve)
- two monthly payment amounts

Then write a function

```
comparePayoffs(amount, rate, monthly1, monthly2)
```

that computes the number of months required to pay off the loan with each monthly payment amount. The interest on the loan balance should compound monthly at a rate of `rate / 100 / 12`. Your function should also plot the loan balances, with each payment amount, over time until both balances reach zero. Then it should print the length of both repayment periods and how much sooner the loan will be paid off if the higher monthly payment is chosen. For example, your output might look like this:

174 ■ Growth and decay

```
Initial balance: 60000
Nominal annual percentage rate: 5
Monthly payment 1: 500
Monthly payment 2: 750
```

If you pay $500.00 per month, the repayment period will be 13 years and 11 months.
If you pay $750.00 per month, the repayment period will be 8 years and 2 months.
If you pay $250.00 more per month, you will repay the loan 5 years and 9 months earlier.

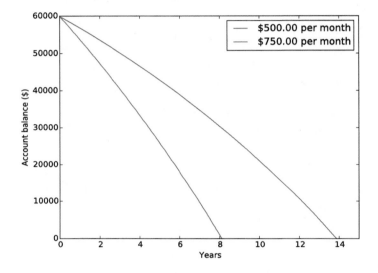

Question 4.2.1 *How long would it take to pay off $20,000 in student loans with a 4% interest rate if you paid $100 per month? Approximately how much would you have to pay per month to pay off the loan in ten years?*

Question 4.2.2 *If you run your program to determine how long it would take to pay off the same loan if you paid only $50 per month, you should encounter a problem. What is it?*

Part 2: How much will I have for retirement?

Your second calculator will compare the retirement nest eggs that result from making two different monthly investments in a retirement fund. Your program should prompt for the following inputs:

- the initial balance in the retirement account
- the current age of the investor
- the desired retirement age of the investor

- the expected nominal annual rate of return on the investment
- two monthly investment amounts

Then write a function

```
compareInvestments(balance, age, retireAge, rate, monthly1, monthly2)
```

that computes the final balance in the retirement account, for each monthly investment, when the investor reaches his or her retirement age. The interest on the current balance should compound monthly at a rate of `rate / 100 / 12`. The function should plot the growth of the retirement account balance for both monthly investment amounts, and then print the two final balances along with the additional amount that results from the higher monthly investment. For example, your output might look like this:

```
Initial balance: 1000
Current age: 20
Retirement age: 70
Nominal annual percentage rate of return: 4.2
Monthly investment 1: 100
Monthly investment 2: 150

The final balance from investing $100.00 per month: $212030.11.
The final balance from investing $150.00 per month: $313977.02.
If you invest $50.00 more per month, you will have $101946.91 more at
    retirement.
```

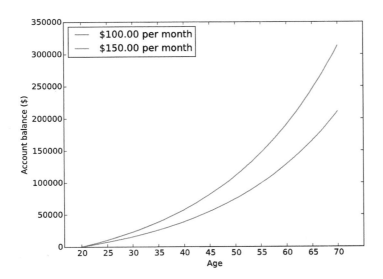

Question 4.2.3 *Suppose you are 30 and, after working for a few years, have managed to save $6,000 for retirement. If you continue to invest $200 per month, how much will you have when you retire at age 72 if your investment grows 3% per year? How much more will you have if you invest $50 more per month?*

Part 3: How long will my retirement savings last?

Your third calculator will initially perform the same computation as your previous calculator, but with only one monthly investment amount. After it computes the final balance in the account, it will estimate how long that nest egg will last into retirement. Your program will need to prompt for the same values as above (but only one monthly investment amount), plus the percentage of the final balance the investor plans to withdraw in the first year after retirement. Then write a function

```
retirement(amount, age, retireAge, rate, monthly, percentWithdraw)
```

that adds the monthly investment amount to the balance in the retirement account until the investor reaches his or her retirement age, and then, after retirement age, withdraws a monthly amount. In the first month after retirement, this amount should be one-twelfth of `percentWithdraw` of the current balance. For every month thereafter, the withdrawal amount should increase according to the rate of inflation, assumed to be 3% annually. Every month, the interest on the current balance should compound at a rate of `rate / 100 / 12`. The function should plot the retirement account balance over time, and then print the age at which the retirement funds run out. For example, your program output might look like this:

```
Initial balance: 10000
Current age: 20
Retirement age: 70
Annual percentage rate of return: 4.2
Monthly investment: 100
Annual withdrawal % at retirement: 4

Your savings will last until you are 99 years and 10 months old.
```

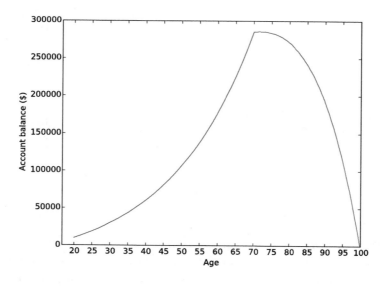

Question 4.2.4 *How long will your retirement savings last if you follow the plan outlined in Question 4.2 (investing $200 per month) and withdraw 4% at retirement?*

*Project 4.3 Market penetration

For this project, we assume that you have read Section 4.4.

In this project, you will write a program that simulates the adoption of a new product in a market over time (i.e., the product's "market penetration") using a difference equation developed by marketing researcher Frank Bass in 1969 [4]. The *Bass diffusion model* assumes that there is one product and a fixed population of N eventual adopters of that product. In the difference equation, $A(t)$ is the fraction (between 0 and 1) of the population that has adopted the new product t weeks after the product launch. Like the difference equations in Section 4.4, the value of $A(t)$ will depend on the value of $A(t-\Delta t)$, the fraction of the population that had adopted the product at the previous time step, $t-\Delta t$. Since t is measured in weeks, Δt is some fraction of a week. The rate of product adoption depends on two factors.

1. The segment of the population that has not yet adopted the product adopts it at a constant adoption rate of r due to chance alone. By the definition of $A(t)$, the fraction of the population that *had* adopted the product at the previous time step is $A(t-\Delta t)$. Therefore, the fraction of the population that *had not yet* adopted the product at the previous time step is $1 - A(t-\Delta t)$. So the fraction of new adopters during week t from this constant adoption rate is

$$r \cdot (1 - A(t-\Delta t)) \cdot \Delta t.$$

2. Second, the rate of adoption is affected by word of mouth within the population. Members of the population who have already adopted the product can influence those who have not. The more adopters there are, the more potential interactions exist between adopters and non-adopters, which boosts the adoption rate. The fraction of all potential interactions that are between adopters and non-adopters in the previous time step is

$$\underbrace{A(t-\Delta t)}_{\text{fraction that are adopters}} \cdot \underbrace{(1 - A(t-\Delta t))}_{\text{fraction that are not adopters}}.$$

The fraction of these interactions that result in a new adoption during one week is called the *social contagion*. We will denote the social contagion by s. So the fraction of new adopters due to social contagion during the time step ending at week t is

$$s \cdot A(t-\Delta t) \cdot (1 - A(t-\Delta t)) \cdot \Delta t.$$

The social contagion measures how successfully adopters are able to convince non-adopters that they should adopt the product. At the extremes, if $s = 1$, then every interaction between a non-adopter and an adopter results in the non-adopter adopting the product. On the other hand, if $s = 0$, then the current adopters cannot convince any non-adopters to adopt the product.

Putting these two parts together, the difference equation for the Bass diffusion model is

$$A(t) = A(t - \Delta t) + \underbrace{r \cdot (1 - A(t - \Delta t)) \cdot \Delta t}_{\substack{\text{fraction of new adopters} \\ \text{from constant rate}}} + \underbrace{s \cdot A(t - \Delta t) \cdot (1 - A(t - \Delta t)) \cdot \Delta t}_{\substack{\text{fraction of new adopters} \\ \text{from social contagion}}}$$

Part 1: Implement the Bass diffusion model

To implement the Bass diffusion model, write a function

`productDiffusion(chanceAdoption, socialContagion, weeks, dt)`

The parameters `chanceAdoption` and `socialContagion` are the values of r and s, respectively. The last two parameters give the number of `weeks` to simulate and the value of Δt to use. Your function should plot two curves, both with time on the x-axis and the proportion of adopters on the y-axis. The first curve is the total fraction of the population that has adopted the product by time t. This is $A(t)$ in the difference equation above. The second curve will be the rate of change of $A(t)$. You can calculate this by the formula

$$\frac{A(t) - A(t - \Delta t)}{\Delta t}.$$

Equivalently, this rate of change can be thought of as the fraction of new adopters at any time step, normalized to a weekly rate, i.e., the fraction of the population added in that time step divided by `dt`.

Write a program that uses your function to simulate a product launch over 15 weeks, using $\Delta t = 0.01$. For this product launch, we will expect that adoption of the product will move slowly without a social effect, but that social contagion will have a significant impact. To model these assumptions, use $r = 0.002$ and $s = 1.03$.

Question 4.3.1 *Describe the picture and explain the pattern of new adoptions and the resulting pattern of total adoptions over the 15-week launch.*

Question 4.3.2 *Now make r very small but leave s the same ($r = 0.00001$, $s = 1.03$), and answer the same question. What kind of market does this represent?*

Question 4.3.3 *Now set r to be 100 times its original value and s to be zero ($r = 0.2$, $s = 0$), and answer the first question again. What kind of market does this represent?*

Part 2: Influentials and imitators

In a real marketplace, some adopters are more influential than others. Therefore, to be more realistic, we will now partition the entire population into two groups called *influentials* and *imitators* [56]. Influentials are only influenced by other influentials, while imitators can be influenced by either group. The numbers of influentials and imitators in the population are N_A and N_B, respectively, so the total population size is $N = N_A + N_B$. We will let $A(t)$ now represent the fraction of the influential population that has adopted the product at time t and let $B(t)$ represent the fraction of the imitator population that has adopted the product at time t.

The adoption rate of the influentials follows the same difference equation as before, except that we will denote the adoption rate and social contagion for the influentials with r_A and s_A.

$$A(t) = A(t - \Delta t) + r_A \cdot (1 - A(t - \Delta t)) \cdot \Delta t + s_A \cdot A(t - \Delta t) \cdot (1 - A(t - \Delta t)) \cdot \Delta t$$

The adoption rate of the imitators will be different because they value the opinions of both influentials and other imitators. Let r_B and s_B represent the adoption rate and social contagion for the imitators. Another parameter, w (between 0 and 1), will indicate how much the imitators value the opinions of the influentials over the other imitators. At the extremes, $w = 1$ means that the imitators are influenced heavily by influentials and not at all by other imitators. On the other hand, $w = 0$ means that they are not at all influenced by influentials, but are influenced by imitators. We will break the difference equation for $B(t)$ into three parts.

1. First, there is a constant rate of adoptions from among the imitators that have not yet adopted, just like the first part of the difference equation for $A(t)$:

$$r_B \cdot (1 - B(t - \Delta t)) \cdot \Delta t$$

2. Second, there is a fraction of the imitators who have not yet adopted who will be influenced to adopt, through social contagion, by influential adopters.

$$w \cdot s_B \cdot \underbrace{A(t - \Delta t)}_{\substack{\text{fraction of} \\ \text{influentials who} \\ \text{have adopted}}} \cdot \underbrace{(1 - B(t - \Delta t))}_{\substack{\text{fraction of} \\ \text{imitators who} \\ \text{have not adopted}}} \cdot \Delta t.$$

Recall from above that w is the extent to which imitators are more likely to be influenced by influentials than other imitators.

3. Third, there is a fraction of the imitators who have not yet adopted who will be influenced to adopt, through social contagion, by other imitators who have already adopted.

$$(1 - w) \cdot s_B \cdot \underbrace{B(t - \Delta t)}_{\substack{\text{fraction of} \\ \text{imitators who} \\ \text{have adopted}}} \cdot \underbrace{(1 - B(t - \Delta t))}_{\substack{\text{fraction of} \\ \text{imitators who} \\ \text{have not adopted}}} \cdot \Delta t.$$

The term $1 - w$ is the extent to which imitators are likely to be influenced by other imitators, compared to influentials.

Putting these three parts together, we have the difference equation modeling the growth of the fraction of imitators who adopt the product.

$$B(t) = B(t - \Delta t) + r_B \cdot (1 - B(t - \Delta t)) \cdot \Delta t$$
$$+ w \cdot s_B \cdot A(t - \Delta t) \cdot (1 - B(t - \Delta t)) \cdot \Delta t$$
$$+ (1 - w) \cdot s_B \cdot B(t - \Delta t) \cdot (1 - B(t - \Delta t)) \cdot \Delta t$$

Now write a function

```
productDiffusion2(inSize, imSize, rIn, sIn, rIm, sIm, weight, weeks, dt)
```

that implements this product diffusion model with influentials and imitators. The parameters are similar to the previous function (but their names have been shortened). The first two parameters are the sizes of the influential and imitator populations, respectively. The third and fourth parameters are the adoption rate (r_A) and social contagion (s_A) for the influentials, respectively. The fifth and sixth parameters are the same values (r_B and s_B) for the imitators. The seventh parameter `weight` is the value of w. Your function should produce two plots. In the first, plot the new adoptions for each group, and the total new adoptions, over time (as before, normalized by dividing by `dt`). In the second, plot the total adoptions for each group, and the total adoptions for both groups together, over time. Unlike in the previous function, plot the *numbers* of adopters in each group rather than the fractions of adopters, so that the different sizes of each population are taken into account. (To do this, just multiply the fraction by the total size of the appropriate population.)

Write a program that uses your function to simulate the same product launch as before, except now there are 600 influentials and 400 imitators in a total population of 1000. The adoption rate and social contagion for the influentials are the same as before ($r_A = 0.002$ and $s_A = 1.03$), but these values are $r_B = 0$ and $s_B = 0.8$ for the imitators. Use a value of $w = 0.6$, meaning that the imitators value the opinions of the influentials over other imitators.

Question 4.3.4 *Describe the picture and explain the pattern of new adoptions and the resulting pattern of total adoptions. Point out any patterns that you find interesting.*

Question 4.3.5 *Now set $w = 0.01$ and rerun the simulation. Describe the new picture and explain how and why this changes the results.*

*Project 4.4 Wolves and moose

For this project, we assume that you have read Section 4.4.

Predator-prey models are commonly used in biology because they can model a wide range of ecological relationships, e.g., wolves and moose, koala and eucalyptus, or humans and tuna. In the simplest incarnation, the livelihood of a population of predators is dependent solely on the availability of a population of prey. The population of prey, in turn, is kept in check by the predators.

In the 1920's, Alfred Lotka and Vito Volterra independently introduced the now-famous Lotka-Volterra equations to model predator-prey relationships. The model consists of a pair of related differential equations that describe the sizes of the two populations over time. We will approximate the differential equations with discrete difference equations. Let's assume that the predators are wolves and the prey are moose. We will represent the sizes of the moose and wolf populations at the end of month t with $M(t)$ and $W(t)$, respectively. The difference equations describing the populations of wolves and moose are:

$$M(t) = M(t - \Delta t) + b_M\, M(t - \Delta t)\, \Delta t - d_M\, W(t - \Delta t)\, M(t - \Delta t)\, \Delta t$$

$$W(t) = W(t - \Delta t) + b_W\, W(t - \Delta t)\, M(t - \Delta t)\, \Delta t - d_W\, W(t - \Delta t)\, \Delta t$$

where

- b_M is the moose birth rate (per month)
- d_M is the moose death rate, or the rate at which a wolf kills a moose that it encounters (per month)
- b_W is the wolf birth rate, or the moose death rate × how efficiently an eaten moose produces a new wolf (per month)
- d_W is the wolf death rate (per month)

Let's look at these equations more closely. In the first equation, the term $b_M\, M(t - \Delta t)\Delta t$ represents the net number of moose births in the last time step, in the absence of wolves, and the term $d_M\, W(t - \Delta t)\, M(t - \Delta t)\Delta t$ represents the number of moose deaths in the last time step. Notice that this term is dependent on both the number of wolves and the number of moose: $W(t - \Delta t)\, M(t - \Delta t)$ is the number of possible wolf-moose encounters and $d_M \Delta t$ is the rate at which a wolf kills a moose that it encounters in a time step of length Δt.

In the second equation, the term $b_W\, W(t - \Delta t)\, M(t - \Delta t)\Delta t$ represents the number of wolf births per month. Notice that this number is also proportional to both the number of wolves and the number of moose, the idea being that wolves will give birth to more offspring when food is plentiful. As described above, b_W is actually based on two quantities, the moose death rate (since wolves have to eat moose to thrive and have offspring) and how efficiently a wolf can use the energy gained by eating a moose to give birth to a new wolf. The term $d_W\, W(t - \Delta t)\Delta t$ represents the net number of wolf deaths per month in the absence of moose.

In this project, you will write a program that uses these difference equations to model the dynamic sizes of a population of wolves and a population of moose over time. There are three parts to the project. In the first part, you will use the Lotka-Volterra model to simulate a baseline scenario. In the second part, you will model the effects that hunting the wolves have on the populations. And, in the third part, you will create a more realistic simulation in which the sizes of the populations are limited by the natural resources available in the area.

182 ■ Growth and decay

Part 1: Implement the Lotka-Volterra model

Write a function in Python

```
PP(preyPop, predPop, dt, months)
```

that simulates this predator prey model using the difference equations above. The parameters `preyPop` and `predPop` are the initial sizes of the prey and predator populations ($M(0)$ and $W(0)$), respectively, `dt` (Δt) is the time interval used in the simulation, and `months` is the number of months (maximum value of t) for which to run the simulation. To cut back on the number of parameters, you can assign constant birth and death rates to local variables inside your function. Start by trying

```
birthRateMoose = 0.5      # b_M
deathRateMoose = 0.02     # d_M
birthRateWolves = 0.005   # b_W = d_M * efficiency of 0.25
deathRateWolves = 0.75    # d_W
```

Your function should plot, using `matplotlib`, the sizes of the wolf and moose populations over time, as the simulation progresses. Write a program that calls your PP function to simulate 500 moose and 25 wolves for 5 years with $dt = 0.01$.

Question 4.4.1 *What happens to the sizes of the populations over time? Why do these changes occur?*

Part 2: Here come the hunters!

Now suppose the wolves begin to threaten local ranchers' livestock (in Wyoming, for example) and the ranchers begin killing the wolves. Simulate this effect by increasing the wolf death rate d_W.

Question 4.4.2 *What is the effect on the moose population?*

Question 4.4.3 *What would the wolf death rate need to be for the wolf population to die out within five years? Note that the death rate can exceed 1. Try increasing the value of d_W slowly and watch what happens. If it seems like you can never kill all the wolves, read on.*

Killing off the wolves appears to be impossible because the equations you are using will never let the value reach zero. (Why?) To compensate, we can set either population to zero when it falls below some threshold, say 1.0. (After all, you can't really have a fraction of a wolf.) To do this, insert the following statements into the body of your `for` loop after you increment the predator and prey populations, and try answering the previous question again.

```
if preyPop < 1.0:
    preyPop = 0.0
if predPop < 1.0:
    predPop = 0.0
```

(Replace `preyPop` and `predPop` with the names you use for the current sizes of the populations.) As we will see shortly, the first two statements will assign 0 to `preyPop` if it is less than 1. The second two statements do the same for `predPop`.

Part 3: Modeling constrained growth

In the simulation so far, we have assumed that a population can grow without bound. For example, if the wolf population died out, the moose population would grow exponentially. In reality, an ecosystem can only support a limited size population due to constraints on space, food, etc. This limit is known as a *carrying capacity*. We can model the moose carrying capacity in a simple way by decreasing the moose birth rate proportionally to the size of the moose population. Specifically, let MCC represent the moose population carrying capacity, which is the maximum number of moose the ecosystem can support. Then change the moose birth term $b_M M(t - \Delta t)\Delta t$ to

$$b_M \left(1 - M(t - \Delta t)/MCC\right) M(t - \Delta t)\Delta t.$$

Notice that now, as the moose population size approaches the carrying capacity, the birth rate slows.

Question 4.4.4 *Why does this change cause the moose birth rate to slow as the size of the moose population approaches the carrying capacity?*

Implement this change to your simulation, setting the moose carrying capacity to 750, and run it again with the original birth and death rates, with 500 moose and 25 wolves, for 10 years.

Question 4.4.5 *How does the result differ from your previous run? What does the result demonstrate? Does the moose population reach its carrying capacity of 750? If not, what birth and/or death rate parameters would need to change to allow this to happen?*

Question 4.4.6 *Reinstate the original birth and death rates, and introduce hunting again; now what would the wolf death rate need to be for the wolf population to die out within five years?*

CHAPTER 5

Forks in the road

> When you come to a fork in the road, take it.
>
> Yogi Berra

So far, our algorithms have been entirely *deterministic*; they have done the same thing every time we execute them with the same inputs. However, the natural world and its inhabitants (including us) are usually not so predictable. Rather, we consider many natural processes to be, to various degrees, *random*. For example, the behaviors of crowds and markets often change in unpredictable ways. The genetic "mixing" that occurs in sexual reproduction can be considered a random process because we cannot predict the characteristics of any particular offspring. We can also use the power of random *sampling* to estimate characteristics of a large population by querying a smaller number of randomly selected individuals.

Most non-random algorithms must also be able to conditionally change course, or select from among a variety of options, in response to input. Indeed, most common desktop applications do nothing unless prompted by a key press or a mouse click. Computer games like racing simulators react to a controller several times a second. The protocols that govern data traffic on the Internet adjust transmission rates continually in response to the perceived level of congestion on the network. In this chapter, we will discover how to design algorithms that can behave differently in response to input, both random and deterministic.

5.1 RANDOM WALKS

Scientists are often interested in simulating random processes to better understand some characteristic of a system. For example, we may wish to estimate the theoretical probability of some genetic abnormality or the expected distance a particle moves in Brownian motion. Algorithms that employ randomness to estimate some value are called *Monte Carlo simulations*, after the famous casino in Monaco. Monte Carlo

> **Box 5.1: Interval notation**
>
> It is customary to represent the interval (i.e., set or range), of *real numbers* between a and b, including a and b, with the notation $[a, b]$. In contrast, the set of *integers* between the integers a and b, including a and b, is denoted $[a..b]$. For example, $[3, 7]$ represents every real number greater than or equal to 3 and less than or equal to 7, while $[3..7]$ represents the integers $3, 4, 5, 6, 7$.
>
> To denote an interval of real numbers between a and b that does not contain an endpoint a or b, we replace the endpoint's square bracket with a parenthesis. So $[a, b)$ is the interval of real numbers between a and b that does contain a but does not contain b. Similarly, $(a, b]$ contains b but not a, and (a, b) contains neither a nor b.

simulations repeat a random experiment many, many times, and average over these trials to arrive at a meaningful result; one run of the simulation, due to its random nature, does not carry any significance.

In 1827, British Botanist Robert Brown, while observing pollen grains suspended in water under his microscope, witnessed something curious. When the pollen grains burst, they emitted much smaller particles that proceeded to wiggle around randomly. This phenomenon, now called *Brownian motion*, was caused by the particles' collisions with the moving water molecules. Brownian motion is now used to describe the motion of any sufficiently small particle (or molecule) in a fluid.

We can model the essence of Brownian motion with a single randomly moving particle in two dimensions. This process is known as a *random walk*. Random walks are also used to model a wide variety of other phenomena such as markets and the foraging behavior of animals, and to sample large social networks. In this section, we will develop a Monte Carlo simulation of a random walk to discover how far away a randomly moving particle is likely to get in a fixed amount of time.

You may have already modeled a simple random walk in Exercise 3.3.15 by moving a turtle around the screen and choosing a random angle to turn at each step. We will now develop a more restricted version of a random walk in which the particle is forced to move on a two-dimensional grid. At each step, we want the particle to move in one of the four cardinal directions, each with equal probability.

To simulate random processes, we need an algorithm or device that produces random numbers, called a *random number generator* (RNG). A computer, as described in Chapter 1, cannot implement a true RNG because everything it does is entirely predictable. Therefore, a computer either needs to incorporate a specialized device that can detect and transmit truly random physical events (like subatomic quantum fluctuations) or simulate randomness with a clever algorithm called a *pseudorandom number generator* (PRNG). A PRNG generates a sequence of numbers that *appear* to be random although, in reality, they are not.

The Python module named `random` provides a PRNG in a function named random. The `random` function returns a pseudorandom number between zero and one, but not including one. For example:

```
>>> import random
>>> random.random()
0.9699738944412686
```

(Your output will differ.) It is convenient to refer to the range of real numbers produced by the `random` function as $[0, 1)$. The square bracket on the left means that 0 is included in the range, and the parenthesis on the right means that 1 is not included in the range. Box 5.1 explains a little more about this so-called *interval notation*, if it is unfamiliar to you.

To randomly move our particle in one of four cardinal directions, we first use the `random` function to assign r to be a random value in $[0, 1)$. Then we divide our space of possible random numbers into four equal intervals, and associate a direction with each one.

1. If r is in $[0, 0.25)$, then move east.

2. If r is in $[0.25, 0.5)$, then move north.

3. If r is in $[0.5, 0.75)$, then move west.

4. If r is in $[0.75, 1.0)$, then move south.

Now let's write an algorithm to take one step of a random walk. We will save the particle's (x, y) location in two variables, `x` and `y` (also called the particle's x and y *coordinates*). To condition each move based on the interval in which r resides, we will use Python's `if` statement. An `if` statement executes a particular sequence of statements only if some condition is true. For example, the following statements assign a pseudorandom value to `r`, and then implement the first case by incrementing `x` when `r` is in $[0, 0.25)$:

```
x = 0
y = 0
r = random.random()
if r < 0.25:          # if r < 0.25,
    x = x + 1         #    move to the east
```

Reflection 5.1 *Why do we not need to also check whether `r` is at least zero?*

An `if` statement is also called a *conditional statement* because, like the `while` loops we saw earlier, they make decisions that are conditioned on a Boolean expression. (Unlike `while` loops, however, an `if` statement is only executed once.) The Boolean expression in this case, `r < 0.25`, is true if `r` is less than 0.25 and false otherwise. If the Boolean expression is true, the statement(s) that are indented beneath the condition are executed. On the other hand, if the Boolean expression is false,

Math symbol	Python symbol
$<$	<
$\leq$	<=
$>$	>
$\geq$	>=
$=$	==
$\neq$	!=

Table 5.1 Python's six comparison operators.

the indented statement(s) are skipped, and the statement following the indented statements is executed next.

In the second case, to check whether `r` is in $[0.25, 0.5)$, we need to check whether `r` is greater than or equal to 0.25 *and* whether `r` is less than 0.5. The meaning of "and" in the previous sentence is identical to the Boolean operator from Section 1.4. In Python, this condition is represented just as you would expect:

```
r >= 0.25 and r < 0.5
```

The `>=` operator is Python's representation of "greater than or equal to" ($\geq$). It is one of six *comparison operators* (or *relational operators*), listed in Table 5.1, some of which have two-character representations in Python. (Note especially that `==` is used to test for equality. We will discuss these operators further in Section 5.4.) Adding this case to the first case, we now have two `if` statements:

```
if r < 0.25:                  # if r < 0.25,
    x = x + 1                 #    move to the east
if r >= 0.25 and r < 0.5:     # if r is in [0.25, 0.5),
    y = y + 1                 #    move to the north
```

Now if `r` is assigned a value that is less than 0.25, the condition in the first `if` statement will be true and `x = x + 1` will be executed. Next, the condition in the second `if` statement will be checked. But since this condition is false, `y` will not be incremented. On the other hand, suppose `r` is assigned a value that is between 0.25 and 0.5. Then the condition in the first `if` statement will be false, so the indented statement `x = x + 1` will be skipped and execution will continue with the second `if` statement. Since the condition in the second `if` statement is true, `y = y + 1` will be executed.

To complete our four-way decision, we can add two more `if` statements:

```
1  if r < 0.25:                   # if r < 0.25,
2      x = x + 1                  #    move to the east
3  if r >= 0.25 and r < 0.5:      # if r is in [0.25, 0.5),
4      y = y + 1                  #    move to the north
5  if r >= 0.5 and r < 0.75:      # if r is in [0.5, 0.75),
6      x = x - 1                  #    move to the west
```

```
7  if r >= 0.75 and r < 1.0:    # if r is in [0.75, 1.0),
8      y = y - 1                #    move to the south
9  print(x, y)                  # executed after the 4 cases
```

There are four possible ways this code could be executed, one for each interval in which `r` can reside. For example, suppose `r` is in the interval $[0.5, 0.75)$. We first execute the `if` statement on line 1. Since the `if` condition is false, the indented statement on line 2, `x = x + 1`, is skipped. Next, we execute the `if` statement on line 3, and test its condition. Since this is also false, the indented statement on line 4, `y = y + 1`, is skipped. We continue by executing the third `if` statement, on line 5. This condition, `r >= 0.5 and r < 0.75`, is true, so the indented statement on line 6, `x = x - 1`, is executed. Next, we continue to the fourth `if` statement on line 7, and test its condition, `r >= 0.75 and r < 1.0`. This condition is false, so execution continues on line 9, after the entire conditional structure, where the values of `x` and `y` are printed. In each of the other three cases, when `r` is in one of the three other intervals, a different condition will be true and the other three will be false. Therefore, exactly one of the four indented statements will be executed for any value of `r`.

Reflection 5.2 *Is this sequence of steps efficient? If not, what steps could be skipped and in what circumstances?*

The code behaves correctly, but it seems unnecessary to test subsequent conditions after we have already found the correct case. If there were many more than four cases, this extra work could be substantial. Here is a more efficient structure:

```
1  if r < 0.25:          # if r < 0.25,
2      x = x + 1         #    move to the east and finish
3  elif r < 0.5:         # otherwise, if r is in [0.25, 0.5),
4      y = y + 1         #    move to the north and finish
5  elif r < 0.75:        # otherwise, if r is in [0.5, 0.75),
6      x = x - 1         #    move to the west and finish
7  elif r < 1.0:         # otherwise, if r is in [0.75, 1.0),
8      y = y - 1         #    move to the south and finish
9  print(x, y)           # executed after each of the 4 cases
```

The keyword `elif` is short for "else if," meaning that the condition that follows an `elif` is checked *only* if no preceding condition was true. In other words, as we sequentially check each of the four conditions, if we find that one is true, then the associated indented statement(s) are executed, and we *skip* the remaining conditions in the group. We were also able to eliminate the lower bound check from each condition (e.g., `r >= 0.25` in the second `if` statement) because, if we get to an `elif` condition, we know that the previous condition was false, and therefore the value of `r` must be greater than or equal to the interval being tested in the previous case.

To illustrate, let's first suppose that the condition in the first `if` statement on line 1 is true, and the indented statement on line 2 is executed. Now none of the remaining `elif` conditions are checked, and execution *skips ahead* to line 9. On the other hand, if the condition in the first `if` statement is false, we know that `r` must be at least 0.25. Next the `elif` condition on line 3 would be checked. If this condition is true, then the indented statement on line 4 would be executed, none of the remaining conditions would be checked, and execution would continue on line 9. If the condition in the second `if` statement is also false, we know that `r` must be at least 0.5. Next, the condition on line 5, `r < 0.75`, would be checked. If it is true, the indented statement on line 6 would be executed, and execution would continue on line 9. Finally, if none of the first three conditions is true, we know that `r` must be at least 0.75, and the `elif` condition on line 7 would be checked. If it is true, the indented statement on line 8 would be executed. And, either way, execution would continue on line 9.

Reflection 5.3 *For each of the four possible intervals to which `r` could belong, how many conditions are checked?*

Reflection 5.4 *Suppose you replace every `elif` in the most recent version above with `if`. What would then happen if `r` had the value 0.1?*

This code can be streamlined a bit more. Since `r` must be in $[0, 1)$, there is no point in checking the last condition, `r < 1.0`. If execution has proceeded that far, `r` *must* be in $[0.75, 1)$. So, we should just execute the last statement, `y = y - 1`, without checking anything. This is accomplished by replacing the last `elif` with an `else` statement:

```
1  if r < 0.25:       # if r < 0.25,
2      x = x + 1      #     move to the east and finish
3  elif r < 0.5:      # otherwise, if r is in [0.25, 0.5),
4      y = y + 1      #     move to the north and finish
5  elif r < 0.75:     # otherwise, if r is in [0.5, 0.75),
6      x = x - 1      #     move to the west and finish
7  else:              # otherwise,
8      y = y - 1      #     move to the south and finish
9  print(x, y)        # executed after each of the 4 cases
```

The `else` signals that, if no previous condition is true, the statement(s) indented under the `else` are to be executed.

Reflection 5.5 *Now suppose you replace every `elif` with `if`. What would happen if `r` had the value 0.1?*

If we (erroneously) replaced the two instances of `elif` above with `if`, then the final `else` would be associated *only* with the last `if`. So if `r` had the value 0.1, all three if conditions would be true and all three of the first three indented statements would be executed. The last indented statement would not be executed because the last if condition was true.

Reflection 5.6 *Suppose we wanted to randomly move a particle on a line instead. Then we only need to check two cases: whether r is less than 0.5 or not. If r is less than 0.5, increment x. Otherwise, decrement x. Write an if/else statement to implement this. (Resist the temptation to look ahead.)*

In situations where there are only two choices, an else can just accompany an if. For example, if wanted to randomly move a particle on a line, our conditional would look like:

```
if r < 0.5:        # if r < 0.5,
    x = x + 1      #    move to the east and finish
else:              # otherwise (r >= 0.5),
    x = x - 1      #    move to the west and finish
print(x)           # executed after the if/else
```

We are now ready to use our if/elif/else conditional structure in a loop to simulate many steps of a random walk on a grid. The randomWalk function below does this, and then returns the distance the particle has moved from the origin.

```
def randomWalk(steps, tortoise):
    """Displays a random walk on a grid.
    
    Parameters:
        steps: the number of steps in the random walk
        tortoise: a Turtle object
    
    Return value: the final distance from the origin
    """
    
    x = 0                      # initialize (x, y) = (0, 0)
    y = 0
    moveLength = 10            # length of a turtle step
    for step in range(steps):
        r = random.random()    # randomly choose a direction
        if r < 0.25:           # if r < 0.25,
            x = x + 1          #    move to the east and finish
        elif r < 0.5:          # otherwise, if r is in [0.25, 0.5),
            y = y + 1          #    move to the north and finish
        elif r < 0.75:         # otherwise, if r is in [0.5, 0.75),
            x = x - 1          #    move to the west and finish
        else:                  # otherwise,
            y = y - 1          #    move to the south and finish
        tortoise.goto(x * moveLength, y * moveLength)   # draw one step
    
    return math.sqrt(x * x + y * y)    # return distance from (0, 0)
```

Figure 5.1 A 1000-step random walk produced by the `randomWalk` function.

To make the grid movement easier to see, we make the turtle move further in each step by multiplying each position by a variable `moveLength`. To try out the random walk, write a `main` function that creates a new `Turtle` object and then calls the `randomWalk` function with 1000 steps. One such run is illustrated in Figure 5.1.

A random walk in Monte Carlo

How far, on average, does a two-dimensional Brownian particle move from its origin in a given number of steps? The answer to this question can, for example, provide insight into the rate at which a fluid spreads or the extent of an animal's foraging territory.

The distance traveled in any one particular random walk is meaningless; the particle may return the origin, walk away from the origin at every step, or do something in between. None of these outcomes tells us anything about the *expected*, or average, behavior of the system. Instead, to model the expected behavior, we need to compute the average distance over many, many random walks. This kind of simulation is called a *Monte Carlo simulation*.

As we will be calling `randomWalk` many times, we would like to speed things up by skipping the turtle visualization of the random walks. We can prevent drawing by incorporating a *flag variable* as a parameter to `randomWalk`. A flag variable takes on a Boolean value, and is used to switch some behavior on or off. In Python, the two possible Boolean values are named `True` and `False` (note the capitalization). In the `randomWalk` function, we will call the flag variable `draw`, and cause its value to influence the drawing behavior with another `if` statement:

```
def randomWalk(steps, tortoise, draw):
    ⋮
    if draw:
        tortoise.goto(x * moveLength, y * moveLength)
    ⋮
```

Now, when we call `randomWalk`, we pass in either `True` or `False` for our third argument. If `draw` is `True`, then `tortoise.goto(···)` will be executed but, if `draw` is `False`, it will be skipped.

To find the average over many trials, we will call our `randomWalk` function repeatedly in a loop, and use an accumulator variable to sum up all the distances.

```
def rwMonteCarlo(steps, trials):
    """A Monte Carlo simulation to find the expected distance
       that a random walk finishes from the origin.

    Parameters:
        steps: the number of steps in the random walk
        trials: the number of random walks

    Return value: the average distance from the origin
    """

    totalDistance = 0
    for trial in range(trials):
        distance = randomWalk(steps, None, False)
        totalDistance = totalDistance + distance
    return totalDistance / trials
```

Notice that we have passed in `None` as the argument for the second parameter (`tortoise`) of `randomWalk`. With `False` being passed in for the parameter `draw`, the value assigned to `tortoise` is never used, so we pass in `None` as a placeholder "dummy value."

Reflection 5.7 *Get ten different estimates of the average distance traveled over five trials of a 500-step random walk by calling* `rwMonteCarlo(500, 5)` *ten times in a loop and printing the result each time. What do you notice? Do you think five trials is enough? Now perform the same experiment with 10, 100, 1000, and 10000 trials. How many trials do you think are sufficient to get a reliable result?*

Ultimately, we want to understand the distance traveled as a function of the number of steps. In other words, if the particle moves n steps, does it travel an average distance of $n/2$ or $n/25$ or $\sqrt{n}$ or $\log_2 n$ or ...? To empirically discover the answer, we need to run the Monte Carlo simulation for many different numbers of steps, and try to infer a pattern from a plot of the results. The following function does this with `steps` equal to 100, 200, 300, ..., `maxSteps`, and then plots the results.

```
import matplotlib.pyplot as pyplot

def plotDistances(maxSteps, trials):
    """Plots the average distances traveled by random walks of
        100, 200, ..., maxSteps steps.

    Parameters:
        maxSteps: the maximum number of steps for the plot
        trials: the number of random walks in each simulation

    Return value: None
    """

    distances = [ ]
    stepRange = range(100, maxSteps + 1, 100)
    for steps in stepRange:
        distance = rwMonteCarlo(steps, trials)
        distances.append(distance)

    pyplot.plot(stepRange, distances, label = 'Simulation')
    pyplot.legend(loc = 'center right')
    pyplot.xlabel('Number of Steps')
    pyplot.ylabel('Distance From Origin')
    pyplot.show()
```

Reflection 5.8 *Run* `plotDistances(1000, 5000)`. *What is your hypothesis for the function approximated by the plot?*

The function we are seeking has actually been mathematically determined, and is approximately $\sqrt{n}$. You can confirm this empirically by plotting this function alongside the simulated results. To do so, insert each of these three statements in their appropriate locations in the `plotDistances` function:

```
y = [ ]    # before loop

y.append(math.sqrt(steps))    # inside loop

pyplot.plot(stepRange, y, label = 'Model Function')    # after loop
```

The result is shown in Figure 5.2. This result tells us that after a random particle moves n unit-length steps, it will be about $\sqrt{n}$ units of distance away from where it started, on average. In any particular instance, however, a particle may be much closer or farther away.

As you discovered in Reflection 5.1, the quality of any Monte Carlo approximation depends on the number of trials. If you call `plotDistances` a few more times with different numbers of trials, you should find that the plot of the simulation results

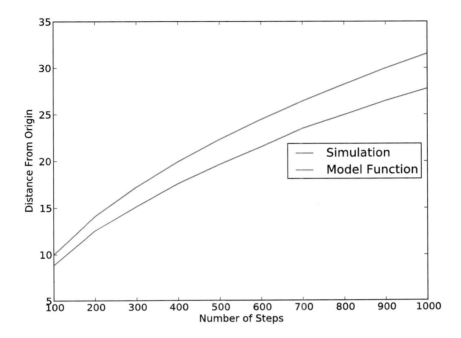

Figure 5.2 Plot generated by `plotDistances(1000, 10000)` with a comparison to the square root function.

gets "smoother" with more trials. But more trials obviously take longer. This is another example of the tradeoffs that we often encounter when solving problems.

Histograms

As we increase the number of trials, our average results become more consistent, but what do the individual trials look like? To find out, we can generate a *histogram* of the individual random walk distances. A histogram for a data set is a bar graph that shows how the items in the data set are distributed across some number of intervals, which are usually called "buckets" or "bins."

The following function is similar to `rwMonteCarlo`, except that it initializes an empty list before the loop, and appends one distance to the list in every iteration. Then it passes this list to the `matplotlib` histogram function `hist`. The first argument to the `hist` function is a list of numbers, and the second parameter is the number of buckets that we want to use.

```
def rwHistogram(steps, trials):
    """Draw a histogram of the given number of trials of
       a random walk with the given number of steps.

    Parameters:
        steps: the number of steps in the random walk
        trials: the number of random walks

    Return value: None
    """

    distances = [ ]
    for trial in range(trials):
        distance = randomWalk(steps, None, False)
        distances.append(distance)

    pyplot.hist(distances, 75)
    pyplot.show()
```

Reflection 5.9 *Call the* `rwHistogram` *function to produce a histogram of 5000 trials of a random walk of 1000 steps. What is the shape of the histogram?*

Exercises

5.1.1. Sometimes we want a random walk to reflect circumstances that bias the probability of a particle moving in some direction (i.e., gravity, water current, or wind). For example, suppose that we need to incorporate gravity, so a movement to the north is modeling a real movement up, away from the force of gravity. Then we might want to decrease the probability of moving north to 0.15, increase the probability of moving south to 0.35, and leave the other directions as they were. Show how to modify the `randomWalk` function to implement this situation.

5.1.2. Suppose the weather forecast calls for a 70% chance of rain. Write a function

  `weather()`

that prints 'RAIN' (or something similar) with probability 0.7, and 'SUN!' otherwise.

Now write another version that snows with probability 0.66, produces a sunny day with probability 0.33, and rains cats and dogs with probability 0.01.

5.1.3. Write a function

  `loaded()`

that simulates the rolling of a single "loaded die" that rolls more 1's and 6's than it should. The probability of rolling each of 1 or 6 should be 0.25. The function should use the `random.random` function and an `if/elif/else` conditional construct to assign a roll value to a variable named `roll`, and then return the value of `roll`.

5.1.4. Write a function

> `roll()`

that simulates the rolling of a single fair die. Then write a function

> `diceHistogram(trials)`

that simulates `trials` rolls of a pair of dice and displays a histogram of the results. (Use a bucket for each of the values 2–12.) What roll is most likely to show up?

5.1.5. The Monty Hall problem is a famous puzzle based on the game show "Let's Make a Deal," hosted, in its heyday, by Monty Hall. You are given the choice of three doors. Behind one is a car, and behind the other two are goats. You pick a door, and then Monty, who knows what's behind all three doors, opens a different one, which always reveals a goat. You can then stick with your original door or switch. What do you do (assuming you would prefer a car)?

We can write a Monte Carlo simulation to find out. First, write a function

> `montyHall(choice, switch)`

that decides whether we win or lose, based on our original door choice and whether we decide to switch doors. Assume that the doors are numbered 0, 1, and 2, and that the car is always behind door number 2. If we originally chose the car, then we lose if we switch but we win if we don't. Otherwise, if we did not originally choose the car, then we win if we switch and lose if we don't. The function should return `True` if we win and `False` if we lose.

Now write a function

> `monteMonty(trials)`

that performs a Monte Carlo simulation with the given number of trials to find the probability of winning if we decide to switch doors. For each trial, choose a random door number (between 0 and 2), and call the `montyHall` function with your choice and `switch = True`. Count the number of times we win, and return this number divided by the number of trials. Can you explain the result?

5.1.6. The value of π can be estimated with Monte Carlo simulation. Suppose you draw a circle on the wall with radius 1, inscribed inside a square with side length 2.

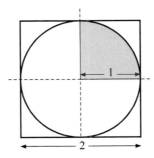

You then close your eyes and throw darts at the circle. Assuming every dart lands inside the square, the fraction of the darts that land in the circle estimates the ratio between the area of the circle and the area of the square. We know that the area of the circle is $C = \pi r^2 = \pi 1^2 = \pi$ and the area of the square is $S = 2^2 = 4$. So the exact ratio is $\pi/4$. With enough darts, f, the fraction (between 0 and 1) that lands in the circle will approximate this ratio: $f \approx \pi/4$, which means that $\pi \approx 4f$.

To make matters a little simpler, we can just throw darts in the upper right quarter of the circle (shaded above). The ratio here is the same: $(\pi/4)/1 = \pi/4$. If we place this quarter circle on x and y axes, with the center of the circle at $(0,0)$, our darts will now all land at points with x and y coordinates between 0 and 1.

Use this idea to write a function

```
montePi(darts)
```

that approximates the value π by repeatedly throwing random virtual darts that land at points with x and y coordinates in $[0,1)$. Count the number that land at points within distance 1 of the origin, and return this fraction.

5.1.7. The Good, The Bad, and The Ugly are in a three-way gun fight (sometimes called a "truel"). The Good always hits his target with probability 0.8, The Bad always hits his target with probability 0.7, and The Ugly always hits his target with probability 0.6. Initially, The Good aims at The Bad, The Bad aims at The Good, and The Ugly aims at The Bad. The gunmen shoot simultaneously. In the next round, each gunman, if he is still standing, aims at his same target, if that target is alive, or at the other gunman, if there is one, otherwise. This continues until only one gunman is standing or all are dead. What is the probability that they all die? What is the probability that The Good survives? What about The Bad? The Ugly? On average, how many rounds are there? Write a function

```
goodBadUgly()
```

that simulates one instance of this three-way gun fight. Your function should return 1, 2, 3, or 0 depending upon whether The Good, The Bad, The Ugly, or nobody is left standing, respectively. Next, write a function

```
monteGBU(trials)
```

that calls your `goodBadUgly` function repeatedly in a Monte Carlo simulation to answer the questions above.

5.1.8. What is printed by the following sequence of statements in each of the cases below? Explain your answers.

```
if votes1 >= votes2:
    print('Candidate one wins!')
elif votes1 <= votes2:
    print('Candidate two wins!')
else:
    print('There was a tie.')
```

(a) votes1 = 184 and votes2 = 206

(b) votes1 = 255 and votes2 = 135

(c) votes1 = 195 and votes2 = 195

5.1.9. There is a problem with the code in the previous exercise. Fix it so that it correctly fulfills its intended purpose.

5.1.10. What is printed by the following sequence of statements in each of the following cases? Explain your answers.

```
majority = (votes1 + votes2 + votes3) / 2
if votes1 > majority:
    print('Candidate one wins!')
if votes2 > majority:
    print('Candidate two wins!')
if votes3 > majority:
    print('Candidate three wins!')
else:
    print('A runoff is required.')
```

(a) votes1 = 5 and votes2 = 5 and votes3 = 5

(b) votes1 = 9 and votes2 = 2 and votes3 = 4

(c) votes1 = 0 and votes2 = 15 and votes3 = 0

5.1.11. Make the code in the previous problem more efficient and fix it so that it fulfills its intended purpose.

5.1.12. What is syntactically wrong with the following sequence of statements?

```
if x < 1:
    print('Something.')
else:
    print('Something else.')
elif x > 3:
    print('Another something.')
```

5.1.13. What is the final value assigned to result after each of the following code segments?

(a)
```
n = 13
result = n
if n > 12:
    result = result + 12
if n < 5:
    result = result + 5
else:
    result = result + 2
```

(b)
```
n = 13
result = n
if n > 12:
    result = result + 12
elif n < 5:
    result = result + 5
else:
    result = result + 2
```

200 ■ Forks in the road

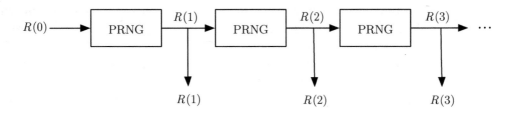

Figure 5.3 An illustration of the operation of a simple PRNG.

5.1.14. Write a function

    computeGrades(scores)

that returns a list of letter grades corresponding to the numerical scores in the list scores. For example, computeGrades([78, 91, 85]) should return the list ['C', 'A', 'B']. To test your function, use lists of scores generated by the following function.

    import random

    def randomScores(n):
        scores = []
        for count in range(n):
            scores.append(random.gauss(80, 10))
        return scores

5.1.15. Determining the number of bins to use in a histogram is part science, part art. If you use too few bins, you might miss the shape of the distribution. If you use too many bins, there may be many empty bins and the shape of the distribution will be too jagged. Experiment with the correct number of bins for 10,000 trials in rwHistogram function. At the extremes, create a histogram with only 3 bins and another with 1,000 bins. Then try numbers in between. What seems to be a good number of bins? (You may also want to do some research on this question.)

*5.2   PSEUDORANDOM NUMBER GENERATORS

Let's look more closely at how a *pseudorandom number generator* (PRNG) creates a sequence of numbers that appear to be random. A PRNG algorithm starts with a particular number called a *seed*. If we let $R(t)$ represent the number returned by the PRNG in step t, then we can denote the seed as $R(0)$. We apply a simple, but carefully chosen, arithmetic function to the seed to get the first number $R(1)$ in our pseudorandom sequence. To get the next number in the pseudorandom sequence, we apply the same arithmetic function to $R(1)$, producing $R(2)$. This process continues for $R(3), R(4), \ldots$, for as long as we need "random" numbers, as illustrated in Figure 5.3.

Reflection 5.10 *Notice that we are computing $R(t)$ from $R(t-1)$ at every step. Does this look familiar?*

If you read Section 4.4, you may recognize this as another example of a difference equation. A difference equation is a function that computes its next value based on its previous value.

A simple PRNG algorithm, known as a *Lehmer pseudorandom number generator*, is named after the late mathematician Derrick Lehmer. Dr. Lehmer taught at the University of California, Berkeley, and was one of the first people to run programs on the ENIAC, the first electronic computer. A Lehmer PRNG uses the following difference equation:

$$R(t) = a \cdot R(t-1) \bmod m$$

where m is a prime number and a is an integer between 1 and $m - 1$. For example, suppose $m = 13$, $a = 5$, and the seed $R(0) = 1$. Then

$$R(1) = 5 \cdot R(0) \bmod 13 = 5 \cdot 1 \bmod 13 = 5 \bmod 13 = 5$$

$$R(2) = 5 \cdot R(1) \bmod 13 = 5 \cdot 5 \bmod 13 = 25 \bmod 13 = 12$$

$$R(3) = 5 \cdot R(2) \bmod 13 = 5 \cdot 12 \bmod 13 = 60 \bmod 13 = 8$$

So this pseudorandom sequence begins $5, 12, 8, \ldots$.

Reflection 5.11 *Compute the next four values in the sequence.*

The next four values are $R(4) = 1$, $R(5) = 5$, $R(6) = 12$, and $R(7) = 8$. So the sequence is now $5, 12, 8, 1, 5, 12, 8, \ldots$.

Reflection 5.12 *Does this sequence look random?*

Unfortunately, our chosen parameters produce a sequence that endlessly repeats the subsequence $1, 5, 12, 8$. This is obviously not very random, but we can fix it by choosing m and a more carefully. We will revisit this in a moment.

Implementation

But first, let's implement the PRNG "black box" in Figure 5.3, based on the Lehmer PRNG difference equation. It is a very simple function!

```
def lehmer(r, m, a):
    """Computes the next pseudorandom number using a Lehmer PRNG.

    Parameters:
        r: the seed or previous pseudorandom number
        m: a prime number
        a: an integer between 1 and m - 1

    Return value: the next pseudorandom number
    """

    return (a * r) % m
```

The parameter `r` represents $R(t-1)$ and the value returned by the function is the single value $R(t)$.

Now, for our function to generate good pseudorandom numbers, we need some better values for m and a. Some particularly good ones were suggested by Keith Miller and Steve Park in 1988:

$$m = 2^{31} - 1 = 2,147,483,647 \text{ and } a = 16,807.$$

A Lehmer generator with these parameters is often called a *Park-Miller psuedorandom number generator*. To create a Park-Miller PRNG, we can call the `lehmer` function repeatedly, each time passing in the previously generated value and the Park-Miller values of m and a, as follows.

```
def randomSequence(length, seed):
    """Returns a list of Park-Miller pseudorandom numbers.

    Parameters:
        length: the number of pseudorandom numbers to generate
        seed: the initial seed

    Return value: a list of Park-Miller pseudorandom numbers
                  with the given length
    """

    r = seed
    m = 2**31 - 1
    a = 16807
    randList = [ ]
    for index in range(length):
        r = lehmer(r, m, a)
        randList.append(r)
    return randList
```

In each iteration, the function appends a new pseudorandom number to the end of the list named `randList`. (This is another list accumulator.) The `randomSequence` function then returns this list of `length` pseudorandom numbers. Try printing the result of `randomSequence(100, 1)`.

Because all of the returned values of `lehmer` are modulo m, they are all in the interval $[0..m-1]$. Since the value of m is somewhat arbitrary, random numbers are usually returned instead as values from the interval $[0,1)$, as Python's `random` function does. This is accomplished by simply dividing each pseudorandom number by m. To modify our `randomSequence` function to returns a list of numbers in $[0,1)$ instead of in $[0..m-1]$, we can simply append `r / m` to `randList` instead of `r`.

Reflection 5.13 *Make this change to the* `randomSequence` *function and call the function again with seeds 3, 4, and 5. Would you expect the results to look similar with such similar seeds? Do they?*

As an aside, you can set the seed used by Python's **random** function by calling the **seed** function. The default seed is based on the current time.

The ability to generate an apparently random, but reproducible, sequence by setting the seed has quite a few practical applications. For example, it is often useful in a Monte Carlo simulation to be able to reproduce an especially interesting run by simply using the same seed. Pseudorandom sequences are also used in electronic car door locks. Each time you press the button on your dongle to unlock the door, it is sending a different random code. But since the dongle and the car are both using the same PRNG with the same seed, the car is able to recognize whether the code is legitimate.

Testing randomness

How can we tell how random a sequence of numbers really is? What does it even mean to be truly "random?" If a sequence is random, then there must be no way to predict or reproduce it, which means that there must be no shorter way to describe the sequence than the sequence itself. Obviously then, a PRNG is not really random at all because it is entirely reproducible and can be described by simple formula. However, for practical purposes, we can ask whether, if we did not know the formula used to produce the numbers, could we predict them, or any patterns in them?

One simple test is to generate a histogram from the list of values.

Reflection 5.14 *Suppose you create a histogram for a list of one million random numbers in $[0, 1)$. If the histogram contains 100 buckets, each representing an interval with length 0.01, about how many numbers should be assigned to each bucket?*

If the list is random, each bucket should contain about 1%, or 10,000, of the numbers. The following function generates such a histogram.

```
import matplotlib.pyplot as pyplot

def histRandom(length):
    """Displays a histogram of numbers generated by the Park-Miller PRNG.

    Parameter:
        length: the number of pseudorandom numbers to generate

    Return value: None
    """

    samples = randomSequence(length, 6)
    pyplot.hist(samples, 100)
    pyplot.show()
```

Reflection 5.15 *Call* `histRandom(100000)`. *Does the* `randomSequence` *function look like it generates random numbers? Does this test guarantee that the numbers are actually random?*

There are several other ways to test for randomness, many of which are more complicated than we can describe here (we leave a few simpler tests as exercises). These tests reveal that the Lehmer PRNG that we described exhibits some patterns that are undesirable in situations requiring higher-quality random numbers. However, the Python `random` function uses one of the best PRNG algorithms known, called the *Mersenne twister*, so you can use it with confidence.

Exercises

5.2.1. We can visually test the quality of a PRNG by using it to plot random points on the screen. If the PRNG is truly random, then the points should be uniformly distributed without producing any noticeable patterns. Write a function

```
testRandom(n)
```

that uses turtle graphics and `random.random` to plot n random points with x and y each in $[0, 1)$. Here is a "skeleton" of the function with some turtle graphics set up for you. Calling the `setworldcoordinates` function redefines the coordinate system so that the lower left corner is $(0, 0)$ and the upper right corner is $(1, 1)$. Use the turtle graphics functions `goto` and `dot` to move the turtle to each point and draw a dot there.

```
def testRandom(n):
    """ your docstring here """

    tortoise = turtle.Turtle()
    screen = tortoise.getscreen()
    screen.setworldcoordinates(0, 0, 1, 1)
    screen.tracer(100)      # only draw every 100 updates
    tortoise.up()
    tortoise.speed(0)

    # draw the points here

    screen.update()         # ensure all updates are drawn
    screen.exitonclick()
```

5.2.2. Repeat Exercise 5.2.1, but use the `lehmer` function instead of `random.random`. Remember that `lehmer` returns an integer in $[0..m-1]$, which you will need to convert to a number in $[0, 1)$. Also, you will need to call `lehmer` twice in each iteration of the loop, so be careful about how you are passing in values of `r`.

5.2.3. Another simple test of a PRNG is to produce many pseudorandom numbers in $[0, 1)$ and make sure their average is 0.5. To test this, write a function

```
avgTest(n)
```

that returns the average of n pseudorandom numbers produced by the `random` function.

*5.3 SIMULATING PROBABILITY DISTRIBUTIONS

Pseudorandom number generators can be used to approximate various *probability distributions* that can be used to simulate natural phenomena. A probability distribution assigns probabilities, which are values in $[0, 1]$, to a set of random *elementary events*. The sum of the probabilities of all of the elementary events must be equal to 1. For example, a weather forecast that calls for a 70% chance of rain is a (predicted) probability distribution: the probability of rain is 0.7 and the probability of no rain is 0.3. Mathematicians usually think of probability distributions as assigning probabilities to numerical values or intervals of values rather than natural events, although we can always define an equivalence between a natural event and a numerical value.

Python provides support for several probability distributions in the `random` module. Here we will just look at two especially common ones: the uniform and normal distributions. Each of these is based on Python's `random` function.

The uniform distribution

The simplest general probability distribution is the *uniform distribution*, which assigns equal probability to every value in a range, or to every equal length interval in a range. A PRNG like the `random` function returns values in $[0, 1)$ according to a uniform distribution. The more general Python function `random.uniform(a, b)` returns a uniformly random value in the interval $[\mathtt{a}, \mathtt{b}]$. In other words, it is equally likely to return any number in $[\mathtt{a}, \mathtt{b}]$. For example,

```
>>> random.uniform(0, 100)
77.10524701669804
```

The normal distribution

Measurements of most everyday and natural processes tend to exhibit a "bell curve" distribution, formally known as a *normal* or *Gaussian* distribution. A normal distribution describes a process that tends to generate values that are centered around a mean. For example, measurements of people's heights and weights tend to fall into a normal distribution, as do the velocities of molecules in an ideal gas.

A normal distribution has two parameters: a mean value and a standard deviation. The standard deviation is, intuitively, the average distance a value is from the mean; a low standard deviation will give values that tend to be tightly clustered around the mean while a high standard deviation will give values that tend to be spread out from the mean.

The normal distribution assigns probabilities to intervals of real numbers in such a way that numbers centered around the mean have higher probability than those that are further from the mean. Therefore, if we are generating random values according to a normal distribution, we are more likely to get values close to the mean than values farther from the mean.

The Python function `random.gauss` returns a value according to a normal distribution with a given mean and standard deviation. For example, the following

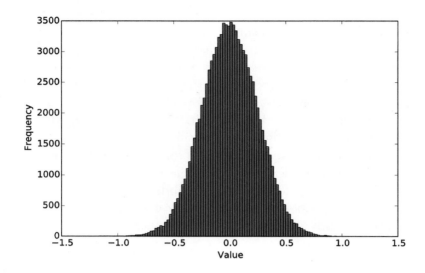

Figure 5.4 A histogram of 100,000 samples from `random.gauss(0, 0.25)`.

returns a value according to the normal distribution with mean 0 and standard deviation 0.25.

```
>>> random.gauss(0, 0.25)
-0.3371607214433552
```

Figure 5.4 shows a histogram*histogram* of 100,000 values returned by `random.gauss(0, 0.25)`. Notice that the "bell curve" of frequencies is centered at the mean of 0.

The central limit theorem

Intuitively, the normal distribution is so common because phenomena that are the sums of many additive random factors tend to be normal. For example, a person's height is the result of several different genes and the person's health, so the distribution of heights in a population tends to be normal. To illustrate this phenomenon, let's create a histogram of the sums of numbers generated by the `random` function. Below, the `sumRandom` function returns the sum of `n` random numbers in the interval $[0, 1)$. The `sumRandomHist` function creates a list named `samples` that contains `trials` of these sums, and then plots a histogram of them.

```
import matplotlib.pyplot as pyplot
import random

def sumRandom(n):
    """Returns the sum of n pseudorandom numbers in [0,1).

    Parameter:
        n: the number of pseudorandom numbers to generate

    Return value: the sum of n pseudorandom numbers in [0,1)
    """

    sum = 0
    for index in range(n):
        sum = sum + random.random()
    return sum

def sumRandomHist(n, trials):
    """Displays a histogram of sums of n pseudorandom numbers.

    Parameters:
        n: the number of pseudorandom numbers in each sum
        trials: the number of sums to generate

    Return value: None
    """

    samples = [ ]
    for index in range(trials):
        samples.append(sumRandom(n))
    pyplot.hist(samples, 100)
    pyplot.show()
```

Reflection 5.16 *Call* sumRandomHist *with 100,000 trials and values of* n *equal to 1, 2, 3, and 10. What do you notice?*

When we call sumRandomHist with n = 1, we get the expected uniform distribution of values, shown in the upper left of Figure 5.5. However, for increasing values of n, we notice that the distribution of values very quickly begins to look like a normal distribution, as illustrated in the rest of Figure 5.5. In addition, the standard deviation of the distribution (intuitively, the width of the "bell") appears to get smaller relative to the mean.

The mathematical name for this phenomenon is the *central limit theorem*. Intuitively, the central limit theorem implies that any large set of independent measurements of a process that is the cumulative result of many random factors will look like a normal distribution. Because a normal distribution is centered around a

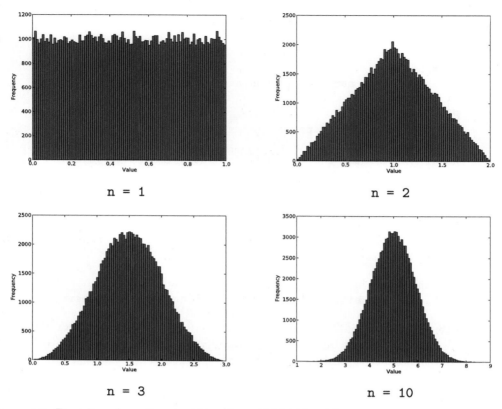

Figure 5.5 Results of `sumRandomHist(n, 100000)` with n = 1, 2, 3, 10.

mean, this also implies that the average of all of the measurements will be close to the true mean.

Reflection 5.17 *Think back to the experiment you ran to answer Reflection 5.9. The shape of the histogram should have resembled a normal distribution. Can you use the central limit theorem to explain why?*

Exercises

5.3.1. A more realistic random walk has the movement in each step follow a *normal distribution*. In particular, in each step, we can change both x and y according to a normal distribution with mean 0. Because the values produced by this distribution will be both positive and negative, the particle can potentially move in any direction. To make the step sizes small, we need to use a small standard deviation, say 0.5:

```
x = x + random.gauss(0, 0.5)
y = y + random.gauss(0, 0.5)
```

Modify the `randomWalk` function so that it updates the position of the particle in this way instead. Then use the `rwMonteCarlo` and `plotDistances` functions to run a Monte Carlo simulation with your new `randomWalk` function. As we did earlier, call `plotDistances(1000, 10000)`. How do your results differ from the original version?

5.3.2. Write a function

   `uniform(a, b)`

that returns a number in the interval $[a, b)$ using only the `random.random` function. (Do not use the `random.uniform` function.)

5.3.3. Suppose we want a pseudorandom integer between 0 and 7 (inclusive). How can we use the `random.random()` and `int` functions to get this result?

5.3.4. Write a function

   `randomRange(a, b)`

that returns a pseudorandom integer in the interval $[a..b]$ using only the `random.random` function. (Do not use `random.randrange` or `random.randint`.)

5.3.5. Write a function

   `normalHist(mean, stdDev, trials)`

that produces a histogram of `trials` values returned by the `gauss` function with the given mean and standard deviation. In other words, reproduce Figure 5.4. (This is very similar to the `sumRandomHist` function.)

5.3.6. Write a function

   `uniformHist(a, b, trials)`

that produces a histogram of `trials` values in the range $[a, b]$ returned by the `uniform` function. In other words, reproduce the top left histogram in Figure 5.5. (This is very similar to the `sumRandomHist` function.)

5.3.7. Write a function

   `plotChiSquared(k, trials)`

that produces a histogram of `trials` values, each of which is the sum of k squares of values given by the `random.gauss` function with mean 0 and standard deviation 1. (This is very similar to the `sumRandomHist` function.) The resulting probability distribution is known as the *chi-squared distribution* (χ^2 distribution) with k degrees of freedom.

5.4 BACK TO BOOLEANS

As we discussed at the beginning of the chapter, virtually every useful algorithm must be able to conditionally change course in response to input. This conditional behavior can be implemented with the same `if` statements that we worked with in Section 5.1. In this section, we will further develop your facility with these conditional statements, and the Boolean logic that drives them. (If your Boolean logic is rusty, you may want to review the concepts in Section 1.4 before continuing.)

In Section 5.1, we used simple `if` statements like

```
if r < 0.5:
    x = x + 1
else:
    x = x - 1
```

to control a random walk based on a random value of **r**. The outcome in this case is based upon the value of the Boolean expression **r < 0.5**. In Python, Boolean expressions evaluate to either the value **True** or the value **False**, which correspond to the binary values 1 and 0, respectively, that we worked with in Section 1.4. The values **True** and **False** can be printed, assigned to variable names, and manipulated just like numeric values. For example, try the following examples and make sure you understand each result.

```
>>> print(0 < 1)
True
>>> name = 'Kermit'
>>> print(name == 'Gonzo')
False
>>> result = 0 < 1
>>> result
True
>>> result and name == 'Kermit'
True
```

The "double equals" (==) operator tests for equality; it has nothing to do with assignment. The Python interpreter will remind you if you mistakenly use a single equals in an **if** statement. For example, try this:

```
>>> if r = 0:
    if r = 0:
         ^
SyntaxError: invalid syntax
```

However, the interpreter will *not* catch the error if you mistakenly use == in an assignment statement. For example, try this:

```
>>> r = 1
>>> r == r + 1    # increment r?
False
```

In a program, nothing will be printed as a result of the second statement, and **r** will not be incremented as expected. So be careful!

As we saw in Section 1.4, Boolean expressions can be combined with the *Boolean operators* (or *logical operators*) **and**, **or**, and **not**. As a reminder, Figure 5.6 contains the truth tables for the three Boolean operators, expressed in Python notation. In the tables, the variable names **a** and **b** represent arbitrary Boolean variables or expressions.

For example, suppose we wanted to determine whether a household has an annual income within some range, say $40,000 to $65,000:

a	b	a and b	a or b	not a
False	False	False	False	True
False	True	False	True	True
True	False	False	True	False
True	True	True	True	False

Figure 5.6 Combined truth table for the three Boolean operators.

```
>>> income = 53000
>>> income >= 40000 and income <= 65000
True
>>> income = 12000
>>> income >= 40000 and income <= 65000
False
```

When 53000 is assigned to `income`, the two Boolean expressions `income >= 40000` and `income <= 65000` are both `True`, so `income >= 40000 and income <= 65000` is also `True`. However, when 12000 is assigned to `income`, `income >= 40000` is `False`, while `income <= 65000` is `True`, so `income >= 40000 and income <= 65000` is now `False`. We can also incorporate this test into a function that simply returns the value of the condition:

```
>>> def middleClass(income):
        return (income >= 40000 and income <= 65000)
>>> middleClass(53000)
True
>>> middleClass(12000)
False
```

Reflection 5.18 *What would happen if we changed* `and` *to* `or` *in the* `middleClass` *function? For what values of* `income` *would the function return* `False`*?*

If we changed the `middleClass` function to be

```
def middleClass(income):
    return (income >= 40000 or income <= 65000)    # WRONG!
```

it would *always* return `True`! Consider the three possible cases:

- If `income` is between 40000 and 65000 (inclusive), it will return `True` because both parts of the `or` expression are `True`.

- If `income < 40000`, it will return `True` because `income <= 65000` is `True`.

- If `income > 65000`, it will return `True` because `income >= 40000` is `True`.

Short circuit evaluation

The **and** and **or** operators exhibit a behavior known as *short circuit evaluation* that comes in handy in particular situations. Since both operands of an **and** expression must be true for the expression to be true, the Python interpreter does not bother to evaluate the second operand in an **and** expression if the first is false. Likewise, since only one operand of an **or** expression must be true for the expression to be true, the Python interpreter does not bother to evaluate the second operand in an **or** expression if the first is true.

For example, suppose we wanted to write a function to decide, in some for-profit company, whether the CEO's compensation divided by the average employees' is at most some "fair" ratio. A simple function that returns the result of this test looks like this:

```
def fair(employee, ceo, ratio):
    """Decide whether the ratio of CEO to employee pay is fair.

    Parameters:
        employee: average employee pay
        ceo: CEO pay
        ratio: the fair ratio

    Return: a Boolean indicating whether ceo / employee is fair
    """

    return (ceo / employee <= ratio)
```

Reflection 5.19 *There is a problem with this function. What is it?*

This function will not always work properly because, if the average employees' compensation equals 0, the division operation will result in an error. Therefore, we have to test whether **employee == 0** before attempting the division and, if so, return **False** (because not paying employees is obviously never fair). Otherwise, we want to return the result of the fairness test. The following function implements this algorithm.

```
def fair(employee, ceo, ratio):
    """ (docstring omitted) """

    if employee == 0:
        result = False
    else:
        result = (ceo / employee <= ratio)
    return result
```

However, with short circuit evaluation, we can simplify the whole function to:

```
def fair(employee, ceo, ratio):
    """ (docstring omitted) """

    return (employee != 0) and (ceo / employee <= ratio)
```

If `employee` is 0, then `(employee != 0)` is `False`, and the function returns `False` *without* evaluating `(ceo / employee <= ratio)`. On the other hand, if `(employee != 0)` is true, then `(ceo / employee <= ratio)` is evaluated, and the return value depends on this outcome. Notice that this would not work if the **and** operator did not use short circuit evaluation because, if `(employee != 0)` were false and then `(ceo / employee <= ratio)` was evaluated, the division would result in a "divide by zero" error!

To illustrate the analogous mechanism with the **or** operator, suppose we wanted to write the function in the opposite way, instead returning `True` if the ratio is unfair. The first version of the function would look like this:

```
def unfair(employee, ceo, ratio):
    """Decide whether the ratio of CEO to employee pay is unfair.

    Parameters:
        employee: average employee pay
        ceo: CEO pay
        ratio: the fair ratio

    Return: a Boolean indicating whether ceo / employee is not fair
    """

    if employee == 0:
        result = True
    else:
        result = (ceo / employee > ratio)
    return result
```

However, taking advantage of short circuit evaluation with the **or** operator, we can simplify the whole function to:

```
def unfair(employee, ceo, ratio):
    """ (docstring omitted) """

    return (employee == 0) or (ceo / employee > ratio)
```

In this case, if `(employee == 0)` is `True`, the whole expression returns `True` *without* evaluating the division test, thus avoiding an error. On the other hand, if `(employee == 0)` is `False`, the division test is evaluated, and the final result is equal to the outcome of this test.

Complex expressions

Some situations require Boolean expressions that are more complex than what we have seen thus far. To illustrate, let's consider how to test whether a particle in a random walk at position (x, y) is in one of the four corners of the screen, as depicted by the shaded regions below.

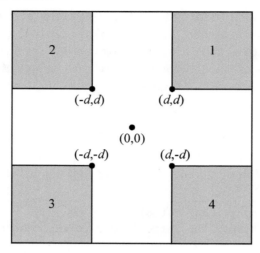

There are (at least) two ways we can write a Boolean expression to test this condition. One way is to test whether the particle is in each corner individually.

1. For the point to be corner 1, `x > d and y > d` must be `True`;
2. for the point to be corner 2, `x < -d and y > d` must be `True`;
3. for the point to be corner 3, `x < -d and y < -d` must be `True`; or
4. for the point to be corner 4, `x > d and y < -d` must be `True`.

Since any one of these can be `True` for our condition to be `True`, the final test, in the form of a function, looks like the following. (The "backslash" (\) character below is the *line continuation character*. It indicates that the line that it ends is continued on the next line. This is sometimes handy for splitting very long lines of code.)

```
>>> def corners(x, y, d):
        return x > d and y > d or x < -d and y > d or \
            x < -d and y < -d or x > d and y < -d
>>> corners(11, -11, 10)
True
>>> corners(11, 0, 10)
False
```

Although our logic is correct, this expression is only correct if the Python interpreter evaluates the **and** operator before it evaluates the **or** operator. Otherwise, if the **or** operator is evaluated first, then the first expression evaluated will be `y > d or x < -d`, shown in red below, which is not what we intended.

	Operators	Description
1.	**	exponentiation (power)
2.	+, -	unary positive and negative, e.g., -(4 * 9)
3.	*, /, //, %	multiplication and division
4.	+, -	addition and subtraction
5.	<, <=, >, >=, !=, ==, in, not in	comparison operators
6.	not	Boolean **not**
7.	and	Boolean **and**
8.	or	Boolean **or**

Table 5.2 Operator precedence, listed from highest to lowest. This is an expanded version of Table 2.1.

> x > d and y > d or x < -d and y > d or x < -d and y < -d or x > d and y < -d

Therefore, understanding the order in which operators are evaluated, known as *operator precedence*, becomes important for more complex conditions such as this. In this case, the expression is correct because **and** has precedence over **or**, as illustrated in Table 5.2. Also notice from Table 5.2 that the comparison operators have precedence over the Boolean operators, which is also necessary for our expression to be correct. If in doubt, you can use parentheses to explicitly specify the intended order. Parentheses would be warranted here in any case, to make the expression more understandable:

```
>>> def corners(x, y, d):
        return (x > d and y > d) or (x < -d and y > d) or \
               (x < -d and y < -d) or (x > d and y < -d)
```

Now let's consider an alternative way to think about the corner condition: for a point to be in one of the corners, it must be true that x must either exceed d or be less than -d and y must either exceed d or be less than -d. The equivalent Boolean expression is enclosed in the following function.

```
>>> def corners2(x, y, d):
        return x > d or x < -d and y > d or y < -d
>>> corners2(11, -11, 10)
True
>>> corners2(11, 0, 10)
True
```

Reflection 5.20 *The second call to* corners2 *gives an incorrect result. Do you see the problem? (You might want to draw a picture.)*

Due to the operator precedence order, this is a situation where parentheses are *required*. Without parentheses, the **and** expression in red below is evaluated first, which is not our intention.

```
x > d or x < -d and y > d or y < -d
```

So the correct implementation is

```
>>> def corners2(x, y, d):
        return (x > d or x < -d) and (y > d or y < -d)
>>> corners2(11, -11, 10)
True
>>> corners2(11, 0, 10)
False
```

*Using truth tables

Reflection 5.21 *How can we confirm that this expression is really correct?*

To confirm that any Boolean expression is correct, we must create a truth table for it, and then confirm that every case matches what we intended. In this expression, there are four separate Boolean "inputs," one for each expression containing a comparison operator. In the truth table, we will represent each of these with a letter to save space:

- `x > d` will be represented by p,
- `x < -d` will be represented by q,
- `y > d` will be represented by r, and
- `y < -d` will be represented by s.

So the final expression we are evaluating is $(p$ `or` $q)$ `and` $(r$ `or` $s)$.

In the truth table below, the first four columns represent our four inputs. With four inputs, we need $2^4 = 16$ rows, one for each possible assignment of truth values to the inputs. There is a trick to quickly writing down all the truth value combinations; see if you can spot it in the first four columns. (We are using `T` and `F` as abbreviations for `True` and `False`. The extra horizontal lines are simply to improve readability.)

x > d	x < -d	y > d	y < -d			
p	q	r	s	p or q	r or s	(p or q) and (r or s)
F	F	F	F	F	F	F
F	F	F	T	F	T	F
F	F	T	F	F	T	F
F	F	T	T	F	T	F
F	T	F	F	T	F	F
F	T	F	T	T	T	T
F	T	T	F	T	T	T
F	T	T	T	T	T	T
T	F	F	F	T	F	F
T	F	F	T	T	T	T
T	F	T	F	T	T	T
T	F	T	T	T	T	T
T	T	F	F	T	F	F
T	T	F	T	T	T	T
T	T	T	F	T	T	T
T	T	T	T	T	T	T

To fill in the fifth column, we only need to look at the first and second columns. Since this is an **or** expression, for each row in the fifth column, we enter a **T** if there is at least one **T** among the first and second columns of that row, or a **F** otherwise. Similarly, to fill in the sixth column, we only look at the third and fourth column. To fill in the last column, we look at the two previous columns, filling in a **T** in each row in which the fifth and sixth columns are *both* **T**, or a **F** otherwise.

To confirm that our expression is correct, we need to confirm that the truth value in each row of the last column is correct with respect to that row's input values. For example, let's look at the fifth, sixth, and eighth rows of the truth table (in red).

In the fifth row, the last column indicates that the expression is **False**. Therefore, it should be the case that a point described by the truth values in the first four columns of that row is *not* in one of the four corners. Those truth values say that **x > d** is false, **x < -d** is true, **y > d** is false, and **y < -d** is false; in other words, **x** is less than **-d** and **y** is between **-d** and **d**. A point within these bounds is not in any of the corners, so that row is correct.

Now let's look at the sixth row, where the final expression is true. The first columns of that row indicate that **x > d** is false, **x < -d** is true, **y > d** is false, and **y < -d** is true; in other words, both **x** and **y** are less than **-d**. A point within these bounds is in the bottom left corner, so that row is also correct.

The eighth row is curious because it states that both **y > d** and **y < -d** are true. But this is, of course, impossible. Because of this, we say that the implied statement is *vacuously true*. In practice, we cannot have such a point, so that row is entirely irrelevant. There are seven such rows in the table, leaving only four other rows that are true, matching the four corners.

Many happy returns

We often come across situations in which we want an algorithm (i.e., a function) to return different values depending on the outcome of a condition. The simplest example is finding the maximum of two numbers `a` and `b`. If `a` is at least as large as `b`, we want to return `a`; otherwise, `b` must be larger, so we return `b`.

```
def max(a, b):
    """Returns the maximum of a and b.

    Parameters:
        a, b: two numbers

    Return value: the maximum value
    """

    if a >= b:
        result = a
    else:
        result = b
    return result
```

We can simplify this function a bit by returning the appropriate value right in the `if/else` statement:

```
def max(a, b):
    """ (docstring omitted)"""

    if a >= b:
        return a
    else:
        return b
```

It may look strange at first to see two `return` statements in one function, but it all makes perfect sense. Recall from Section 3.5 that `return` *both* ends the function *and* assigns the function's return value. So this means that at most one `return` statement can ever be executed in a function. In this case, if `a >= b` is true, the function ends and returns the value of `a`. Otherwise, the function executes the `else` clause, which returns the value of `b`.

The fact that the function ends if `a >= b` is true means that we can simplify it even further: if execution continues past the `if` part of the `if/else`, it *must* be the case that `a >= b` is false. So the `else` is extraneous; the function can be simplified to:

```
def max(a, b):
    """ (docstring omitted)"""

    if a >= b:
        return a
    return b
```

This same principle can be applied to situations with more than two cases. Suppose we wrote a function to convert a percentage grade to a grade point (i.e., GPA) on a 0–4 scale. A natural implementation of this might look like the following:

```
def assignGP(score):
    """Returns the grade point equivalent of score.

    Parameter:
        score: a score between 0 and 100

    Return value: the equivalent grade point value
    """

    if score >= 90:
        return 4
    elif score >= 80:
        return 3
    elif score >= 70:
        return 2
    elif score >= 60:
        return 1
    else:
        return 0
```

Reflection 5.22 *Why do we not need to check upper bounds on the scores in each case? In other words, why does the second condition not need to be* score >= 80 and score < 90*?*

Suppose score was 92. In this case, the first condition is true, so the function returns the value 4 and ends. Execution never proceeds past the statement return 4. For this reason, the "el" in the next elif is extraneous. In other words, because execution would never have made it there if the previous condition was true, there is no need to tell the interpreter to skip testing this condition if the previous condition was true.

Now suppose score was 82. In this case, the first condition would be false, so we continue on to the first elif condition. Because we got to this point, we already know that score < 90 (hence the omission of that check). The first elif condition is true, so we immediately return the value 3. Since the function has now completed, there is no need for the "el" in the second `elif` either. In other words, because execution would never have made it to the second `elif` if either of the previous conditions were true, there is no need to skip testing this condition if a previous condition was true. In fact, we can remove the "el"s from all of the `elif`s, and the final `else`, with no loss in efficiency at all.

```
def assignGP(score):
    """ (docstring omitted) """

    if score >= 90:
        return 4
    if score >= 80:
        return 3
    if score >= 70:
        return 2
    if score >= 60:
        return 1
    return 0
```

Some programmers find it clearer to leave the `elif` statements in, and that is fine too. We will do it both ways in the coming chapters. But, as you begin to see more algorithms, you will probably see code like this, and so it is important to understand why it is correct.

Exercises

5.4.1. Write a function

> password()

that asks for a username and a password. It should return **True** if the username is entered as `alan.turing` and the password is entered as `notTouring`, and return **False** otherwise.

5.4.2. Suppose that in a game that you are making, the player wins if her score is at least 100. Write a function

> hasWon(score)

that returns **True** if she has won, and **False** otherwise.

5.4.3. Suppose you have designed a sensor that people can wear to monitor their health. One task of this sensor will be to monitor body temperature: if it falls outside the range 97.9° F to 99.3° F, the person may be getting sick. Write a function

> monitor(temperature)

that takes a temperature (in Fahrenheit) as a parameter, and returns **True** if `temperature` falls in the healthy range and **False** otherwise.

5.4.4. Write a function

>   `winner(score1, score2)`

> that returns 1 or 2, indicating whether the winner of a game is Player 1 or Player 2. The higher score wins and you can assume that there are no ties.

5.4.5. Repeat the previous exercise, but now assume that ties are possible. Return 0 to indicate a tie.

5.4.6. Your firm is looking to buy computers from a distributor for $1500 per machine. The distributor will give you a 5% discount if you purchase more than 20 computers. Write a function

>   `cost(quantity)`

> that takes as a parameter the `quantity` of computers you wish to buy, and returns the cost of buying them from this distributor.

5.4.7. Repeat the previous exercise, but now add three additional parameters: the cost per machine, the number of computers necessary to get a discount, and the discount.

5.4.8. The speeding ticket fine in a nearby town is $50 plus $5 for each mph over the posted speed limit. In addition, there is an extra penalty of $200 for all speeds above 90 mph. Write a function

>   `fine(speedLimit, clockedSpeed)`

> that returns the fine amount (or 0 if `clockedSpeed` ≤ `speedLimit`).

5.4.9. Write a function

>   `gradeRemark()`

> that prompts for a grade, and then returns the corresponding remark (as a string) from the table below:

Grade	Remark
96-100	Outstanding
90-95	Exceeds expectations
80-89	Acceptable
1-79	Trollish

5.4.10. Write a function that takes two integer values as parameters and returns their sum if they are not equal and their product if they are.

5.4.11. Write a function

>   `amIRich(amount, rate, years)`

> that accumulates interest on `amount` dollars at an annual rate of `rate` percent for a number of `years`. If your final investment is at least double your original amount, return `True`; otherwise, return `False`.

5.4.12. Write a function

> `max3(a, b, c)`

the returns the maximum value of the parameters a, b, and c. Be sure to test it with many different numbers, including some that are equal.

5.4.13. Write a function

> `shipping(amount)`

that returns the shipping charge for an online retailer based on a purchase of amount dollars. The company charges a flat rate of $6.95 for purchases up to $100, plus 5% of the amount over $100.

5.4.14. Starting with two positive integers a and b, consider the sequence in which the next number is the digit in the ones place of the sum of the previous two numbers. For example, if $a = 1$ and $b = 1$, the sequence is

$$1, 1, 2, 3, 5, 8, 3, 1, 4, 5, 9, 4, 3, 7, 0, \ldots$$

Write a function

> `mystery(a, b)`

that returns the length of the sequence when the last two numbers repeat the values of a and b for the first time. (When $a = 1$ and $b = 1$, the function should return 62.)

5.4.15. The Chinese zodiac relates each year to an animal in a twelve-year cycle. The animals for the most recent years are given below.

Year	Animal	Year	Animal
2004	Monkey	2010	Tiger
2005	Rooster	2011	Rabbit
2006	Dog	2012	Dragon
2007	Pig	2013	Snake
2008	Rat	2014	Horse
2009	Ox	2015	Goat

Write a function

> `zodiac(year)`

that takes as a parameter a four-digit year (this could be any year in the past or future) and returns the corresponding animal as a string.

5.4.16. A year is a leap year if it is divisible by four, unless it is a century year in which case it must be divisible by 400. For example, 2012 and 1600 are leap years, but 2011 and 1800 are not. Write a function

> `leap(year)`

that returns a Boolean value indicating whether the year is a leap year.

5.4.17. Write a function

> `even(number)`

that returns `True` if `number` is even, and `False` otherwise.

5.4.18. Write a function

> `between(number, low, high)`

that returns `True` if `number` is in the interval $[\text{low}, \text{high}]$ (between (or equal to) `low` and `high`), and `False` otherwise.

5.4.19. Write a function

> `justone(a, b)`

that returns `True` if exactly one (but not both) of the numbers a or b is 10, and `False` otherwise.

5.4.20. Write a function

> `roll()`

that simulates rolling two of the loaded dice implemented in Exercise 5.1.3 (by calling the function `loaded`), and returns `True` if the sum of the dice is 7 or 11, or `False` otherwise.

5.4.21. The following function returns a Boolean value indicating whether an integer `number` is a perfect square. Rewrite the function in one line, taking advantage of the short-circuit evaluation of `and` expressions.

```
def perfectSquare(number):
    if number < 0:
        return False
    else:
        return math.sqrt(number) == int(math.sqrt(number))
```

5.4.22. Consider the `rwMonteCarlo` function on Page 193. What will the function return if `trials` equals 0? What if `trials` is negative? Propose a way to deal with these issues by adding statements to the function.

5.4.23. Write a Boolean expression that is `True` if and only if the point (x, y) resides in the shaded box below (including its boundary).

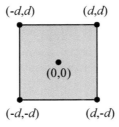

5.4.24. Use a truth table to show that the Boolean expressions

> `(x > d and y > d) or (x < -d and y > d)`

and

$$(y > d) \text{ and } (x > d \text{ or } x < -d)$$

are equivalent.

5.4.25. Write a function

`drawRow(tortoise, row)`

that uses turtle graphics to draw one row of an 8 × 8 red/black checkerboard. If the value of `row` is even, the row should start with a red square; otherwise, it should start with a black square. You may want to use the `drawSquare` function you wrote in Exercise 3.3.5. Your function should only need one `for` loop and only need to draw one square in each iteration.

5.4.26. Write a function

`checkerBoard(tortoise)`

that draws an 8 × 8 red/black checkerboard, using the function you wrote in Exercise 5.4.25.

5.5 A GUESSING GAME

In this section, we will practice a bit more writing algorithms that contain complex conditional constructs and `while` loops by designing an algorithm that lets us play the classic "I'm thinking of a number" game with the computer. The computer (really the PRNG) will "think" of a number between 1 and 100, and we will try to guess it. Here is a function to play the game:

```
import random

def guessingGame(maxGuesses):
    """Plays a guessing game.  The human player tries to guess
       the computer's number from 1 to 100.

    Parameter:
        maxGuesses: the maximum number of guesses allowed

    Return value: None
    """

    secretNumber = random.randrange(1, 101)
    for guesses in range(maxGuesses):
        myGuess = input('Please guess my number: ')
        myGuess = int(myGuess)

        if myGuess == secretNumber:       # win
            print('You got it!')
        else:                              # try again
            print('Nope.  Try again.')
```

The **randrange** function returns a random integer that is at least the value of its first argument, but less than its second argument (similar to the way the **range** function interprets its arguments). After the function chooses a random integer between 1 and 100, it enters a **for** loop that will allow us to guess up to **maxGuesses** times. The function prompts us for our guess with the **input** function, and then assigns the response to **myGuess** as a string. Because we want to interpret the response as an integer, we use the **int** function to convert the string. Once it has **myGuess**, the function uses the **if/else** statement to tell us whether we have guessed correctly. After this, the loop will give us another guess, until we have used them all up.

Reflection 5.23 *Try playing the game by calling* **guessingGame(20)**. *Does it work? Is there anything we still need to work on?*

You may have noticed three issues:

1. After we guess correctly, unless we have used up all of our guesses, the loop iterates again and gives us another guess. Instead, we want the function to end at this point.

2. It would be much friendlier for the game to tell us whether an incorrect guess is too high or too low.

3. If we do not guess correctly in at most **maxGuesses** guesses, the last thing we see is `'Nope. Try again.'` before the function ends. But there is no opportunity to try again; instead, it should tell us that we have lost.

We will address these issues in order.

1. Ending the game if we win

Our current **for** loop, like all **for** loops, iterates a prescribed number of times. Instead, we want the loop to only iterate for as long as we need another guess. So this is another instance that calls for a **while** loop. Recall that a **while** loop iterates while some Boolean condition is true. In this case, we need the loop to iterate while we have not guessed the secret number, in other words, while **myGuess != secretNumber**. But we also need to stop when we have used up all of our guesses, counted in our current function by the index variable **guesses**. So we also want our **while** loop to iterate while **guesses < maxGuesses**. Since both of these conditions must be true for us to keep iterating, our desired **while** loop condition is:

```
while (myGuess != secretNumber) and (guesses < maxGuesses):
```

Because we are replacing our **for** loop with this **while** loop, we will now need to manage the index variable manually. We do this by initializing **guesses = 0** before the loop and incrementing **guesses** in the body of the loop. Here is the function with these changes:

```
def guessingGame(maxGuesses):
    """ (docstring omitted) """

    secretNumber = random.randrange(1, 101)
    myGuess = 0
    guesses = 0
    while (myGuess != secretNumber) and (guesses < maxGuesses):
        myGuess = input('Please guess my number: ')
        myGuess = int(myGuess)
        guesses = guesses + 1          # increment # of guesses
        if myGuess == secretNumber:    # win
            print('You got it!')
        else:                          # try again
            print('Nope.  Try again.')
```

Reflection 5.24 *Notice that we have also included* `myGuess = 0` *before the loop. Why do we bother to assign a value to* `myGuess` *before the loop? Is there anything special about the value 0? (Hint: try commenting it out.)*

If we comment out `myGuess = 0`, we will see the following error on the line containing the `while` loop:

`UnboundLocalError: local variable 'myGuess' referenced before assignment`

This error means that we have referred to an unknown variable named `myGuess`. The name is unknown to the Python interpreter because we had not defined it before it was first referenced in the `while` loop condition. Therefore, we need to initialize `myGuess` before the `while` loop, so that the condition makes sense the first time it is tested. To make sure the loop iterates at least once, we need to initialize `myGuess` to a value that cannot be the secret number. Since the secret number will be at least 1, 0 works for this purpose. This logic can be generalized as one of two rules to always keep in mind when designing an algorithm with a `while` loop:

1. Initialize the condition before the loop. Always make sure that the condition makes sense and will behave in the intended way the first time it is tested.

2. In each iteration of the loop, work toward the condition eventually becoming false. Not doing so will result in an infinite loop.

To ensure that our condition will eventually become false, we need to understand when this happens. For the `and` expression in the `while` loop to become false, either `(myGuess != secretNumber)` must be false or `(guesses < maxGuesses)` must be false.

Reflection 5.25 *How do the statements in the body of the loop ensure that eventually* `(myGuess != secretNumber)` *or* `(guesses < maxGuesses)` *will be false?*

Prompting for a new guess creates the opportunity for the first part to become false, while incrementing **guesses** ensures that the second part will eventually become false. Therefore, we cannot have an infinite loop.

This reasoning is enshrined in the first of *De Morgan's laws*, named after 19th century British mathematician Augustus De Morgan. De Morgan's two laws are:

1. **not** (a **and** b) = **not** a **or not** b

2. **not** (a **or** b) = **not** a **and not** b

You may recognize these from Exercises 1.4.10 and 1.4.11. The first law says that a **and** b is false if either a is false or b is false. The second law says that a **or** b is false if both a is false and b is false. Applied to our case, the first law tells us that the negation of our `while` loop condition is

```
(myGuess == secretNumber) or (guesses >= maxGuesses)
```

This is just a more formal way of deriving what we already concluded.

2. Friendly hints

Inside the loop, we currently handle two cases: (1) we win, and (2) we do not win but get another guess. To be friendlier, we should split the second case into two: (2a) our guess was too low, and (2b) our guess was too high. We can accomplish this by replacing the not-so-friendly `print('Nope. Try again.')` with another `if/else` that decides between the two new subcases:

```
if myGuess == secretNumber:                 # win
    print('You got it!')
else:                                       # try again
    if myGuess < secretNumber:              # too low
        print('Too low. Try again.')
    else:                                   # too high
        print('Too high. Try again.')
```

Now, if `myGuess == secretNumber` is false, we execute the first `else` clause, the body of which is the new `if/else` construct. If `myGuess < secretNumber` is true, we print that the number is too low; otherwise, we print that the number is too high.

Reflection 5.26 *Do you see a way in which the conditional construct above can be simplified?*

The conditional construct above is really just equivalent to a decision between three disjoint possibilities: (a) the guess is equal to the secret number, (b) the guess is less than the secret number, or (c) the guess is greater than the secret number. In other words, it is equivalent to:

```
    if myGuess == secretNumber:                # win
        print('You got it!')
    elif myGuess < secretNumber:                # too low
        print('Too low.  Try again.')
    else:                                       # too high
        print('Too high.  Try again.')
```

3. Printing a proper win/lose message

Now inside the loop, we currently handle three cases: (a) we win, and (b) our guess is too low and we get another guess, and (c) our guess is too high and we get another guess. But we are missing the case in which we run out of guesses. In this situation, we want to print something like `'Too bad. You lose.'` instead of one of the "try again" messages.

Reflection 5.27 *How can we augment the `if/elif/else` statement so that it correctly handles all four cases?*

In the game, if we incorrectly use our last guess, then two things must be true just before the `if` condition is tested: first, `myGuess` is not equal to `secretNumber` and, second, `guesses` is equal to `maxGuesses`. So we can incorporate this condition into the `if/elif/else`:

```
    if (myGuess != secretNumber) and (guesses == maxGuesses): # lose
        print('Too bad.  You lose.')
    elif myGuess == secretNumber:               # win
        print('You got it!')
    elif myGuess < secretNumber:                # too low
        print('Too low.  Try again.')
    else:                                       # too high
        print('Too high.  Try again.')
```

Notice that we have made the previous first `if` condition into an `elif` statement because we only want one of the four messages to be printed. However, here is an alternative structure that is more elegant:

```
    if myGuess == secretNumber:                 # win
        print('You got it!')
    elif guesses == maxGuesses:                 # lose
        print('Too bad.  You lose.')
    elif myGuess < secretNumber:                # too low
        print('Too low.  Try again.')
    else:                                       # too high
        print('Too high.  Try again.')
```

By placing the new condition second, we can leverage the fact that, if we get to the first `elif`, we already know that `myGuess != secretNumber`. Therefore, we do not need to include it explicitly.

There is a third way to handle this situation that is perhaps even more elegant. Notice that both of the first two conditions are going to happen at most once, and at the end of the program. So it might make more sense to put them *after* the loop. Doing so also exhibits a nice parallel between these two events and the two parts of the `while` loop condition. As we discussed earlier, the negation of the `while` loop condition is

```
(myGuess == secretNumber) or (guesses >= maxGuesses)
```

So when the loop ends, at least one of these two things is true. Notice that these two events are exactly the events that define a win or a loss: if the first part is true, then we won; if the second part is true, we lost. So we can move the win/loss statements after the loop, and decide which to print based on which part of the `while` loop condition became false:

```
if myGuess == secretNumber:           # win
    print('You got it!')
else:                                 # lose
    print('Too bad.  You lose.')
```

In the body of the loop, with these two cases gone, we will now need to check if we still get another guess (mirroring the `while` loop condition) before we print one of the "try again" messages:

```
if (myGuess != secretNumber) and (guesses < maxGuesses):
    if myGuess < secretNumber:            # too low
        print('Too low.  Try again.')
    else:                                 # too high
        print('Too high.  Try again.')
```

Reflection 5.28 *Why is it not correct to combine the two `if` statements above into a single statement like the following?*

```
if (myGuess != secretNumber) and (guesses < maxGuesses) \
                       and (myGuess < secretNumber):
    print('Too low.  Try again.')
else:
    print('Too high.  Try again.')
```

Hint: what does the function print when `guesses < maxGuesses` *is false and* `myGuess < secretNumber` *is true?*

These changes are incorporated into the final game that is shown below.

```
import random

def guessingGame(maxGuesses):
    """ (docstring omitted) """

    secretNumber = random.randrange(1, 101)
    myGuess = 0
    guesses = 0
    while (myGuess != secretNumber) and (guesses < maxGuesses):
        myGuess = input('Please guess my number: ')
        myGuess = int(myGuess)
        guesses = guesses + 1        # increment # of guesses used

        if (myGuess != secretNumber) and (guesses < maxGuesses):
            if myGuess < secretNumber:          # guess is too low
                print('Too low.  Try again.')   #    give a hint
            else:                               # guess is too high
                print('Too high.  Try again.')  #    give a hint

    if myGuess == secretNumber:      # win
        print('You got it!')
    else:                            # lose
        print('Too bad.  You lose.')

def main():
    guessingGame(10)

main()
```

As you play the game, think about what the best strategy is. How many guesses do different strategies require? Exercise 5.5.7 asks you to write a Monte Carlo simulation to compare three different strategies for playing the game.

Exercises

5.5.1. Write a function

```
ABC()
```

that prompts for a choice of A, B, or C and uses a `while` loop to keep prompting until it receives the string 'A', 'B', or 'C'.

5.5.2. Write a function

```
numberPlease()
```

that prompts for an integer between 1 and 100 (inclusive) and uses a `while` loop to keep prompting until it receives a number within this range.

5.5.3. Write a function

> `differentNumbers()`

that prompts for two different numbers. The function should use a `while` loop to keep prompting for a pair of numbers until the two numbers are different.

5.5.4. Write a function

> `rockPaperScissorsLizardSpock(player1, player2)`

that decides who wins in a game of rock-paper-scissors-lizard-Spock[1]. Each of `player1` and `player2` is a string with value 'rock', 'paper', 'scissors', 'lizard', or 'Spock'. The function should return the value 1 if player 1 wins, -1 if player 2 wins, or 0 if they tie. Test your function by playing the game with the following main program:

```
def main():
    player1 = input('Player1: ')
    player2 = input('Player2: ')

    outcome = rockPaperScissorsLizardSpock(player1, player2)

    if outcome == 1:
        print('Player 1 wins!')
    elif outcome == -1:
        print('Player 2 wins!')
    else:
        print('Player 1 and player 2 tied.')
```

5.5.5. Write a function

> `yearsUntilDoubled(amount, rate)`

that returns the number of years until `amount` is doubled when it earns the given rate of interest, compounded annually. Use a `while` loop.

5.5.6. The *hailstone numbers* are a sequence of numbers generated by the following simple algorithm. First, choose any positive integer. Then, repeatedly follow this rule: if the current number is even, divide it by two; otherwise, multiply it by three and add one. For example, suppose we choose the initial integer to be 3. Then this algorithm produces the following sequence:

$$3, 10, 5, 16, 8, 4, 2, 1, 4, 2, 1, 4, 2, 1 \ldots$$

For every initial integer ever tried, the sequence always reaches one and then repeats the sequence 4, 2, 1 forever after. Interestingly, however, no one has ever *proven* that this pattern holds for every integer!

Write a function

> `hailstone(start)`

---
[1] See http://en.wikipedia.org/wiki/Rock-paper-scissors-lizard-Spock for the rules.

that prints the hailstone number sequence starting from the parameter `start`, until the value reaches one. Your function should return the number of integers in your sequence. For example, if `n` were 3, the function should return 8. (Use a `while` loop.)

5.5.7. In this exercise, we will design a Monte Carlo simulation to compare the effectiveness of three strategies for playing the guessing game. Each of these strategies will be incorporated into the guessing game function we designed in this chapter, but instead of checking whether the player wins or loses, the function will continue until the number is guessed, and then return the number of guesses used. We will also make the maximum possible secret number a parameter, so that we can compare the results for different ranges of secret numbers.

The first strategy is to just make a random guess each time, ignoring any previous guesses:

```python
def guessingGame1(maxNumber):
    """Play the guessing game by making random guesses."""

    secretNumber = random.randrange(1, maxNumber + 1)
    myGuess = 0
    guesses = 0
    while (myGuess != secretNumber):
        myGuess = random.randrange(1, maxNumber + 1)
        guesses = guesses + 1

    return guesses
```

The second strategy is to incrementally try every possible guess from 1 to 100, thereby avoiding any duplicates:

```python
def guessingGame2(maxNumber):
    """Play the guessing game by making incremental guesses."""

    secretNumber = random.randrange(1, maxNumber + 1)
    myGuess = 0
    guesses = 0
    while (myGuess != secretNumber):
        myGuess = myGuess + 1
        guesses = guesses + 1

    return guesses
```

Finally, the third strategy uses the outcomes of previous guesses to narrow in on the secret number:

```
def guessingGame3(maxNumber):
    """Play the guessing game intelligently by narrowing in
       on the secret number."""

    secretNumber = random.randrange(1, maxNumber + 1)
    myGuess = 0
    low = 1
    high = maxNumber
    guesses = 0
    while (myGuess != secretNumber):
        myGuess = (low + high) // 2
        guesses = guesses + 1

        if myGuess < secretNumber:
            low = myGuess + 1
        elif myGuess > secretNumber:
            high = myGuess - 1

    return guesses
```

Write a Monte Carlo simulation to compare the expected (i.e., average) behavior of these three strategies. Use a sufficiently high number of trials to get consistent results. Similarly to what we did in Section 5.1, run your simulation for a range of maximum secret numbers, specifically $5, 10, 15, \ldots, 100$, and plot the average number of guesses required by each strategy for each maximum secret number. (The x-axis of your plot will be the maximum secret number and the y-axis will be the average number of guesses.) Explain the results. In general, how many guesses on average do you think each strategy requires to guess a secret number between 1 and n?

5.6 SUMMARY

In previous chapters, we designed deterministic algorithms that did the same thing every time we executed them, if we gave them the same inputs (i.e., arguments). Giving those algorithms different arguments, of course, could change their behavior, whether it be drawing a different size shape, modeling a different population, or experimenting with a different investment scenario. In this chapter, we started to investigate a new class of algorithms that can change their behavior "on the fly," so to speak. These algorithms all make choices using *Boolean logic*, the same Boolean logic on which computers are fundamentally based. By combining *comparison operators* and *Boolean operators*, we can characterize just about any decision. By incorporating these Boolean expressions into *conditional statements* (`if/elif/else`) and *conditional loops* (`while`), we vastly increase the diversity of algorithms that we can design. These are fundamental techniques that we will continue to use and develop over the next several chapters, as we start to work with textual and numerical data that we read from files and download from the web.

5.7 FURTHER DISCOVERY

This chapter's epigraph is a famous "Yogiism," from Hall of Fame catcher, coach, and manager Yogi Berra [6].

If you would like to learn more about Robert Brown's experiments, and the history and science behind them, visit the following web site, titled "What Brown Saw and You Can Too."

http://physerver.hamilton.edu/Research/Brownian/index.html

The Drunkard's Walk by Leonard Mlodinow [34] is a very accessible book about how randomness and chance affect our lives. For more information about generating random numbers, and the differences between PRNGs and true random number generators, visit

https://www.random.org/randomness/ .

The Park-Miller random number generator is due to Keith Miller and the late Steve Park [37].

The Roper Center for Public Opinion Research, at the University of Connecticut, maintains some helpful educational resources about random sampling and errors in the context of public opinion polling at

http://www.ropercenter.uconn.edu/education.html .

5.8 PROJECTS

Project 5.1 The magic of polling

> According to the latest poll, the president's job approval rating is at 45%, with a margin of error of ±3%, based on interviews with approximately 1,500 adults over the weekend.

We see news headlines like this all the time. But how can a poll of 1,500 randomly chosen people claim to represent the opinions of millions in the general population? How can the pollsters be so certain of the margin of error? In this project, we will investigate how well random sampling can really estimate the characteristics of a larger population. We will assume that we know the true percentage of the overall population with some characteristic or belief, and then investigate how accurate a much smaller poll is likely to get.

Suppose we know that 30% of the national population agrees with the statement, "Animals should be afforded the same rights as human beings." Intuitively, this means that, if we randomly sample ten individuals from this population, we should, on average, find that three of them agree with the statement and seven do not. But does it follow that every poll of ten randomly chosen people will mirror the percentage of the larger population? Unlike a Monte Carlo simulation, a poll is taken just once (or maybe twice) at any particular point in time. To have confidence in the poll results, we need some assurance that the results would not be drastically different if the poll had queried a different group of randomly chosen individuals. For example, suppose you polled ten people and found that two agreed with the statement, then polled ten more people and found that seven agreed, and then polled ten more people and found that all ten agreed. What would you conclude? There is too much variation for this poll to be credible. But what if we polled more than ten people? Does the variation, and hence the trustworthiness, improve?

In this project, you will write a program to investigate questions such as these and determine empirically how large a sample needs to be to reliably represent the sentiments of a large population.

1. Simulate a poll

In conducting this poll, the pollster asks each randomly selected individual whether he or she agrees with the statement. We know that 30% of the population does, so there is a 30% chance that each individual answers "yes." To simulate this polling process, we can iterate over the number of individuals being polled and count them as a "yes" with probability 0.3. The final count at the end of the loop, divided by the number of polled individuals, gives us the poll result. Implement this simulation by writing a function

```
poll(percentage, pollSize)
```

that simulates the polling of `pollSize` individuals from a large population in which the given `percentage` (between 0 and 100) will respond "yes." The function should return the percentage (between 0 and 100) of the poll that actually responded "yes." Remember that the result will be different every time the function is called. Test your function with a variety of poll sizes.

2. Find the polling extremes

To investigate how much variation there can be in a poll of a particular size, write a function

```
pollExtremes(percentage, pollSize, trials)
```

that builds a list of `trials` poll results by calling `poll(percentage, pollSize)` `trials` times. The function should return the minimum and maximum percentages in this list. For example, if five trials give the percentages [28, 35, 31, 24, 31], the function should return the minimum 24 and maximum 35. If the list of poll results is named `pollResults`, you can return these two values with

```
return min(pollResults), max(pollResults)
```

Test your function with a variety of poll sizes and numbers of trials.

3. What is a sufficient poll size?

Next, we want to use your previous functions to investigate how increasing poll sizes affect the variation of the poll results. Intuitively, the more people you poll, the more accurate the results should be. Write a function

```
plotResults(percentage, minPollSize, maxPollSize, step, trials)
```

that plots the minimum and maximum percentages returned by calling the function `pollExtremes(percentage, pollSize, trials)` for values of `pollSize` ranging from `minPollSize` to `maxPollSize`, in increments of `step`. For each poll size, call your `pollExtremes` function with

```
low, high = pollExtremes(percentage, pollSize, trials)
```

and then append the values of `low` and `high` each to its own list for the plot. Your function should return the margin of error for the largest poll, defined to be (`high` - `low`) / 2. The poll size should be on the x-axis of your plot and the percentage should be on the y-axis. Be sure to label both axes.

Question 5.1.1 *Assuming that you want to balance a low margin of error with the labor involved in polling more people, what is a reasonable poll size? What margin of error does this poll size give?*

Write a `main` function (if you have not already) that calls your `plotResults` function to investigate an answer to this question. You might start by calling it with `plotResults(30, 10, 1000, 10, 100)`.

4. Does the error depend on the actual percentage?

To investigate this question, write another function

```
plotErrors(pollSize, minPercentage, maxPercentage, step, trials)
```

that plots the margin of error in a poll of `pollSize` individuals, for actual percentages ranging from `minPercentage` to `maxPercentage`, in increments of `step`. To find the margin of error for each poll, call the `pollExtremes` function as above, and compute (`high` - `low`) / 2. In your plot, the percentage should be on the x-axis and the margin of error should be on the y-axis. Be sure to label both axes.

Question 5.1.2 *Does your answer to the previous part change if the actual percentage of the population is very low or very high?*

You might start to investigate this question by calling the function with `plotErrors(1500, 10, 80, 1, 100)`.

Project 5.2 Escape!

In some scenarios, movement of the "particle" in a random walk is restricted to a bounded region. But what if there is a small opening through which the particle might escape or disappear? How many steps on average does it take the particle to randomly come across this opening and escape? This model, which has become known as the *narrow escape problem*, could represent a forager running across a predator on the edge its territory, an animal finding an unsecured gate in a quarantined area, a molecule finding its way through a pore in the cell membrane, or air molecules in a hot room escaping through an open door.

1. Simulate the narrow escape

Write a function

    escape(openingDegrees, tortoise, draw)

that simulates the narrow escape problem in a circle with radius 1 and an opening of `openingDegrees` degrees. In the circle, the opening will be between 360 − `openingDegrees` and 360 degrees, as illustrated below.

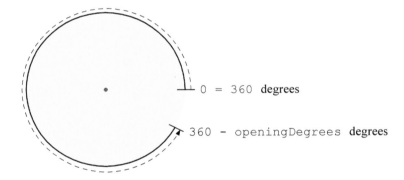

The particle should follow a *normally distributed* random walk, as described in Exercise 5.3.1. The standard deviation of the normal distribution needs to be quite small for the particle to be able to find small openings. A value of $\pi/128$ is suggested in [7]. Since we are interested in the number of steps taken by the particle (instead of the distance traveled, as before), the number of steps will need to be incremented in each iteration of the loop. When the particle hits a wall, it should "bounce" back to its previous position. Since the particle is moving within a circle, we can tell if it hits a wall by comparing its distance from the origin to the radius of the circle. If the particle moves out to the distance of the wall, but is within the angle of the opening, the loop should end, signaling the particle's escape.

Finding the current angle of the particle with respect to the origin requires some trigonometry. Since we know the x and y coordinates of the particle, the angle can be found by computing the arctangent of y/x: $\tan^{-1}(y/x)$. However, this will cause a problem with $x = 0$, so we need to check for that possibility and fudge the value of x a bit. Also, the Python arctangent ($\tan^{-1}$) function, `math.atan`, always returns an angle between $-\pi/2$ and $\pi/2$ radians (between −90 and 90 degrees), so the result needs to be adjusted to be between 0 and 360 degrees. The following function handles this for you.

```
def angle(x, y):
    if x == 0:          # avoid dividing by zero
        x = 0.001
    angle = math.degrees(math.atan(y / x))
    if angle < 0:
        if y < 0:
            angle = angle + 360    # quadrant IV
        else:
            angle = angle + 180    # quadrant II
    elif y < 0:
        angle = angle + 180        # quadrant III
    return angle
```

Below you will find a "skeleton" of the `escape` function with the loop and drawing code already written. Drawing the partial circle is handled by the function `setupWalls` below. Notice that the function uses a `while` loop with a Boolean *flag variable* named `escaped` controlling the iteration. The value of `escaped` is initially `False`, and your algorithm should set it to `True` when the particle escapes. Most, *but not all*, of the remaining code is needed in the `while` loop.

```
def escape(openingDegrees, tortoise, draw):
    x = y = 0                       # initialize (x, y) = (0, 0)
    radius = 1                      # moving in unit radius circle
    stepLength = math.pi / 128      # std dev of each step

    if draw:
        scale = 300                 # scale up drawing
        setupWalls(tortoise, openingDegrees, scale, radius)

    escaped = False                 # has particle escaped yet?
    while not escaped:

        # random walk and detect wall and opening here

        if draw:
            tortoise.goto(x * scale, y * scale)   # move particle

    if draw:
        screen = tortoise.getscreen()   # update screen to compensate
        screen.update()                 #   for high tracer value
```

```
def setupWalls(tortoise, openingDegrees, scale, radius):
    screen = tortoise.getscreen()
    screen.mode('logo')                 # east is 0 degrees
    screen.tracer(5)                    # speed up drawing

    tortoise.up()                       # draw boundary with
    tortoise.width(0.015 * scale)       #    shaded background
    tortoise.goto(radius * scale, 0)
    tortoise.down()
    tortoise.pencolor('lightyellow')
    tortoise.fillcolor('lightyellow')
    tortoise.begin_fill()
    tortoise.circle(radius * scale)
    tortoise.end_fill()
    tortoise.pencolor('black')
    tortoise.circle(radius * scale, 360 - openingDegrees)
    tortoise.up()
    tortoise.home()

    tortoise.pencolor('blue')           # particle is a blue circle
    tortoise.fillcolor('blue')
    tortoise.shape('circle')
    tortoise.shapesize(0.75, 0.75)

    tortoise.width(1)                   # set up for walk
    tortoise.pencolor('green')
    tortoise.speed(0)
    tortoise.down()                     # comment this out to hide trail
```

2. Write a Monte Carlo simulation

Write a function

   escapeMonteCarlo(openingDegrees, trials)

that returns the average number of steps required, over the given number of trials, to escape with an opening of openingDegrees degrees. This is very similar to the rwMonteCarlo function from Section 5.1.

3. Empirically derive the function

Write a function

   plotEscapeSteps(minOpening, maxOpening, openingStep, trials)

that plots the average number of steps required, over the given number of trials, to escape openings with widths ranging from minOpening to maxOpening degrees, in increments of openingStep. (The x-axis values in your plot are the opening widths and y-axis values are the average number of steps required to escape.) This is very similar to the plotDistances function from Section 5.1.

Plot the average numbers of steps for openings ranging from 10 to 180 degrees, in 10-degree steps, using at least 1,000 trials to get a smooth curve. As this number of trials will take a few minutes to complete, start with fewer trials to make sure your simulation is working properly.

In his undergraduate thesis at the University of Pittsburgh, Carey Caginalp [7] mathematically derived a function that describes these results. In particular, he proved that the expected time required by a particle to escape an opening width of α degrees is

$$T(\alpha) = \frac{1}{2} - 2\ln\left(\sin\frac{\alpha}{4}\right).$$

Plot this function in the same graph as your empirical results. You will notice that the $T(\alpha)$ curve is considerably below the results from your simulation, which has to do with the step size that we used (i.e., the value of stepLength in the escape function). To adjust for this step size, multiply each value returned by the escapeMonteCarlo function by the square of the step size $((\pi/128)^2)$ before you plot it. Once you do this, the results from your Monte Carlo simulation will be in the same *time* units used by Caginalp, and should line up closely with the mathematically derived result.

CHAPTER 6

Text, documents, and DNA

> So, here's what I can say: the Library of Congress has more than 3 petabytes of digital collections. What else I can say with all certainty is that by the time you read this, all the numbers — counts and amount of storage — will have changed.
>
> Leslie Johnston, former Chief of Repository Development, Library of Congress
Blog post (2012)

> The roughly 2000 sequencing instruments in labs and hospitals around the world can collectively sequence 15 quadrillion nucleotides per year, which equals about 15 petabytes of compressed genetic data. A petabyte is 2^{50} bytes, or in round numbers, 1000 terabytes. To put this into perspective, if you were to write this data onto standard DVDs, the resulting stack would be more than 2 miles tall. And with sequencing capacity increasing at a rate of around three- to fivefold per year, next year the stack would be around 6 to 10 miles tall. At this rate, within the next five years the stack of DVDs could reach higher than the orbit of the International Space Station.
>
> Michael C. Schatz and Ben Langmead
The DNA Data Deluge (2013)

NEW technologies are producing data at an incredible rate, sometimes faster than it can be stored and analyzed. By some accounts, Google plans to have scanned every book in existence by the end of this decade. Sequencing of the approximately three billion base pairs of the human genome took 10 years when first published in 2000, but can now be done in less than a week. Social scientists currently have at their disposal a steady stream of data from online social networks. Analyses of data of these magnitudes, most of which is stored as text, requires computational tools. In this chapter, we will discuss the fundamental algorithmic techniques for developing these tools. We will look at how text is represented in a computer, how to read text from both files and the web, and develop algorithms to process and analyze textual data.

6.1 COUNTING WORDS

As we have already seen, we can store text as a sequence of characters called a *string*. A string constant (also called a *string literal*), such as `'You got it!'`, is enclosed in either single quotes (`'`) or double quotes (`"`). We have also seen that string values can be assigned to variable names, passed as parameters, and compared to each other just like numbers, as illustrated in the following silly functions.

```
def roar(n):
    return 'r' + ('o' * n) + ('a' * n) + 'r!'

def speak(animal):
    if animal == 'cat':
        word = 'meow.'
    elif animal == 'dog':
        word = 'woof.'
    else:
        word = roar(10)

    print('The', animal, 'says', word)

speak('monkey')
```

Reflection 6.1 *What is printed by the code above? Make sure you understand why.*

Like a `Turtle`, a string is an object, an instance of a class named `str`. As such, a string is also another example of an abstract data type. Recall that an abstract data type hides the implementation details of how its data is stored, allowing a programmer to interact with it through a set of functions called methods. As we will discuss in Section 6.3, strings are actually stored as binary sequences, yet we can interact with them in a very natural way.

There are several methods available for strings.[1] For example, the `upper()` method returns a new string with all the lower case characters in the old string capitalized. For example,

```
>>> pirate = 'Longbeard'
>>> newName = pirate.upper()
>>> newName
'LONGBEARD'
```

Remember that, when using methods, the name of the object, in this case the string object named `pirate`, precedes the name of the method.

Smaller strings contained within larger strings are called *substrings*. We can use the operators `in` and `not in` to test whether a string contains a particular substring. For example, the conditions in each of the following `if` statements are true.

---

[1] See Appendix B.6 for a list.

```
>>> if 'bear' in pirate:
...     print('There is a bear in Longbeard.')
...
There is a bear in Longbeard.
>>> if 'beer' not in pirate:
...     print('There is no beer in Longbeard.')
...
There is no beer in Longbeard.
```

Two particularly useful methods that deal with substrings are **replace** and **count**. The **replace** method returns a new string with all occurrences of one substring replaced with another substring. For example,

```
>>> newName = pirate.replace('Long', 'Short')
>>> newName
'Shortbeard'
>>> quote = 'Yo-ho-ho, and a bottle of rum!'
>>> quote2 = quote.replace(' ', '')
>>> quote2
'Yo-ho-ho,andabottleofrum!'
```

The second example uses the **replace** method to delete spaces from a string. The second parameter to **replace**, quotes with nothing in between, is called the *empty string*, and is a perfectly legal string containing zero characters.

Reflection 6.2 *Write a statement to replace all occurrences of* brb *in a text with* be right back.

```
>>> txt = 'Hold on brb'
>>> txt.replace('brb', 'be right back')
'Hold on be right back'
```

The **count** method returns the number of occurrences of a substring in a string. For example,

```
>>> pirate.count('ear')
1
>>> pirate.count('eye')
0
```

We can count the number of words in a text by counting the number of "whitespace" characters, since almost every word must be followed by a space of some kind. Whitespace characters are spaces, tabs, and newline characters, which are the hidden characters that mark the end of a line of text. Tab and newline characters are examples of non-printable *control characters*. A tab character is denoted by the two characters \t and a newline character is denoted by the two characters \n. The **wordCount1** function below uses the **count** method to return the number of whitespace characters in a string named **text**.

```python
def wordCount1(text):
    """Approximate the number of words in a string by counting
        the number of spaces, tabs, and newlines.

    Parameter:
        text: a string object

    Return value: the number of spaces, tabs and newlines in text
    """

    return text.count(' ') + text.count('\t') + text.count('\n')

def main():
    shortText = 'This is not long.  But it will do. \n' + \
                'All we need is a few sentences.'
    wc = wordCount1(shortText)
    print(wc)

main()
```

In the string `shortText` above, we explicitly show the space characters (␣) and break it into two parts because it does not fit on one line.

Reflection 6.3 *In a Python shell, try entering*

```
print('Two tabs\t\there.  Two newlines\n\nhere.')
```

to see the effects of the `\t` *and* `\n` *characters.*

Reflection 6.4 *What answer does the* `wordCount1` *function give in the example above? Is this the correct answer? If not, why not?*

The `wordCount1` function returns 16, but there are actually only 15 words in `shortText`. We did not get the correct answer because there are two spaces after the first period, a space and newline character after the second period, and nothing after the last period. How can we fix this? If we knew that the text contained no newline characters, and was perfectly formatted with one space between words, two spaces between sentences, and no space at the end, then we could correct the above computation by subtracting the number of instances of two spaces in the text and adding one for the last word:

```python
def wordCount2(text):
    """ (docstring omitted) """

    return text.count(' ') - text.count('  ') + 1
```

But most text is not perfectly formatted, so we need a more sophisticated approach.

Reflection 6.5 *If you were faced with a long text, and could only examine one character at a time, what algorithm would you use to count the number of words?*

As a starting point, let's think about how the `count` method must work. To count the number of instances of some character in a string, the `count` method must examine each character in the string, and increment an accumulator variable for each one that matches.

To replicate this behavior, we can iterate over a string with a `for` loop, just as we have iterated over `ranges` of integers. For example, the following `for` loop iterates over the characters in the string `shortText`. Insert this loop into the `main` function above to see what it does.

```
for character in shortText:
    print(character)
```

In this loop, each character in the string is assigned to the index variable `character` in order. To illustrate this, we have the body of the loop simply print the character assigned to the index variable. Given the value assigned to `shortText` in the `main` function above, this loop will print

```
T
h
i
⋮    (middle omitted)
e
s
.
```

To count whitespace characters, we can use the same loop but, in the body of the loop, we need to increment a counter each time the value of `character` is equal to a whitespace character:

```
def wordCount3(text):
    """ (docstring omitted) """

    count = 0
    for character in text:
        if character == ' ' or character == '\t' or character == '\n':
            count = count + 1
    return count
```

Reflection 6.6 *What happens if we replace the* or *operators with* and *operators?*

Alternatively, this `if` condition can be simplified with the `in` operator:

```
if character in ' \t\n':
```

Reflection 6.7 *What answer does the* `wordCount3` *function give when it is called with the parameter* `shortText` *that we defined above?*

When we call this function with `shortText`, we get the same answer that we did with `wordCount1` (16) because we have simply replicated that function's behavior.

Reflection 6.8 *Now that we have a baseline word count function, how can we make it more accurate on text with extra spaces?*

To improve upon `wordCount3`, we need to count only the *first* whitespace or newline character in any sequence of such characters.

Reflection 6.9 *If the value of* `character` *is a space, tab, or newline, how can we tell if it is the* first *in a sequence of such characters? (Hint: if it is the first, what must the previous character not be?)*

If the value of `character` is a whitespace or newline character, we know it is the first in a sequence if the previous character was *not* one of these characters. So we need to also keep track of the previous character and check its value in the `if` statement. If we use a variable named `prevCharacter` to store this value, we would need to change the `if` statement to

```
if character in ' \t\n' and prevCharacter not in ' \t\n':
    count = count + 1
```

Reflection 6.10 *Where should we assign* `prevCharacter` *a value?*

We need to assign the current value of `character` to `prevCharacter` at the end of each iteration because this current value of `character` will be the previous character during the next iteration. With these pieces added, our revised `wordCount4` looks like this:

```
def wordCount4(text):
    """ (docstring omitted) """

    count = 0
    prevCharacter = ' '
    for character in text:
        if character in ' \t\n' and prevCharacter not in ' \t\n':
            count = count + 1
        prevCharacter = character
    return count
```

The one remaining piece, included above, is to initialize `prevCharacter` to something before the loop so that the `if` condition makes sense in the first iteration.

Reflection 6.11 *What happens if we do not initialize* `prevCharacter` *before the loop? Why did we initialize* `prevCharacter` *to a space? Does its initial value matter?*

Let's consider two possibilities for the first character in `text`: either it is a whitespace character, or it is not. If the first character in `text` is not a whitespace or newline character (as would normally be the case), then the first part of the `if` condition (`character in ' \t\n'`) will be false. Therefore, the initial value of `prevCharacter` does not matter. On the other hand, if the first value assigned to `character` *is* a whitespace or newline character, then the first part of the `if` condition will be true. But we want to make sure that this character does *not* count as ending a word. Setting `prevCharacter` to a space initially will ensure that the second part of the `if` condition (`prevCharacter not in ' \t\n'`) is initially false, and prevent `count` from being incremented.

Reflection 6.12 *Are there any other values that would work for the initial value of* `prevCharacter`*?*

Finally, we need to deal with the situation in which the text does not end with a whitespace character. In this case, the final word would not have been counted, so we need to increment `count` by one.

Reflection 6.13 *How can we tell if the last character in* `text` *is a whitespace or newline character?*

Since `prevCharacter` will be assigned the last character in `text` after the loop completes, we can check its value after the loop. If it is a whitespace character, then the last word has already been counted; otherwise, we need to increment `count`. So the final function looks like this:

```
def wordCount5(text):
    """Count the number of words in a string.

    Parameter:
        text: a string object

    Return value: the number of words in text
    """

    count = 0
    prevCharacter = ' '
    for character in text:
        if character in ' \t\n' and prevCharacter not in ' \t\n':
            count = count + 1
        prevCharacter = character
    if prevCharacter not in ' \t\n':
        count = count + 1
    return count
```

Although our examples have used short strings, our function will work on any size string we want. In the next section, we will see how to read an entire text file or web page into a string, and then use our word count function, unchanged, to count the words in the file. In later sections, we will design similar algorithms to carry out more sophisticated analyses of text files containing entire books and long sequences of DNA.

Exercises

6.1.1. Write a function

`twice(text)`

that returns the string `text` repeated twice, with a space in between. For example, `twice('bah')` should return the string `'bah bah'`.

6.1.2. Write a function

`repeat(text, n)`

that returns a string that is `n` copies of the string `text`. For example, `repeat('AB', 3)` should return the string `'ABABAB'`.

6.1.3. Write a function

`vowels(word)`

that uses the `count` method to return the number of vowels in the string `word`. (Note that `word` may contain upper and lower case letters.)

6.1.4. Write a function

`nospaces(sentence)`

that uses the `replace` string method to return a version of the string `sentence` in which all the spaces have been replaced by the underscore (_) character.

6.1.5. Write a function

`txtHelp(txt)`

that returns an expanded version of the string `txt`, which may contain texting abbreviations like "brb" and "lol." Your function should expand at least four different texting abbreviations. For example, `txtHelp('imo u r lol brb')` might return the string `'in my opinion you are laugh out loud be right back'`.

6.1.6. Write a function

`letters(text)`

that prints the characters of the string `text`, one per line. For example `letters('abc')` should print

    a
    b
    c

6.1.7. Write a function

   count(text, letter)

   that returns the number of occurrences of the one-character string named letter in the string text, *without* using the count method. (Use a for loop instead.)

6.1.8. Write a function

   vowels(word)

   that returns the same result as Exercise 6.1.3 *without* using the count method. (Use a for loop instead.)

6.1.9. Write a function

   sentences(text)

   that returns the number of sentences in the string text. A sentence may end with a period, question mark, or exclamation point. It may also end with multiple punctuation marks, such as '!!!' or '?!'.

6.1.10. Write a function

   nospaces(sentence)

   that returns the same result as Exercise 6.1.4 *without* using the replace method. (Hint: use a for loop and the string concatenation operator (+) inside the loop to build up a new string. We will see more examples like this in Section 6.3.)

6.1.11. When bits are sent across a network, they are sometimes corrupted by interference or errors. Adding some redundant bits to the end of the transmission can help to detect these errors. The simplest error detection algorithm is known as *parity checking*. A bit string has *even parity* if it has an even number of ones, and *odd parity* otherwise. In an even parity scheme, the sender adds a single bit to the end of the bit string so that the final bit string has an even number of ones. For example, if we wished to send the data 1101011, would actually send 11010111 instead so that the bit string has an even number of ones. If we wished to send 1101001 instead, we would actually send 11010010. The receiver then checks whether the received bit string has even parity; if it does not, then an error must have occurred so it requests a retransmission.

   (a) Parity can only detect very simple errors. Give an example of an error that cannot be detected by an even parity scheme.

   (b) Propose a solution that would detect the example error you gave above.

   (c) In the next two problems, we will pretend that bits are sent as strings (they are not; this would be terribly inefficient). Write a function

      evenParity(bits)

      that uses the count method to return True if the string bits has even parity and False otherwise. For example, evenParity('110101') should return True and evenParity('110001') should return False.

   (d) Now write the evenParity function without using the count method.

   (e) Write a function

```
            makeEvenParity(bits)
```
that returns a string consisting of `bits` with one additional bit concatenated so that the returned string has even parity. Your function should call your `evenParity` function. For example, `makeEvenParity('110101')` should return `'1101010'` and `makeEvenParity('110001')` should return `'1100011'`.

6.2 TEXT DOCUMENTS

In real applications, we read text from *files* stored on a hard drive or a flash drive. Like everything else in a computer system, files are sequences of bits. But we interact with files as abstractions, electronic documents containing information such as text, spreadsheets, or images. These abstractions are mediated by a part of the operating system called a *file system*. The file system organizes files in folders in a hierarchical fashion, such as in the simplified view of a Mac OS X file system in Figure 6.1. Hierarchical structures like this are called *trees*. The top of the tree is called the *root* and branches are called *children*. Elements of the tree with no children are called *leaves*. In a file system, a folder can have children, consisting of files or other folders residing inside it, whereas files are always leaves.

Below the root in this figure is a folder named `Users` where every user of the computer has a *home folder* labeled with his or her name, say `george`. In the picture above, this home folder contains two subfolders named `CS 111` and `HIST 216`. We can represent the location of a file with the *path* one must follow to get there from the root. For example, the path to the file `dna.py` is `/Users/george/CS 111/dna.py`. The initial slash character represents the root, whereas the others simply divide layers. Any path without the first slash character is considered to be a *relative path*, relative to the current *working directory* set by the operating system. For example, if the current working directory were `/Users/george`, then `dna.py` could be specified with the relative path `CS 111/dna.py`.

Reading from text files

In Python, a file is represented by an abstraction called a *file object*. We associate a file with a file object by using the `open` function, which takes a file name or a path as an argument and returns a file object. For example, the following statement associates the file object named `inputFile` with the file named `mobydick.txt` in the current working directory.[2]

```
    inputFile = open('mobydick.txt', 'r')
```

The second argument to `open` is the *mode* to use when working with the file; `'r'` means that we want to read from the file. By default, every file is assumed to contain text.

---

[2]This file can be obtained from the book's website or from Project Gutenberg at `http://www.gutenberg.org/files/2701/2701.txt`.

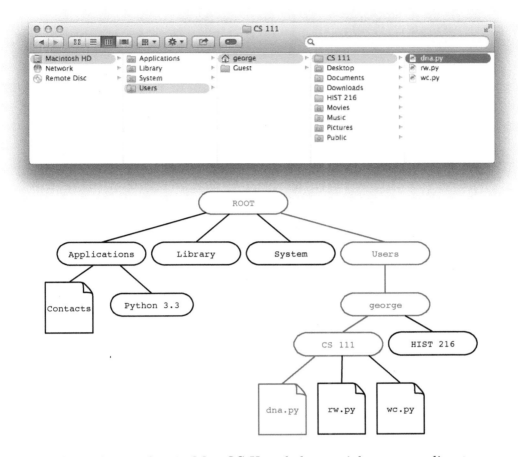

Figure 6.1 A Finder window in Mac OS X and the partial corresponding tree representation. Ovals represent folders and rectangles represent files.

The **read** method of a file object reads the entire contents of a file into a string. For example, the following statement would read the entire text from the file into a string assigned to the variable named **text**.

```
text = inputFile.read()
```

When we are finished with a file, it is important that we close it. Closing a file signals to the operating system that we are done using it, and ensures that any memory allocated to the file by the program is released. To close a file, we simply use the file object's **close** method:

```
inputFile.close()
```

Let's look at an example that puts all of this together. The following function reads in a file with the given file name and returns the number of words in the file using our **wordCount5** function.

```
def wcFile(fileName):
    """Return the number of words in the file with the given name.

    Parameter:
        fileName: the name of a text file

    Return value: the number of words in the file
    """

    textFile = open(fileName, 'r', encoding = 'utf-8')
    text = textFile.read()
    textFile.close()

    return wordCount5(text)
```

The optional `encoding` parameter to the `open` function indicates how the bits in the file should be interpreted (we will discuss what UTF-8 is in Section 6.3).

Reflection 6.14 *How many words are there in the file* `mobydick.txt`*?*

Now suppose we want to print a text file, formatted with line numbers to the left of each line. A "line" is defined to be a sequence of characters that end with a newline ('\n') character. Since we need to print a line number to the left of every line, this problem would be much more easily solved if we could read in a file one line at a time. Fortunately, we can. In the same way that we can iterate over a range of integers or the characters of a string, we can iterate over the lines in a file. When we use a file object as the sequence in a `for` loop, the index variable is assigned a string containing each line in the file, one line per iteration. For example, the following code prints each line in the file object named `textFile`:

```
for line in textFile:
    print(line)
```

In each iteration of this loop, the index variable named `line` is assigned the next line in the file, which is then printed in the body of the loop. We can easily extend this idea to our line-printing problem:

```
def lineNumbers(fileName):
    """Print the contents of the file with the given name
       with each line preceded by a line number.

    Parameter:
        fileName: the name of a text file

    Return value: None
    """

    textFile = open(fileName, 'r', encoding = 'utf-8')
    lineCount = 1
    for line in textFile:
        print('{0:<5} {1}'.format(lineCount, line[:-1]))
        lineCount = lineCount + 1
    textFile.close()
```

The `lineNumbers` function combines an accumulator with a `for` loop that reads the text file line by line. After the file is opened, the accumulator variable `lineCount` is initialized to one. Inside the loop, each line is printed using a format string that precedes the line with the current value of `lineCount`. At the end of the loop, the accumulator is incremented and the loop repeats.

Reflection 6.15 *Why does the function print* `line[:-1]` *instead of* `line` *in each iteration? What happens if you replace* `line[:-1]` *with* `line`*?*

Reflection 6.16 *How would the output change if* `lineCount` *was instead incremented in the loop before calling* `print`*?*

Reflection 6.17 *How many lines are there in the file* `mobydick.txt`*?*

Writing to text files

We can also create new files or write to existing ones. For example, instead of printing a file formatted with line numbers, we might want to create a new version of the file that includes line numbers. To write text to a new file, we first have to open it using `'w'` ("write") mode:

```
newTextFile = open('newfile.txt', 'w')
```

Opening a file in this way will create a new file named `newfile.txt`, if a file by that name does not exist, or overwrite the file by that name if it does exist. (So be careful!) To append to the end of an existing file, use the `'a'` ("append") mode instead. Once the file is open, we can write text to it using the `write` method:

```
newTextFile.write('Hello.\n')
```

The `write` method does not write a newline character at the end of the string by default, so we have to include one explicitly, if one is desired.

The following function illustrates how we can modify the `lineNumbers` function so that it writes the file with line numbers directly to another file instead of printing it.

```
def lineNumbersFile(fileName, newFileName):
    """Write the contents of the file fileName to the file
        newFileName, with each line preceded by a line number.

    Parameters:
        fileName: the name of a text file
        newFileName: the name of the output text file

    Return value: None
    """

    textFile = open(fileName, 'r', encoding = 'utf-8')
    newTextFile = open(newFileName, 'w')
    lineCount = 1
    for line in textFile:
        newTextFile.write('{0:<5} {1}\n'.format(lineCount, line[:-1]))
        lineCount = lineCount + 1
    textFile.close()
    newTextFile.close()
```

Remember to always close the new file when you are done. Closing a file to which we have written ensures that the changes have actually been written to the drive. To improve efficiency, an operating system does not necessarily write text out to the drive immediately. Instead, it usually waits until a sufficient amount builds up, and then writes it all at once. Therefore, if you forget to close a file and your computer crashes, your program's last writes may not have actually been written. (This is one reason why we sometimes have trouble with corrupted files after a computer crash.)

Reading from the web

The ability to read files allows us to conveniently apply the functions that we have already developed to real data that we download from the web. It would be even more convenient, however, if we could read some data *directly* from the web rather than having to manually download it and then read it with our program. Conveniently, we can do that in Python too.

The `urllib.request` module contains a function named `urlopen` that returns a file object abstraction for a web page that is similar to the one returned by the `open` function. The `urlopen` function takes as a parameter the address of the web page we wish to access. A web page address is formally known as a *URL*, short for *Uniform Resource Locator*, and normally begins with the prefix `http://` ("http" is short for "hypertext transfer protocol"). For example, to open the main web page at `python.org`, we can do the following:

```
>>> import urllib.request as web
>>> webpage = web.urlopen('http://python.org')
```

Once we have the file object, we can read from it just as we did earlier.

```
>>> text = webpage.read()
>>> webpage.close()
```

Behind the scenes, the Python interpreter communicates with a web server over the Internet to read this web page. But thanks to the magic of abstraction, we did not have to worry about any of those details.

Because the `urlopen` function does not accept an encoding parameter, the `read` function cannot tell how the text of the web page is encoded. Therefore, `read` returns a `bytes` object instead of a string. A `bytes` object contains a sequence of raw bytes that are not interpreted in any particular way. To convert the `bytes` object to a string before we print it, we can use the `decode` method, as follows:

```
>>> print(text.decode('utf-8'))
<!doctype html>
    ⋮
```

(Again, we will see what UTF-8 is in Section 6.3.) What you see when you print `text` is HTML (short for hypertext markup language) code for the `python.org` home page. HTML is the language in which most web pages are written.

We can download data files from the web in the same way if we know the correct URL. For example, the U.S. Food and Drug Administration (FDA) lists recalls of products that it regulates at `http://www.fda.gov/Safety/Recalls/default.htm`. The raw data behind this list is also available, so we can write a program that reads and parses it to gather statistics about recalls over time. Let's first take a look at what the data looks like. Because the file is quite long, we will initially print only the first twenty lines by calling the `readline` method, which just reads a single line into a string, in a `for` loop:

```
>>> url = 'http://www.fda.gov/DataSets/Recalls/RecallsDataSet.xml'
>>> webpage = web.urlopen(url)
>>> for i in range(20):
...     line = webpage.readline()
...     print(line.decode('utf-8'))
...
```

This loop prints something like the following.

```
<?xml version="1.0" encoding="UTF-8" ?>

<RECALLS_DATA>
<PRODUCT>
    <DATE>Mon, 11 Aug 2014 00:00:00 -0400</DATE>
    <BRAND_NAME><![CDATA[Good Food]]></BRAND_NAME>
    <PRODUCT_DESCRIPTION><![CDATA[Carob powder]]></PRODUCT_DESCRIPTION>
    <REASON><![CDATA[Salmonella]]></REASON>
    <COMPANY><![CDATA[Goodfood Inc.]]></COMPANY>
    <COMPANY_RELEASE_LINK>http://www.fda.gov/Safety/Recalls/ucm40969.htm
        </COMPANY_RELEASE_LINK>
    <PHOTOS_LINK></PHOTOS_LINK>
</PRODUCT>
    ⋮
</RECALLS_DATA>
```

This file is in a common format called XML (short for e<u>x</u>tensible <u>m</u>arkup <u>l</u>anguage). In XML, data elements are enclosed in matching pairs of *tags*. In this file, each product recall is enclosed in a pair of `<PRODUCT>` ⋯ `</PRODUCT>` tags. Within that element are other elements enclosed in matching tags that give detailed information about the product.

Reflection 6.18 *Look at the (fictitious) example product above enclosed in the `<PRODUCT>` ⋯ `</PRODUCT>` tags. What company made the product? What is the product called? Why was it recalled? When was it recalled?*

Before we can do anything useful with the data in this file, we need to be able to identify individual product recall descriptions. Notice that the file begins with a header line that describes the version of XML and the text encoding. Following some blank lines, we next notice that all of the recalls are enclosed in a matching pair of `<RECALLS_DATA>` ⋯ `</RECALLS_DATA>` tags. So to get to the first product, we need to read until we find the `<RECALLS_DATA>` tag. We can do this with a `while` loop:

```
url = 'http://www.fda.gov/DataSets/Recalls/RecallsDataSet.xml'
webpage = web.urlopen(url)

line = ''
while line[:14] != '<RECALLS_DATA>':
    line = webpage.readline()
    line = line.decode('utf-8')
```

Implicit in the file object abstraction is a *file pointer* that keeps track of the position of the next character to be read. So this `while` loop helpfully moves the file pointer to the beginning of the line after the `<RECALLS_DATA>` tag.

Reflection 6.19 *Why do we have to initialize `line` before the loop?*

Reflection 6.20 *Why will* `line != '<RECALLS_DATA>'` *not work as the condition in the* `while` *loop? (What control character is "hiding" at the end of that line?)*

To illustrate how we can identify a product element, let's write an algorithm to print just the first one. We can then put this code into a loop to print all of the product elements, or modify the code to do something more interesting with the data. Assuming that the file pointer is now pointing to the beginning of a `<PRODUCT>` tag, we want to print lines until we find the matching `</PRODUCT>` tag. This can be accomplished with another `while` loop:

```
line = webpage.readline()          # read the <PRODUCT> line
line = line.decode('utf-8')
while line[:10] != '</PRODUCT>':   # while we don't see </PRODUCT>
    print(line.rstrip())           # print the line
    line = webpage.readline()      # read the next line
    line = line.decode('utf-8')
print(line.rstrip())               # print the </PRODUCT> line
```

The two statements before the `while` loop read and decode one line of XML data, so that the condition in the `while` loop makes sense when it is first tested. The `while` loop then iterates while we have not read the `</PRODUCT>` tag that marks the end of the product element. Inside the loop, we print the line and then read the next line of data. Since the loop body is not executed when `line[:10] == '</PRODUCT>'`, we add another call to the `print` function after the loop to print the closing tag.

Reflection 6.21 *Why do we call* `line.rstrip()` *before printing each line? (What happens if we omit the call to* **rstrip***?)*

Reflection 6.22 *What would happen if we did not call* `readline` *inside the loop?*

Finally, let's put this loop inside another loop to print and count all of the product elements. We saw earlier that the product elements are enclosed within `<RECALLS_DATA>` ⋯ `</RECALLS_DATA>` tags. Therefore, we want to print product elements while we do not see the closing `</RECALLS_DATA>` tag.

```
def printProducts():
    """Print the products on the FDA recall list.

    Parameters: none

    Return value: None
    """

    url = 'http://www.fda.gov/DataSets/Recalls/RecallsDataSet.xml'
    webpage = web.urlopen(url)

    line = ''
```

```
        while line[:14] != '<RECALLS_DATA>':    # read past headers
            line = webpage.readline()
            line = line.decode('utf-8')

        productNum = 1
        line = webpage.readline()                # read first <PRODUCT> line
        line = line.decode('utf-8')
        while line[:15] != '</RECALLS_DATA>':    # while more products
            print(productNum)
            while line[:10] != '</PRODUCT>':     # print one product element
                print(line.rstrip())
                line = webpage.readline()
                line = line.decode('utf-8')
            print(line.rstrip())
            productNum = productNum + 1
            line = webpage.readline()            # read next <PRODUCT> line
            line = line.decode('utf-8')

        webpage.close()
```

The new **while** loop and the new statements that manage the accumulator to count the products are marked in red. A loop within a loop, often called a *nested loop*, can look complicated, but it helps to think of the previously written inner **while** loop (the five black statements in the middle) as a functional abstraction that prints one product. The outer loop repeats this segment while we do not see the final </RECALLS_DATA> tag. The condition of the outer loop is initialized by reading the first <PRODUCT> line before the loop, and the loop moves toward the condition becoming false by reading the next <PRODUCT> (or </RECALLS_DATA>) line at the end of the loop.

Although our function only prints the product information, it provides a framework in which to do more interesting things with the data. Exercise 6.5.7 in Section 6.5 asks you to use this framework to compile the number of products recalled for a particular reason in a particular year.

Exercises

6.2.1. Modify the `lineNumbers` function so that it only prints a line number on every tenth line (for lines 1, 11, 21, ...).

6.2.2. Write a function

wcWeb(url)

that reads a text file from the web at the given URL and returns the number of words in the file using the final `wordCount5` function from Section 6.1. You can test your function on books from Project Gutenberg at http://www.gutenberg.org. For any book, choose the "Plain Text UTF-8" or ASCII version. In either case, the file should end with a .txt file extension. For example,

```
wcWeb('http://www.gutenberg.org/cache/epub/98/pg98.txt')
```

should return the number of words in *A Tale of Two Cities* by Charles Dickens. You can also access a mirrored copy of *A Tale of Two Cities* from the book web site at

```
http://discovercs.denison.edu/chapter6/ataleoftwocities.txt
```

6.2.3. Write a function

```
wcLines(fileName)
```

that uses `wordCount5` function from Section 6.1 to print the number of words in each line of the file with the given file name.

6.2.4. Write a function

```
pigLatinDict(fileName)
```

that prints the Pig Latin equivalent of every word in the dictionary file with the given file name. (See Exercise 6.3.3.) Assume there is exactly one word on each line of the file. Start by testing your function on small files that you create. An actual dictionary file can be found on most Mac OS X and Linux computers at `/usr/share/dict/words`. There is also a dictionary file available on the book web site.

6.2.5. Repeat the previous exercise, but have your function write the results to a new file instead, one Pig Latin word per line. Add a second parameter for the name of the new file.

6.2.6. Write a function

```
strip(fileName, newFileName)
```

that creates a new version of the file with the given `fileName` in which all whitespace characters (' ', '\n', and '\t') have been removed. The second parameter is the name of the new file.

6.3 ENCODING STRINGS

Because everything in a computer is stored in binary, strings must be also. Although it is not necessary to understand every detail of their implementation at this point, some insight is very helpful.

Indexing and slicing

A string is a sequence of characters that are stored in contiguous memory cells. We can conveniently illustrate the situation in this way:

```
pirate ──▶ | L | o | n | g | b | e | a | r | d |
             0   1   2   3   4   5   6   7   8
            -9  -8  -7  -6  -5  -4  -3  -2  -1
```

A reference to the entire string of nine characters is assigned to the name `pirate`. Each character in the string is identified by an *index* that indicates its position. Indices always start at 0. We can access a character directly by referring to its index in square brackets following the name of the string. For example,

```
>>> pirate = 'Longbeard'
>>> pirate[0]
'L'
>>> pirate[2]
'n'
```

Notice that each character is itself represented as a single-character string in quotes. As indicated in the figure above, we can also use *negative indexing*, which starts from the end of the string. For example, in the figure, `pirate[2]` and `pirate[-7]` refer to the same character.

Reflection 6.23 *Using negative indexing, how can we always access the last character in a string, regardless of its length?*

The length of a string is returned by the `len` function:

```
>>> len(pirate)
9
>>> pirate[len(pirate) - 1]
'd'
>>> pirate[-1]
'd'
>>> pirate[len(pirate)]
IndexError: string index out of range
```

Reflection 6.24 *Why does the last statement above result in an error?*

Notice that `len` is *not* a method and that it returns the number of characters in the string, *not* the index of the last character. The positive index of the last character in a string is always the length of the string minus one. As shown above, referring to an index that does not exist will give an *index error*.

If we need to find the index of a particular character or substring in a string, we can use the `find` method. But the `find` method only returns the position of the *first* occurrence. For example,

```
>>> pirate2 = 'Willie Stargell'
>>> pirate2.find('gel')
11
>>> pirate2.find('ll')
2
>>> pirate2.find('jelly')
-1
```

In the last example, the substring `'jelly'` was not found, so `find` returned −1.

Reflection 6.25 *Why does −1 make sense as a return value that means "not found?"*

Creating modified strings

Suppose, in the string assigned to `pirate` (still `'Longbeard'`), we want to change the character at index 5 to an `'o'`. It seems natural that we could use indexing and assignment to do this:

```
>>> pirate[5] = 'o'
TypeError: 'str' object does not support item assignment
```

However, in Python, we cannot do so. Strings are *immutable*, meaning they cannot be changed in place. Although this may seem like an arbitrary (and inconvenient) decision, there are good reasons for it. The primary one is efficiency of both space and time. The memory to store a string is allocated when the string is created. Later, the memory immediately after the string may be used to store some other values. If strings could be lengthened by adding more characters to the end, then, when this happens, a larger chunk of memory would need to be allocated, and the old string would need to be copied to the new space. The illustration below depicts what might need to happen *if* we were allowed to add an `'s'` to the end of `pirate`. In this example, the three variable names refer to three different adjacent chunks of a computer's memory: `answer` refers to the value 42, `pirate` refers to the string `'Longbeard'`, and `fish` refers to the value 17.

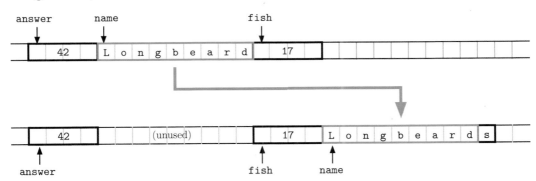

Since the value 17 is already stored immediately after `'Longbeard'`, we cannot make the string any longer. Instead, a larger chunk of memory must be allocated after 17. Then the old string would need to be copied there, and the `'s'` added. This is time consuming, and leaves an empty space in memory that may or may not be used in the future. Continuously adding characters would repeat this process.

Instead, to change a character (or group of characters) in a string, we must create a new string based on the original one. The easiest way is to use *slicing*. A slice is simply a substring of a string. For example, if `pirate` refers to the string `'Longbeard'`, then `pirate[4:8]` refers to the string `'bear'`. The first index in the slice is the index of the first character in the slice and the second index is the position of the character *just past* the last character in the slice. This works just like the `range` function.

We often want a slice at the beginning or end of a string. In these cases, we can omit the first and/or last indices in the slice notation. For example,

```
>>> pirate[:4]
'Long'
>>> pirate[4:]
'beard'
>>> pirate[:]
'Longbeard'
>>> pirate[:-1]
'Longbear'
```

The second to last expression creates a copy of the string `pirate`. Since the index −1 refers to the index of the last character in the string, the last example gives us the slice that includes everything up to, but not including, the last character.

Reflection 6.26 *How would you create a slice of* `pirate` *that contains all but the first character? How about a slice of* `pirate` *that evaluates to* `'ear'`*?*

Let's now return to our original question: how can we change the `'e'` in `'Longbeard'` to an `'o'`? To accomplish this with slicing, we need to assign the concatenation of three strings to `pirate`: the slice of `pirate` before the `'e'`, an `'o'`, and the slice of `pirate` after the `'e'`:

```
>>> pirate = pirate[:5] + 'o' + pirate[6:]
>>> pirate
'Longboard'
```

A more general technique for creating a copy, or a modified version, of a string is to iterate over it and, for each character, concatenate the character, or a modified version of it, to the end of a new string. The following function implements the simplest example of this idea.

```
def copy(text):
    """Return a copy of text.

    Parameter:
        text: a string object

    Return value: a copy of text
    """

    newText = ''
    for character in text:
        newText = newText + character
    return newText
```

This technique is really just another version of an accumulator, and is very similar to the list accumulators that we have been using for plotting. To illustrate how this works, suppose `text` was `'abcd'`. The table below illustrates how each value changes in each iteration of the loop.

Iteration	character	newText
1	'a'	'' + 'a' → 'a'
2	'b'	'a' + 'b' → 'ab'
3	'c'	'ab' + 'c' → 'abc'
4	'd'	'abc' + 'd' → 'abcd'

In the first iteration, the first character in `text` is assigned to `character`, which is 'a'. Then `newText` is assigned the concatenation of the current value of `newText` and `character`, which is '' + 'a', or 'a'. In the second iteration, `character` is assigned 'b', and `newText` is assigned the concatenation of the current value of `newText` and `character`, which is 'a' + 'b', or 'ab'. This continues for two more iterations, resulting in a value of `newText` that is identical to the original `text`.

This *string accumulator* technique can form the basis of all kinds of transformations to text. For example, let's write a function that removes all whitespace from a string. This is very similar to what we did above, except that we only want to concatenate the character if it is not a whitespace character.

```
def noWhitespace(text):
    """Return a version of text without whitespace.

    Parameter:
        text: a string object

    Return value: a version of text without whitespace
    """

    newText = ''
    for character in text:
        if character not in ' \t\n':
            newText = newText + character
    return newText
```

Some of the exercises below ask you to write similar functions to modify text in a variety of ways.

Encoding characters

As we alluded to earlier, in a computer's memory, each of the characters in a string must be encoded in binary in some way. Up until recently, most English language text was encoded in a format known as *ASCII*.[3] ASCII assigns each character a 7-bit code, and is therefore able to represent 2^7 = 128 different characters. In memory, each ASCII character is stored in one byte, with the leftmost bit of the byte being a 0. So a string is stored as a sequence of bytes, which can also be represented as a

---
[3] ASCII is an acronym for American Standard Code for Information Interchange.

Control characters	Space	Punctuation characters	Digits	Punctuation characters	Upper case letters	Punctuation characters	Lower case letters	Punctuation characters	Delete
0 31	32	33 47	48 57	58 64	65 90	91 96	97 122	123 126	127

Figure 6.2 An overview of the organization (not to scale) of the ASCII character set (and the Basic Latin segment of the Unicode character set) with decimal code ranges. For the complete Unicode character set, refer to http://unicode.org.

sequence of numbers between 0 and 127. For example, the string 'Longbeard' is stored in memory in ASCII as

```
01001100011011110110111001100111011000100110010101100001011100100110 0100
    76        111       110       103       98        101       97        114       100
     L         o         n         g         b         e         a         r         d
```

The middle row contains the decimal equivalents of the binary codes. Figure 6.2 illustrates an overview of the organization of the ASCII character set.

The ASCII character set has been largely supplanted, including in Python, by an international standard known as *Unicode*. Whereas ASCII only provides codes for Latin characters, Unicode encodes over 100,000 different characters from more than 100 languages, using up to 4 bytes per character. A Unicode string can be encoded in one of three ways, but is most commonly encoded using a variable-length system called *UTF-8*. Conveniently, UTF-8 is backwards-compatible with ASCII, so each character in the ASCII character set is encoded in the same 1-byte format in UTF-8. In Python, we can view the Unicode code (in decimal) for any character using the ord function. (ord is short for "ordinal.") For example,

```
>>> ord('L')
76
```

The chr function is the inverse of ord; given a Unicode code, chr returns the corresponding character.

```
>>> chr(76)
'L'
```

We can use the ord and chr functions to convert between letters and numbers. For example, when we print a numeric value on the screen using the print function, each digit in the number must be converted to its corresponding character to be displayed. In other words, the value 0 must be converted to the character '0', the value 1 must be converted to the character '1', etc. The Unicode codes for the digit characters are conveniently sequential, so the code for any digit character is equal to the code for '0', which is ord('0'), plus the value of the digit. For example,

```
>>> ord('2')
50
>>> ord('0') + 2
50
```

Therefore, for any one-digit integer, we can get the corresponding character by passing the value plus ord('0') into the chr function. For example:

```
>>> chr(ord('0') + 2)
'2'
```

The following function generalizes this idea for any one-digit integer named digit:

```
def digit2String(digit):
    """Converts an integer digit to its string representation.

    Parameter:
        digit: an integer in 0, 1, ..., 9

    Return value:
        the corresponding character '0', '1', ..., '9'
        or None if a non-digit is passed as an argument
    """

    if (digit < 0) or (digit > 9):
        return None
    return chr(ord('0') + digit)
```

If digit is not the value of a decimal digit, we need to recognize that and do something appropriate. In this case, we simply return None.

We can use a similar idea to convert a character into a number. Suppose we want to convert a letter of the alphabet into an integer representing its position. In other words, we want to convert the character 'A' or 'a' to 1, 'B' or 'b' to 2, etc. Like the characters for the digits, the codes for the upper case and lower case letters are in consecutive order. Therefore, for an upper case letter, we can subtract the code for 'A' from the code for the letter to get the letter's offset relative to 'A'. Similarly, we can subtract the code for 'a' from the code for a lower case letter to get the lower case letter's offset with respect to 'a'. Try it yourself:

```
>>> ord('D') - ord('A')
3
```

Since this gives us one less than the value we want, we can simply add one to get the correct position for the letter.

```
>>> ord('D') - ord('A') + 1
4
```

The following function uses if statements to handle three possible cases.

> **Box 6.1: Compressing text files**
>
> If a text file is stored in UTF-8 format, then each character is represented by an eight-bit code, requiring one byte of storage per character. For example, the file `mobydick.txt` contains about 1.2 million characters, so it requires about 1.2 MB of disk space. But text files can usually be modified to use far less space, without losing any information. Suppose that a text file contains upper and lower case letters, plus whitespace and punctuation, for a total of sixty unique characters. Since $2^6 = 64$, we can adopt an alternative encoding scheme in which each of these sixty unique characters is represented by a six-bits code instead. By doing so, the text file will use only 6/8 = 75% of the space.
>
> The *Huffman coding* algorithm can do even better by creating a *prefix code*, in which shorter codes are used for more frequent characters. As a simple example, suppose a 23,000-character text file contains only five unique characters: A, C, G, N, and T with frequencies of 5, 6, 4, 3, and 5 thousand, respectively. Using the previous fixed-length scheme, we could devise a three-bit code for these characters and use only $3 \cdot 23,000 = 69,000$ bits instead of the original $8 \cdot 23,000 = 184,000$ bits. But, by using a prefix code that assigns the more frequent characters A, C, and T to shorter two-bit codes (A = 10, C = 00, and T = 11) and the less frequent characters G and N to three-bit codes (G = 010 and N = 011), we can store the file in
>
> $$2 \cdot 5,000 + 2 \cdot 6,000 + 3 \cdot 4,000 + 3 \cdot 3,000 + 2 \cdot 5,000 = 53,000$$
>
> bits instead. This is called a prefix code because no code is a prefix of another code, which is essential for decoding the file.
>
> An alternative compression technique, used by the *Lempel-Ziv-Welch algorithm*, replaces repeated strings of characters with fixed-length codes. For example, in the string `CANTNAGATANCANCANNAGANT`, the repeated sequences `CAN` and `NAG` might each be represented with its own code.

```python
def letter2Index(letter):
    """Returns the position of a letter in the alphabet.

    Parameter:
        letter: an upper case or lower case letter

    Return value: the position of a letter in the alphabet
    """

    if (letter >= 'A') and (letter <= 'Z'):     # upper case
        return ord(letter) - ord('A') + 1
    if (letter >= 'a') and (letter <= 'z'):     # lower case
        return ord(letter) - ord('a') + 1
    return None                                 # non-letter
```

Notice above that we can compare characters in the same way we compare numbers. The values being compared are actually the Unicode codes of the characters, but since the letters and numbers are in consecutive order, the comparisons follow alphabetical order. We can also compare longer strings in the same way.

```
>>> 'cat' < 'dog'
True
>>> 'cat' < 'catastrophe'
True
>>> 'Cat' < 'cat'
True
>>> '1' < 'one'
True
```

Reflection 6.27 *Why are the expressions* `'Cat' < 'cat'` *and* `'1' < 'one'` *both* **True**? *Refer to Figure 6.2.*

Exercises

6.3.1. Suppose you have a string stored in a variable named `word`. Show how you would print

   (a) the string length

   (b) the first character in the string

   (c) the third character in the string

   (d) the last character in the string

   (e) the last three characters in the string

   (f) the string consisting of the second, third, and fourth characters

   (g) the string consisting of the fifth, fourth, and third to last characters

   (h) the string consisting of all but the last character

6.3.2. Write a function

   `username(first, last)`

   that returns a person's username, specified as the last name followed by an underscore and the first initial. For example, `username('martin', 'freeman')` should return the string `'freeman_m'`.

6.3.3. Write a function

   `piglatin(word)`

   that returns the pig latin equivalent of the string `word`. Pig latin moves the first character of the string to the end, and follows it with `'ay'`. For example, pig latin for `'python'` is `'ythonpay'`.

6.3.4. Suppose

> quote = 'Well done is better than well said.'

(The quote is from Benjamin Franklin.) Use slicing notation to answer each of the following questions.

(a) What slice of `quote` is equal to `'done'`?

(b) What slice of `quote` is equal to `'well said.'`?

(c) What slice of `quote` is equal to `'one is bet'`?

(d) What slice of `quote` is equal to `'Well do'`?

6.3.5. Write a function

> noVowels(text)

that returns a version of the string `text` with all the vowels removed. For example, `noVowels('this is an example.')` should return the string `'ths s n xmpl.'`.

6.3.6. Suppose you develop a code that replaces a string with a new string that consists of all the even indexed characters of the original followed by all the odd indexed characters. For example, the string `'computers'` would be encoded as `'cmuesoptr'`. Write a function

> encode(word)

that returns the encoded version of the string named `word`.

6.3.7. Write a function

> decode(codeword)

that reverses the process from the `encode` function in the previous exercise.

6.3.8. Write a function

> daffy(word)

that returns a string that has Daffy Duck's lisp added to it (Daffy would pronounce the 's' sound as though there was a 'th' after it). For example, `daffy("That's despicable!")` should return the string `"That'sth desthpicable!"`.

6.3.9. Suppose you work for a state in which all vehicle license plates consist of a string of letters followed by a string of numbers, such as `'ABC 123'`. Write a function

> randomPlate(length)

that returns a string representing a randomly generated license plate consisting of `length` upper case letters followed by a space followed by `length` digits.

6.3.10. Write a function

> int2String(n)

that converts a positive integer value `n` to its string equivalent, *without* using the `str` function. For example, `int2String(1234)` should return the string `'1234'`. (Use the `digit2String` function from this section.)

Encoding strings ■ 269

6.3.11. Write a function

    `reverse(text)`

that returns a copy of the string `text` in reverse order.

6.3.12. When some people get married, they choose to take the last name of their spouse or hyphenate their last name with the last name of their spouse. Write a function

    `marriedName(fullName, spouseLastName, hyphenate)`

that returns the person's new full name with hyphenated last name if `hyphenate` is `True` or the person's new full name with the spouse's last name if `hyphenate` is `False`. The parameter `fullName` is the person's current full name in the form `'Firstname Lastname'` and the parameter `spouseLastName` is the spouse's last name. For example, `marriedName('Jane Doe', 'Deer', True)` should return the string `'Jane Doe-Deer'` and `marriedName('Jane Doe', 'Deer', False)` should return the string `'Jane Deer'`.

6.3.13. Write a function

    `letter(n)`

that returns the nth capital letter in the alphabet, using the `chr` and `ord` functions.

6.3.14. Write a function

    `value(digit)`

that returns the integer value corresponding to the parameter string `digit`. The parameter will contain a single character `'0'`, `'1'`, ..., `'9'`. Use the `ord` function. For example, `value('5')` should return the integer value 5.

6.3.15. Using the `chr` and `ord` functions, we can convert a numeric exam score to a letter grade with fewer `if`/`elif`/`else` statements. In fact, for any score between 60 and 99, we can do it with one expression using `chr` and `ord`. Demonstrate this by replacing `SOMETHING` in the function below.

```
def letterGrade(grade):
    if grade >= 100:
        return 'A'
    if grade > 59:
        return SOMETHING
    return 'F'
```

6.3.16. Write a function

    `captalize(word)`

that returns a version of the string `word` with the first letter capitalized. (Note that the word may already be capitalized!)

6.3.17. Similar to parity in Exercise 6.1.11, a *checksum* can be added to more general strings to detect errors in their transmission. The simplest way to compute a checksum for a string is to convert each character to an integer representing its position in the alphabet (a = 0, b = 1, ..., z = 25), add all of these integers, and then convert the sum back to a character. For simplicity, assume that the string contains only lower case letters. Because the sum will likely be greater than 25, we will need to convert the sum to a number between 0 and 25 by finding the remainder modulo 26. For example, to find the checksum character for the string 'snow', we compute (18 + 13 + 14 + 22) mod 26 (because s = 18, n = 13, o = 14 and w = 22), which equals 67 mod 26 = 15. Since 15 = p, we add 'p' onto the end of 'snow' when we transmit this sequence of characters. The last character is then checked on the receiving end. Write a function

> checksum(word)

that returns word with the appropriate checksum character added to the end. For example, checksum('snow') should return 'snowp'. (Hint: use chr and ord.)

6.3.18. Write a function

> checksumCheck(word)

that determines whether the checksum character at the end of word (see Exercise 6.3.17) is correct. For example, checksumCheck('snowp') should return True, but checksumCheck('snowy') should return False.

6.3.19. Julius Caesar is said to have sent secret correspondence using a simple encryption scheme that is now known as the Caesar cipher. In the Caesar cipher, each letter in a text is replaced by the letter some fixed distance, called the *shift*, away. For example, with a shift of 3, A is replaced by D, B is replaced by E, etc. At the end of the alphabet, the encoding wraps around so that X is replaced by A, Y is replaced by B, and Z is replaced by C. Write a function

> encipher(text, shift)

that returns the result of encrypting text with a Caesar cypher with the given shift. Assume that text contains only upper case letters.

6.3.20. Modify the encipher function from the previous problem so that it either encrypts or decrypts text, based on the value of an additional Boolean parameter.

6.4 LINEAR-TIME ALGORITHMS

It is tempting to think that our one-line wordCount1 function from the beginning of Section 6.1 must run considerably faster than our final wordCount5 function. However, from our discussion of how the count method must work, it is clear that both functions must examine every character in the text. So the truth is more complicated.

Recall from Section 1.3 that we analyze the time complexity of algorithms by counting the number of elementary steps that they require, in relation to the size of the input. An elementary step takes the same amount of time every time we execute it, no matter what the input is. We can consider basic arithmetic, assignment, and comparisons between numbers and individual characters to be valid elementary steps.

Reflection 6.28 *Suppose that the variable names* `word1` *and* `word2` *are both assigned string values. Should a comparison between these two strings, like* `word1 < word2`, *also count as an elementary step?*

The time complexity of string operations must be considered carefully. To determine whether `word1 < word2`, the first characters must be compared, and then, if the first characters are equal, the second characters must be compared, etc. until either two characters are not the same or we reach the end of one of the strings. So the total number of individual character comparisons required to compare `word1` and `word2` depends on the values assigned to those variable names.

Reflection 6.29 *If* `word1` = 'python' *and* `word2` = 'rattlesnake', *how many character comparisons are necessary to conclude that* `word1 < word2` *is true? What if* `word1` = 'rattlesnake' *and* `word2` = 'rattlesnakes'?

In the first case, only a comparison of the first characters is required to determine that the expression is true. However, in the second case, or if the two strings are the same, we must compare every character to yield an answer. Therefore, assuming one string is not the empty string, the minimum number of comparisons is 1 and the maximum number of comparisons is n, where n is the length of the shorter string. (Exercise 6.4.1 more explicitly illustrates how a string comparison works.)

Put another way, the *best-case* time complexity of a string comparison is constant because it does not depend on the input size, and the *worst-case* time complexity for a string comparison is directly proportional to n, the length of the shorter string.

Reflection 6.30 *Do you think the best-case time complexity or the worst-case time complexity is more representative of the true time complexity in practice?*

We are typically much more interested in the worst-case time complexity of an algorithm because it describes a guarantee on how long an algorithm can take. It also tends to be more representative of what happens in practice. Intuitively, the average number of character comparisons in a string comparison is about half the length of the shorter string, or $n/2$. Since $n/2$ is directly proportional to n, it is more similar to the worst-case time complexity than the best-case.

Reflection 6.31 *What if one or both of the strings are constants? For example, how many character comparisons are required to evaluate* 'buzzards' == 'buzzword'?

Because both values are constants instead of variables, we know that this comparison *always* requires five character comparisons (to find that the fifth characters in the strings are different). In other words, the number of character comparisons required by this string comparison is independent of the input to any algorithm containing it. Therefore, the time complexity is constant.

Let's now return to a comparison of the `wordCount1` and `wordCount5` functions. The final `wordCount5` function is reproduced below.

```
 1  def wordCount5(text):
 2      """ (docstring omitted) """
 3
 4      count = 0
 5      prevCharacter = ' '
 6      for character in text:
 7          if character in ' \t\n' and prevCharacter not in ' \t\n':
 8              count = count + 1
 9          prevCharacter = character
10      if prevCharacter not in ' \t\n':
11          count = count + 1
12      return count
```

Each of the first two statements in the function, on lines 4 and 5, is an elementary step because its time does not depend on the value of the input, `text`. These statements are followed by a `for` loop on lines 6–9.

Reflection 6.32 *Suppose that* `text` *contains n characters (i.e.,* `len(text)` *is n). In terms of n, how many times does the* `for` *loop iterate?*

The loop iterates n times, once for each character in `text`. In each of these iterations, a new character from `text` is implicitly assigned to the index variable `character`, which we should count as one elementary step per iteration. Then the comparison on line 7 is executed. Since both `character` and `prevCharacter` consist of a single character, and each of these characters is compared to the three characters in the string ` \t\n`, there are at most six character comparisons here in total. Although `count` is incremented on line 8 only when this condition is true, we will assume that it happens in every iteration because we are interested in the worst-case time complexity. So this increment and the assignment on line 9 add two more elementary steps to the body of the loop, for a total of nine. Since these nine elementary steps are executed once for each character in the string `text`, the total number of elementary steps in the loop is $9n$. Finally, the comparison on line 10 counts as three more elementary steps, the increment on line 11 counts as one, and the `return` statement on line 12 counts as one. Adding all of these together, we have a total of $9n + 7$ elementary steps in the worst case.

Now let's count the number of elementary steps required by the `wordCount1` and `wordCount3` functions. Although they did not work as well, they provide a useful comparison. The `wordCount1` function is reproduced below.

```
def wordCount1(text):
    """ (docstring omitted) """

    return text.count(' ') + text.count('\t') + text.count('\n')
```

The `wordCount1` function simply calls the `count` method three times to count the number of spaces, tabs, and newlines in the parameter `text`. But, as we saw in the previous section, each of these method calls hides a `for` loop that is comparing every character in `text` to the character argument that is passed into `count`. We can estimate the time complexity of `wordCount1` by looking at `wordCount3`, reproduced below, which we developed to mimic the behavior of `wordCount1`.

```
def wordCount3(text):
    """ (docstring omitted) """

    count = 0
    for character in text:
        if character in ' \t\n':
            count = count + 1
    return count
```

Compared to `wordCount5`, we see that `wordCount3` lacks three comparisons and an assignment statement in the body of the `for` loop. Therefore, the `for` loop performs a total of $5n$ elementary steps instead of $9n$. Outside the `for` loop, `wordCount3` has only two elementary steps, for a total of $5n+2$. In summary, according to our analysis, the time complexities of all three functions grow at a rate that is proportional to n, but `wordCount1` and `wordCount3` should be about $(9n+7)/(5n+2) \approx 9/5$ times faster than `wordCount5`.

To see if this analysis holds up in practice, we timed `wordCount1`, `wordCount3`, and `wordCount5` on increasingly long portions of the text of Herman Melville's *Moby Dick*. The results of this experiment are shown in Figure 6.3. We see that, while `wordCount1` was much faster, all of the times grew at a rate proportional to n. In particular, on the computer on which we ran this experiment, `wordCount5` required about $51n$ nanoseconds, `wordCount3` required about $41.5n$ nanoseconds, and `wordCount1` required about $2.3n$ nanoseconds for every value of n. So `wordCount1` was about 22.5 times faster than `wordCount5` for every single value of n. (On another computer, the ratio might be different.) The time of `wordCount3` fell in the middle, but was closer to `wordCount5`.

Reflection 6.33 *Our analysis predicted that* `wordCount1` *would be about 9/5 times faster than* `wordCount5` *and about the same as* `wordCount3`. *Why do you think* `wordCount1` *was so much faster than both of them?*

This experiment confirmed that the time complexities of all three functions are proportional to n, the length of the input. But it contradicted our expectation about the magnitude of the constant ratio between their time complexities. This is not surprising at all. The discrepancy is due mainly to three factors. First, we missed some hidden elementary steps in our analyses of `wordCount5` and `wordCount3`. For example, we did not count the `and` operation in line 7 of `wordCount5` or the time it takes to "jump" back up to the top of the loop at the end of an iteration.

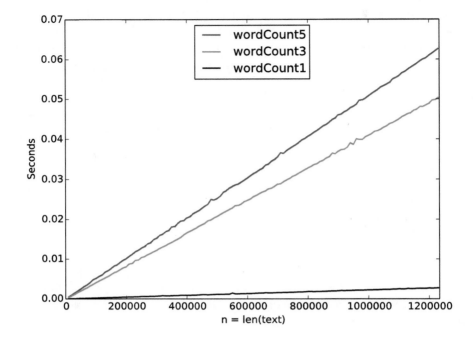

Figure 6.3 An empirical comparison of the time complexities of the `wordCount1`, `wordCount3`, and `wordCount5` functions.

Second, we did not take into account that different types of elementary steps take different amounts of real time. Third, and most importantly here, there are various optimizations that are hidden in the implementations of all built-in Python functions and methods, like `count`, that make them faster than our Python code. At our level of abstraction, these optimizations are not available.

Asymptotic time complexity

Technical factors like these can make the prediction of actual running times very difficult. But luckily, it turns out that these factors are not nearly as important as the *rates* at which the time complexities grow with increasing input size. And we are *really* only interested in the rate at which an algorithm's time complexity grows as the input gets *very large*. Another name for this concern is *scalability*; we are concerned with how well an algorithm will cope with inputs that are scaled up to larger sizes. Intuitively, this is because virtually all algorithms are going to be very fast when the input sizes are small. (Indeed, even the running time of `wordCount5` on the entire text of *Moby Dick*, about 1.2 million characters, was less than one-tenth of a second.) However, when input sizes are large, differences in time complexity can become quite significant.

This more narrow concern with growth rates allows us to simplify the time

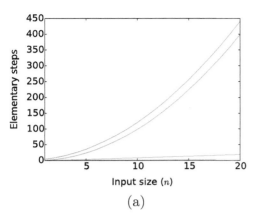
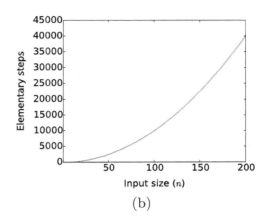

Figure 6.4 Two views of the time complexities $n^2 + 2n + 2$ (blue), n^2 (green), and n (red).

complexity expressions of algorithms to include just those terms that have an impact on the growth rate as the input size grows large. For example, we can simplify $9n + 7$ to just n because, as n gets very large, $9n + 7$ and n grow at the same rate. In other words, the constants 9 and 7 in the expression $9n + 7$ do not impact the rate at which the expression grows when n gets large. This is illustrated in the following table, which shows what happens to each expression as n increases by factors of 10.

n	↑ factor	$9n + 7$	↑ factor
10	—	97	—
100	10	907	9.35052
1,000	10	9,007	9.93054
10,000	10	90,007	9.99301
100,000	10	900,007	9.99930
1,000,000	10	9,000,007	9.99993

The second column of each row shows the ratio of the value of n in that row to the value of n in the previous row, always a factor of ten. The fourth column shows the same ratios for $9n + 7$, which get closer and closer to 10, as n gets larger. In other words, the values of n and $9n + 7$ grow at essentially the same rate as n gets larger.

We call this idea the *asymptotic time complexity*. So the asymptotic time complexity of each of `wordCount1`, `wordCount3`, and `wordCount5` is n. *Asymptotic* refers to our interest in arbitrarily large (i.e., infinite) input sizes. An asymptote, a line that an infinite curve gets arbitrarily close to, but never actually touches, should be familiar if you have taken some calculus. By saying that $9n + 7$ is asymptotically n, we are saying that $9n + 7$ is really the same thing as n, as our input size parameter n approaches infinity. Algorithms with asymptotic time complexity equal to n are said to have *linear time complexity*. We also say that each is a *linear-time algorithm*. In general, any algorithm that performs a constant number of operations on each element in a list or string has linear time complexity. All of our word count algorithms, and the string comparison that we started with, have linear time complexity.

Reflection 6.34 *The* MEAN *algorithm on Page 10 also has linear time complexity. Can you explain why?*

As another example, consider an algorithm that requires $n^2 + 2n + 2$ steps when its input has size n. Since n^2 grows faster than both $2n$ and 2, we can ignore both of those terms, and say that this algorithm's asymptotic time complexity is n^2. This is illustrated in Figure 6.4. When we view $n^2 + 2n + 2$ and n^2 together for small values of n in Figure 6.4(a), it appears as if $n^2 + 2n + 2$ is diverging from n^2, but when we "zoom out," as in Figure 6.4(b), we see that the functions are almost indistinguishable. We can also see that both algorithms have time complexities that are significantly different from that of a linear-time algorithm. Algorithms that have asymptotic time complexity proportional to n^2 are said to have *quadratic time complexity*. Quadratic-time algorithms usually pass over a string or list a number of times that is proportional to the length of the string or list. An example of a quadratic-time algorithm is one that sorts a list by first doing one pass to find the smallest element, then doing another pass to find the second smallest element, ..., continuing until it has found the nth smallest element. Since this algorithm has performed n passes over the list, each of which requires n steps, it has asymptotic time complexity n^2. (We will look at this algorithm more closely in Chapter 11.)

As we will see, algorithms can have a variety of different asymptotic time complexities. Algorithms that simply perform arithmetic on a number, like converting a temperature from Celsius to Fahrenheit are said to have *constant time complexity* because they take the same amount of time regardless of the input. At the other extreme, suppose you worked for a shipping company and were tasked with dividing a large number of assorted crates into two equal-weight groups. It turns out that there is, in general, no better algorithm than the one that simply tries all possible ways to divide the crates. Since there are 2^n possible ways to do this with n crates, this algorithm has *exponential time complexity*. As we will explore further in Section 11.5, exponential-time algorithms are so slow that they are, practically speaking, almost useless.

Exercises

6.4.1. The following function more explicitly illustrates how a string comparison works "behind the scenes." The `steps` variable counts how many individual character comparisons are necessary.

```
def compare(word1, word2):
    index = 0
    steps = 3   # number of comparisons
    while index < len(word1) and \
          index < len(word2) and \
          word1[index] == word2[index]:   # character comparison
        steps = steps + 3
        index = index + 1

    if index == len(word1) and index == len(word2):    # case 1: ==
        print(word1, 'and', word2 , 'are equal.')
        steps = steps + 2
    elif index == len(word1) and index < len(word2):   # case 2: <
        print(word1, 'comes before', word2)
        steps = steps + 2
    elif index == len(word2) and index < len(word1):   # case 3: >
        print(word1, 'comes after', word2)
        steps = steps + 2
    elif word1[index] < word2[index]:                  # case 4: <
        print(word1, 'comes before', word2)
        steps = steps + 1
    else:   # word2[index] < word1[index]:             # case 5: >
        print (word1, 'comes after', word2)
        steps = steps + 1

    print('Number of comparisons =', steps)
```

The variable `index` keeps track of the position in the two strings that is currently being compared. The `while` loop increments `index` while the characters at position `index` are the same. The `while` loop ends when one of three things happens:

(a) `index` reaches the end of `word1`,

(b) `index` reaches the end of `word2`, or

(c) `word1[index]` does not equal `word2[index]`.

If `index` reaches all the way to the end of `word1` *and* all the way to the end of `word2`, then the two strings have the same length and all of their characters along the way were equal, which means that the two strings must be equal (case 1). If `index` reaches all the way to the end of `word1`, but `index` is still less than the length of `word2`, then `word1` must be a prefix of `word2` and therefore `word1` < `word2` (case 2). If `index` reaches all the way to the end of `word2`, but `index` is still less than the length of `word1`, then we have the opposite situation and therefore `word1` > `word2` (case 3). If `index` did not reach the end of either string, then a mismatch between characters must have occurred, so we need to compare the last characters again to figure out which was less (cases 4 and 5).

The variable `steps` counts the total number of comparisons in the function. This includes both character comparisons and comparisons involving the length of the strings. The value of `steps` is incremented by 3 before the 3 comparisons in each iteration of the `while` loop, and by 1 or 2 more in each case after the loop.

Experiment with this function and answer the following questions.

   (a) How many comparisons happen in the `compare` function when
      i. `'canny'` is compared to `'candidate'`?
      ii. `'canny'` is compared to `'danny'`?
      iii. `'canny'` is compared to `'canny'`?
      iv. `'can'` is compared to `'canada'`?
      v. `'canoeing'` is compared to `'canoe'`?

   (b) Suppose `word1` and `word2` are the same n-character string. How many comparisons happen when `word1` is compared to `word2`?

   (c) Suppose `word1` (with m characters) is a prefix of `word2` (with n characters). How many comparisons happen when `word1` is compared to `word2`?

   (d) The value of `steps` actually overcounts the number of comparisons in some cases. When does this happen?

6.4.2. For each of the following code snippets, think carefully about how it must work, and then indicate whether it represents a linear-time algorithm, a constant-time algorithm, or something else. Each variable name refers to a string object with length n. Assume that any operation on a single character is one elementary step.

   (a) `name = name.upper()`
   (b) `name = name.find('x')`
   (c) `name = 'accident'.find('x')`
   (d) `newName = name.replace('a', 'e')`
   (e) `newName = name + 'son'`
   (f) `newName = 'jack' + 'son'`
   (g) `index = ord('H') - ord('A') + 1`
   (h) `for character in name:`
       `    print(character)`
   (i) `for character in 'hello':`
       `    print(character)`
   (j) `if name == newName:`
       `    print('yes')`
   (k) `if name == 'hello':`
       `    print('yes')`
   (l) `if 'x' in name:`
       `    print('yes')`
   (m) `for character in name:`
       `    x = name.find(character)`

6.4.3. What is the asymptotic time complexity of an algorithm that requires each of the following numbers of elementary steps? Assume that n is the length of the input in each case.

    (a) $7n - 4$

    (b) 6

    (c) $3n^2 + 2n + 6$

    (d) $4n^3 + 5n + 2^n$

    (e) $n \log_2 n + 2n$

6.4.4. Suppose that two algorithms for the same problem require $12n$ and n^2 elementary steps. On a computer capable of executing 1 billion steps per second, how long will each algorithm take (in seconds) on inputs of size $n = 10$, 10^2, 10^4, 10^6, and 10^9? Is the algorithm that requires n^2 steps ever faster than the algorithm that requires $12n$ steps?

6.5 ANALYZING TEXT

In this section, we will illustrate more algorithmic techniques that we can use to analyze large texts. We will first generalize our word count algorithm from Section 6.1, and then discuss how we can search for particular words and display their context in the form of a *concordance*.

Counting and searching

In Section 6.1, we counted words in a text by iterating over the characters in the string looking for whitespace. We can generalize this technique to count the number of occurrences of any character in a text. Notice the similarity of the following algorithm to the `wordCount3` function on Page 245:

```
def count1(text, target):
    """Count the number of target characters in text.

    Parameters:
        text: a string object
        target: a single-character string object

    Return value: the number of occurrences of target in text
    """

    count = 0
    for character in text:
        if character == target:
            count = count + 1
    return count
```

Reflection 6.35 *Assuming* `target` *is a letter, how can we modify the function to count both lower and upper case instances of* `target`*? (Hint: see Appendix B.6.)*

Reflection 6.36 *If we allow* `target` *to contain more than one character, how can we count the number of occurrences of any character in* `target`*? (Hint: the word "in" is the key.)*

What if we want to generalize `count1` to find the number of occurrences of any particular word or substring in a text?

Reflection 6.37 *Can we use the same* `for` *loop that we used in the* `count1` *function to count the number of occurrences of a string containing more than one character?*

Iterating over the characters in a string only allows us to "see" one character at a time, so we only have one character at a time that we can compare to a target string in the body of the loop. Instead, we need to compare the target string to all multi-character substrings with the same length in `text`. For example, suppose we want to search for the target string `'good'` in a larger string named `text`. Then we need to check whether `text[0:4]` is equal to `'good'`, then whether `text[1:5]` is equal to `'good'`, then whether `text[2:6]` is equal to `'good'`, etc. More concisely, for all values of `index` equal to 0, 1, 2, ..., we need to test whether `text[index:index + 4]` is equal to `'good'`. In general, for all values of `index` equal to 0, 1, 2, ..., we need to test whether `text[index:index + len(target)]` is equal to `target`. To examine these slices, we need a `for` loop that iterates over every *index* of `text`, rather than over the characters in `text`.

Reflection 6.38 *How can we get a list of every index in a string?*

The list of indices in a string named `text` is 0, 1, 2, ..., `len(text) - 1`. This is precisely the list of integers given by `range(len(text))`. So our desired `for` loop looks like the following:

```
def count(text, target):
    """Count the number of target strings in text.

    Parameters:
        text: a string object
        target: a string object

    Return value: the number of occurrences of target in text
    """

    count = 0
    for index in range(len(text)):
        if text[index:index + len(target)] == target:
            count = count + 1
    return count
```

Let's look at how `count` works when we call it with the following arguments:[4]

```
result = count('Diligence is the mother of good luck.', 'the')
```

If we "unwind" the loop, we find that the statements executed in the body of the loop are equivalent to:

```
if 'Dil' == 'the':      # compare text[0:3] to 'the'
    count = count + 1   # not executed; count is still 0
if 'ili' == 'the':      # compare text[1:4] to 'the'
    count = count + 1   # not executed; count is still 0
if 'lig' == 'the':      # compare text[2:5] to 'the'
    count = count + 1   # not executed; count is still 0
         ⋮
if ' th' == 'the':      # compare text[12:15] to 'the'
    count = count + 1   # not executed; count is still 0
if 'the' == 'the':      # compare text[13:16] to 'the'
    count = count + 1   # count is now 1
if 'he ' == 'the':      # compare text[14:17] to 'the'
    count = count + 1   # not executed; count is still 1
         ⋮
if 'oth' == 'the':      # compare text[18:21] to 'the'
    count = count + 1   # not executed; count is still 1
if 'the' == 'the':      # compare text[19:22] to 'the'
    count = count + 1   # count is now 2
         ⋮
if 'k.' == 'the':       # compare text[35:38] to 'the'
    count = count + 1   # not executed; count is still 2
if '.' == 'the':        # compare text[36:39] to 'the'
    count = count + 1   # not executed; count is still 2
return count            # return 2
```

Notice that the last two comparisons can never be true because the strings corresponding to `text[35:38]` and `text[36:39]` are too short. Therefore, we never need to look at a slice that starts after `len(text) - len(target)`. To eliminate these needless comparisons, we could change the range of indices to

```
range(len(text) - len(target) + 1)
```

Reflection 6.39 *What is returned by a slice of a string that starts beyond the last character in the string (e.g., `'good'[4:8]`)? What is returned by a slice that starts before the last character but extends beyond the last character (e.g., `'good'[2:10]`)?*

Iterating over the indices of a string is an alternative to iterating over the characters. For example, the `count1` function could alternatively be written as:

---

[4]"Diligence is the mother of good luck." is from *The Way to Wealth* (1758) by Benjamin Franklin.

```
def count1(text, target):
    """ (docstring omitted) """

    count = 0
    for index in range(len(text)):
        if text[index] == target:
            count = count + 1
    return count
```

Compare the two versions of this function, and make sure you understand how the two approaches perform exactly the same comparisons in their loops.

There are some applications in which iterating over the string indices is necessary. For example, consider a function that is supposed to return the index of the first occurrence of a particular character in a string. If the function iterates over the characters in the string, it would look like this:

```
def find1(text, target):
    """Find the index of the first occurrence of target in text.

    Parameters:
        text: a string object to search in
        target: a single-character string object to search for

    Return value: the index of the first occurrence of target in text
    """

    for character in text:
        if character == target:
            return ???          # return the index of character?
    return -1
```

This is just like the first version of `count1`, except we want to return the index of `character` when we find that it equals `target`. But when this happens, we are left without a satisfactory return value because we do not know the index of `character`! Instead, consider a version that iterates over the indices of the string:

```
def find1(text, target):
    """ (docstring omitted) """

    for index in range(len(text)):
        if text[index] == target:
            return index        # return the index of text[index]
    return -1
```

Now when we find that `text[index] == target`, we know that the desired character is at position `index`, and we can return that value.

Reflection 6.40 *What is the purpose of* `return -1` *at the end of the function? Under what circumstances is it executed? Why is the following alternative implementation incorrect?*

```
def find1BAD(text, target):
    """ (docstring omitted) """

    for index in range(len(text)):
        if text[index] == target:
            return index
        else:                  # THIS IS
            return -1          # INCORRECT!
```

Let's look at the `return` statements in the correct `find1` function first. If `text[index] == target` is true for some value of `index` in the `for` loop, then the `find1` function is terminated by the `return index` statement. In this case, the loop never reaches its "natural" conclusion and the `return -1` statement is never reached. Therefore, the `return -1` statement in the `find1` function is only executed if no match for `target` is found in `text`.

In the incorrect `find1BAD` function, the `return -1` is misplaced because it causes the function to *always* return during the first iteration of the loop! When 0 is assigned to `index`, if `text[index] == target` is true, then the value 0 will be returned. Otherwise, the value −1 will be returned. Either way, the next iteration of the `for` loop will never happen. The function will *appear* to work correctly if `target` happens to match the first character in `text` or if `target` is not found in `text`, but it will be incorrect if `target` only matches some later character in `text`. Since the function does not work for *every* input, it is incorrect overall.

Just as we generalized the `count1` function to `count` by using slicing, we can generalize `find1` to a function `find` that finds the first occurrence of any substring.

```
def find(text, target):
    """Find the index of the first occurrence of target in text.

    Parameters:
        text: a string object to search in
        target: a string object to search for

    Return value: the index of the first occurrence of target in text
    """

    for index in range(len(text) - len(target) + 1):
        if text[index:index + len(target)] == target:
            return index
    return -1
```

Notice how similar the following algorithm is to that of `find1` and that, if `len(target)` is 1, `find` does the same thing as `find1`. Like the `wordCount5` function from Section 6.1, the `count1` and `find1` functions implement linear-time algorithms. To see why, let's look at the `find1` function more closely. As we did before, let n represent the length of the input `text`. The second input to `find1`, `target`, has length one. So the total size of the input is $n + 1$. In the `find1` function, the most frequent elementary step is the comparison in the `if` statement inside the `for` loop. Because the function ends when `target` is found in the string `text`, the worst-case (i.e., maximum number of comparisons) occurs when `target` is not found. In this case, there are n comparisons, one for every character in `text`. Since the number of elementary steps is asymptotically the same as the input size, `find1` is a linear-time algorithm. For this reason, the algorithmic technique used by `find1` is known as a *linear search* (or *sequential search*). In Chapter 8, we will see an alternative search algorithm that is much faster, but it can only be used in cases where the data is maintained in sorted order.

Reflection 6.41 *What are the time complexities of the more general* `count` *and* `find` *functions? Assume* `text` *has length n and* `target` *has length m. Are these also linear-time algorithms?*

A concordance

Now that we have functions to count and search for substrings in text, we can apply these to whatever text we want, including long text files. For example, we can write an interactive program to find the first occurrence of any desired word in *Moby Dick*:

```
def main():
    textFile = open('mobydick.txt', 'r', encoding = 'utf-8')
    text = textFile.read()
    textFile.close()

    word = input('Search for: ')
    while word != 'q':
        index = find(text, word)
        print(word, 'first appears at position', index)
        word = input('Search for: ')
```

But, by exploiting `find` as a functional abstraction, we can develop algorithms to glean even more useful information about a text file. One example is a *concordance*, an alphabetical listing of all the words in a text, with the context in which each appears. A concordance for the works of William Shakespeare can be found at http://www.opensourceshakespeare.org/concordance/.

The simplest concordance just lists all of the lines in the text that contain each word. We can create a concordance entry for a single word by iterating over each line in a text file, and using our `find` function to decide whether the line contains the given word. If it does, we print the line.

Reflection 6.42 *If we call the `find` function to search for a word, how do we know if it was found?*

The following function implements the algorithm to print a single concordance entry:

```
def concordanceEntry(fileName, word):
    """Print all lines in a file containing the given word.

    Parameters:
        fileName: the name of the text file as a string
        word: the word to search for

    Return value: None
    """

    textFile = open(fileName, 'r', encoding = 'utf-8')
    for line in textFile:
        found = find(line, word)
        if found >= 0:                  # found the word in line
            print(line.rstrip())
    textFile.close()
```

When we call the `concordanceEntry` function on the text of *Moby Dick*, searching for the word "lashed," we get 14 matches, the first 6 of which are:

```
things not properly belonging to the room, there was a hammock lashed
blow her homeward; seeks all the lashed sea's landlessness again;
sailed with. How he flashed at me!--his eyes like powder-pans! is he
I was so taken all aback with his brow, somehow. It flashed like a
with storm-lashed guns, on which the sea-salt cakes!
to the main-top and firmly lashed to the lower mast-head, the strongest
```

It would be easier to see where "lashed" appears in each line if we could line up the words like this:

```
... belonging to the room, there was a hammock lashed
              blow her homeward; seeks all the lashed sea's ...
                               sailed with. How he flashed at me!...
... all aback with his brow, somehow. It flashed like a
                                    with storm-lashed guns, ...
           to the main-top and firmly lashed to the ...
```

Reflection 6.43 *Assume that each line in the text file is at most 80 characters long. How many spaces do we need to print before each line to make the target words line up? (Hint: use the value of `found`.)*

> **Box 6.2: Natural language processing**
>
> Researchers in the field of *natural language processing* seek to not only search and organize text, but to develop algorithms that can "understand" and respond to it, in both written and spoken forms. For example, Google Translate (`http://translate.google.com`) performs automatic translation from one language to another in real time. The "virtual assistants" that are becoming more prevalent on commercial websites seek to understand your questions and provide useful answers. Cutting edge systems seek to derive an understanding of immense amounts of unstructured data available on the web and elsewhere to answer open-ended questions. If these problems interest you, you might want to look at `http://www.nltk.org` to learn about the *Natural Language Toolkit (NLTK)*, a Python module that provides tools for natural language processing. An associated book is available at `http://nltk.org/book`.

In each line in which the word is found, we know it is found at the index assigned to `found`. Therefore, there are `found` characters before the word in that line. We can make the ends of the target words line up at position 80 if we preface each line with (80 - len(word) - found) spaces, by replacing the call to `print` with:

```
space = ' ' * (80 - len(word) - found)
print(space + line.rstrip())
```

Finally, these passages are not very useful without knowing where in the text they belong. So we should add line numbers to each line of text that we print. As in the `lineNumbers` function from Section 6.2, this is accomplished by incorporating an accumulator that is incremented every time we read a line. When we print a line, we format the line number in a field of width 6 to maintain the alignment that we introduced previously.

```
def concordanceEntry(fileName, word):
    """ (docstring omitted) """

    text = open(fileName, 'r', encoding = 'utf-8')
    lineCount = 1
    for line in text:
        found = find(line, word)
        if found >= 0:                      # found the word in line
            space = ' ' * (80 - len(word) - found)
            print('{0:<6}'.format(lineCount) + space + line.rstrip())
        lineCount = lineCount + 1
    text.close()
```

There are many more enhancements we can make to this function, some of which we leave as exercises.

Section 6.7 demonstrates how these algorithmic techniques can also be applied to problems in genomics, the field of biology that studies the function and structure of the DNA in living cells.

Exercises

6.5.1. For each of the following `for` loops, write an equivalent loop that iterates over the indices of the string `text` instead of the characters.

(a) ```
for character in text:
    print(character)
```

(b) ```
newText = ''
for character in text:
    if character != ' ':
        newText = newText + character
```

(c) ```
for character in text[2:10]:
    if character >= 'a' and character <= 'z':
        print(character)
```

(d) ```
for character in text[1:-1]:
    print(text.count(character))
```

6.5.2. Describe what is wrong with the syntax of each the following blocks of code, and show how to fix it. Assume that a string value has been previously assigned to `text`.

(a) ```
for character in text:
    caps = caps + character.upper()
```

(b) ```
while answer != 'q':
    answer = input('Word? ')
    print(len(answer))
```

(c) ```
for index in range(text):
    if text[index] != ' ':
        print(index)
```

6.5.3. Write a function

   prefixes(word)

that prints all of the prefixes of the given word. For example, `prefixes('cart')` should print

```
c
ca
car
cart
```

6.5.4. Modify the `find` function so that it only finds instances of `target` that are whole words.
```

6.5.5. Enhance the `concordanceEntry` function in each of the following ways:

(a) Modify the function so that it matches instances of the target word regardless of whether the case of the letters match. For example:

```
MOBY DICK; OR THE WHALE
                The pale Usher--threadbare in coat ...
... embellished with all the gay flags of all
...    in hand to school others, and to teach them by what
```

(b) In the line that is printed for each match, display `word` in all caps. For example:

```
... any whale could so SMITE his stout sloop-of-war
...    vessel, so as to SMITE down some of the spars and
```

(c) Use the modified `find` function from Exercise 6.5.4 so that each target word is matched only if it is a complete word.

(d) The current version of `concordanceEntry` will only identify the first instance of a word on each line. Modify it so that it will display a new context line for every instance of the target word in every line of the text. For example, "ship" appears twice in the line, "upon the ship, than to rejoice that the ship had so victoriously gained," so the function should print:

```
                            upon the SHIP, than to ...
upon the ship, than to rejoice that the SHIP had so ...
```

You may want to create modified versions of the `count` and `find` functions.

6.5.6. Write a function

```
concordance(dictFileName, textFileName)
```

that uses the dictionary with the given file name and the `concordanceEntry` function to print a complete concordance for the text with the given file name. (See Exercise 6.2.4 for information on how to get a dictionary file.) The function should list the context of every word in the text in alphabetical order.

This is a very inefficient way to compile a concordance for two reasons. First, it would be far more efficient to only search for words in the text instead of every word in the dictionary. Second, it is inefficient to completely search the text file from the beginning for every new word. (For these reasons, do not try to print a complete concordance. Limit the number of words that you read from the dictionary.) We will revisit this problem again in Chapter 11 and develop a more efficient algorithm.

6.5.7. In this exercise, you will augment the `printProducts` function from Section 6.2 so that it finds the number of products recalled for a particular reason in a particular year.

(a) Write a function

```
findReason(product)
```

that returns a string containing the year and reason for the recall described by the given product recall element. For example, if `product` were equal to the string

```
'<PRODUCT>
  <DATE>Mon, 19 Oct 2009 00:00:00 -0400</DATE>
  <BRAND_NAME><![CDATA[Good food]]></BRAND_NAME>
  <PRODUCT_DESCRIPTION><![CDATA[Cake]]></PRODUCT_DESCRIPTION>
  <REASON><![CDATA[Allergen]]></REASON>
  <COMPANY><![CDATA[Good food Inc.]]></COMPANY>
  <COMPANY_RELEASE_LINK> ... </COMPANY_RELEASE_LINK>
  <PHOTOS_LINK></PHOTOS_LINK>
</PRODUCT>'
```

then the function should return the string `'2009 Allergen'`.

(b) Modify the `printProducts` function from Section 6.2 to create a new function

```
recalls(reason, year)
```

that returns the number of recalls issued in the given year for the given reason. To do this, the new function should, instead of printing each product recall element, create a string containing the product recall element (like the one above), and then pass this string into the `findReason` function. (Replace each `print(line.rstrip())` with a string accumulator statement and eliminate the statements that implement product numbering. Do not delete anything else.) With each result returned by `findReason`, increment a counter if it contains the `reason` and `year` that are passed in as parameters.

6.6 COMPARING TEXTS

There have been many methods developed to measure similarity between texts, most of which are beyond the scope of this book. But one particular method, called a *dot plot* is both accessible and quite powerful. In a dot plot, we associate one text with the x-axis of a plot and another text with the y-axis. We place a dot at position (x, y) if the character at index x in the first text is the same as the character at index y in the second text. In this way, a dot plot visually illustrates the similarity between the two texts.

Let's begin by writing an algorithm that places dots only where characters at the same indices in the two texts match. Consider the following two sentences:

Text 1: `Peter Piper picked a peck of pickled peppers.`
Text 2: `Peter Pepper picked a peck of pickled capers.`

Our initial algorithm will compare `P` in text 1 to `P` in text 2, then `e` in text 1 to `e` in text 2, then `t` in text 1 to `t` in text 2, etc. Although this algorithm, shown below, must iterate over both strings at the same time, and compare the two strings at each position, it requires only one loop because we always compare the strings at the same index.

290 ■ Text, documents, and DNA

```
import matplotlib.pyplot as pyplot

def dotplot1(text1, text2):
    """Display a simplified dot plot comparing two equal-length strings.

    Parameters:
        text1: a string object
        text2: a string object

    Return value: None
    """

    text1 = text1.lower()
    text2 = text2.lower()
    x = []
    y = []
    for index in range(len(text1)):
        if text1[index] == text2[index]:
            x.append(index)
            y.append(index)
    pyplot.scatter(x, y)              # scatter plot
    pyplot.xlim(0, len(text1))        # x axis covers entire text1
    pyplot.ylim(0, len(text2))        # y axis covers entire text2
    pyplot.xlabel(text1)
    pyplot.ylabel(text2)
    pyplot.show()
```

Reflection 6.44 *What is the purpose of the calls to the* `lower` *method?*

Reflection 6.45 *Why must we iterate over the indices of the strings rather than the characters in the strings?*

Every time two characters are found to be equal in the loop, the index of the matching characters is added to both a list of x-coordinates and a list of y-coordinates. These lists are then plotted with the `scatter` function from `matplotlib`. The `scatter` function simply plots points without lines attaching them. Figure 6.5 shows the result for the two sequences above.

Reflection 6.46 *Look at Figure 6.5. Which dots correspond to which characters? Why are there only dots on the diagonal?*

We can see that, because this function only recognizes matches at the same index and most of the identical characters in the two sentences do not line up perfectly, this function does not reveal the true degree of similarity between them. But if we were to simply insert two gaps into the strings, the character-by-character comparison would be quite different:

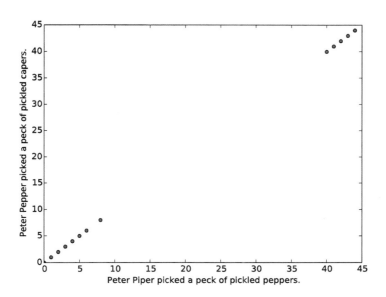

Figure 6.5 Output from the `dotplot1` function.

Text 1: `Peter Pip er picked a peck of pickled peppers.`
Text 2: `Peter Pepper picked a peck of pickled ca pers.`

A real dot plot compares every character in one sequence to every character in the other sequence. This means that we want to compare `text1[0]` to `text2[0]`, then `text1[0]` to `text2[1]`, then `text1[0]` to `text2[2]`, etc., as illustrated below:

```
            0 1 2 3 4 5 6 7 8 9 ...
    text1:  Peter  Piper  picked  a  peck  of  pickled  peppers.
            
    text2:  Peter  Pepper  picked  a  peck  of  pickled  capers.
            0 1 2 3 4 5 6 7 8 9 ...
```

After we have compared `text1[0]` to all of the characters in `text2`, we need to repeat this process with `text1[1]`, comparing `text1[1]` to `text2[0]`, then `text1[1]` to `text2[1]`, then `text1[1]` to `text2[2]`, etc., as illustrated below:

```
            0 1 2 3 4 5 6 7 8 9 ...
    text1:  Peter  Piper  picked  a  peck  of  pickled  peppers.
            
    text2:  Peter  Pepper  picked  a  peck  of  pickled  capers.
            0 1 2 3 4 5 6 7 8 9 ...
```

In other words, for each value of `index`, we want to compare `text1[index]` to every base in `text2`, not just to `text2[index]`. To accomplish this, we need to replace the `if` statement in `dotplot1` with another `for` loop:

```
import matplotlib.pyplot as pyplot

def dotplot(text1, text2):
    """Display a dot plot comparing two strings.

    Parameters:
        text1: a string object
        text2: a string object

    Return value: None
    """

    text1 = text1.lower()
    text2 = text2.lower()
    x = []
    y = []
    for index in range(len(text1)):
        for index2 in range(len(text2)):
            if text1[index] == text2[index2]:
                x.append(index)
                y.append(index2)
    pyplot.scatter(x, y)
    pyplot.xlim(0, len(text1))
    pyplot.ylim(0, len(text2))
    pyplot.xlabel(text1)
    pyplot.ylabel(text2)
    pyplot.show()
```

With this change inside the first `for` loop, each character `text1[index]` is compared to every character in `text2`, indexed by the index variable `index2`, just as in the illustrations above. If a match is found, we append `index` to the `x` list and `index2` to the `y` list.

Reflection 6.47 *Suppose we pass in 'plum' for text1 and 'pea' for text2. Write the sequence of comparisons that would be made in the body of the for loop. How many comparisons are there?*

There are $4 \cdot 3 = 12$ total comparisons because each of the four characters in 'plum'

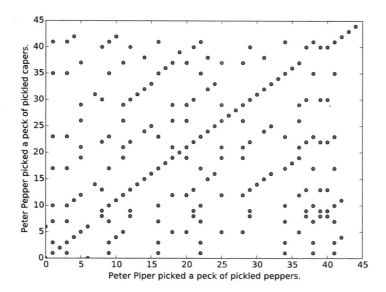

Figure 6.6 Output from the revised `dotplot` function.

is compared to each of the three characters in `'pea'`. These twelve comparisons, in order, are:

1. `text1[0] == text2[0]` or `'p' == 'p'`
2. `text1[0] == text2[1]` or `'p' == 'e'`
3. `text1[0] == text2[2]` or `'p' == 'a'`
4. `text1[1] == text2[0]` or `'l' == 'p'`
5. `text1[1] == text2[1]` or `'l' == 'e'`
6. `text1[1] == text2[2]` or `'l' == 'a'`
7. `text1[2] == text2[0]` or `'u' == 'p'`
8. `text1[2] == text2[1]` or `'u' == 'e'`
9. `text1[2] == text2[2]` or `'u' == 'a'`
10. `text1[3] == text2[0]` or `'m' == 'p'`
11. `text1[3] == text2[1]` or `'m' == 'e'`
12. `text1[3] == text2[2]` or `'m' == 'a'`

Notice that when `index` is 0, the inner `for` loop runs through all the values of `index2`. Then the inner loop finishes, also finishing the body of the outer `for` loop, and `index` is incremented to 1. The inner loop then runs through all of its values again, and so on.

Figure 6.6 shows the plot from the revised version of the function. Because the two strings share many characters, there are quite a few matches, contributing to a "noisy" plot. But the plot does now pick up the similarity in the strings, illustrated by the dots along the main diagonal.

We can reduce the "noise" in a dot plot by comparing substrings instead of

294 ■ Text, documents, and DNA

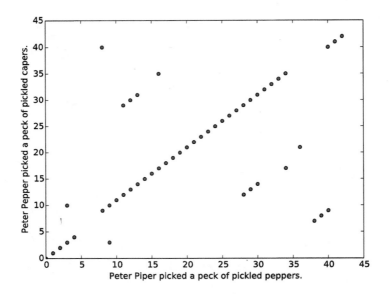

Figure 6.7   Output from the `dotplot` function from Exercise 6.6.5 (3-grams).

individual characters. In textual analysis applications, substrings with length n are known as n-grams. When $n > 1$, there are many more possible substrings, so fewer matches tend to exist. Exercise 6.6.5 asks you to generalize this dot plot function so that it compares all n-grams in the two texts instead of all single characters. Figure 6.7 shows the result of this function with $n = 3$.

An interesting application of dot plots is in plagiarism detection. For example, Figure 6.8 shows a dot plot comparison between the following two passages. The first is from page 23 of *The Age of Extremes: A History of the World, 1914–1991* by Eric Hobsbawm [19]:

> All this changed in 1914. The First World War involved all major powers and indeed all European states except Spain, the Netherlands, the three Scandinavian countries and Switzerland. What is more, troops from the world overseas were, often for the first time, sent to fight and work outside their own regions.

The second is a fictitious attempt to plagiarize Hobsbawm's passage:

> All of this changed in 1914. World War I involved all major powers and all European states except the three Scandinavian countries, Spain, the Netherlands, and Switzerland. In addition, troops from the world overseas were sent to fight and work outside their own regions.

Reflection 6.48 *Just by looking at Figure 6.8, would you conclude that the passage had been plagiarized? (Think about what a dotplot comparing two random passages would look like.)*

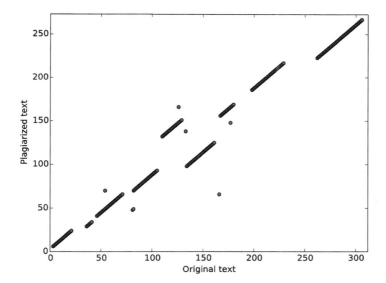

Figure 6.8 A dot plot comparing 6-grams from an original and a plagiarized passage.

Exercises

6.6.1. What is printed by the following nested loop?

```
text = 'imho'
for index1 in range(len(text)):
    for index2 in range(index1, len(text)):
        print(text[index1:index2 + 1])
```

6.6.2. Write a function

```
difference(word1, word2)
```

that returns the first index at which the strings word1 and word2 differ. If the words have different lengths, and the shorter word is a prefix of the longer word, the function should return the length of the shorter word. If the two words are the same, the function should return −1. Do this without directly testing whether word1 and word2 are equal.

6.6.3. Hamming distance, defined to be the number of bit positions that are different between two bit strings, is used to measure the error that is introduced when data is sent over a network. For example, suppose we sent the bit sequence 011100110001 over a network, but the destination received 011000110101 instead. To measure the transmission error, we can find the Hamming distance between the two sequences by lining them up as follows:

Sent: 011100110001
Received: 011000110101

Since the bit sequences are different in two positions, the Hamming distance is 2. Write a function

```
hamming(bits1, bits2)
```

that returns the Hamming distance between the two given bit strings. Assume that the two strings have the same length.

6.6.4. Repeat Exercise 6.6.3, but make it work correctly even if the two strings have different lengths. In this case, each "missing" bit at the end of the shorter string counts as one toward the Hamming distance. For example, `hamming('000', '10011')` should return 3.

6.6.5. Generalize the `dotplot` function so that it compares n-grams instead of individual characters.

6.6.6. One might be able to gain insight into a text by viewing the frequency with which a word appears in "windows" of some length over the length of the text. Consider the very small example below, in which we are counting the frequency of the "word" a in windows of size 4 in the string `'abracadabradab'`:

```
          2                 2
       ┌─────┐           ┌─────┐
     a b r a c a d a b r a d a b
     └───┘         └───┘
       2             1
```

In this example, the window skips ahead 3 characters each time. So the four windows' frequencies are $2, 2, 1, 2$, which can be plotted like this:

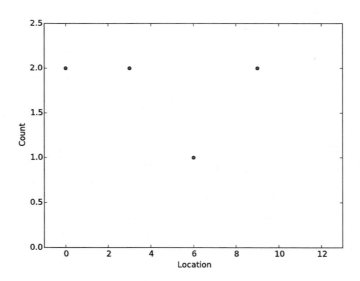

The numbers on the x-axis are indices of the beginnings of each window. Write a function

```
wordFrequency(fileName, word, windowSize, skip)
```

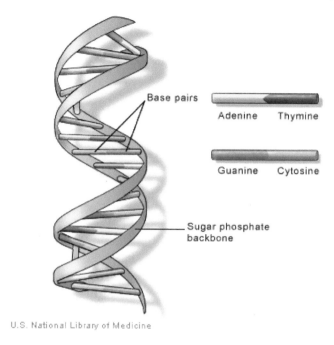

Figure 6.9 An illustration of a DNA double helix.

that displays a plot like this showing the frequencies of the string `word` in windows of size `windowSize`, where the windows skip ahead by `skip` indices each time. The text to analyze should be read in from the file with the given `fileName`.

*6.7   GENOMICS

Every living cell contains molecules of DNA (deoxyribonucleic acid) that encode the genetic blueprint of the organism. Decoding the information contained in DNA, and understanding how it is used in all the processes of life, is an ongoing grand challenge at the center of modern biology and medicine. Comparing the DNA of different organisms also provides insight into evolution and the tree of life. The lengths of DNA molecules and the sheer quantity of DNA data that have been read from living cells and recorded in text files require that biologists use computational methods to answer this challenge.

A genomics primer

As illustrated in Figure 6.9, DNA is a long double-stranded molecule in the shape of a double helix. Each strand is a chain of smaller units called *nucleotides*. A nucleotide consists of a sugar (deoxyribose), a phosphate group, and one of four nitrogenous

298 ■ Text, documents, and DNA

bases: adenine (A), thymine (T), cytosine (C), or guanine (G). Each nucleotide in a molecule can be represented by the letter corresponding to the base it contains, and an entire strand can be represented by a string of characters corresponding to its sequence of nucleotides. For example, the following string represents a small DNA molecule consisting of an adenine nucleotide followed by a guanine nucleotide followed by a cytosine nucleotide, etc.:

```
>>> dna = 'agcttttcattctgactg'
```

(The case of the characters is irrelevant; some sequence repositories use lower case and some use upper case.) Real DNA sequences are stored in large text files; we will look more closely at these later.

A base on one strand is connected via a hydrogen bond to a complementary base on the other strand. C and G are complements, as are A and T. A base and its connected complement are called a *base pair*. The two strands are "antiparallel" in the sense that they are traversed in opposite directions by the cellular machinery that copies and reads DNA. On each strand, DNA is read in an "upstream-to-downstream" direction, but the "upstream" and "downstream" ends are reversed on the two strands. For reasons that are not relevant here, the "upstream" end is called 5' (read "five prime") and the downstream end is called 3' (read "three prime"). For example, the sequence in the top strand below, read from 5' to 3', is `AGCTT...CTG`, while the bottom strand, called its *reverse complement*, also read 5' to 3', is `CAGTC...GCT`.

Reflection 6.49 *When DNA sequences are stored in text files, only the sequence on one strand is stored. Why?*

RNA (ribonucleic acid) is a similar molecule, but each nucleotide contains a different sugar (ribose instead of deoxyribose), and a base named uracil (U) takes the place of thymine (T). RNA molecules are also single-stranded. As a result, RNA tends to "fold" when complementary bases on the same strand pair with each other. This folded structure forms a unique three-dimensional shape that plays a significant role in the molecule's function.

Some regions of a DNA molecule are called *genes* because they encode genetic information. A gene is *transcribed* by an enzyme called *RNA polymerase* to produce a complementary molecule of RNA. For example, the DNA sequence `5'-GACTGAT-3'` would be transcribed into the RNA sequence `3'-CUGACUA-5'`. If the RNA is a messenger RNA (mRNA), then it contains a *coding region* that will ultimately be used to build a protein. Other RNA products of transcription, called *RNA genes*, are not translated into proteins, and are often instead involved in regulating whether genes are transcribed or translated into proteins. Upstream from the transcribed

	U	C	A	G	
U	UUU Phe F UUC Phe F UUA Leu L UUG Leu L	UCU Ser S UCC Ser S UCA Ser S UCG Ser S	UAU Tyr Y UAC Tyr Y UAA Stop * UAG Stop *	UGU Cys C UGC Cys C UGA Stop * UGG Trp W	U C A G
C	CUU Leu L CUC Leu L CUA Leu L CUG Leu L	CCU Pro P CCC Pro P CCA Pro P CCG Pro P	CAU His H CAC His H CAA Gln Q CAG Gln Q	CGU Arg R CGC Arg R CGA Arg R CGG Arg R	U C A G
A	AUU Ile I AUC Ile I AUA Ile I AUG Met M	ACU Thr T ACC Thr T ACA Thr T ACG Thr T	AAU Asn N AAC Asn N AAA Lys K AAG Lys K	AGU Ser S AGC Ser S AGA Arg R AGG Arg R	U C A G
G	GUU Val V GUC Val V GUA Val V GUG Val V	GCU Ala A GCC Ala A GCA Ala A GCG Ala A	GAU Asp D GAC Asp D GAA Glu E GAG Glu E	GGU Gly G GGC Gly G GGA Gly G GGG Gly G	U C A G

Table 6.1 The standard genetic code that translates between codons and amino acids. For each codon, both the three letter code and the single letter code are shown for the corresponding amino acid.

sequence is a *promoter* region that binds to RNA polymerase to initiate transcription. Upstream from there is often a *regulatory region* that influences whether or not the gene is transcribed.

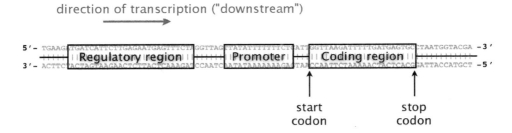

The coding region of a mRNA is *translated* into a protein by a molecular machine called a *ribosome*. A ribosome reads the coding region in groups of three nucleotides called *codons*. In prokaryotes (e.g., bacteria), the coding region begins with a particular *start codon* and ends with one of three *stop codons*. The ribosome translates each mRNA codon between the start and stop codons into an amino acid, according to the *genetic code*, shown in Table 6.1. It is this sequence of amino acids that comprises the final protein. The genes of eukaryotes (organisms whose cells contain a nucleus) are more complicated, partially because they are interspersed with untranslated regions called *introns* that must be spliced out before translation can occur.

Reflection 6.50 *What amino acid sequence is represented by the mRNA sequence* `CAU UUU GAG`*?*

Reflection 6.51 *Notice that, in the genetic code (Table 6.1), most amino acids are represented by several similar codons. Keeping in mind that nucleotides can mutate over time, what evolutionary advantage might this hold?*

The complete sequence of an organism's DNA is called its *genome*. The size of a genome can range from 10^5 base pairs (bp) in the simplest bacteria to 1.5×10^{11} bp in some plants. The human genome contains about 3.2×10^9 (3.2 billion) bp. Interestingly, the size of an organism's genome does not necessarily correspond to its complexity; plants have some of the largest genomes, far exceeding the size of the human genome.

The subfield of biology that studies the structure and function of genomes is called *genomics*. To better understand a genome, genomicists ask questions such as:

- What is the frequency of each base in the genome? What is the frequency of each codon? For each amino acid, is there a bias toward particular codons?

- Where are the genes and what are their functions?

- How similar are two sequences? Sequence comparison can be used to determine whether two genes have a shared ancestry, called *homology*. Sequence comparison between homologous sequences can also give clues to an unidentified gene's function.

- How are a set of organisms related evolutionarily? We can use sequence comparison to build a *phylogenetic tree* that specifies genetic relations between species.

- Scientists have discovered that mammalian genomes are largely reshuffled collections of similar blocks, called *synteny blocks*. What is the most likely sequence of rearrangement events to have changed one genome into the other?

- What genes are regulated together? Identifying groups of genes that are regulated in the same way can lead to insights into genes' functions, especially those related to disease.

Because genomes are so large, questions like these can only be answered computationally. In the next few pages, we will look at how the methods and techniques from previous sections can be used to answer some of the questions above. We leave many additional examples as exercises. We will begin by working with small sequences; later, we will discuss how to read some longer sequences from files and the web.

Basic DNA analysis

We can use some of the techniques from the previous section to answer questions in the first bullet above. For example, bacterial genomes are sometimes categorized according to their ratio of G and C bases to A and T bases. The fraction of bases that are G or C is called the genome's *GC content*. There are a few ways that we can compute the GC content. First, we could simply use the `count` method:

```
>>> gc = (dna.count('c') + dna.count('g')) / len(dna)
>>> gc
0.3888888888888889
```

Or, we could use a `for` loop like the `count1` function in the previous section:

```
def gcContent(dna):
    """Compute the GC content of a DNA sequence.

    Parameter:
        dna: a string representing a DNA sequence

    Return value: the GC content of the DNA sequence
    """

    dna = dna.lower()
    count = 0
    for nt in dna:             # nt is short for "nucleotide"
        if nt in 'cg':
            count = count + 1
    return count / len(dna)
```

Reflection 6.52 *Why do we convert* `dna` *to lower case at the beginning of the function?*

Because DNA sequences can be in either upper or lower case, we need to account for both possibilities when we write a function. But rather than check for both possibilities in our functions, it is easier to just convert the parameter to either upper or lower case at the beginning.

To gather statistics about the codons in genes, we need to count the number of non-overlapping occurrences of any particular codon. This is very similar to the `count` function from the previous section, except that we need to increment the index variable by three in each step:

```
def countCodon(dna, codon):
    """Find the number of occurrences of a codon in a DNA sequence.

    Parameters:
        dna: a string representing a DNA sequence
        codon: a string representing a codon

    Return value: the number of occurrences of the codon in dna
    """

    count = 0
    for index in range(0, len(dna) - 2, 3):
        if dna[index:index + 3] == codon:
            count = count + 1
    return count
```

Reflection 6.53 *Why do we subtract 2 from* `len(dna)` *above?*

Transforming sequences

When DNA sequences are stored in databases, only the sequence of one strand is stored. But genes and other features may exist on either strand, so we need to be able to derive the reverse complement sequence from the original. To first compute the complement of a sequence, we can use a string accumulator, but append the complement of each base in each iteration.

```
def complement(dna):
    """Return the complement of a DNA sequence.

    Parameter:
        dna: a string representing a DNA sequence

    Return value: the complement of the DNA sequence
    """

    dna = dna.lower()
    compdna = ''
    for nt in dna:
        if nt == 'a':
            compdna = compdna + 't'
        elif nt == 'c':
            compdna = compdna + 'g'
        elif nt == 'g':
            compdna = compdna + 'c'
        else:
            compdna = compdna + 'a'
    return compdna
```

Genomics ■ 303

To turn our complement into a reverse complement, we need to write an algorithm to reverse a string of DNA. Let's look at two different ways we might do this. First, we could iterate over the original string in reverse order and append the characters to the new string in that order.

Reflection 6.54 *How can we iterate over indices in reverse order?*

We can iterate over the indices in reverse order by using a `range` that starts at `len(dna) - 1` and goes down to, but not including, −1 using a step of −1. A function that uses this technique follows.

```
def reverse(dna):
    """Return the reverse of a DNA sequence.

    Parameter:
        dna: a string representing a DNA sequence

    Return value: the reverse of the DNA sequence
    """

    revdna = ''
    for index in range(len(dna) - 1, -1, -1):
        revdna = revdna + dna[index]
    return revdna
```

A more elegant solution simply iterates over the characters in `dna` in the normal order, but *prepends* each character to the `revdna` string.

```
def reverse(dna):
    """ (docstring omitted) """

    revdna = ''
    for nt in dna:
        revdna = nt + revdna
    return revdna
```

To see how this works, suppose `dna` was `'agct'`. The table below illustrates how each value changes in each iteration of the loop.

Iteration	nt	new value of revdna
1	'a'	'a' + '' → 'a'
2	'g'	'g' + 'a' → 'ga'
3	'c'	'c' + 'ga' → 'cga'
4	't'	't' + 'cga' → 'tcga'

Finally, we can put these two pieces together to create a function for reverse complement:

```
def reverseComplement(dna):
    """Return the reverse complement of a DNA sequence.

    Parameter:
        dna: a string representing a DNA sequence

    Return value: the reverse complement of the DNA sequence
    """

    return reverse(complement(dna))
```

This function first computes the `complement` of `dna`, then calls `reverse` on the result. We can now, for example, use the `reverseComplement` function to count the frequency of a particular codon on both strands of a DNA sequence:

```
countForward = countCodon(dna, 'atg')
countBackward = countCodon(reverseComplement(dna), 'atg')
```

Comparing sequences

Measuring the similarity between DNA sequences, called *comparative genomics*, has become an important area in modern biology. Comparing the genomes of different species can provide fundamental insights into evolutionary relationships. Biologists can also discover the function of an unknown gene by comparing it to genes with known functions in evolutionarily related species.

Dot plots are used heavily in computational genomics to provide visual representations of how similar large DNA and amino sequences are. Consider the following two sequences. The `dotplot1` function in Section 6.5 would show pairwise matches between the nucleotides in black only.

<pre>
            Sequence 1:   agctttgcattctgacag
            Sequence 2:   accttttaattctgtacag
</pre>

But notice that the last four bases in the two sequences are actually the same; if you insert a gap in the first sequence, above the last T in the second sequence, then the number of differing bases drops from eight to four, as illustrated below.

<pre>
            Sequence 1:   agctttgcattctg-acag
            Sequence 2:   accttttaattctgtacag
</pre>

Evolutionarily, if the DNA of the closest common ancestor of these two species contained a T in the location of the gap, then we interpret the gap as a deletion in the first sequence. Or, if the closest common ancestor did not have this T, then

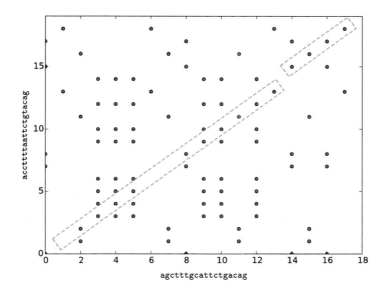

Figure 6.10 A dot plot comparing individual bases in `agctttgcattctgacag` and `accttttaattctgtacag`. The dots representing the main alignment are highlighted.

we interpret the gap as an insertion in the second sequence. These insertions and deletions, collectively called *indels*, are common; therefore, sequence comparison algorithms must take them into account. So, just as the first `dotplot1` function did not adequately represent the similarity between texts, neither does it for DNA sequences. Figure 6.10 shows a complete dot plot, using the `dotplot` function from Section 6.5, for the sequences above.

As we saw in the previous section, dot plots tend to be more useful when we reduce the "noise" by instead comparing subsequences of a given length, say ℓ, within a sliding window. Because there are 4^ℓ different possible subsequences with length ℓ, fewer matches are likely. Dot plots are also more useful when comparing sequences of amino acids. Since there are 20 different amino acids, we tend to see less noise in the plots. For example, Figure 6.11 shows a dot plot for the following hypothetical small proteins. Each letter corresponds to a different amino acid. (See Table 6.1 for the meanings of the letters if you are interested.)

```
seq1 = 'PDAQNPDMSFFKMLFLPESARWIQRTHGKNS'
seq2 = 'PDAQNPDMPLFLMKFFSESARWIQRTHGKNS'
```

Notice how the plot shows an *inversion* in one sequence compared the other, highlighted in red above.

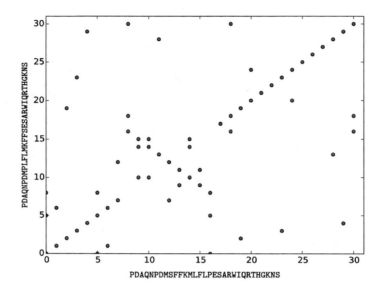

Figure 6.11 A dot plot comparing two hypothetical short proteins PDAQNPDMSFFKMLFLPESARWIQRTHGKNS and PDAQNPDMPLFLMKFFSESARWIQRTHGKNS.

Reading sequence files

DNA sequences are archived and made publicly available in a database managed by the National Center for Biotechnology Information (NCBI), a division of the National Institutes of Health (NIH). Each sequence in the database has an associated identifier known as an *accession number*. For example, the accession number for the first segment of the Burmese python (*Python molurus bivittatus*) genome is AEQU02000001. You can find this data by visiting the NCBI web site at http://www.ncbi.nlm.nih.gov, entering the accession number in the search box, and selecting "Nucleotide" from the drop-down menu.

Genomic data is commonly stored in a format called *FASTA* [38]. A FASTA file begins with a header preceded by a greater-than sign (>), followed by several lines of sequence data. Each line of sequence is terminated with a newline character. For example, the FASTA file for the first segment of the Burmese python genome looks like this:

```
>gi|541020484|gb|AEQU02000001.1| Python bivittatus ...
AGGCCTGGCGCAATATGGTTCATGGGGTCACGAAGAGTCGGACACGACTTAACGACTAAACAACAATTAA
TTCTAAACCCTAGCTATCTGGGTTGCTTCCCTACATTATATTTCCGTGAATAAATCTCACTAAACTCAGA
AGGATTTACTTTTGAGTGAACACTCATGATCAAGAACTCAGAAAACTGAAAGGATTCGGGAATAGAAGTG
TTATTTCAGCCCTTTCCAATCTTCTAAAAAGCTGACAGTAATACCTTTTTTTGTTTTAAAAATTTAAAAA
GAGCACTCACATTTATGCTACAGAATGAGTCTGTAACAGGGAAGCCAGAAGGGAAAAAGACTGAAACGTA
AACCAAGGACATAATGCAGGGATCATAATTTCTAGTTGGCAAGACTAAAGTCTATACATGTTTTATTGAA
AAACCACTGCTATTTGCTCTTACAAGAACTTATCCCTCCAGAAAAAGTAGCACTGGCTTGATACTGCCTG
CTCTAACTCAACAGTTTCAGGAAGTGCTGGCAGCCCCTCCTAACACCATCTCAGCTAGCAAGACCAAGTT
GTGCAAGTCCACTCTATCTCATAACAAAACCTCTTTAATTAAACTTGAAAGCCTAGAGCTTTTTTACCT
TGGTTCTTCAAGTCTTCGTTTGTACAATATGGGCCAATGTCATAGCTTGGTGTATTAGTATTATTATTGT
TACTGTTGGTTACACAGTCAGTCAGATGTTTAGACTGGATTTGACATTTTTACCCATGTATCGAGTCCTT
CCCAAGGACCTGGGATAGGCAGATGTTGGTGTTTGATGGTGTTAAAGGT
```

To use this DNA sequence in a program, we need to extract the unadorned sequence into a string, removing the header and newline characters. A function to do this is very similar to those in Section 6.2.

```
def readFASTA(filename):
    """Read a FASTA file containing a single sequence and return
       the sequence as a string.

    Parameter:
        filename: a string representing the name of a FASTA file

    Return value: a string containing the sequence in the FASTA file
    """

    inputFile = open(filename, 'r', encoding = 'utf-8')
    header = inputFile.readline()
    dna = ''
    for line in inputFile:
        dna = dna + line[:-1]
    inputFile.close()
    return dna
```

After opening the FASTA file, we read the header line with the `readline` function. We will not need this header, so this `readline` just serves to move the file pointer past the first line, to the start of the actual sequence. To read the rest of the file, we iterate over it, line by line. In each iteration, we append the line, without the newline character at the end, to a growing string named `dna`. There is a link on the book website to the FASTA file above so that you can try out this function.

We can also directly access FASTA files by using a URL that sends a query to NCBI. The URL below submits a request to the NCBI web site to download the FASTA file for the first segment of the Burmese python genome (with accession number AEQU02000001).

```
http://eutils.ncbi.nlm.nih.gov/entrez/eutils/efetch.fcgi?db=nuccore
    &id=AEQU02000001&rettype=fasta&retmode=text
```

To retrieve a different sequence, we would just need to insert a different accession number (in red). The following function puts all of the pieces together.

```
import urllib.request as web

def getFASTA(id):
    """Fetch the DNA sequence with the given id from NCBI and return
        it as a string.

    Parameter:
        id: the identifier of a DNA sequence

    Return value: a string representing the sequence with the given id
    """

    prefix1 = 'http://eutils.ncbi.nlm.nih.gov/entrez/eutils/efetch.fcgi'
    prefix2 = '?db=nuccore&id='
    suffix = '&rettype=fasta&retmode=text'
    url = prefix1 + prefix2 + id + suffix
    readFile = web.urlopen(url)
    header = readFile.readline()
    dna = ''
    for line in readFile:
        line = line[:-1]
        dna = dna + line.decode('utf-8')
    readFile.close()
    return dna
```

The function first creates a string containing the URL for the accession number passed in parameter `id`. The first common parts of the URL are assigned to the `prefix` variables, and the last part of the URL is assigned to `suffix`. We construct the custom URL by concatenating these pieces with the accession number. We then open the URL and use the code from the previous `readFASTA` function to extract the DNA sequence and return it as a string.

Reflection 6.55 *Use the* `getFASTA` *function to read the first segment of the Burmese python genome (accession number AEQU02000001).*

We have just barely touched on the growing field of computational genomics. The following exercises explore some additional problems and provide many opportunities to practice your algorithm design and programming skills.

Exercises

6.7.1. Write a function

    countACG(dna)

that returns the fraction of nucleotides in the given DNA sequence that are not T. Use a `for` loop that iterates over the characters in the string.

6.7.2. Repeat the previous exercise, but use a `for` loop that iterates over the indices of the string instead.

6.7.3. Write a function

    printCodons(dna)

that prints the codons, starting at the left end, in the string `dna`. For example, `printCodons('ggtacactgt')` would print

    ggt
    aca
    ctg

6.7.4. Write a function

    findCodon(dna, codon)

that returns the index of the first occurrence of the given `codon` in the string `dna`. Unlike in the `countCodon` function, increment the loop index by 1 instead of 3, so you can find the codon at any position.

6.7.5. Write a function

    findATG(dna)

that builds up a list of indices of *all* of the positions of the codon `ATG` in the given DNA string. Do not use the built-in `index` method or `find` method. Constructing the list of indices is very similar to what we have done several times in building lists to plot.

6.7.6. Since codons consist of three bases each, transcription can occur in one of three independent *reading frames* in any sequence of DNA by starting at offsets 0, 1, and 2 from the left end. For example, if our DNA sequence was `ggtacactgtcat`, the codons in the three reading frames are:

    RF 1:   ggt aca ctg tca t
    RF 2: g gta cac tgt cat
    RF 3: gg tac act gtc at

Write a function

    printCodonsAll(dna)

that uses your function from Exercise 6.7.3 to print the codons in all three reading frames.

6.7.7. Most vertebrates have much lower density of CG pairs (called *CG dinucleotides*) than would be expected by chance. However, they often have relatively high concentrations upstream from genes (coding regions). For this reason, finding these so-called "CpG islands" is often a good way to find putative sites of genes. (The "p" between C and G denotes the phosphodiester bond between the C and G, versus a hydrogen bond across two complementary strands.) Without using the built-in count method, write a function

   CpG(dna)

that returns the fraction of dinucleotides that are cg. For example, if dna were atcgttcg, then the function should return 0.5 because half of the sequence is composed of CG dinucleotides.

6.7.8. A *microsatellite* or *simple sequence repeat (SSR)* is a repeat of a short sequence of DNA. The repeated sequence is typically 2–4 bases in length and can be repeated 10–100 times or more. For example, cacacacaca is a short SSR of ca, a very common repeat in humans. Microsatellites are very common in the DNA of most organisms, but their lengths are highly variable within populations because of replication errors resulting from "slippage" during copying. Comparing the distribution of length variants within and between populations can be used to determine genetic relationships and learn about evolutionary processes.
Write a function

   ssr(dna, repeat)

that returns the length (number of repeats) of the first SSR in dna that repeats the parameter repeat. If repeat is not found, return 0. Use the find method of the string class to find the first instance of repeat in dna.

6.7.9. Write another version of the ssr function that finds the length of the *longest* SSR in dna that repeats the parameter repeat. Your function should repeatedly call the ssr function in the previous problem. You will probably want to use a while loop.

6.7.10. Write a third version of the ssr function that uses the function in the previous problem to find the longest SSR of *any dinucleotide* (sequence with length 2) in dna. Your function should return the longest repeated dinucleotide.

6.7.11. Write a function

   palindrome(dna)

that returns True if dna is the same as its reverse complement, and False otherwise. (Note that this is different from the standard definition of palindrome.) For example, gaattc is a palindrome because it and its reverse complement are the same. These sequences turn out to be very important because certain restriction enzymes target specific palindromic sequences and cut the DNA molecule at that location. For example, the EcoR1 restriction enzyme cuts DNA of the bacteria *Escherichia coli* at the sequence gaattc in the following way:

```
5'--gaattc--3'                5'--g   \ aattc--3'
    ||||||          ⇒             |    \    |
3'--cttaag--5'                3'--cttaa \   g--5'
```

6.7.12. Write a function

    `dna2rna(dna)`

that returns returns a copy of `dna` with every T replaced by a U.

6.7.13. Write a function

    `transcribe(dna)`

that returns the RNA equivalent of the reverse complement of `dna`. (Utilize previously written functions to accomplish this in one line.)

6.7.14. Write a function

    `clean(dna)`

that returns a new DNA string in which every character that is not an A, C, G, or T is replaced with an N. For example, `clean('goat')` should return the string `'gnat'`.

6.7.15. When DNA molecules are sequenced, there are often ambiguities that arise in the process. To handle these situations, there are also "ambiguous symbols" defined that code for one of a set of bases.

Symbol	Possible Bases	Symbol	Possible Bases
R	A G	B	C G T
Y	C T	D	A G T
K	G T	H	A C T
M	A C	V	A C G
S	C G	N	A C G T
W	A T		

Write a function

    `fix(dna)`

that returns a DNA string in which each ambiguous symbol is replaced with one of the possible bases it represents, each with equal probability. For example, if an R exists in `dna`, it should be replaced with either an A with probability 1/2 or a G with probability 1/2.

6.7.16. Write a function

    `mark(dna)`

that returns a new DNA string in which every start codon ATG is replaced with >>>, and every stop codon (TAA, TAG, or TGA) is replaced with <<<. Your loop should increment by 3 in each iteration so that you are only considering non-overlapping codons. For example, `mark('ttgatggagcattagaag')` should return the string `'ttg>>>gagcat<<<aag'`.

6.7.17. The accession number for the hepatitis A virus is `NC_001489`. Write a program that uses the `getFASTA` function to get the DNA sequence for hepatitis A, and then the `gcContent` function from this section to find the GC content of hepatitis A.

6.7.18. The DNA of the hepatitis C virus encodes a single long protein that is eventually processed by enzymes called proteases to produce ten smaller proteins. There are seven different types of the virus. The accession numbers for the proteins produced by type 1 and type 2 are NP_671491 and YP_001469630. Write a program that uses the getFASTA function and your dot plot function from Exercise 6.6.5 to produce a dot plot comparing these two proteins using a window of size 4.

6.7.19. *Hox* genes control key aspects of embryonic development. Early in development, the embryo consists of several segments that will eventually become the main axis of the head-to-tail body plan. The *Hox* genes dictate what body part each segment will become. One particular *Hox* gene, called *Hox A1*, seems to control development of the hindbrain. Write a program that uses the getFASTA function and your dot plot function from Exercise 6.6.5 to produce a dot plot comparing the human and mouse *Hox A1* genes, using a window of size 8. The human *Hox A1* gene has accession number U10421.1 and the mouse *Hox A1* gene has accession number NM_010449.4.

6.8 SUMMARY

Text is stored as a sequence of bytes, which we can read into one or more strings. The most fundamental string algorithms have one of the following structures:

```
for character in text:
    # process character

for index in range(len(text)):
    # process text[index]
```

In the first case, consecutive characters in the string are assigned to the `for` loop index variable `character`. In the body of the loop, each character can then be examined individually. In the second case, consecutive integers from the list `[0, 1, 2, ..., len(text) - 1]`, which are precisely the indices of the characters in the string, are assigned to the `for` loop index variable `index`. In this case, the algorithm has more information because, not only can it access the character at `text[index]`, it also knows where that character resides in the string. The first choice tends to be more elegant, but the second choice is necessary when the algorithm needs to know the index of each character, or if it needs to process slices of the string, which can only be accessed with indices.

We called one special case of these loops a *string accumulator*:

```
newText = ''
for character in text:
    newText = newText + _____
```

Like an integer accumulator and a list accumulator, a string accumulator builds its result cumulatively in each iteration of the loop. Because strings are immutable, a string accumulator must create a new string in each iteration that is composed of the old string with a new character concatenated.

Algorithms like these that perform one pass over their string parameters and execute a constant number of elementary steps per character are called *linear-time algorithms* because their number of elementary steps is proportional to the length of the input string.

In some cases, we need to compare every character in one string to every character in a second string, so we need a nested loop like the following:

```
for index1 in range(len(text1)):
    for index2 in range(len(text2)):
        # process text1[index1] and text2[index2]
```

If both strings have length n, then a nested loop like this constitutes a *quadratic-time algorithm* with time complexity n^2 (as long as the body of the loop is constant-time) because every one of n characters in the first string is compared to every one of n characters in the second string. We will see more loops like this in later chapters.

6.9 FURTHER DISCOVERY

The first of the two epigraphs at the beginning of this chapter is from the following blog post by Leslie Johnston, the former Chief of the Repository Development Center at the Library of Congress. She is currently Director of Digital Preservation at The National Archives.

```
http://blogs.loc.gov/digitalpreservation/2012/04/
a-library-of-congress-worth-of-data-its-all-in-how-you-define-it/
```

The second epigraph is from an article titled "The DNA Data Deluge" by Michael C. Schatz and Ben Langmead [46], which can be found at

`http://spectrum.ieee.org/biomedical/devices/the-dna-data-deluge` .

The Keyword in Context (KWIC) indexing system, also known as a permuted index, is similar to a concordance. In a KWIC index, every word in the title of an article appears in the index in the context in which it appears.

If you are interested in learning more about computational biology, two good places to start are *The Mathematics of Life* [53] by Ian Stewart and *Natural Computing* [51] by Dennis Shasha and Cathy Lazere. The latter book has a wider focus than just computational biology.

6.10 PROJECTS

Project 6.1 Polarized politics

The legislative branch of the United States government, with its two-party system, goes through phases in which the two parties work together productively to pass laws, and other phases in which partisan lawmaking is closer to the norm. In this project, we will use data available on the website of the Clerk of the U.S. House of Representatives (`http://clerk.house.gov/legislative/legvotes.aspx`) to analyze the voting behavior and polarization of the U.S. House over time.

Part 1: Party line votes

We will call a vote *party line* when at least half of the voting members of one party vote one way and at least half of the voting members of the other party vote the other way. In the House, when the vote of every member is recorded, it is called a *roll call* vote. Individual roll call votes are available online at URLs that look like the following:

```
http://clerk.house.gov/evs/<year>/roll<number>.xml
```

The placeholder `<year>` is replaced with the year of the vote and the placeholder `<number>` is replaced with the roll call number. For example, the results of roll call vote 194 from 2010 (the final vote on the Affordable Care Act) are available at

```
http://clerk.house.gov/evs/2010/roll194.xml
```

If the roll call number has fewer than three digits, the number is filled with zeros. For example, the results of roll call vote 11 from 2010 are available at

```
http://clerk.house.gov/evs/2010/roll011.xml
```

Run the following program to view one of these files:

```
import urllib.request as web

def main():
    url = 'http://clerk.house.gov/evs/2010/roll194.xml'
    webpage = web.urlopen(url)
    for line in webpage:
        line = line.decode('utf-8')
        print(line.rstrip())
    webpage.close()

main()
```

You will notice many XML elements contained in matching

```
<recorded-vote> ... </recorded-vote>
```

tags. Each of these elements is a vote of one member of the House.
To begin, write a function

```
partyLine(year, number)
```

that determines whether House roll call vote `number` from the given `year` was a party line vote. Use only the recorded vote elements, and no other parts of the XML files. Do not use any built-in string methods. Be aware that some votes are recorded as Yea/Nay and some are recorded as Aye/No. Do not count votes that are recorded in any other way (e.g., "Present" or "Not Voting"), even toward the total numbers of votes. Test your function thoroughly before moving to the next step.

Question 6.1.1 *What type of data should this function return? (Think about making your life easier for the next function.)*

Question 6.1.2 *Was the vote on the Affordable Care Act a party line vote?*

Question 6.1.3 *Choose another vote that you care about. Was it a party line vote?*

Second, write another function

`countPartyLine(year, maxNumber)`

that uses the `partyLine` function to return the fraction of votes that were party line votes during the given `year`. The parameter `maxNumber` is the number of the last roll call vote of the year.

Finally, write a function

`plotPartyLine()`

that plots the fractions of party line votes for the last 20 years. To keep things simple, you may assume that there were 450 roll call votes each year. If you would prefer to count all of the roll call votes, here are the numbers for each of the last 20 years:

Year	Number	Year	Number
2013	641	2003	677
2012	659	2002	484
2011	949	2001	512
2010	664	2000	603
2009	991	1999	611
2008	690	1998	547
2007	1186	1997	640
2006	543	1996	455
2005	671	1995	885
2004	544	1994	507

Question 6.1.4 *What fraction of votes from each of years 1994 to 2013 went along party lines?*

Note that collecting this data may take a long time, so make sure that your functions are correct for a single year first.

Question 6.1.5 *Describe your plot. Has American politics become more polarized over the last 20 years?*

Question 6.1.6 *Many news outlets report on the issue of polarization. Find a news story about this topic online, and compare your results to the story. It might be helpful to think about the motivations of news outlets.*

Part 2: Blue state, red state

We can use the roll call vote files to infer other statistics as well. For this part, write a function

```
stateDivide(state)
```

that plots the number of Democratic and Republican representatives for the given `state` every year for the last 20 years. Your plot should have two different curves (one for Democrat and one for Republican). You may also include curves for other political parties if you would like. Test your function on several different states.

Call your `stateDivide` function for at least five states, including your home state.

Question 6.1.7 *Describe the curves you get for each state. Interpret your results in light of the results from Part 1. Has the House of Representatives become more polarized?*

*Project 6.2 Finding genes

This project assumes that you have read Section 6.7.

In prokaryotic organisms, the coding regions of genes are usually preceded by the start codon ATG, and terminated by one of three stop codons, TAA, TAG, and TGA. A long region of DNA that is framed by start and stop codons, with no stop codons in between, is called an *open reading frame* (or ORF). Searching for sufficiently long ORFs is an effective (but not fool-proof) way to locate putative genes.

For example, the rectangle below marks a particular open reading frame beginning with the start codon ATG and ending with the stop codon TAA. Notice that there are no other stop codons in between.

```
ggcgg atgaaacgcattagcaccaccattaccaccaccatcaccattaccacaggtaa gttc
```

Not every ORF corresponds to a gene. Since there are $4^3 = 64$ possible codons and 3 different stop codons, one is likely to find a stop codon approximately every $64/3 \approx 21$ codons in a random stretch of DNA. Therefore, we are really only interested in ORFs that have lengths substantially longer than this. Since such long ORFs are unlikely to occur by chance, there is good reason to believe that they may represent a coding region of a gene.

Not all open reading frames on a strand of DNA will be aligned with the left (5') end of the sequence. For instance, in the example above, if we only started searching for open reading frames in the codons aligned with the left end — `ggc`, `gga`, `tga`, etc. — we would have missed the boxed open reading frame. To find all open reading frames in a strand of DNA, we must look at the codon sequences with offsets 0, 1, and 2 from the left end. (See Exercise 6.7.6 for another example.) The codon sequences with these offsets are known as the three *forward reading frames*. In the example above, the three forward reading frames begin as follows:

Reading frame 0: ggc gga tga aac ...

Reading frame 1: gcg gat gaa acg ...

Reading frame 2: cgg atg aaa cgc ...

Because DNA is double stranded, with a reverse complement sequence on the other strand, there are also three *reverse reading frames*. In this strand of DNA, the reverse reading frames would be:

Reverse reading frame 0: gaa ctt acc tgt ...

Reverse reading frame 1: aac tta cct gtg ...

Reverse reading frame 2: act tac ctg tgg ...

Because genomic sequences are so long, finding ORFs must be performed computationally. For example, consider the following sequence of 1,260 nucleotides from an *Escherichia coli* (or *E. coli*) genome, representing only about 0.03% of the complete genome. (This number represents about 0.00005% of the human genome.) Can you find an open reading frame in this sequence?

```
agcttttcattctgactgcaacgggcaatatgtctctgtgtggattaaaaaaagagtgtctgatagcagc
ttctgaactggttacctgccgtgagtaaattaaaattttattgacttaggtcactaaatactttaaccaa
tataggcatagcgcacagacagataaaattacagagtacacaacatccatgaaacgcattagccaccacc
attaccaccaccatcaccattaccacaggtaacggtgcgggctgacgcgtacaggaaacacagaaaaaag
cccgcacctgacagtgcgggctttttttttcgaccaaaggtaacgaggtaacaaccatgcgagtgttgaag
ttcggcggtacatcagtggcaaatgcagaacgttttctgcggttgccgatattctggaaagcaatgcca
ggcaggggcaggtggccaccgtcctctctgccccgccaaaatcaccaaccatctggtagcgatgattga
aaaaaccattagcggtcaggatgctttacccaatatcagcgatgccgaacgtattttgccgaacttctg
acgggactcgccgccgcccagccgggatttccgctggcacaattgaaactttcgtcgaccaggaatttg
cccaaataaaacatgtcctgcatggcatcagtttgttggggcagtgcccggatagcatcaacgctgcgct
gatttgccgtggcgagaaaatgtcgatcgccattatggccggcgtgttagaagcgcgtggtcacaacgtt
accgttatcgatccggtcgaaaaactgctggcagtgggtcattacctcgaatctaccgttgatattgctg
aatccaccgcgtattgcggcaagccgcattccggctgaccacatggtgctgatggctggtttcactgc
cggtaatgaaaaggcgagctggtggttctgggacgcaacggttccgactactccgctgcggtgctggcg
gcctgtttacgcgccgattgttgcgagatctggacggatgttgacggtgtttatacctgcgatccgcgtc
aggtgcccgatgcgaggttgttgaagtcgatgtcctatcaggaagcgatggagcttcttacttcggcgc
taaagttcttcaccccgcaccattaccccatcgcccagttccagatcccttgcctgattaaaaatacc
ggaaatccccaagcaccaggtacgctcattggtgccagccgtgatgaagacgaattaccggtcaagggca
```

This is clearly not a job for human beings. But, if you spent enough time, you might spot an ORF in reading frame 2 between positions 29 and 97, an ORF in reading frame 0 between positions 189 and 254, and another ORF in reading frame 0 between positions 915 and 1073 (to name just a few). These are highlighted in red below.

```
agcttttcattctgactgcaacgggcaatATGTCTCTGTGTGGATTAAAAAAAGAGTGTCTGATAGCAGC
TTCTGAACTGGTTACCTGCCGTGAGTAAattaaaattttattgacttaggtcactaaatactttaaccaa
tataggcatagcgcacagacagataaaaattacagagtacacaacatccATGAAACGCATTAGCACCACC
ATTACCACCACCATCACCATTACCACAGGTAACGGTGCGGGCTGAcgcgtacaggaaacacagaaaaaag
cccgcacctgacagtgcgggcttttttttcgaccaaaggtaacgaggtaacaaccatgcgagtgttgaag
ttcggcggtacatcagtggcaaatgcagaacgttttctgcgggttgccgatattctggaaagcaatgcca
ggcaggggcaggtggccaccgtcctctctgccccgccaaaatcaccaaccatctggtagcgatgattga
aaaaccattagcggtcaggatgctttacccaatatcagcgatgccgaacgtattttgccgaacttctg
acgggactcgccgccgcccagccgggatttccgctggcacaattgaaaactttcgtcgaccaggaatttg
cccaaataaaacatgtcctgcatggcatcagtttgttggggcagtgcccggatagcatcaacgctgcgct
gatttgccgtggcgagaaaatgtcgatcgccattatggccggcgtgttagaagcgcgtggtcacaacgtt
accgttatcgatccggtcgaaaaactgctggcagtgggtcattacctcgaatctaccgttgatattgctg
aatccacccgccgtattgcggcaagccgcattccggctgaccacatggtgctgatggctggtttcactgc
cggtaATGAAAAAGGCGAGCTGGTGGTTCTGGGACGCAACGGTTCCGACTACTCCGCTGCGGTGCTGGCG
GCCTGTTTACGCGCCGATTGTTGCGAGATCTGGACGGATGTTGACGGTGTTTATACCTGCGATCCGCGTC
AGGTGCCCGATGCGAGGTTGTTGAagtcgatgtcctatcaggaagcgatggagcttttcttacttcggcgc
taaagttcttcaccccccgcaccattaccccatcgcccagttccagatcccttgcctgattaaaaatacc
ggaaatccccaagcaccaggtacgctcattggtgccagccgtgatgaagacgaattaccggtcaagggca
```

Of these three, only the second is actually a known gene.

Part 1: Find ORFs

For the first part of this project, we will read a text file containing a long DNA sequence, write this sequence to a turtle graphics window, and then draw bars with intervals of alternating colors over the sequence to indicate the locations of open reading frames. In the image below, which represents just a portion of a window, the three bars represent the three forward reading frames with reading frame 0 at the bottom. Look closely to see how the blue bars begin and end on start and stop codons (except for the top bar which ends further to the right). The two shorter blue bars are actually unlikely to be real genes due to their short length.

To help you get started and organize your project, you can download a "skeleton" of the program from the book web page. In the program, the `viewer` function sets up the turtle graphics window, writes the DNA sequence at the bottom (one character per x coordinate), and then calls the two functions that you will write. The `main` function reads a long DNA sequence from a file and into a string variable, and then calls `viewer`.

To display the open reading frames, you will write the function

```
orf1(dna, rf, tortoise)
```

to draw colored bars representing open reading frames in one forward reading frame with offset `rf` (0, 1, or 2) of string `dna` using the turtle named `tortoise`. This function will be called three times with different values of `rf` to draw the three reading frames. As explained in the skeleton program, the drawing function is already written; you just have to change colors at the appropriate times before calling it. Hint: Use a Boolean variable `inORF` in your `for` loop to keep track of whether you are currently in an ORF.

Also on the book site are files containing various size prefixes of the genome of a particular strain of *E. coli*. Start with the smaller files.

Question 6.2.1 *How long do you think an open reading frame should be for us to consider it a likely gene?*

Question 6.2.2 *Where are the likely genes in the first 10,000 bases of the E. coli genome?*

Part 2: GC content

The GC content of a particular stretch of DNA is the ratio of the number of C and G bases to the total number of bases. For example, in the following sequence

    TCTACGACGT

the GC content is 0.5 because 5/10 of the bases are either C or G. Because the GC content is usually higher around coding sequences, this can also give clues about gene location. (This is actually more true in vertebrates than in bacteria like *E. coli*.) In long sequences, GC content can be measured over "windows" of a fixed size, as in the example we discussed back in Section 1.3. For example, in the tiny example above, if we measure GC content over windows of size 4, the resulting ratios are

    TCTACGACGT

    TCTA   0.25
     CTAC   0.50
      TACG   0.50
       ACGA   0.50
        CGAC   0.75
         GACG   0.75
          ACGT 0.50

We can then plot this like so:

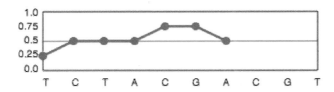

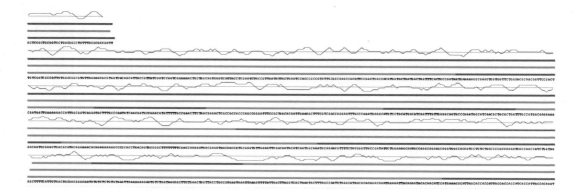

Figure 6.12 The finished product.

Finish writing the function

```
gcFreq(dna, window, tortoise)
```

that plots the GC frequency of the string `dna` over windows of size `window` using the turtle named `tortoise`. Plot this in the same window as the ORF bars. As explained in the skeleton code, the plotting function is already written; you just need to compute the correct GC fractions. As we discussed in Section 1.3, you should not need to count the GC content anew for each window. Once you have counted the G and C bases for the first window, you can incrementally modify this count for the subsequent windows. The final display should look something like Figure 6.12.

CHAPTER 7
Designing programs

From then on, when anything went wrong with a computer, we said it had bugs in it.

Admiral Grace Hopper
upon pulling a two-inch moth from the Harvard Mark I computer in 1945

Every hour of planning saves about a day of wasted time and effort.

Steve McConnell
Software Project Survival Guide (1988)

IT should go without saying that we want our programs to be correct. That is, we want our algorithms to produce the correct output for every possible input, and we want our implementations to be faithful to the design of our algorithms. Unfortunately however, in practice, perfectly correct programs are virtually nonexistent. Due to their complexity, virtually every significant piece of software has errors, or "bugs" in it. But there are techniques that we can use to increase the *likelihood* that our functions and programs are correct. First, we can make sure that we thoroughly understand the problem we are trying to solve and spend quality time designing a solution, well before we start typing any code. Second, we can think carefully about the requirements on our inputs and function parameters, and enforce these requirements to head off problems down the road. And third, we can adopt a thorough testing strategy to make sure that our programs work correctly. In this chapter, we will introduce the basics of these three practices, and encourage you to follow them hereafter. Following these more methodical practices can actually save time in the long run by preventing errors and more easily discovering those that do creep into your programs.

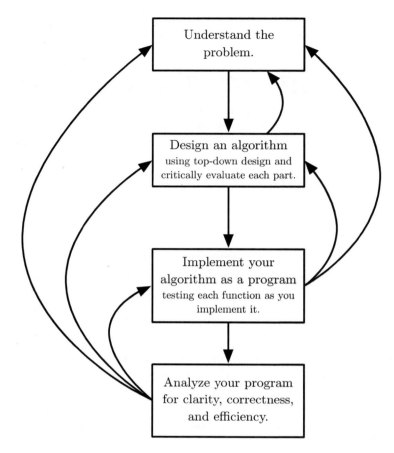

Figure 7.1 A simplified view of the problem solving process.

7.1 HOW TO SOLVE IT

Designing and implementing a computational solution to a problem is a multistep process requiring planning, reflection and revision, not unlike the writing process. This is especially true for large projects involving many people, for which there are established guidelines and formal processes, the primary concerns of a field called *software engineering*. In this book, we are working with relatively small problems, so less formality is necessary. However, it is still very important to develop good habits, represented by the process illustrated in Figure 7.1.

This process, and the title of this section, come from the book *How to Solve It* by mathematician George Polya [41]. Originally published in 1945, Polya's problem solving framework has withstood the test of time. Polya outlined a four step process:

1. *Understand the problem. What is the unknown [output]? What are the data [inputs]? What is the condition [linking the input to the output]?*

2. *Devise a plan to solve the problem.*

3. *Carry out the plan, checking each step.*

4. *Look back. Check the result. Can you derive the result differently?*

These four steps, with some modifications, can be applied just as well to computational problem solving. In this section, we will outline these four steps, and illustrate the process with the following example problem.

Given a text, we would like to measure the distribution of a particular word by breaking the text into "windows" of some length and computing how many times the word appears in each window. The output will be a plot of these frequencies across all of the windows, as well as the average frequency per window.

Understand the problem

As we discussed in the first pages of this book, every useful computation solves a problem with an input and output. Therefore, the first step is to understand the inputs to the problem, the desired outputs, and the connections between them. In addition, we should ask whether there are any restrictions or special conditions relating to the inputs or outputs. If something is unclear, seek clarification right away; it is much better to do so immediately than to wait until you have spent a lot of time working on the wrong problem! It is a good idea at this point to re-explain the problem back to the poser, either orally or in writing. The feedback you get from this exercise might identify points of misunderstanding. You might include a picture, depending upon the type of problem. In addition, work out some examples by hand to make sure you understand all of the requirements and subtleties involved.

The inputs to our problem are a text, a word, and a window length. The outputs are a plot of the frequencies of the word in each window of the text and the average frequency across all of the windows.

Reflection 7.1 *What more do we need to know about the input and output to better understand this problem?*

Where should the text come from? Should we prompt for the word and window length? Should the windows overlap? What should the plot look like? Since we would like to be able to analyze large texts, let's read the text in from a file. But we can prompt for the name of the file, the word and the window length. For simplicity, we will not have overlapping windows. Here is a sketch of the desired plot:

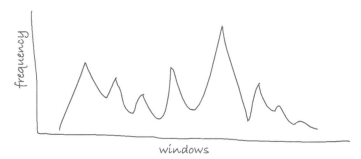

324 ■ Designing programs

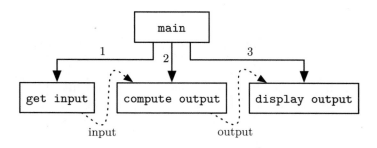

Figure 7.2 A generic top-down design.

The windows over the text will be represented on the x-axis, perhaps by the starting index of each window. The y-axis will correspond to the number of times the word occurs in each window.

Design an algorithm

Once we think we understand the problem to be solved, the next step is to design an algorithm. An important part of this process is to identify smaller *subproblems* that can be solved individually, each with its own function. Once written, these functions can be used as functional abstractions to solve the overall problem. We saw an example of this in the flower-drawing program illustrated in Figure 3.7. This technique is called *top-down design* because it involves starting from the problem to be solved, and then breaking this problem into smaller pieces. Each of these pieces might also be broken down into yet smaller pieces, until we arrive at subproblems that are relatively straightforward to solve. A generic picture of a top-down design is shown in Figure 7.2. The generic program is broken into three main subproblems (i.e., functions). First, a function gets the input. Second, a function computes the output from the input. Third, another function displays the output. The dotted lines represent the flow of information through return values and parameters. The function that gets the input returns it to the main program, which then passes it into the function that computes the output. This function then returns the output to the main program, which passes it into the function that displays it. In Python, this generic program looks like the following:

```
def main():
    input = getInput()
    output = compute(input)
    displayOutput(output)

main()
```

Moving on from this generic structure, the next step is to break these three subproblems into smaller subproblems, as needed, depending on the particular problem. Identifying these subproblems/functions can be as much an art as a science, but here are a few guidelines to keep in mind:

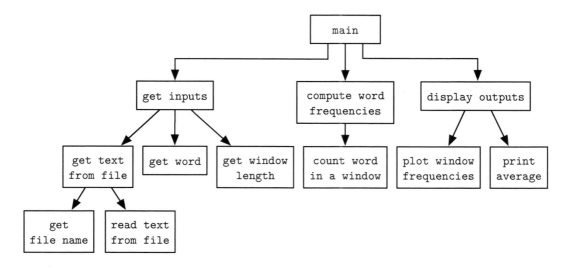

Figure 7.3 A top-down design for the word frequency problem.

1. A subproblem/function should accomplish something relatively small, and make sense standing on its own.

2. Functions should be written for subproblems that may be called upon frequently, perhaps with different arguments.

3. A function should generally fit on a page or, in many cases, less.

4. The **main** function should be short, generally serving only to set up the program and call other functions that carry out the work.

You may find that, during this process, new questions arise about the problem, sending you briefly back to step one. This is fine; it is better to discover this immediately and not later in the process. The problem solving process, like the writing process, is inherently cyclic and fluid, as illustrated in Figure 7.1.

Reflection 7.2 *What steps should an algorithm follow to solve our problem?*

First, we can specialize the generic design by specifying what each of the three main subproblems should do.

1. Get inputs: prompt for a file name, a word, and a window length. Read the text from the file.

2. Compute: a list of word frequencies and the average word frequency.

3. Display outputs: plot the frequencies and print the average frequency.

Since each of these subproblems has multiple parts, we can break each subproblem down further:

1. Get inputs.
   (a) Get the text.
      i. Prompt for and get a file name.
      ii. Read the text from the file with that file name.
   (b) Prompt for and get a word.
   (c) Prompt for and get a window length.
2. Compute a list of word frequencies and the average word frequency.
   (a) In each window, count the frequency of the word.
3. Display outputs:
   (a) Plot the frequencies.
   (b) Print the average frequency.

This outline is represented visually in Figure 7.3. We have broken the input subproblem into three parts, one for each input that we need. The text input part is further broken down into getting the file name and then reading the text from that file. Next, in the computation subproblem, we will need to count the words in each of the windows that we construct, so it will make sense to create another function that counts the words in each of those windows. Finally, we break the output subproblem into two parts, one for the plot and one for the average frequency.

Next, before we implement our top-down design as a program, we need to design an algorithm for each "non-trivial" subproblem. Each of the input and output subproblems is pretty straightforward, but the main computation subproblem is not. It is convenient at this stage to write our algorithm semi-formally in what is known as *pseudocode*. Pseudocode does not have an exact definition, but is somewhere between English and a programming language. When we write algorithms in pseudocode, we need to be guided by our understanding of what can and cannot be done in a programming language, but not necessarily write out every detail. For example, in the algorithm that will find the frequency of the word in every window, we know that we will have to somehow iterate over these windows using a loop, but we do not necessarily have to decide at this point the precise form of that loop. Here is one possible algorithm written in a Python-like pseudocode. We have colored it blue to remind you that it is not an actual Python function.

```
getWordCounts(text, word, window_length)
    total_word_count = 0
    total_window_count = 0
    word_count_list = [ ]
    for each non-overlapping window:
        increment total_window_count
        count = instances of word in window
        add count to total_word_count
        append count to word_count_list
    return (total_word_count / total_window_count) and word_count_list
```

In the algorithm, we are maintaining three main values. The total number of instances of the word that we have found (**total_word_count**) and the total number of windows that we have processed (**total_window_count**) are needed to compute the average count per window. The list of word counts (**word_count_list**), which stores one word count per window, is needed for the plot. The loop iterates over all of the windows and, for each one, finds the count of the word in that window and uses it to update the total word count and the word count list. At the end, we return the average word count and the list of word counts.

Although analyzing your algorithm is the last step in Polya's four-step process, we really should not wait until then to consider the efficiency, clarity, and correctness of our algorithms. At each stage, we need to critically examine our work and look for ways to improve it. For example, if our windows overlapped, we could make the algorithm above more efficient by basing the count in each window on the count in the previous window, as we did in Section 1.3. Notice that this improvement could be made before we implement any code.

Implement your algorithm as a program

Only once your program and function designs are complete should you begin writing your program. This point cannot be emphasized enough. Beginning programmers often rush to the keyboard prematurely, only to find that, after many hours of work, they really do not understand the problem and/or how to solve it. Although it often appears as if getting right to the program will save time, in the long run, the very opposite tends to happen. Expert programmers will sometimes prototype specific pieces of code in very long programs earlier in the process to test whether new ideas are feasible. However, beginning programmers, writing smaller programs like the ones in this book, are always better off with a tentative solution on paper before they start writing any code.

When you do sit down to write your program, your first task should be to lay out your functions, including **main**, with their docstrings. Because the Python interpreter will complain if your function does not contain any statements, you can initially place the "dummy" statement **pass** in the body of your functions. In our design, we have decided to group all of the input in one function because it will be so little code. Also, we have not created a separate function for printing the average count per window because it will be just a single line.

```
def getInputs():
    """Prompt for the name of a text file, a word to analyze, and a
       window length.  Read and return the text, word, and window length.

    Parameters: none

    Return value: a text, word, and window length
    """

    pass
```

```
def count(text, target):
    """Count the number of target strings in text.

    Parameters:
        text: a string object
        target: a string object

    Return value: the number of times target occurs in text
    """

    pass

def getWordCounts(text, word, windowLength):
    """Find the number of times word occurs in each window in the text.

    Parameters:
        text: a string containing a text
        word: a string
        windowLength: the integer length of the windows

    Return values: average count per window and list of window counts
    """

    pass

def plotWordCounts(wordCounts):
    """Plot a list of word frequencies.

    Parameter:
        wordCounts: a list of word frequencies

    Return value: None
    """

    pass

def main():
    pass

    # call getInputs
    # call getWordCounts
    # call plotWordCounts
    # print average count per window

main()
```

In Section 7.2, we will discuss a more formal method for thinking about and enforcing requirements for your parameters and return values. Requirements for parameters are called *preconditions* and requirements for return values (and side effects) are called *postconditions*. For example, in the `getWordCounts` function, we are assuming that `text` and `target` are strings, and that `windowLength` is a positive integer, but we do not check whether they actually are. If we mistakenly pass a negative number in for `windowLength`, an error is likely to ensue. A precondition would state formally that `windowLength` must be a positive integer and make sure that the function does not proceed if it is not.

Once your functions exist, you can begin to implement them, one at a time. Comment your code as you go. While you are working on a function, test it often by calling it from **main** and running the program. By doing so, you can ensure that you always have working code that is free of syntax errors. If you are running your program often, then finding syntax errors when they do crop up will be much easier because they must exist in the last few lines that you wrote.

Once you are pretty sure that a function works, you can initiate a more formal testing process. This is the subject of Section 7.3. You must make sure that each function is working properly in isolation before moving on to the next one. Once you are sure that it works, a function becomes a functional abstraction that you don't need to think about any more, making your job easier! Once each of your functions is working properly, you can assemble your complete program.

For example, we might start by implementing the **getInputs** function:

```
def getInputs():
    """ (docstring omitted) """

    fileName = input('Text file name: ')                    # get file name

    textFile = open(fileName, 'r', encoding = 'utf-8')     # read file
    text = textFile.read()
    textFile.close()

    word = input('Word: ')                                  # get word
    windowLength = int(input('Window length: '))            # get window length

    return text, word, windowLength
```

Now we need to test this function on its own before continuing to the next one. Call the function from **main** and just print the results to make sure it is working correctly.

```
def main():
    testText, testWord, testLength = getInputs()
    print(testText, testWord, testLength)

main()
```

If there were syntax errors or something did not work the way you expected, it is important to fix it now. This process continues for each of the remaining three functions. Then the final **main** function ties them all together:

```
def main():
    text, word, windowLength = getInputs()
    average, wordCounts = getWordCounts(text, word, windowLength)
    plotWordCounts(wordCounts)
    print(word, 'occurs on average', average, 'times per window.')
```

You can find a finished version of the program on the book web site. An example of a plot generated by the program is shown in Figure 7.4.

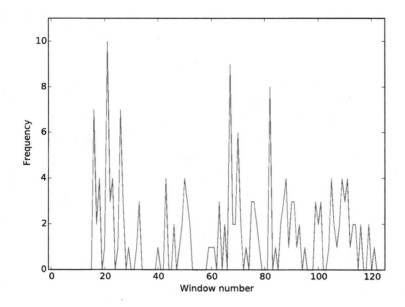

Figure 7.4 Word frequency scatter plot for the word "Pequod" in the text of *Moby Dick* (window length = 10,000).

Analyze your program for clarity, correctness, and efficiency

Once you are done with your program, it is time to reflect on your work as a whole.

- Does your program work properly?

    As you did for each function, test your program with many different inputs, especially ones that seem difficult or that you think might break your program. It is important to test with inputs that are representative of the entire range of possibilities. For example, if your input is a number, try both negative and positive integers and floating point numbers, and zero. If some inputs are not allowed, this should be stated explicitly in the docstring(s), as we will discuss further in the next section.

- Is your program as efficient as it could be?

    Is your program doing too much work? Is there a way to streamline it? As our programs get more complicated, we will see some examples of how to bring down the time complexity of algorithms.

- Is your program easy to read? Is it commented properly? Is there anything extraneous that could be omitted?

    Refer back to Section 3.4 for commenting and program structure guidelines.

- Are there chunks of code that should be made into functions to improve readability?

You may very well decide in this process that revisions are necessary, sending you back up to an earlier step. Again, this is completely normal and encouraged. No one ever designs a perfect solution on a first try!

Keep these steps and Figure 7.1 in mind as we move forward. We will gradually begin to solve more complex problems in which these steps will be crucial to success.

*7.2 DESIGN BY CONTRACT

There are often subtleties in a problem definition that are easily overlooked, leading to serious problems later. In this section, we will follow a more rigorous and nuanced approach to thinking about the input and output of a problem, and where things might go awry in an algorithm. We will start by looking at some small functions from Section 6.3, and then revisit the program from the previous section.

Preconditions and postconditions

Part of understanding the input to a problem is understanding any restrictions on it. For example, consider the `digit2String` function from Section 6.3.

```
def digit2String(digit):
    """Converts an integer digit to its corresponding string
       representation.

    Parameter:
        digit: an integer in 0, 1, ..., 9

    Return value:
        the corresponding character '0', '1', ..., '9'
        or None if a non-digit is passed as an argument
    """

    if (digit < 0) or (digit > 9):
        return None
    return chr(ord('0') + digit)
```

Recall that this function is meant to convert an integer digit between 0 and 9 to its associated character. For example, if the input were 4, the output should be the string '4'. We noted when we designed this function that inputs less than 0 or greater than 9 would not work, so we returned None in those cases. The requirement that digit must be integer between 0 and 9 is called a *precondition*. A precondition for a function is something that must be true when the function is called for it to behave correctly. Put another way, we do not guarantee that the function will give the correct output when the precondition is false.

On the other hand, the *postcondition* for a function is a statement of what must be true when the function finishes. A postcondition usually specifies what the function returns and what, if any, side effects it has. Recall that a side effect occurs when a global variable is modified or some other event in a function modifies a global resource. For example, calls to `print` and modifications of files can be considered side effects because they have impacts outside of the function itself.

Reflection 7.3 *What is the postcondition of the* `digit2String` *function?*

The postcondition for the `digit2String` function should state that the function returns a string representation of its input.

The use of preconditions and postconditions is called *design by contract* because the precondition and postcondition establish a contract between the function designer and the function caller. The function caller understands that the precondition must be met before the function is called and the function designer guarantees that, if the precondition is met, the postcondition will also be met.

Because they describe the input and output of a function, and therefore how the function can be used, preconditions and postconditions should be included in the docstring for a function. If they include information about the parameters and return value, they can replace those parts of the docstring. For example, a new docstring for `digit2String` would look like:

```
def digit2String(digit):
    """Converts an integer digit to its corresponding string
        representation.

    Precondition: digit is an integer between 0 and 9

    Postcondition: returns the string representation of digit, or
                   None if digit is not an integer between 0 and 9
    """
```

Checking parameters

The `digit2String` function partially "enforces" the precondition by returning `None` if `digit` is not between 0 and 9. But there is another part of the precondition that we do not handle.

Reflection 7.4 *What happens if the argument passed in for* `digit` *is not an integer?*

Calling `digit2String('cookies')` results in the following error:

```
TypeError: unorderable types: str() < int()
```

This error happens in the `if` statement when the value of `digit`, in this case `'cookies'`, is compared to 0. To avoid this rather opaque error message, we should make sure that `digit` is an integer at the beginning of the function. We can do so using the `isinstance` function, which takes two arguments: a variable name (or value) and the name of a class. The function returns `True` if the variable or value refers to an object (i.e., instance) of the class, and `False` otherwise. For example,

```
>>> isinstance(5, int)
True
>>> isinstance(5, float)
False
>>> s = 'a string'
>>> isinstance(s, str)
True
>>> isinstance(s, float)
False
```

We can use `isinstance` in the `if` statement to make `digit2String` more robust:

```
def digit2String(digit):
    """ (docstring omitted) """

    if not isinstance(digit, int) or (digit < 0) or (digit > 9):
        return None
    return chr(ord('0') + digit)
```

Now let's consider how we might use this function in a computation. For example, we might want to use `digit2String` in the solution to Exercise 6.3.10 to convert a positive integer named `value` to its corresponding string representation:

```
def int2String(value):
    """Convert an integer into its corresponding string representation.

    Parameter:
        value: an integer

    Return value: the string representation of value
    """

    intString = ''
    while value > 0:
        digit = value % 10
        value = value // 10
        intString = digit2String(digit) + intString
    return intString
```

In each iteration of the `while` loop, the rightmost digit of `value` is extracted and `value` is divided by ten, the digit is converted to a string using `digit2String`, and the result is prepended to the beginning of a growing string representation.

Reflection 7.5 *What happens if* `value` *is a floating point number, such as* `123.4`?

If `value` is `123.4`, then the first value of `digit` will be `3.4`. Because this is not an integer, `digit2String(digit)` will return `None`. This, in turn, will cause an error when we try to prepend `None` to the string:

```
TypeError: unsupported operand type(s) for +: 'NoneType' and 'str'
```

Because we clearly stated in the docstring of `digit2String` that the function would behave this way, by the principle of design by contract, the `int2String` function, as the caller of `digit2String`, is responsible for catching this possibility and handling it appropriately. In particular, it needs to check whether `digit2String` returns `None` and, if it does, do something appropriate. For example, we could just abort and return `None`:

```
def int2String(value):
    """ (docstring omitted) """

    intString = ''
    while value > 0:
        digit = value % 10
        value = value // 10
        digitStr = digit2String(digit)
        if digitStr != None:
            intString = digit2Str + intString
        else:
            return None
    return intString
```

Reflection 7.6 *What are some alternatives to this function returning* `None` *in the event that* `value` *is a floating point number?*

Alternatively, we could return a string representing the integer portion of `value`. In other words, if `value` were assigned `123.4`, it could return the string `'123'`. But this would need to be stated clearly in the postcondition of the function. Exercises 7.2.1 and 7.2.2 ask you to write a precondition and postcondition for this function, and then modify the function so that it works correctly for all inputs.

Assertions

A stricter alternative is to display an error and abort the entire program if a parameter violates a precondition. This is precisely what happened above when the Python interpreter displayed a `TypeError` and aborted when the + operator tried to concatenate `None` and a string. These now-familiar `TypeError` and `ValueError` messages are called *exceptions*. A built-in function "raises an exception" when something goes wrong and the function cannot continue. When a function raises an exception, it does not return normally. Instead, execution of the function ends at the moment the exception is raised and execution instead continues in part of the Python interpreter called an *exception handler*. By default, the exception handler prints an error message and aborts the entire program.

It is possible for our functions to also raise `TypeError` and `ValueError` exceptions, but the mechanics of how to do so are beyond the scope of this book. We will instead consider just one particularly simple type of exception called an `AssertionError`, which may be raised by an `assert` statement. An `assert` statement tests a Boolean condition, and raises an `AssertionError` if the condition is false. If the condition is true, the `assert` statement does nothing. For example,

```
>>> value = -1
>>> assert value < 0
>>> assert value > 0
AssertionError
>>> assert value > 0, 'value must be positive'
AssertionError: value must be positive
```

The first `assert` statement above does nothing because the condition being asserted is `True`. But the condition in the second `assert` statement is `False`, so an `AssertionError` exception is raised. The third `assert` statement demonstrates that we can also include an informative message to accompany an `AssertionError`.

We can replace the `if` statement in our `digit2String` function with assertions to catch both types of errors that we discussed previously:

```
def digit2String(digit):
    """ (docstring omitted) """

    assert isinstance(digit, int), 'digit must be an integer'
    assert (digit >= 0) and (digit <= 9), 'digit must be in [0..9]'

    return chr(ord('0') + digit)
```

If `digit` is not an integer, then the first `assert` statement will display

```
AssertionError: digit must be an integer
```

and abort the program. If `digit` *is* an integer (and therefore gets past the first assertion) but is not in the correct range, the second `assert` statement will display

```
AssertionError: digit must be in [0..9]
```

and abort the program.

Reflection 7.7 *Call this modified version of* `digit2String` *from the* `int2String` *function. What happens now when you call* `int2String(123.4)`?

Note that, since the `assert` statement aborts the entire program, it should only be used in circumstances in which there is no other reasonable course of action. But the definition of "reasonable" usually depends on the circumstances.

For another example, let's look at the `count` and `getWordCounts` functions that we discussed in the previous section. We did not implement the functions there, so they are shown below. We have added to the `count` function a precondition and postcondition, and an assertion that tests the precondition.

```python
def count(text, target):
    """Count the number of target strings in text.

    Precondition: text and target are string objects

    Postcondition: returns number of occurrences of target in text
    """
    assert isinstance(text, str) and isinstance(target, str), \
           'arguments must be strings'
    count = 0
    for index in range(len(text) - len(target) + 1):
        if text[index:index + len(target)] == target:
            count = count + 1
    return count

def getWordCounts(text, word, windowLength):
    """Find the number of times word occurs in each window in the text.

    Parameters:
        text: a string containing a text
        word: a string
        windowLength: the integer length of the windows

    Return values: average count per window and list of window counts
    """
    wordCount = 0
    windowCount = 0
    wordCounts = []
    for index in range(0, len(text) - windowLength + 1, windowLength):
        window = text[index:index + windowLength]
        windowCount = windowCount + 1
        wordsInWindow = count(window, word)
        wordCount = wordCount + wordsInWindow
        wordCounts.append(wordsInWindow)
    return wordCount / windowCount, wordCounts
```

Design by contract ■ 337

Reflection 7.8 *What are a suitable precondition and postcondition for the* getWordCounts *function? Should the precondition state anything about the values of* text *or* word*? (What happens if* text *is too short?)*

The precondition should state that both text and word are strings, and wordLength is a positive integer. In addition, special attention must be paid to the contents of text. If the length of text is smaller than windowLength, then there will be no iterations of the loop. In this case, windowCount will remain at zero and the division in the return statement will result in an error. Depending on the circumstances, we might also want to prescribe some restrictions on the value of word but, in general, this is not necessary for the function to work correctly. Incorporating these requirements, a suitable precondition might look like this:

```
Precondition: text and word are string objects and
              windowLength is a positive integer and
              text contains at least windowLength characters
```

Assuming the precondition is met, the postcondition is relatively simple:

```
Postcondition: returns a list of word frequencies and the
               average number of occurrences of word per window
```

Not satisfying any one of the precondition requirements will break the getWordCounts function, so let's check that the requirements are met with assert statements. First, we can check that text and word are strings with the following assert statement:

```
assert isinstance(text, str) and isinstance(word, str), \
       'first two arguments must be strings'
```

We can check that windowLength is a positive integer with:

```
assert isinstance(windowLength, int) and windowLength > 0, \
       'window length must be a positive integer'
```

To ensure that we avoid dividing by zero, we need to assert that the text is at least as long as windowLength:

```
assert len(text) >= windowLength, \
       'window length must be shorter than the text'
```

Finally, it is not a bad idea to add an extra assertion just before the division at the end of the function:

```
assert windowCount > 0, 'no windows were found'
```

Although our previous assertions should prevent this assertion from ever failing, being a defensive programmer is always a good idea. Incorporating these changes into the function looks like this:

```
def getWordCounts(text, word, windowLength):
    """Find the number of times word occurs in each window in the text.

    Precondition: text and word are string objects and
                  windowLength is a positive integer and
                  text contains at least windowLength characters

    Postcondition: returns a list of word frequencies and the
                   average number of occurrences of word per window
    """
    assert isinstance(text, str) and isinstance(word, str), \
        'first two arguments must be strings'

    assert isinstance(windowLength, int) and windowLength > 0, \
        'window length must be a positive integer'

    assert len(text) >= windowLength, \
        'window length must be shorter than the text'

    wordCount = 0
    windowCount = 0
    wordCounts = []
    for index in range(0, len(text) - windowLength + 1, windowLength):
        window = text[index:index + windowLength]
        windowCount = windowCount + 1
        wordsInWindow = count(window, word)
        wordCount = wordCount + wordsInWindow
        wordCounts.append(wordsInWindow)

    assert windowCount > 0, 'no windows were found'

    return wordCount / windowCount, wordCounts
```

In the next section, we will discuss a process for more thoroughly testing these functions once we think they are correct.

Exercises

7.2.1. Write a suitable precondition and postcondition for the `int2String` function.

7.2.2. Modify the `int2String` function with `if` statements so that it correctly satisfies the precondition and postcondition that you wrote in the previous exercise, and works correctly for all possible inputs.

7.2.3. Modify `int2String` function with `assert` statements so that it correctly satisfies the precondition and postcondition that you wrote in Exercise 7.2.1, and works correctly for all possible inputs.

For each of the following functions from earlier chapters, (a) write a suitable precondition and postcondition, and (b) add `assert` *statements to enforce your precondition. Be sure to include an informative error message with each* `assert` *statement.*

7.2.4.
```
def volumeSphere(radius):
    return (4 / 3) * math.pi * (radius ** 3)
```

7.2.5.
```
def fair(employee, ceo, ratio):
    return (ceo / employee <= ratio)
```

7.2.6.
```
def windChill(temp, wind):
    chill = 13.12 + 0.6215 * temp + \
            (0.3965 * temp - 11.37) * wind ** 0.16
    return round(chill)
```
Note that wind chill is only defined for temperatures at or below 10° C and wind speeds above 4.8 km/h.

7.2.7.
```
def plot(tortoise, n):
    for x in range(-n, n + 1):
        tortoise.goto(x, x * x)
```

7.2.8.
```
def pond(years):
    population = 12000
    print('Year  Population')
    print('{0:<4}  {1:>9.2f}'.format(0, population))
    for year in range(1, years + 1):
        population = 1.08 * population - 1500
        print('{0:<4}  {1:>9.2f}'.format(year, population))
    return population
```

7.2.9.
```
def decayC14(originalAmount, years, dt):
    amount = originalAmount
    k = -0.00012096809434
    numIterations = int(years / dt) + 1
    for i in range(1, numIterations):
        amount = amount + k * amount * dt
    return amount
```

7.2.10.
```
def argh(n):
    return 'A' + ('r' * n) + 'gh!'
```

7.2.11.
```
def reverse(word):
    newstring = ''
    for index in range(len(word) - 1, -1, -1):
        newstring = newstring + word[index]
    return newstring
```

7.2.12.
```
def find(text, target):
    for index in range(len(text) - len(target) + 1):
        if text[index:index + len(target)] == target:
            return index
    return -1
```

7.2.13.
```
def lineNumbers(fileName):
    textFile = open(fileName, 'r', encoding = 'utf-8')
    lineCount = 1
    for line in textFile:
        print('{0:<5} {1}'.format(lineCount, line[:-1]))
        lineCount = lineCount + 1
    textFile.close()
```

To determine whether a file exists, you can use:

```
import os.path
assert os.path.isfile(fileName)
```

Then, to make sure you can read the file, you can use:

```
import os
assert os.access(fileName, os.R_OK)
```

The value `os.R_OK` is telling the function to determine whether reading the file is OK.

*7.3 TESTING

Once we have carefully defined and implemented a function, we need to test it thoroughly. You have surely been testing your functions all along, but now that we have written some more involved ones, it is time to start taking a more deliberate approach.

It is very important to test each function that we write *before* we move on to other functions. The process of writing a program should consist of multiple iterations of *design—implement—test*, as we illustrated in Figure 7.1 in Section 7.1. Once you come up with an overall design and identify what functions you need, you should follow this *design—implement—test* process for each function individually. If you do not follow this advice and instead test everything for the first time when you think you are done, it will likely be very hard to discern where your errors are.

The only way to really ensure that a function is correct is to either mathematically prove it is correct or test it with every possible input (i.e., every possible parameter value). But, since both of these strategies are virtually impossible in all but the most trivial situations, the best we can do is to test our functions with a variety of carefully chosen inputs that are representative of the entire range of possibilities. In large software companies, there are dedicated teams whose sole jobs are to design and carry out tests of individual functions, the interplay between functions, and the overall software project.

Unit testing

We will group our tests for each function in what is known as a *unit test*. The "unit" in our case will be an individual function, but in general it may be any block of code with a specific purpose. Each unit test will itself be a function, named `test_` followed by the name of the function that we are testing. For example, our unit test function for the `count` function will be named `test_count`. Each unit test function will contain several individual tests, each of which will `assert` that calling the function with a particular set of parameters returns the correct answer.

To illustrate, we will design unit tests for the `count` and `getWordCounts` functions from the previous section, reproduced below in a simplified form. We have omitted docstrings and the messages attached to the `assert` statements, simplified `getWordCounts` slightly by removing the list of window frequencies, and simplified the `main` function by removing the plot.

```
def getInputs():
    :    # code omitted

def count(text, target):
    assert isinstance(text, str) and isinstance(target, str)

    count = 0
    for index in range(len(text) - len(target) + 1):
        if text[index:index + len(target)] == target:
            count = count + 1
    return count

def getWordCounts(text, word, windowLength):
    assert isinstance(text, str) and isinstance(word, str)
    assert isinstance(windowLength, int) and windowLength > 0
    assert len(text) >= windowLength

    wordCount = 0
    windowCount = 0
    for index in range(0, len(text) - windowLength + 1, windowLength):
        window = text[index:index + windowLength]
        windowCount = windowCount + 1
        wordsInWindow = count(window, word)
        wordCount = wordCount + wordsInWindow

    assert windowCount > 0
    return wordCount / windowCount

def main():
    text, word, windowLength = getInputs()
    average = getWordCounts(text, word, windowLength)
    print(word, 'occurs on average', average, 'times per window.')

main()
```

We will place all of the unit tests for a program in a separate file to reduce clutter in our program file. If the program above is saved in a file named `wordcount.py`, then the unit test file will be named `test_wordcount.py`, and have the following structure:

```
from wordcount import count, getWordCounts

def test_count():
    """Unit test for count"""

    # tests of count here

    print('Passed all tests of count!')

def test_getWordCounts():
    """Unit test for getWordCounts"""

    # tests of getWordCounts here

    print('Passed all tests of getWordCounts!')

def test():
    test_count()
    test_getWordCounts()

test()
```

The first line imports the two functions that we are testing from `wordcount.py` into the global namespace of the test program. Recall from Section 3.6 that a normal `import` statement creates a new namespace containing all of the functions from an imported module. Instead, the

```
from <module name> import <function names>
```

form of the `import` statement imports functions into the *current* namespace. The advantage is that we do not have to preface every call to `count` and `getWordCounts` with the name of the module. In other words, we can call

```
count('This is fun.', 'is')
```

instead of

```
wordcount.count('This is fun.', 'is')
```

The `import` statement in the program is followed by the unit test functions and a main `test` function that calls all of these unit tests.

Regression testing

When we test our program, we will call `test()` instead of individual unit test functions. Besides being convenient, this technique has the advantage that, when we test new functions, we also re-test previously tested functions. If we make changes to any one function in a program, we want to *both* make sure that this change worked *and* make sure that we have not inadvertently broken something that was working earlier. This idea is called *regression testing* because we are making sure that our program has not *regressed* to an earlier error state.

Before we design the actual unit tests, there is one more technical matter to deal with. Although we have not thought about it this way, the `import` statement both imports names from another file *and* executes the code in that file. Therefore, when `test_wordcount.py` imports the functions from `wordcount.py` into the global namespace, the code in `wordcount.py` is executed. This means that `main` will be called, and therefore the entire program in `wordcount.py` will be executed, before `test()` is called. But we only want to execute `main()` when we run `wordcount.py` as the main program, not when we import it for testing. To remedy this situation, we need to place the call to `main()` in `wordcount.py` inside the following conditional statement:

```
if __name__ == '__main__':
    main()
```

You may recall from Section 3.6 that `__name__` is the name of the current module, which is assigned the value `'__main__'` if the module is run as the main program. When `wordcount.py` is imported into another module instead, `__name__` is set to `'wordcount'`. So this statement executes `main()` only if the module is run as the main program.

Designing unit tests

Now that we have the infrastructure in place, let's design the unit test for the `count` function. We will start with a few easy tests:

```
def test_count():
    quote = 'Diligence is the mother of good luck.'
    assert count(quote, 'the') == 2
    assert count(quote, 'them') == 0

    print('Passed all tests of count!')
```

Reflection 7.9 *What is printed by the* `assert` *statements in the* `test_count` *function?*

If the `count` function is working correctly, *nothing* should be printed by the `assert` statements. On the other hand, if the `count` function were to fail one of the tests, the program would abort at that point with an `AssertionError` exception. To see this, change the second assertion to (incorrectly) read `assert count(quote, 'them') == 1` instead. Then rerun the program. You should see

> **Box 7.1: Unit testing frameworks**
>
> Python also provides two modules to specifically facilitate unit testing. The `doctest` module provides a way to incorporate tests directly into function docstrings. The `unittest` module provides a fully featured unit testing framework that automates the unit and regression testing process, and is more similar to the approach we are taking in this section. If you would like to learn more about these tools, visit
>
> http://docs.python.org/3/library/development.html

```
Traceback (most recent call last):

  ⋮ (order of function calls displayed here)

    assert count(quote, 'them') == 1
AssertionError
```

This error tells you which assertion failed so that you can track down the problem. (In this case, there is no problem; change the assertion back to the correct version.)

On their own, the results of these two tests do not provide enough evidence to show that the `count` function is correct. As we noted earlier, we need to choose a variety of tests that are representative of the entire range of possibilities. The input that we use for a particular test is called a *test case*. We can generally divide test cases into three categories:

1. *Common cases.* First, test the function on several straightforward inputs to make sure that its basic functionality is intact. Be sure to choose test cases that result in a range of different answers. For example, for the `count` function, we might create the following tests (which include the two above):

    ```
    quote = 'Diligence is the mother of good luck.'
    assert count(quote, 'the') == 2
    assert count(quote, 'the ') == 1
    assert count(quote, 'them') == 0
    assert count(quote, ' ') == 6
    ```

   Notice that we have chosen inputs that result in a variety of results, including the case in which the target string is not found.

2. *Boundary cases.* A *boundary case* is an input that rests on a boundary of the range of legal inputs. In the case of the `count` function, the input is two strings, so we want to test the function on strings with the minimum possible lengths and on strings with lengths that are close to the minimum. (If there were a maximum length, we would want to test that too.)

   The minimum length string is the empty string (''). So we need to design test cases for the `count` function in which either parameter is the empty string. We will start with cases in which `target` is the empty string.

Reflection 7.10 *What should the correct answer be for* `count(quote, '')`? *In other words, how many times does the empty string occur in a non-empty string?*

This question is similar to asking how many zeros there are in 10. In mathematics, this number (10/0) is undefined, and Python gives an error. Should counting the number of empty strings in a string also result in an error? Or perhaps the answer should be 0; in other words, there are no empty strings in a non-empty string.

Let's suppose that we want the function to return zero if `target` is the empty string. So we add the following three test cases to the `test_count` function:

```
assert count(quote, '') == 0
assert count('a', '') == 0
assert count('', '') == 0
```

Notice that we passed in strings with three different lengths for `text`.

Reflection 7.11 *Does the* `count` *function pass these tests? What does* `count(quote, '')` *actually return? Look at the code and try to understand why. For comparison, what does the built-in string method* `quote.count('')` *return?*

The return values of both `count(quote, '')` and `quote.count('')` are 38, which is `len(quote) + 1`, so the tests fail. This happens because, when `target` is the empty string, every test of the `if` condition, which evaluates to

```
if text[index:index] == '':
```

is true. Since the loop iterates `len(text) + 1` times, each time incrementing `count`, `len(text) + 1` is assigned to `count` at the end.

This is a situation in which testing has turned up a case that we might not have noticed otherwise. Since our expectation differs from the current behavior, we need to modify the `count` function by inserting an `if` statement before the `for` loop:

```
def count(text, target):
    assert isinstance(text, str) and isinstance(target, str)

    if target == '':
        return 0

    count = 0
    for index in range(len(text) - len(target) + 1):
        if text[index:index + len(target)] == target:
            count = count + 1
    return count
```

With this change, all of the tests should pass.

Now what if `text` is the empty string? It seems reasonable to expect that, if `text` is empty, then `count` should always return zero. So we add two more test cases, one with a long `target` and the other with a short `target`:

```
assert count('', quote) == 0
assert count('', 'a') == 0
```

Reflection 7.12 *What does* `count('', quote)` *actually return?*

In this case, both tests pass. If `text` is the empty string and `target` contains at least one character, then the range of the `for` loop is empty and the function returns zero.

In addition to tests with the empty string, we should add a few tests in which both strings are close to the empty string:

```
assert count(quote, 'e') == 4
assert count('e', 'e') == 1
assert count('e', 'a') == 0
```

Another kind of boundary case involves situations in which an algorithm is searching near the beginning or end of a string (or, in the next chapter, a list). For example, we should make sure that our function correctly counts substrings that appear at the beginning and end of `text`:

```
assert count(quote, 'D') == 1
assert count(quote, 'Di') == 1
assert count(quote, 'Dx') == 0
assert count(quote, '.') == 1
assert count(quote, 'k.') == 1
assert count(quote, '.x') == 0
```

Notice that we have also included two cases in which the target is close to something at the beginning and end, but does not actually match.

3. *Corner cases.* A *corner case* is any other kind of rare input that might cause the function to break. For the `count` function, our boundary cases took care of most of these. But two other unusual cases to check might be if the `text` and `target` are the same, and if `text` is shorter than `target`:

   ```
   assert count(quote, quote) == 1
   assert count('the', quote) == 0
   ```

   Thinking up pathological corner cases is an acquired skill that comes with experience. Many companies pay top dollar for programmers whose sole job is to discover corner cases that break their software!

Putting all of these test cases together, our unit test for the `count` function looks like this:

```
def test_count():
    quote = 'Diligence is the mother of good luck.'
    assert count(quote, 'the') == 2        # common cases
    assert count(quote, 'the ') == 1
    assert count(quote, 'them') == 0
    assert count(quote, ' ') == 6
    assert count(quote, '') == 0            # boundary cases
    assert count('a', '') == 0
    assert count('', '') == 0
    assert count('', quote) == 0
    assert count('', 'a') == 0
    assert count(quote, 'e') == 4
    assert count('e', 'e') == 1
    assert count('e', 'a') == 0
    assert count(quote, 'D') == 1
    assert count(quote, 'Di') == 1
    assert count(quote, 'Dx') == 0
    assert count(quote, '.') == 1
    assert count(quote, 'k.') == 1
    assert count(quote, '.x') == 0
    assert count(quote, quote) == 1         # corner cases
    assert count('the', quote) == 0

    print('Passed all tests of count!')
```

Testing floating point values

Once we are sure that the `count` function works correctly, we can go on to implement and test the `getWordCounts` function. The beginning of a unit test for this function might look like the following. (We will leave the remaining tests as an exercise.)

```
def test_getWordCounts():
    text = 'Call me Ishmael. Some years ago--never mind how long \
precisely--having little or no money in my purse, and nothing \
particular to interest me on shore, I thought I would sail about \
a little and see the watery part of the world.  It is a way I have \
of driving off the spleen and regulating the circulation.'

    assert getWordCounts(text, 'the', 20) == 4 / 15
    assert getWordCounts(text, 'spleen', 20) == 1 / 15

    print('Passed all tests of getWordCounts!')
```

Reflection 7.13 *Why did we use* 4 / 15 *and* 1 / 15 *in the tests above instead of something like* 0.267 *and* 0.067?

Using these floating point approximations of 4/15 and 1/15 would cause the assertions to fail. For example, try this:

```
>>> assert 4 / 15 == 0.267    # fails
```

In general, you should *never* test for equality between floating point numbers. There are two reasons for this. First, the value you use may not accurately represent the correct value that you are testing against. This was the case above. To get `assert 4 / 15 == 0.267` to pass, you would have to add a lot more digits to the right of the decimal point (e.g., 0.26666 ⋯ 66). But, even then, the number of digits in the value of 4 / 15 may depend on your specific computer system, so even using more digits is a bad idea. Second, as we discussed in Sections 2.2 and 4.4, floating point numbers have finite precision and are therefore approximations. For example, consider the following example from Section 4.4.

```
sum = 0
for index in range(1000000):
    sum = sum + 0.0001
assert sum == 100.0
```

This loop adds one ten-thousandths one million times, so the answer should be one hundred, as reflected in the `assert` statement. However, the `assert` fails because the value of `sum` is actually slightly greater than 100 due to rounding errors. To deal with this inconvenience, we should always test floating point values within a range instead. For example, the `assert` statement above should be replaced by

```
assert sum > 99.99 and sum < 100.01
```

or

```
assert sum > 99.99999 and sum < 100.00001
```

The size of the range that you test will depend on the accuracy that is necessary in your particular application.

Let's apply this idea to the familiar `volumeSphere` function:

```
def volumeSphere(radius):
    return (4 / 3) * math.pi * (radius ** 3)
```

To generate some test cases for this function, we would figure out what the answers should be for a variety of different values of `radius` and then write `assert` statements for each of these test cases. For example, the volume of a sphere with radius 10 is about 4188.79. So our `assert` statement should look something like

```
assert volumeSphere(10) > 4188.789 and volumeSphere(10) < 4188.791
```

Exercises

7.3.1. Finish the unit test for the `getWordCounts` function. Be sure to include both boundary and corner cases.

7.3.2. Design a thorough unit test for the `digit2String` function from Section 7.2.

7.3.3. Design a thorough unit test for the `int2String` function from Section 7.2.

7.3.4. Design a thorough unit test for the `assignGP` function below.

```
def assignGP(score):
    if score >= 90:
        return 4
    if score >= 80:
        return 3
    if score >= 70:
        return 2
    if score >= 60:
        return 1
    return 0
```

7.3.5. Design a thorough unit test for the `volumeSphere` function in Exercise 7.2.4.

7.3.6. Design a thorough unit test for the `windChill` function in Exercise 7.2.6.

7.3.7. Design a thorough unit test for the `decayC14` function in Exercise 7.2.9.

7.3.8. Design a thorough unit test for the `reverse` function in Exercise 7.2.11.

7.3.9. Design a thorough unit test for the `find` function in Exercise 7.2.12.

7.4 SUMMARY

Designing correct algorithms and programs requires following a careful, reflective process. We outlined an adaptation of Polya's *How to Solve It* with four main steps:

1. Understand the problem.

2. Design an algorithm, starting from a top-down design.

3. Implement your algorithm as a program.

4. Analyze your program for clarity, correctness, and efficiency.

It is important to remember, however, that designing programs is not a linear process. For example, sometimes after we start programming (step 3), we notice a more efficient way to organize our functions and return to step 2. And, while testing a function (step 4), we commonly find errors that need to be corrected, returning us to step 3. So treat these four steps as guidelines, but always allow some flexibility.

In the last two sections, we formalized the process of designing and testing our functions. In the design process, we introduced *design by contract* using *preconditions* and *postconditions*, and the use of `assert` statements to enforce them. Finally, we introduced *unit testing* and *regression testing* as means for discovering errors in your functions and programs. In the remaining chapters, we encourage you to apply design by contract and unit testing to your projects. Although this practice requires more time up front, it virtually always ends up saving time overall because less time is wasted chasing down hard-to-find errors.

7.5 FURTHER DISCOVERY

The first epigraph is from an April 1984 *Time* magazine article titled "The Wizard inside the Machine" about Admiral Grace Hopper [55]. The second epigraph is from the *Software Project Survival Guide* by Steve McConnell [32] (p. 36).

Grace Hopper was an extraordinarily influential figure in the early development of programming languages and compilers. The original room-sized computers of the 1940's and 1950's had to be programmed directly in binary code by highly trained experts. But, by developing early compilers, Admiral Hopper demonstrated that programming could be made more user-friendly. Grace Hopper earned a Ph.D. in mathematics from Yale University in 1934, and taught at Vassar College until 1943, when she left to join the U.S. Naval Reserve. After World War II, she worked with early computers for both industry and the military. She retired from the Navy in 1986 with the rank of Rear Admiral. You can find a detailed biography at

`http://www.history.navy.mil/bios/hopper_grace.htm` .

A biographical video was also recently produced for the web site `fivethirtyeight.com`. You can watch it at

`http://fivethirtyeight.com/features/the-queen-of-code/` .

CHAPTER 8

Data analysis

"Data! Data! Data!" he cried impatiently. "I can't make bricks without clay."

Sherlock Holmes
The Adventure of the Copper Beeches (1892)

I<small>N</small> Chapter 6, we designed algorithms to analyze and manipulate text, which is stored as a sequence of characters. In this chapter, we will design algorithms to process and learn from more general collections of data. The problems in this chapter involve earthquake measurements, SAT scores, isotope ratios, unemployment rates, meteorite locations, consumer demand, river flow, and more. Data sets such as these have become a (if not, *the*) vital component of many scientific, non-profit, and commercial ventures. Many of these now employ experts in *data science* and/or *data mining* who use advanced techniques to transform data into valuable information to guide the organization.

To solve these problems, we need an abstract data type (ADT) in which to store a collection of data. The simplest and most intuitive such abstraction is a *list*, which is simply a sequence of items. In previous chapters, we discovered how to generate Python `list` objects with the `range` function and how to accumulate lists of coordinates to visualize in plots. To solve the problems in this chapter, we will also grow and shrink lists, and modify and rearrange their contents, without having to worry about where or how they are stored in memory.

8.1 SUMMARIZING DATA

A list is represented as a sequence of items, separated by commas, and enclosed in square brackets (`[ ]`). Lists can contain any kind of data we want; for example:

352 ■ Data analysis

```
>>> sales = [32, 42, 11, 15, 58, 44, 16]
>>> unemployment = [0.082, 0.092, 0.091, 0.063, 0.068, 0.052]
>>> votes = ['yea', 'yea', 'nay', 'yea', 'nay']
>>> points = [[2, 1], [12, 3], [6, 5], [3, 14]]
```

The first example is a list representing hundreds of daily sales, the second example is a list of the 2012 unemployment rates of the six largest metropolitan areas, and the third example is a list of votes of a five-member board. The last example is a list of (x, y) coordinates, each of which is represented by a two-element list. Although they usually do not in practice, lists can also contain items of different types:

```
>>> crazy = [15, 'gtaac', [1, 2, 3], max(4, 14)]
>>> crazy
[15, 'gtaac', [1, 2, 3], 14]
```

Since lists are sequences like strings, they can also be indexed and sliced. But now indices refer to list elements instead of characters and slices are sublists instead of substrings:

```
>>> sales[1]
42
>>> votes[:3]
['yea', 'yea', 'nay']
```

Now suppose that we are running a small business, and we need to get some basic descriptive statistics about last week's daily sales, starting with the average (or mean) daily sales for the week. Recall from Section 1.2 that, to find the mean of a list of numbers, we need to first find their sum by iterating over the list. Iterating over the values in a list is essentially identical to iterating over the characters in a string, as illustrated below.

```
def mean(data):
    """Compute the mean of a list of numbers.

    Parameter:
        data: a list of numbers

    Return value: the mean of the numbers in data
    """
    sum = 0
    for item in data:
        sum = sum + item
    return sum / len(data)
```

In each iteration of the `for` loop, `item` is assigned the next value in the list named `data`, and then added to the running sum. After the loop, we divide the sum by the length of the list, which is retrieved with the same `len` function we used on strings.

Reflection 8.1 *Does this work when the list is empty?*

If `data` is the empty list (`[ ]`), then the value of `len(data)` is zero, resulting in a "division by zero" error in the `return` statement. We have several options to deal with this. First, we could just let the error happen. Second, if you read Section 7.2, we could use an `assert` statement to print an error message and abort. Third, we could detect this error with an `if` statement and return something that indicates that an error occurred. In this case, we will adopt the last option by returning `None` and indicating this possibility in the docstring.

```
 1  def mean(data):
 2      """Compute the mean of a non-empty list of numbers.
 3
 4      Parameter:
 5          data: a list of numbers
 6
 7      Return value: the mean of numbers in data or None if data is empty
 8      """
 9
10      if len(data) == 0:
11          return None
12
13      sum = 0
14      for item in data:
15          sum = sum + item
16      return sum / len(data)
```

This `for` loop is yet another example of an accumulator, and is virtually identical to the `countLinks` function that we developed in Section 4.1. To illustrate what is happening, suppose we call `mean` from the following `main` function.

```
def main():
    sales = [32, 42, 11, 15, 58, 44, 16]
    averageSales = mean(sales)
    print('Average daily sales were', averageSales)

main()
```

The call to `mean(sales)` above will effectively execute the following sequence of statements inside the `mean` function. The changing value of `item` assigned by the `for` loop is highlighted in red. The numbers on the left indicate which line in the `mean` function is being executed in each step.

```
14   sum = 0                      # sum is initialized
15   item = 32                    # for loop assigns 32 to item
16   sum = sum + item             # sum is assigned 0 + 32 = 32
15   item = 42                    # for loop assigns 42 to item
```

```
16    sum = sum + item          # sum is assigned 32 + 42 = 74
          ⋮
15    item = 16                 # for loop assigns 16 to item
16    sum = sum + item          # sum is assigned 202 + 16 = 218
17    return sum / len(data)    # returns 218 / 7 ≈ 31.14
```

Reflection 8.2 *Fill in the missing steps above to see how the function arrives at a sum of 218.*

The mean of a data set does not adequately describe it if there is a lot of variability in the data, i.e., if there is no "typical" value. In these cases, we need to accompany the mean with the *variance*, which is measure of how much the data varies from the mean. Computing the variance is left as Exercise 8.1.10.

Now let's think about how to find the minimum and maximum sales in the list. Of course, it is easy for us to just look at a short list like the one above and pick out the minimum and maximum. But a computer does not have this ability. Therefore, as you think about these problems, it may be better to think about a very long list instead, one in which the minimum and maximum are not so obvious.

Reflection 8.3 *Think about how you would write an algorithm to find the minimum value in a long list. (Similar to a running sum, keep track of the current minimum.)*

As the hint suggests, we want to maintain the current minimum while we iterate over the list with a `for` loop. When we examine each item, we need to test whether it is smaller than the current minimum. If it is, we assign the current item to be the new minimum. The following function implements this algorithm.

```
def min(data):
    """Compute the minimum value in a non-empty list of numbers.

    Parameter:
        data: a list of numbers

    Return value: the minimum value in data or None if data is empty
    """

    if len(data) == 0:
        return None

    minimum = data[0]
    for item in data[1:]:
        if item < minimum:
            minimum = item
    return minimum
```

Before the loop, we initialize `minimum` to be the first value in the list, using indexing. Then we iterate over the slice of remaining values in the list. In each iteration, we compare the current value of `item` to `minimum` and, if `item` is smaller than `minimum`, update `minimum` to the value of `item`. At the end of the loop, `minimum` is assigned the smallest value in the list.

Reflection 8.4 *If the list [32, 42, 11, 15, 58, 44, 16] is assigned to* `data`*, then what are the values of* `data[0]` *and* `data[1:]` *?*

Let's look at a small example of how this function works when we call it with the list containing only the first four numbers from the list above: [32, 42, 11, 15]. The function begins by assigning the value 32 to `minimum`. The first value of `item` is 42. Since 42 is not less than 32, `minimum` remains unchanged. In the next iteration of the loop, the third value in the list, 11, is assigned to `item`. In this case, since 11 is less than 32, the value of `minimum` is updated to 11. Finally, in the last iteration of the loop, `item` is assigned the value 15. Since 15 is greater than 11, `minimum` is unchanged. At the end, the function returns the final value of `minimum`, which is 11. A function to compute the maximum is very similar, so we leave it as an exercise.

Reflection 8.5 *What would happen if we iterated over* `data` *instead of* `data[1:]` *? Would the function still work?*

If we iterated over the entire list instead, the first comparison would be useless (because `item` and `minimum` would be the same) so it would be a little less efficient, but the function would still work fine.

Now what if we also wanted to know on which day of the week the minimum sales occurred? To answer this question, assuming we know how indices correspond to days of the week, we need to find the index of the minimum value in the list. As we learned in Chapter 6, we need to iterate over the indices in situations like this:

```
def minDay(data):
    """Compute the index of the minimum value in a non-empty list.

    Parameter:
        data: a list of numbers

    Return value: the index of the minimum value in data
                  or -1 if data is empty
    """

    if len(data) == 0:
        return -1

    minIndex = 0
    for index in range(1, len(data)):
        if data[index] < data[minIndex]:
            minIndex = index
    return minIndex
```

356 ■ Data analysis

This function performs almost exactly the same algorithm as our previous `min` function, but now each value in the list is identified by `data[index]` instead of `item` and we remember the index of current minimum in the loop instead of the actual minimum value.

Reflection 8.6 *How can we modify the `minDay` function to return a day of the week instead of an index, assuming the sales data starts on a Sunday.*

One option would be to replace `return minIndex` with `if/elif/else` statements, like the following:

```
if minIndex == 0:
    return 'Sunday'
elif minIndex == 1:
    return 'Monday'
    ⋮
else:
    return 'Saturday'
```

But a more clever solution is to create a list of the days of the week that are in the same order as the sales data. Then we can simply use the value of `minIndex` as an index into this list to return the correct string.

```
days = ['Sunday', 'Monday', 'Tuesday', 'Wednesday', 'Thursday',
        'Friday', 'Saturday']
return days[minIndex]
```

There are many other descriptive statistics that we can use to summarize the contents of a list. The following exercises challenge you to implement some of them.

Exercises

8.1.1. Suppose a list is assigned to the variable name `data`. Show how you would

  (a) print the length of `data`

  (b) print the third element in `data`

  (c) print the last element in `data`

  (d) print the last three elements in `data`

  (e) print the first four elements in `data`

  (f) print the list consisting of the second, third, and fourth elements in `data`

8.1.2. In the `mean` function, we returned `None` if `data` was empty. Show how to modify the following `main` function so that it properly tests for this possibility and prints an appropriate message.

```
def main():
    someData = getInputFromSomewhere()
    average = mean(someData)
    print('The mean value is', average)
```

8.1.3. Write a function

>   sumList(data)

that returns the sum of all of the numbers in the list data. For example, sumList([1, 2, 3]) should return 6.

8.1.4. Write a function

>   sumOdds(data)

that returns the sum of only the odd integers in the list data. For example, sumOdds([1, 2, 3]) should return 4.

8.1.5. Write a function

>   countOdds(data)

that returns the number of odd integers in the list data. For example, countOdds([1, 2, 3]) should return 2.

8.1.6. Write a function

>   multiples5(data)

that returns the number of multiples of 5 in a list of integers. For example, multiples5([5, 7, 2, 10]) should return 2.

8.1.7. Write a function

>   countNames(words)

that returns the number of capitalized names in the list of strings named words. For example, countNames(['Fili', 'Oin', 'Thorin', 'and', 'Bilbo', 'are', 'characters', 'in', 'a', 'book', 'by', 'Tolkien']) should return 5.

8.1.8. The percentile associated with a particular value in a data set is the number of values that are less than or equal to it, divided by the total number of values, times 100. Write a function

>   percentile(data, value)

that returns the percentile of value in the list named data.

8.1.9. Write a function

>   meanSquares(data)

that returns the mean (average) of the squares of the numbers in a list named data.

8.1.10. Write a function

>   variance(data)

that returns the variance of a list of numbers named data. The variance is defined to be the mean of the squares of the numbers in the list minus the square of the mean of the numbers in the list. In your implementation, call your function from Exercise 8.1.9 and the mean function from this section.

8.1.11. Write a function

> `max(data)`

that returns the maximum value in the list of numbers named **data**. Do not use the built-in **max** function.

8.1.12. Write a function

> `shortest(words)`

that returns the shortest string in a list of strings named **words**. In case of ties, return the first shortest string. For example, `shortest(['spider', 'ant', 'beetle', 'bug'])` should return the string `'ant'`.

8.1.13. Write a function

> `span(data)`

that returns the difference between the largest and smallest numbers in the list named **data**. Do not use the built-in **min** and **max** functions. (But you may use your own functions.) For example, `span([9, 4, 2, 1, 7, 7, 3, 2])` should return 8.

8.1.14. Write a function

> `maxIndex(data)`

that returns the *index* of the maximum item in the list of numbers named **data**. Do not use the built-in **max** function.

8.1.15. Write a function

> `secondLargest(data)`

that returns the second largest number in the list named **data**. Do not use the built-in **max** function. (But you may use your **maxIndex** function from Exercise 8.1.14.)

8.1.16. Write a function

> `search(data, target)`

that returns **True** if the **target** is in the list named **data**, and **False** otherwise. Do not use the **in** operator to test whether an item is in the list. For example, `search(['Tris', 'Tobias', 'Caleb'], 'Tris')` should return **True**, but `search(['Tris', 'Tobias', 'Caleb'], 'Peter')` should return **False**.

8.1.17. Write a function

> `search(data, target)`

that returns the index of **target** if it is found in the list named **data**, and -1 otherwise. Do not use the **in** operator or the **index** method to test whether items are in the list. For example, `search(['Tris', 'Tobias', 'Caleb'], 'Tris')` should return 0, but `search(['Tris', 'Tobias', 'Caleb'], 'Peter')` should return -1.

8.1.18. Write a function

> `intersect(data1, data2)`

that returns `True` if the two lists named `data1` and `data2` have any common elements, and `False` otherwise. (Hint: use your `search` function from Exercise 8.1.16.) For example, `intersect(['Katniss', 'Peeta', 'Gale'], ['Foxface', 'Marvel', 'Glimmer'])` should return `False`, but `intersect(['Katniss', 'Peeta', 'Gale'], ['Gale', 'Haymitch', 'Katniss'])` should return `True`.

8.1.19. Write a function

> `differ(data1, data2)`

that returns the first index at which the two lists `data1` and `data2` differ. If the two lists are the same, your function should return -1. You may assume that the lists have the same length. For example, `differ(['CS', 'rules', '!'], ['CS', 'totally', 'rules!'])` should return the index 1.

8.1.20. A *checksum* is a digit added to the end of a sequence of data to detect error in the transmission of the data. (This is a generalization of *parity* from Exercise 6.1.11 and similar to Exercise 6.3.17.) Given a sequence of decimal digits, the simplest way to compute a checksum is to add all of the digits and then find the remainder modulo 10. For example, the checksum for the sequence 48673 is 8 because (4 + 8 + 6 + 7 + 3) mod 10 = 28 mod 10 = 8. When this sequence of digits is transmitted, the checksum digit will be appended to the end of the sequence and checked on the receiving end. Most numbers that we use on a daily basis, including credit card numbers and ISBN numbers, as well as data sent across the Internet, include checksum digits. (This particular checksum algorithm is not particularly good, so "real" checksum algorithms are a bit more complicated; see, for example, Exercise 8.1.21.)

(a) Write a function

> `checksum(data)`

that computes the checksum digit for the list of integers named `data` and returns the list with the checksum digit added to the end. For example, `checksum([4, 8, 6, 7, 3])` should return the list [4, 8, 6, 7, 3, 8].

(b) Write a function

> `check(data)`

that returns a Boolean value indicating whether the last integer in `data` is the correct checksum value.

(c) Demonstrate a transmission (or typing) error that could occur in the sequence [4, 8, 6, 7, 3, 8] (the last digit is the checksum digit) that would *not* be detected by the `check` function.

8.1.21. The Luhn algorithm is the standard algorithm used to validate credit card numbers and protect against accidental errors. Read about the algorithm online, and then write a function

```
validateLuhn(number)
```

that returns `True` if the number if valid and `False` otherwise. The `number` parameter will be a list of digits. For example, to determine if the credit card number 4563 9601 2200 1999 is valid, one would call the function with the parameter [4, 5, 6, 3, 9, 6, 0, 1, 2, 2, 0, 0, 1, 9, 9, 9]. (Hint: use a `for` loop that iterates in reverse over the indices of the list.)

8.2 CREATING AND MODIFYING LISTS

We often want to create or modify data, rather than just summarize it. We have already seen some examples of this with the list accumulators that we have used to build lists of data to plot. In this section, we will revisit some of these instances, and then build upon them to demonstrate some of the nuts and bolts of working with data stored in lists. Several of the projects at the end of this chapter can be solved using these techniques.

List accumulators, redux

We first encountered lists in Section 4.1 in a slightly longer version of the following function.

```
import matplotlib.pyplot as pyplot

def pond(years):
    """ (docstring omitted) """

    population = 12000
    populationList = [population]
    for year in range(1, years + 1):
        population = 1.08 * population - 1500
        populationList.append(population)
    pyplot.plot(range(years + 1), populationList)
    pyplot.show()
    return population
```

In the first red statement, the list named `populationList` is initialized to the single-item list [12000]. In the second red statement, inside the loop, each new population value is appended to the list. Finally, the list is plotted with the `pyplot.plot` function.

We previously called this technique a *list accumulator*, due to its similarity to integer accumulators. List accumulators can be applied to a variety of problems. For example, consider the `find` function from Section 6.5. Suppose that, instead of returning only the index of the first instance of the target string, we wanted to return a list containing the indices of all the instances:

```
def find(text, target):
    """ (docstring omitted) """
    indexList = []
    for index in range(len(text) - len(target) + 1):
        if text[index:index + len(target)] == target:
            indexList.append(index)
    return indexList
```

In this function, just as in the `pond` function, we initialize the list before the loop, and append an index to the list inside the loop (wherever we find the target string). At the end of the function, we return the list of indices. The function call

```
find('Well done is better than well said.', 'ell')
```

would return the list `[1, 26]`. On the other hand, the function call

```
find('Well done is better than well said.', 'Franklin')
```

would return the empty list `[]` because the condition in the `if` statement is never true.

This list accumulator pattern is so common that there is a shorthand for it called a *list comprehension*. You can learn more about list comprehensions by reading the optional material at the end of this section, starting on Page 368.

Lists are mutable

We can modify an existing list with `append` because, unlike strings, lists are *mutable*. In other words, the components of a list can be changed. We can also change individual elements in a list. For example, if we need to change the second value in the list of unemployment rates, we can do so:

```
>>> unemployment = [0.082, 0.092, 0.091, 0.063, 0.068, 0.052]
>>> unemployment[1] = 0.062
>>> unemployment
[0.082, 0.062, 0.091, 0.063, 0.068, 0.052]
```

We can change individual elements in a list because each of the elements is an independent reference to a value, like any other variable name. We can visualize the original `unemployment` list like this:

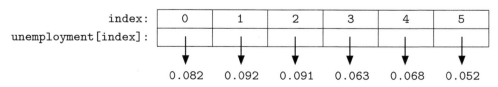

So the value 0.082 is assigned to the name `unemployment[0]`, the value 0.092 is assigned to the name `unemployment[1]`, etc. When we assigned a new value to `unemployment[1]` with `unemployment[1] = 0.062`, we were simply assigning a new value to the name `unemployment[1]`, like any other assignment statement:

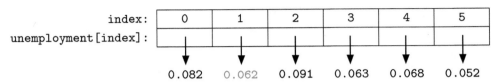

Suppose we wanted to adjust all of the unemployment rates in this list by subtracting one percent from each of them. We can do this with a `for` loop that iterates over the indices of the list.

```
>>> for index in range(len(unemployment)):
        unemployment[index] = unemployment[index] - 0.01
>>> unemployment
[0.072, 0.052, 0.081, 0.053, 0.058, 0.042]
```

This `for` loop is simply equivalent to the following six statements:

```
unemployment[0] = unemployment[0] - 0.01
unemployment[1] = unemployment[1] - 0.01
unemployment[2] = unemployment[2] - 0.01
unemployment[3] = unemployment[3] - 0.01
unemployment[4] = unemployment[4] - 0.01
unemployment[5] = unemployment[5] - 0.01
```

Reflection 8.7 *Is it possible to achieve the same result by iterating over the values in the list instead? In other words, does the following `for` loop accomplish the same thing? (Try it.) Why or why not?*

```
for rate in unemployment:
    rate = rate - 0.01
```

This loop does *not* modify the list because `rate`, which is being modified, is not a name in the list. So, although the value assigned to `rate` is being modified, the list itself is not. For example, at the beginning of the first iteration, 0.082 is assigned to `rate`, as illustrated below.

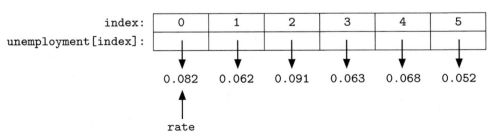

Then, when the modified value `rate - 0.01` is assigned to `rate`, this only affects `rate`, not the original list, as illustrated below.

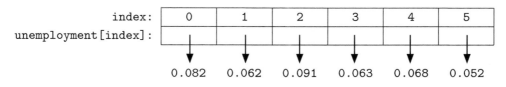

List parameters are mutable too

Now let's put the correct loop above in a function named `adjust` that takes a list of unemployment rates as a parameter. We can then call the function with an actual list of unemployment rates that we wish to modify:

```
def adjust(rates):
    """Subtract one percent (0.01) from each rate in a list.

    Parameter:
        rates: a list of numbers representing rates (percentages)

    Return value: None
    """

    for index in range(len(rates)):
        rates[index] = rates[index] - 0.01

def main():
    unemployment = [0.053, 0.071, 0.065, 0.074]
    adjust(unemployment)
    print(unemployment)

main()
```

The list named `unemployment` is assigned in the `main` function and then passed in for the parameter `rates` to the `adjust` function. Inside the `adjust` function, every value in `rates` is decremented by 0.01. What effect, if any, does this have on the list assigned to `unemployment`? To find out, we need to look carefully at what happens when the function is called.

Right after the assignment statement in the `main` function, the situation looks like the following, with the variable named `unemployment` in the `main` namespace assigned the list [0.053, 0.071, 0.065, 0.074].

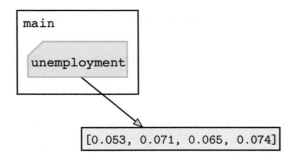

Now recall from Section 3.5 that, when an argument is passed to a function, it is *assigned* to its associated parameter. Therefore, immediately after the `adjust` function is called from `main`, the parameter `rates` is assigned the same list as `unemployment`:

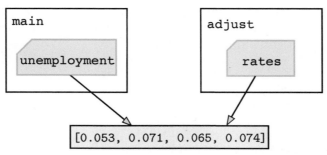

After `adjust` executes, 0.01 has been subtracted from each value in `rates`, as the following picture illustrates.

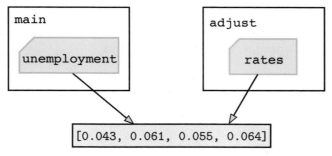

But notice that, since the same list is assigned to `unemployment`, these changes will also be reflected in the value of `unemployment` back in the `main` function. In other words, after the `adjust` function returns, the picture looks like this:

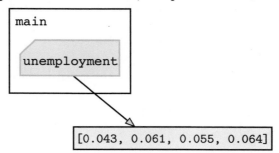

So when `unemployment` is printed at the end of `main`, the adjusted list [0.043, 0.061, 0.055, 0.064] will be displayed.

Reflection 8.8 *Why does the argument's value change in this case when it did not in the parameter passing examples in Section 3.5? What is different?*

The difference here is that lists are mutable. When you pass a mutable value as an argument, any changes to the associated formal parameter inside the function will be reflected in the value of the argument. Therefore, when we pass a list as an argument to a function, the values in the list can be changed inside the function.

What if we did *not* want to change the argument to the `adjust` function (i.e., `unemployment`), and instead return a *new* adjusted list? One alternative, illustrated in the function below, would be to make a copy of `rates`, using the list method `copy`, and then modify this copy instead.

```python
def adjust(rates):
    """ (docstring omitted) """

    ratesCopy = rates.copy()
    for index in range(len(ratesCopy)):
        ratesCopy[index] = ratesCopy[index] - 0.01
    return ratesCopy
```

The `copy` method creates an independent copy of the list in memory, and returns a reference to this new list so that it can be assigned to a variable name (in this case, `ratesCopy`). There are other solutions to this problem as well, which we leave as exercises.

Tuples

Python offers another list-like object called a *tuple*. A tuple works just like a list, with two substantive differences. First, a tuple is enclosed in parentheses instead of square brackets. Second, a tuple is immutable. For example, as a tuple, the `unemployment` data would look like (0.053, 0.071, 0.065, 0.074).

Tuples can be used in place of lists in situations where the object being represented has a fixed length, and individual components are not likely to change. For example, colors are often represented by their (red, green, blue) components (see Box 3.2) and two-dimensional points by (x, y).

```
>>> point = (4, 2)
>>> point
(4, 2)
>>> green = (0, 255, 0)
```

Reflection 8.9 *Try reassigning the first value in* `point` *to 7. What happens?*

```
>>> point[0] = 7
TypeError: 'tuple' object does not support item assignment
```

Tuples are more memory efficient than lists because extra memory is set aside in a list for a few appends before more memory must be allocated.

List operators and methods

Two operators that we used to create new strings can also be used to create new lists. The repetition operator * creates a new list that is built from repeats of the contents of a smaller list. For example:

```
>>> empty = [0] * 5
>>> empty
[0, 0, 0, 0, 0]
>>> ['up', 'down'] * 4
['up', 'down', 'up', 'down', 'up', 'down', 'up', 'down']
```

The concatenation operator + creates a new list that is the result of "sticking" two lists together. For example:

```
>>> unemployment = [0.082, 0.092, 0.091, 0.063, 0.068, 0.052]
>>> unemployment = unemployment + [0.087, 0.101]
>>> unemployment
[0.082, 0.092, 0.091, 0.063, 0.068, 0.052, 0.087, 0.101]
```

Notice that the concatenation operator combines two lists to create a new list, whereas the `append` method *inserts* a new element into an existing list as the last element. In other words,

```
unemployment = unemployment + [0.087, 0.101]
```

accomplishes the same thing as the two statements

```
unemployment.append(0.087)
unemployment.append(0.101)
```

However, using concatenation actually creates a new list that is then assigned to `unemployment`, whereas using `append` modifies an existing list. So using `append` is usually more efficient than concatenation if you are just adding to the end of an existing list.

Reflection 8.10 *How do the results of the following two statements differ? If you want to add the number* `0.087` *to the end of the list, which is correct?*

```
unemployment.append(0.087)
```

and

```
unemployment.append([0.087])
```

The `list` class has several useful methods in addition to `append`. We will use many of these in the upcoming sections to solve a variety of problems. For now, let's look at four of the most common methods: `sort`, `insert`, `pop`, and `remove`.

The `sort` method simply sorts the elements in a list in increasing order. For example, suppose we have a list of SAT scores that we would like to sort:

```
>>> scores = [620, 710, 520, 550, 640, 730, 600]
>>> scores.sort()
>>> scores
[520, 550, 600, 620, 640, 710, 730]
```

Note that the `sort` and `append` methods, as well as `insert`, `pop` and `remove`, do not return new lists; instead they modify the lists in place. In other words, the following is a mistake:

```
>>> scores = [620, 710, 520, 550, 640, 730, 600]
>>> newScores = scores.sort()
```

Reflection 8.11 *What is the value of* `newScores` *after we execute the statement above?*

Printing the value of `newScores` reveals that it refers to the value `None` because `sort` does not return anything (meaningful). However, `scores` was modified as we expected:

```
>>> newScores
>>> scores
[520, 550, 600, 620, 640, 710, 730]
```

The `sort` method will sort any list that contains comparable items, including strings. For example, suppose we have a list of names that we want to be in alphabetical order:

```
>>> names = ['Eric', 'Michael', 'Connie', 'Graham']
>>> names.sort()
>>> names
['Connie', 'Eric', 'Graham', 'Michael']
```

Reflection 8.12 *What happens if you try to sort a list containing items that cannot be compared to each other? For example, try sorting the list* `[3, 'one', 4, 'two']`.

The `insert` method inserts an item into a list at a particular index. For example, suppose we want to insert new names into the sorted list above to maintain alphabetical order:

```
>>> names.insert(3, 'John')
>>> names
['Connie', 'Eric', 'Graham', 'John', 'Michael']
>>> names.insert(0, 'Carol')
>>> names
['Carol', 'Connie', 'Eric', 'Graham', 'John', 'Michael']
```

The first parameter of the `insert` method is the index where the inserted item will reside *after* the insertion.

The `pop` method is the inverse of `insert`; `pop` deletes the list item at a given index and returns the deleted value. For example,

```
>>> inMemoriam = names.pop(3)
>>> names
['Carol', 'Connie', 'Eric', 'John', 'Michael']
>>> inMemoriam
'Graham'
```

If the argument to `pop` is omitted, `pop` deletes and returns the last item in the list.

The `remove` method also deletes an item from a list, but takes the *value* of an item as its parameter rather than its position. If there are multiple items in the list with the given value, the `remove` method only deletes the first one. For example,

```
>>> names.remove('John')
>>> names
['Carol', 'Connie', 'Eric', 'Michael']
```

Reflection 8.13 *What happens if you try to remove* `'Graham'` *from* `names` *now?*

*List comprehensions

As we mentioned at the beginning of this section, the list accumulator pattern is so common that there is a shorthand for it called a *list comprehension*. A list comprehension allows us to build up a list in a single statement. For example, suppose we wanted to create a list of the first 15 even numbers. Using a `for` loop, we can construct the desired list with:

```
evens = [ ]
for i in range(15):
    evens.append(2 * i)
```

An equivalent list comprehension looks like this:

```
evens = [2 * i for i in range(15)]
```

The first part of the list comprehension is an expression representing the items we want in the list. This is the same as the expression that would be passed to the `append` method if we constructed the list the "long way" with a `for` loop. This expression is followed by a `for` loop clause that specifies the values of an index variable for which the expression should be evaluated. The `for` loop clause is also identical to the `for loop` that we would use to construct the list the "long way." This correspondence is illustrated below:

> **Box 8.1: NumPy arrays**
>
> NumPy is a Python module that provides a different list-like class named `array`. (Because the `numpy` module is required by the `matplotlib` module, you should already have it installed.) Unlike a list, a NumPy `array` is treated as a mathematical *vector*. There are several different ways to create a new array. We will only illustrate two:
>
> ```
> >>> import numpy
> >>> a = numpy.array([1, 2, 3, 4, 5])
> >>> print(a)
> [1 2 3 4 5]
> >>> b = numpy.zeros(5)
> >>> print(b)
> [ 0.  0.  0.  0.  0.]
> ```
>
> In the first case, `a` was assigned an `array` created from a list of numbers. In the second, `b` was assigned an `array` consisting of 5 zeros. One advantage of an `array` over a `list` is that arithmetic operations and functions are applied to each of an `array` object's elements individually. For example:
>
> ```
> >>> print(a * 3)
> [ 3  6  9 12 15]
> >>> c = numpy.array([3, 4, 5, 6, 7])
> >>> print(a + c)
> [ 4  6  8 10 12]
> ```
>
> There are also many functions and methods available to `array` objects. For example:
>
> ```
> >>> print(c.sum())
> 25
> >>> print(numpy.sqrt(c))
> [ 1.73205081  2.          2.23606798  2.44948974  2.64575131]
> ```
>
> An `array` object can also have more than one dimension, as we will discuss in Chapter 9. If you are interested in learning more about NumPy, visit `http://www.numpy.org`.

```
evens = [ ]
for i in range(15):
    evens.append(2 * i)

evens = [2 * i for i in range(15)]
```

List comprehensions can also incorporate `if` statements. For example, suppose we wanted a list of the first 15 even numbers that are not divisible by 6. A `for` loop to create this list would look just like the previous example, with an additional `if` statement that checks that `2 * i` is not divisible by 6 before appending it:

```
evens = [ ]
for i in range(15):
    if 2 * i % 6 != 0:
        evens.append(2 * i)
```

This can be reproduced with a list comprehension that looks like this:

```
evens = [2 * i for i in range(15) if 2 * i % 6 != 0]
```

The corresponding parts of this loop and list comprehension are illustrated below:

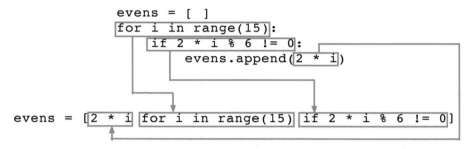

In general, the initial expression in a list comprehension can be followed by any sequence of `for` and `if` clauses that specify the values for which the expression should be evaluated.

Reflection 8.14 *Rewrite the* `find` *function on Page 361 using a list comprehension.*

The `find` function can be rewritten with the following list comprehension.

```
def find(text, target):
    """ (docstring omitted) """

    return [index for index in range(len(text) - len(target) + 1)
            if text[index:index + len(target)] == target]
```

Look carefully at the corresponding parts of the original loop version and the list comprehension version, as we did above.

Exercises

8.2.1. Show how to add the string `'grapes'` to the end of the following list using both concatenation and the `append` method.

```
fruit = ['apples', 'pears', 'kiwi']
```

8.2.2. Write a function

```
squares(n)
```

that returns a list containing the squares of the integers 1 through n. Use a `for` loop.

8.2.3. (This exercise assumes that you have read Section 6.7.) Write a function

> `getCodons(dna)`

that returns a list containing the codons in the string `dna`. Your algorithm should use a `for` loop.

8.2.4. Write a function

> `square(data)`

that takes a list of numbers named `data` and squares each number in `data` in place. The function should not return anything. For example, if the list `[4, 2, 5]` is assigned to a variable named `numbers` then, after calling `square(numbers)`, `numbers` should have the value `[16, 4, 25]`.

8.2.5. Write a function

> `swap(data, i, j)`

that swaps the positions of the items with indices `i` and `j` in the list named `data`.

8.2.6. Write a function

> `reverse(data)`

that reverses the list `data` in place. Your function should not return anything. (Hint: use the `swap` function you wrote above.)

8.2.7. Suppose you are given a list of `'yea'` and `'nay'` votes. Write a function

> `winner(votes)`

that returns the majority vote. For example, `winner(['yea', 'nay', 'yea'])` should return `'yea'`. If there is a tie, return `'tie'`.

8.2.8. Write a function

> `delete(data, index)`

that returns a new list that contains the same elements as the list `data` except for the one at the given `index`. If the value of `index` is negative or exceeds the length of `data`, return a copy of the original list. Do not use the `pop` method. For example, `delete([3, 1, 5, 9], 2)` should return the list `[3, 1, 9]`.

8.2.9. Write a function

> `remove(data, value)`

that returns a new list that contains the same elements as the list `data` except for those that equal `value`. Do not use the builit-in `remove` method. Note that, unlike the built-in `remove` method, your function should remove *all* items equal to `value`. For example, `remove([3, 1, 5, 3, 9], 3)` should return the list `[1, 5, 9]`.

8.2.10. Write a function

> `centeredMean(data)`

that returns the average of the numbers in `data` with the largest and smallest numbers removed. You may assume that there are at least three numbers in the list. For example, `centeredMean([2, 10, 3, 5])` should return 4.

8.2.11. On Page 365, we showed one way to write the `adjust` function so that it returned an adjusted list rather than modifying the original list. Give another way to accomplish the same thing.

8.2.12. Write a function

> `shuffle(data)`

that shuffles the items in the list named `data` in place, without using the `shuffle` function from the `random` module. Instead, use the `swap` function you wrote in Exercise 8.2.5 to swap 100 pairs of randomly chosen items. For each swap, choose a random index for the first item and then choose a greater random index for the second item.

8.2.13. We wrote the following loop on Page 362 to subtract 0.01 from each value in a list:

```
for index in range(len(unemployment)):
    unemployment[index] = unemployment[index] - 0.01
```

Carefully explain why the following loop does *not* accomplish the same thing.

```
for rate in unemployment:
    rate = rate - 0.01
```

8.2.14. Write a function

> `smooth(data, windowLength)`

that returns a new list that contains the values in the list `data`, averaged over a window of the given length. This is the problem that we originally introduced in Section 1.3.

8.2.15. Write a function

> `median(data)`

that returns the median number in a list of numbers named `data`.

8.2.16. Consider the following alphabetized grocery list:

> `groceries = ['cookies', 'gum', 'ham', 'ice cream', 'soap']`

Show a sequence of calls to list methods that insert each of the following into their correct alphabetical positions, so that the final list is alphabetized:

(a) `'jelly beans'`
(b) `'donuts'`
(c) `'bananas'`
(d) `'watermelon'`

Next, show a sequence of calls to the `pop` method that delete each of the following items from the final list above.

(a) `'soap'`

(b) `'watermelon'`

(c) `'bananas'`

(d) `'ham'`

8.2.17. Given n people in a room, what is the probability that at least one pair of people shares a birthday? To answer this question, first write a function

`sameBirthday(people)`

that creates a list of `people` random birthdays and returns `True` if two birthdays are the same, and `False` otherwise. Use the numbers 0 to 364 to represent 365 different birthdays. Next, write a function

`birthdayProblem(people, trials)`

that performs a Monte Carlo simulation with the given number of `trials` to approximate the probability that, in a room with the given number of `people`, two people share a birthday.

8.2.18. Write a function that uses your solution to the previous problem to return the smallest number of people for which the probability of a pair sharing a birthday is at least 0.5.

8.2.19. Rewrite the `squares` function from Exercise 8.2.2 using a list comprehension.

8.2.20. (This exercise assumes that you have read Section 6.7.) Rewrite the `getCodons` function from Exercise 8.2.3 using a list comprehension.

8.3 FREQUENCIES, MODES, AND HISTOGRAMS

In addition to the mean and range of a list of data, which are single values, we might want to learn how data values are distributed throughout the list. The number of times that a value appears in a list is called its *frequency*. Once we know the frequencies of values in a list, we can represent them visually with a *histogram* and find the most frequent value(s), called the *mode(s)*.

Tallying values

To get a sense of how we might compute the frequencies of values in a list, let's consider the ocean buoy temperature readings that we first encountered in Chapter 1. As we did then, let's simplify matters by just considering one week's worth of data:

`temperatures = [18.9, 19.1, 18.9, 19.0, 19.3, 19.2, 19.3]`

Now we want to iterate over the list, keeping track of the frequency of each value that we see. We can imagine using a simple table for this purpose. After we see the first value in the list, 18.9, we create an entry in the table for 18.9 and mark its frequency with a tally mark.

Temperature:	18.9
Frequency:		

The second value in the list is 19.1, so we create another entry and tally mark.

Temperature:	18.9	19.1
Frequency:				

The third value is 18.9, so we add another tally mark to the 18.9 column.

Temperature:	18.9	19.1
Frequency:	||		

The fourth value is 19.0, so we create another entry in the table with a tally mark.

Temperature:	18.9	19.1	19.0
Frequency:	||				

Continuing in this way with the rest of the list, we get the following final table.

Temperature:	18.9	19.1	19.0	19.2	19.3
Frequency:	||							||

Or, equivalently:

Temperature:	18.9	19.1	19.0	19.2	19.3
Frequency:	2	1	1	1	2

Now to find the mode, we find the maximum frequency in the table, and then return a list of the values with this maximum frequency: [18.9, 19.3].

Dictionaries

Notice how the frequency table resembles the picture of a list on Page 361, except that the indices are replaced by temperatures. In other words, the frequency table looks like a generalized list in which the indices are replaced by values that we choose. This kind of abstract data type is called a *dictionary*. In a dictionary, each index is replaced with a more general *key*. Unlike a list, in which the indices are implicit, a dictionary in Python (called a `dict` object) must define the correspondence between a key and its value explicitly with a `key:value` pair. To differentiate it from a list, a dictionary is enclosed in curly braces ({ }). For example, the frequency table above would be represented in Python like this:

```
>>> frequency = {18.9: 2, 19.1: 1, 19.0: 1, 19.2: 1, 19.3: 2}
```

The first pair in **frequency** has key 18.9 and value 2, the second item has key 19.1 and value 1, etc.

If you type the dictionary above in the Python shell, and then display it, you might notice something unexpected:

```
>>> frequency
{19.0: 1, 19.2: 1, 18.9: 2, 19.1: 1, 19.3: 2}
```

The items appear in a different order than they were originally. This is okay because a dictionary is an *unordered* collection of pairs. The displayed ordering has to do with the way dictionaries are implemented, using a structure called a *hash table*. If you are interested, you can learn a bit more about hash tables in Box 8.2.

Each entry in a dictionary object can be referenced using the familiar indexing notation, but using a key in the square brackets instead of an index. For example:

```
>>> frequency[19.3]
2
>>> frequency[19.1]
1
```

As we alluded to above, the model of a dictionary in memory is similar to a list (except that there is no significance attached to the ordering of the key values):

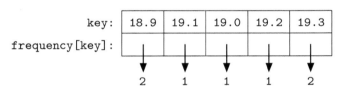

Each entry in the dictionary is a reference to a value in the same way that each entry in a list is a reference to a value. So, as with a list, we can change any value in a dictionary. For example, we can increment the value associated with the key 19.3:

```
>>> frequency[19.3] = frequency[19.3] + 1
```

Now let's use a dictionary to implement the algorithm that we developed above to find frequencies and the mode(s) of a list. To begin, we will create an empty dictionary named **frequency** in which to record our tally marks:

```
def mode(data):
    """Compute the mode of a non-empty list.
    
    Parameter:
        data: a list of items
    
    Return value: the mode of the items in data
    """
    
    frequency = { }
```

> **Box 8.2: Hash tables**
>
> A Python dictionary is implemented with a structure called a *hash table*. A hash table contains a fixed number of indexed *slots* in which the `key:value` pairs are stored. The slot assigned to a particular `key:value` pair is determined by a *hash function* that "translates" the key to a slot index. The picture below shows how some items in the `frequency` dictionary might be placed in an underlying hash table.
>
>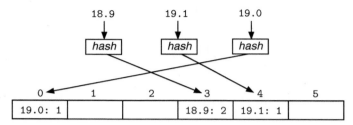
>
> In this illustration, the hash function associates the pair `18.9: 2` with slot index 3, `19.1: 1` with slot index 4, and `19.0: 1` with slot index 0.
>
> The underlying hash table allows us to access a value in a dictionary (e.g., `frequency[18.9]`) or test for inclusion (e.g., `key in frequency`) in a constant amount of time because each operation only involves a hash computation and then a direct access (like indexing in a string or a list). In contrast, if the pairs were stored in a list, then the list would need to be searched (in linear time) to perform these operations.
>
> Unfortunately, this constant-time access could be foiled if a key is mapped to an occupied slot, an event called a *collision*. Collisions can be resolved by using adjacent slots, using a second hash function, or associating a list of items with each slot. A good hash function tries to prevent collisions by assigning slots in a seemingly random manner, so that keys are evenly distributed in the table and similar keys are not mapped to the same slot. Because hash functions tend to be so good, we can still consider an average dictionary access to be a constant-time operation, or one elementary step, even with collisions.

Each entry in this dictionary will have its key equal to a unique item in `data` and its value equal to the item's frequency count. To tally the frequencies, we need to iterate over the items in `data`. As in our tallying algorithm, if there is already an entry in `frequency` with a key equal to the item, we will increment the item's associated value; otherwise, we will create a new entry with `frequency[item] = 1`. To differentiate between the two cases, we can use the `in` operator: `item in frequency` evaluates to `True` if there is a key equal to `item` in the dictionary named `frequency`.

```
for item in data:
    if item in frequency:     # item is already a key in frequency
        frequency[item] = frequency[item] + 1   # count the item
    else:
        frequency[item] = 1   # create a new entry item: 1
```

The next step is to find the maximum frequency in the dictionary. Since the frequencies are the values in the dictionary, we need to extract a list of the values, and then find the maximum in this list. A list of the values in a dictionary object is returned by the `values` method, and a list of the keys is returned by the `keys` method. For example, if we already had the complete `frequency` dictionary for the example above, we could extract the values with

```
>>> frequency.values()
dict_values([1, 1, 2, 1, 2])
```

The `values` method returns a special kind of object called a `dict_values` object, but we want the values in a list. To convert the `dict_values` object to a list, we simply use the `list` function:

```
>>> list(frequency.values())
[1, 1, 2, 1, 2]
```

We can get a list of keys from a dictionary in a similar way:

```
>>> list(frequency.keys())
[19.0, 19.2, 18.9, 19.1, 19.3]
```

The resulting lists will be in whatever order the dictionary happened to be stored, but the `values` and `keys` methods are guaranteed to produce lists in which every `key` in a `key:value` pair is in the same position in the list of keys as its corresponding `value` is in the list of values.

Returning to the problem at hand, we can use the `values` method to retrieve the list of frequencies, and then find the maximum frequency in that list:

```
frequencyValues = list(frequency.values())
maxFrequency = max(frequencyValues)
```

Finally, to find the mode(s), we need to build a list of all the items in `data` with frequency equal to `maxFrequency`. To do this, we iterate over all the keys in `frequency`, appending to the list of modes each key that has a value equal to `maxFrequency`. Iterating over the keys in `frequency` is done with the familiar `for` loop. When the `for` loop is done, we return the list of modes.

```
modes = [ ]
for key in frequency:
    if frequency[key] == maxFrequency:
        modes.append(key)

return modes
```

With all of these pieces in place, the complete function looks like the following.

378 ■ Data analysis

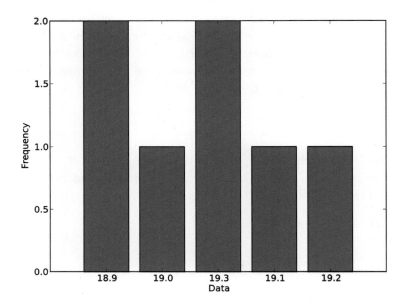

Figure 8.1 A histogram displaying the frequency of each temperature reading in the list [18.9, 19.1, 18.9, 19.0, 19.3, 19.2, 19.3].

```
def mode(data):
    """ (docstring omitted) """

    frequency = { }

    for item in data:
        if item in frequency:
            frequency[item] = frequency[item] + 1
        else:
            frequency[item] = 1

    frequencyValues = list(frequency.values())
    maxFrequency = max(frequencyValues)

    modes = [ ]
    for key in frequency:
        if frequency[key] == maxFrequency:
            modes.append(key)

    return modes
```

As a byproduct of computing the mode, we have also done all of the work necessary to create a *histogram* for the values in `data`. The histogram is simply a vertical bar chart with the keys on the x-axis and the height of the bars equal to the frequency of each key. We leave the creation of a histogram as Exercise 8.3.7. As an example, Figure 8.1 shows a histogram for our `temperatures` list.

We can use dictionaries for a variety of purposes beyond counting frequencies. For example, as the name suggests, dictionaries are well suited for handling translations. The following dictionary associates a meaning with each of three texting abbreviations.

```
>>> translations = {'lol': 'laugh out loud', 'u': 'you', 'r': 'are'}
```

We can find the meaning of `lol` with

```
>>> translations['lol']
'laugh out loud'
```

There are more examples in the exercises that follow.

Exercises

8.3.1. Write a function

```
printFrequencies(frequency)
```

that prints a formatted table of frequencies stored in the dictionary named `frequency`. The key values in the table must be listed in sorted order. For example, for the dictionary in the text, the table should look something like this:

```
Key     Frequency
18.9    2
19.0    1
19.1    1
19.2    1
19.3    2
```

(Hint: iterate over a sorted list of keys.)

8.3.2. Write a function

```
wordFrequency(text)
```

that returns a dictionary containing the frequency of each word in the string `text`. For example, `wordFrequency('I am I.')` should return the dictionary `{'I': 2, 'am': 1}`. (Hint: the `split` and `strip` string methods will be useful; see Appendix B.)

8.3.3. The *probability mass function* (PMF) of a data set gives the probability of each value in the set. A dictionary representing the PMF is a frequency dictionary with each frequency value divided by the total number of values in the original data set. For example, the probabilities for the values represented in the table in Exercise 8.3.1 are shown below.

Key	Probability
18.9	2/7
19.0	1/7
19.1	1/7
19.2	1/7
19.3	2/7

Write a function

`pmf(frequency)`

that returns a dictionary containing the PMF of the frequency dictionary passed as a parameter.

8.3.4. Write a function

`wordFrequencies(fileName)`

that prints an alphabetized table of word frequencies in the text file with the given `fileName`.

8.3.5. Write a function

`firstLetterCount(words)`

that takes as a parameter a list of strings named `words` and returns a dictionary with lower case letters as keys and the number of words in `words` that begin with that letter (lower or upper case) as values. For example, if the list is ['ant', 'bee', 'armadillo', 'dog', 'cat'], then your function should return the dictionary {'a': 2, 'b': 1, 'c': 1, 'd': 1}.

8.3.6. Similar to the Exercise 8.3.5, write a function

`firstLetterWords(words)`

that takes as a parameter a list of strings named `words` and returns a dictionary with lower case letters as keys. But now associate with each key the *list of the words* in `words` that begin with that letter. For example, if the list is ['ant', 'bee', 'armadillo', 'dog', 'cat'], then your function should return the dictionary {'a': ['ant', 'armadillo'], 'b': ['bee'], 'c': ['cat'], 'd': ['dog']}.

8.3.7. Write a function

`histogram(data)`

that displays a histogram of the values in the list `data` using the `bar` function from `matplotlib`. The `matplotlib` function `bar(x, heights, align = 'center')` draws a vertical bar plot with bars of the given `heights`, centered at the x-coordinates in x. The `bar` function defines the widths of the bars with respect to the range of values on the x-axis. Therefore, if `frequency` is the name of your dictionary, it is best to pass in `range(len(frequency))`, instead of `list(frequency.keys())`, for x. Then label the bars on the x-axis with the `matplotlib` function `xticks(range(len(frequency)), list(frequency.keys()))`.

8.3.8. Write a function

> `bonus(salaries)`

that takes as a parameter a dictionary named `salaries`, with names as keys and salaries as values, and increases the salary of everyone in the dictionary by 5%.

8.3.9. Write a function

> `updateAges(names, ages)`

that takes as parameters a list of `names` of people whose birthday it is today and a dictionary named `ages`, with names as keys and ages as values, and increments the age of each person in the dictionary whose birthday is today.

8.3.10. Write a function

> `seniorList(students, year)`

that takes as a parameter a dictionary named `students`, with names as keys and class years as values, and returns a list of names of students who are graduating in `year`.

8.3.11. Write a function

> `createDictionary()`

that creates a dictionary, inserts several English words as keys and the Pig Latin (or any other language) translations as values, and then returns the completed dictionary.

Next write a function

> `translate()`

that calls your `createDictionary` function to create a dictionary, and then repeatedly asks for a word to translate. For each entered word, it should print the translation using the dictionary. If a word does not exist in the dictionary, the function should say so. The function should end when the word `quit` is typed.

8.3.12. Write a function

> `txtTranslate(word)`

that uses a dictionary to return the English meaning of the texting abbreviation `word`. Incorporate translations for at least ten texting abbreviations. If the abbreviation is not in the dictionary, your function should return a suitable string message instead. For example, `txtTranslate('lol')` might return `'laugh out loud'`.

8.3.13. Write a function

> `login(passwords)`

that takes as a parameter a dictionary named `passwords`, with usernames as keys and passwords as values, and repeatedly prompts for a username and password until a valid pair is entered. Your function should continue to prompt even if an invalid username is entered.

8.3.14. Write a function

> `union(dict1, dict2)`

that returns a new dictionary that contains all of the entries of the two dictionaries `dict1` and `dict2`. If the dictionaries share a key, use the value in the first dictionary. For example, `union({'pies': 3, 'cakes': 5}, {'cakes': 4, 'tarts': 2}` should return the dictionary `{'pies': 3, 'cakes': 5, 'tarts': 2}`.

8.3.15. The *Mohs hardness scale* rates the hardness of a rock or mineral on a 10-point scale, where 1 is very soft (like talc) and 10 is very hard (like diamond). Suppose we have a list such as

> ```
> rocks = [('talc', 1), ('lead', 1.5), ('copper', 3),
>          ('nickel', 4), ('silicon', 6.5), ('emerald', 7.5),
>          ('boron', 9.5), ('diamond', 10)]
> ```

where the first element of each tuple is the name of a rock or mineral, and the second element is its hardness. Write a function

> `hardness(rocks)`

that returns a dictionary organizing the rocks and minerals in such a list into four categories: soft (1–3), medium (3.1–5), hard (5.1–8), and very hard (8.1–10). For example, given the list above, the function would return the dictionary

> ```
> {'soft': ['talc', 'lead', 'copper'],
>  'medium': ['nickel'],
>  'hard': ['silicon', 'emerald'],
>  'very hard': ['boron', 'diamond']}
> ```

8.3.16. (This exercise assumes that you have read Section 6.7.) Rewrite the `complement` function on page 302 using a dictionary. (Do not use any `if` statements.)

8.3.17. (This exercise assumes that you have read Section 6.7.) Suppose we have a set of homologous DNA sequences with the same length. A *profile* for these sequences contains the frequency of each base in each position. For example, suppose we have the following five sequences, lined up in a table:

	G	G	T	T	C
	G	A	T	T	A
	G	C	A	T	A
	C	A	A	T	C
	G	C	A	T	A
A:	0	2	3	0	3
C:	1	2	0	0	2
G:	4	1	0	0	0
T:	0	0	2	5	0
	G	A	A	T	A	← consensus sequence

Their profile is shown below the sequences. The first column of the profile indicates that there is one sequence with a `C` in its first position and four sequences with a `G` in their first position. The second column of the profile shows that there are two sequences with `A` in their second position, two sequences with `C` in their second position, and one sequence with `G` in its second position. And so on. The *consensus sequence* for a set of sequences has in each position the most common base in the profile. The consensus for this list of sequences is shown below the profile.

A profile can be implemented as a list of 4-element dictionaries, one for each column. A consensus sequence can then be constructed by finding the base with the maximum frequency in each position. In this exercise, you will build up a function to find a consensus sequence in four parts.

(a) Write a function

```
profile1(sequences, index)
```

that returns a dictionary containing the frequency of each base in position `index` in the list of DNA sequences named `sequences`. For example, suppose we pass in the list ['GGTTC', 'GATTA', 'GCATA', 'CAATC', 'GCATA'] for `sequences` (the sequences from the example above) and the value 2 for `index`. Then the function should return the dictionary {'A': 3, 'C': 0, 'G': 0, 'T': 2}, equivalent to the third column in the profile above.

(b) Write a function

```
profile(sequences)
```

that returns a list of dictionaries representing the profile for the list of DNA sequences named `sequences`. For example, given the list of sequences above, the function should return the list

```
[{'A': 0, 'C': 1, 'G': 4, 'T': 0},
 {'A': 2, 'C': 2, 'G': 1, 'T': 0},
 {'A': 3, 'C': 0, 'G': 0, 'T': 2},
 {'A': 0, 'C': 0, 'G': 0, 'T': 5},
 {'A': 3, 'C': 2, 'G': 0, 'T': 0}]
```

Your `profile` function should call your `profile1` function in a loop.

(c) Write a function

```
maxBase(freqs)
```

that returns the base with the maximum frequency in a dictionary of base frequencies named `freqs`. For example,

```
maxBase({'A': 0, 'C': 1, 'G': 4, 'T': 0})
```

should return 'G'.

(d) Write a function

```
consensus(sequences)
```

that returns the consensus sequence for the list of DNA sequences named `sequences`. Your `consensus` function should call your `profile` function and also call your `maxBase` function in a loop.

8.4 READING TABULAR DATA

In Section 6.2, we read one-dimensional text from files and designed algorithms to analyze the text in various ways. A lot of data, however, is maintained in the form of two-dimensional tables instead. For example, the U.S. Geological Survey (USGS) maintains up-to-the-minute tabular data about earthquakes happening around the world at `http://earthquake.usgs.gov/earthquakes/feed/v1.0/csv.php`. Earthquakes typically occur on the boundaries between tectonic plates. Therefore, by extracting this data into a usable form, and then plotting the locations of the earthquakes with `matplotlib`, we can visualize the locations of the earth's tectonic plates.

The USGS earthquake data is available in many formats, the simplest of which is called CSV, short for "comma-separated values." The first few columns of a USGS data file look like the following.

```
time,latitude,longitude,depth,mag,magType,...
2013-09-24T20:01:22.700Z,40.1333,-123.863,29,1.8,Md,...
2013-09-24T18:57:59.000Z,59.8905,-151.2392,56.4,2.5,Ml,...
2013-09-24T18:51:19.100Z,37.3242,-122.1015,0.3,1.8,Md,...
2013-09-24T18:40:09.100Z,34.3278,-116.4663,8.5,1.2,Ml,...
2013-09-24T18:20:06.300Z,35.0418,-118.3227,1.3,1.4,Ml,...
2013-09-24T18:09:53.700Z,32.0487,-115.0075,28,3.2,Ml,...
```

As you can see, CSV files contain one row of text per line, with columns separated by commas. The first row contains the names of the fifteen columns in this file, only six of which are shown here. Each additional row consists of fifteen columns of data from one earthquake. The first earthquake in this file was first detected at 20:01 UTC (Coordinated Universal Time) on 2013-09-24 at 40.1333 degrees latitude and −123.863 degrees longitude. The earthquake occurred at a depth of 29 km and had magnitude 1.8.

The CSV file containing data about all of the earthquakes in the past month is available on the web at the URL

`http://earthquake.usgs.gov/earthquakes/feed/v1.0/summary/all_month.csv`

(If you have trouble with this file, you can try smaller files by replacing `all_month` with `2.5_month` or `4.5_month`. The numbers indicate the minimum magnitude of the earthquakes included in the file.)

Reflection 8.15 *Enter the URL above into a web browser to see the data file for yourself. About how many earthquakes were recorded in the last month?*

We can read the contents of CSV files in Python using the same techniques that we used in Section 6.2. We can either download and save this file manually (and then read the file from our program), or we can download it directly from our program using the `urllib.request` module. We will use the latter method.

To begin our function, we will open the URL, and then read the header row containing the column names, as follows. (We do not actually need the header row; we just need to get past it to get to the data.)

```
import urllib.request as web

def plotQuakes():
    """Plot the locations of all earthquakes in the past month.

    Parameters: None

    Return value: None
    """

    url = 'http://earthquake.usgs.gov/...'  # see above for full URL
    quakeFile = web.urlopen(url)
    header = quakeFile.readline()
```

To visualize fault boundaries with `matplotlib`, we will need all the longitude (x) values in one list and all the latitude (y) values in another list. In our plot, we will color the points according to their depths, so we will also need to extract the depths of the earthquakes in a third list. To maintain an association between the latitude, longitude, and depth of a particular earthquake, we will need these lists to be *parallel* in the sense that, at any particular index, the longitude, latitude, and depth at that index belong to the same earthquake. In other words, if we name these three lists `longitudes`, `latitudes`, and `depths`, then for any value of `index`, `longitudes[index]`, `latitudes[index]`, and `depths[index]` belong to the same earthquake.

In our function, we next initialize these three lists to be empty and begin to iterate over the lines in the file:

```
longitudes = []
latitudes = []
depths = []
for line in quakeFile:
    line = line.decode('utf-8')
```

To extract the necessary information from each line of the file, we can use the `split` method from the string class. The `split` method splits a string at every instance of a given character, and returns a list of strings that result. For example, `'this;is;a;line'.split(';')` returns the list of strings `['this', 'is', 'a', 'line']`. (If no argument is given, it splits at strings of whitespace characters.) In this case, we want to split the string `line` at every comma:

```
row = line.split(',')
```

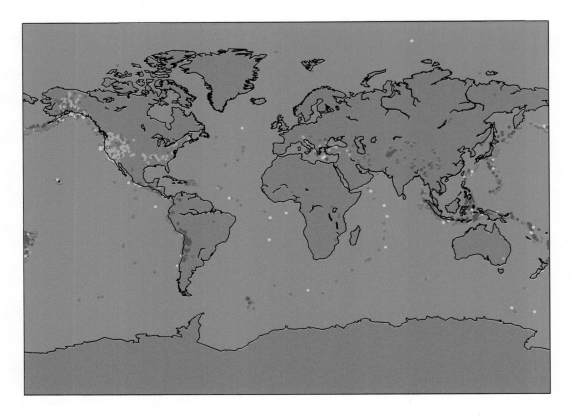

Figure 8.2 Plotted earthquake locations with colors representing depths (yellow dots are shallower, red dots are medium, and blue dots are deeper).

The resulting list named `row` will have the time of the earthquake at index 0, the latitude at index 1, the longitude at index 2 and the depth at index 3, followed by 11 additional columns that we will not use. Note that each of these values is a string, so we will need to convert each one to a number using the `float` function. After converting each value, we append it to its respective list:

```
latitudes.append(float(row[1]))
longitudes.append(float(row[2]))
depths.append(float(row[3]))
```

Finally, after the loop completes, we close the file.

```
quakeFile.close()
```

Once we have the data from the file in this format, we can plot the earthquake epicenters, and depict the depth of each earthquake with a color. Shallow (less than 10 km deep) earthquakes will be yellow, medium depth (between 10 and 50 km) earthquakes will be red, and deep earthquakes (greater than 50 km deep) will be blue. In `matplotlib`, we can color each point differently by passing the `scatter` function a list of colors, one for each point. To create this list, we iterate over the `depths` list and, for each depth, append the appropriate color string to another list named `colors`:

```
colors = []
for depth in depths:
    if depth < 10:
        colors.append('yellow')
    elif depth < 50:
        colors.append('red')
    else:
        colors.append('blue')
```

Finally, we can plot the earthquakes in their respective colors:

```
pyplot.scatter(longitudes, latitudes, 10, color = colors)
pyplot.show()
```

In the call to `scatter` above, the third argument is the square of the size of the point marker, and `color = colors` is a *keyword argument*. (We also saw keyword arguments briefly in Section 4.1.) The name `color` is the name of a parameter of the `scatter` function, for which we are passing the argument `colors`. (We will only use keyword arguments with `matplotlib` functions, although we could also use them in our functions if we wished to do so.)

The complete `plotQuakes` function is shown below.

```
import urllib.request as web
import matplotlib.pyplot as pyplot

def plotQuakes():
    """Plot the locations of all earthquakes in the past month.

    Parameters: None

    Return value: None
    """

    url = 'http://earthquake.usgs.gov/...' # see above for full URL
    quakeFile = web.urlopen(url)
    header = quakeFile.readline()

    longitudes = []
    latitudes = []
```

```
        depths = []
        for line in quakeFile:
            line = line.decode('utf-8')
            row = line.split(',')
            latitudes.append(float(row[1]))
            longitudes.append(float(row[2]))
            depths.append(float(row[3]))
        quakeFile.close()

        colors = []
        for depth in depths:
            if depth < 10:
                colors.append('yellow')
            elif depth < 50:
                colors.append('red')
            else:
                colors.append('blue')

        pyplot.scatter(longitudes, latitudes, 10, color = colors)
        pyplot.show()
```

The plotted earthquakes are shown in Figure 8.2 over a map background. (Your plot will not show the map, but if you would like to add it, look into the `Basemap` class from the `mpl_toolkits.basemap` module.) Geologists use illustrations like Figure 8.2 to infer the boundaries between tectonic plates.

Reflection 8.16 *Look at Figure 8.2 and try to identify the different tectonic plates. Can you infer anything about the way neighboring plates interact from the depth colors?*

For example, the ring of red and yellow dots around Africa encloses the African plate, and the dense line of blue and red dots northwest of Australia delineates the boundary between the Eurasian and Australian plates. The depth information gives geologists information about the types of the boundaries and the directions in which the plates are moving. For example, the shallow earthquakes on the west coast of North America mark a divergent boundary in which the plates are moving away from each other, while the deeper earthquakes in the Aleutian islands near Alaska mark a subduction zone on a convergent boundary where the Pacific plate to the south is diving underneath the North American plate to the north.

Exercises

8.4.1. Modify `plotQuakes` so that it also reads earthquake magnitudes into a list, and then draws larger circles for higher magnitude earthquakes. The sizes of the circles can be changed by passing a list of sizes, similar to the list of colors, as the third argument to the `scatter` function.

8.4.2. Modify the function from Exercise 8.3.5 so that it takes a file name as a parameter and uses the words from this file instead. Test your function using the SCRABBLE® dictionary on the book web site.[1]

8.4.3. Modify the function from Exercise 8.3.13 so that it takes a file name as a parameter and creates a username/password dictionary with the usernames and passwords in that file before it starts prompting for a username and password. Assume that the file contains one username and password per line, separated by a space. There is an example file on the book web site.

8.4.4. Write a function

```
plotPopulation()
```

that plots the world population over time from the tab-separated data file on the book web site named `worldpopulation.txt`. To read a tab-separated file, split each line with `line.split('\t')`. These figures are U.S. Census Bureau midyear population estimates from 1950–2050. Your function should create two plots. The first shows the years on the x-axis and the populations on the y-axis. The second shows the years on the x-axis and the annual growth rate on the y-axis. The growth rate is the difference between this year's population and last year's population, divided by last year's population. Be sure to label your axes in both plots with the `xlabel` and `ylabel` functions.

What is the overall trend in world population growth? Do you have any hypotheses regarding the most significant spikes and dips?

8.4.5. Write a function

```
plotMeteorites()
```

that plots the location (longitude and latitude) of every known meteorite that has fallen to earth, using a tab-separated data file from the book web site named `meteoritessize.txt`. Split each line of a tab-separated file with `line.split('\t')`. There are large areas where no meteorites have apparently fallen. Is this accurate? Why do you think no meteorites show up in these areas?

8.4.6. Write a function

```
plotTemps()
```

that reads a CSV data file from the book web site named `madison_temp.csv` to plot several years of monthly minimum temperature readings from Madison, Wisconsin. The temperature readings in the file are integers in tenths of a degree Celsius and each date is expressed in YYYYMMDD format. Rather than putting every date in a list for the x-axis, just make a list of the years that are represented in the file. Then plot the data and put a year label at each January tick with

```
pyplot.plot(range(len(minTemps)), minTemps)
pyplot.xticks(range(0, len(minTemps), 12), years)
```

The first argument to the `xticks` function says to only put a tick at every twelfth x value, and the second argument supplies the list of years to use to label those ticks. It will be helpful to know that the data file starts with a January 1 reading.

---

[1] SCRABBLE® is a registered trademark of Hasbro Inc.

8.4.7. Write a function

   `plotZebras()`

that plots the migration of seven Burchell's zebra in northern Botswana. (The very interesting story behind this data can be found at `https://www.movebank.org/node/11921`.) The function should read a CSV data file from the book web site named `zebra.csv`. Each line in the data file is a record of the location of an individual zebra at a particular time. Each individual has a unique identifier. There are over 50,000 records in the file. For each record, your function should extract the individual identifier (column index 9), longitude (column index 3), and latitude (column index 4). Store the data in two dictionaries, each with a key equal to an identifier. The values in the two dictionaries should be the longitudes and latitudes, respectively, of all tracking events for the zebra with that identifier. Plot the locations of the seven zebras in seven different colors to visualize their migration patterns.

How can you determine from the data which direction the zebras are migrating?

8.4.8. On the book web site, there is a tab-separated data file named `education.txt` that contains information about the maximum educational attainment of U.S. citizens, as of 2013. Each non-header row contains the number of people in a particular category (in thousands) that have attained each of fifteen different educational levels. Look at the file in a text editor (*not* a spreadsheet program) to view its contents. Write a function

   `plotEducation()`

that reads this data and then plots separately (but in one figure) the educational attainment of all males, females, and both sexes together over 18 years of age. The x-axis should be the fifteen different educational attainment levels and the y axis should be the *percentage* of each group that has attained that level. Notice that you will only need to extract three lines of the data file, skipping over the rest. To label the ticks on the x-axis, use the following:

```
pyplot.xticks(range(15), titles[2:], rotation = 270)
pyplot.subplots_adjust(bottom = 0.45)
```

The first statement labels the x ticks with the educational attainment categories, rotated 270 degrees. The second statement reserves 45% of the vertical space for these x tick labels. Can you draw any conclusions from the plot about relative numbers of men and women who pursue various educational degrees?

*8.5   DESIGNING EFFICIENT ALGORITHMS

For nearly all of the problems we have encountered so far, there has been only one algorithm, and its design has been relatively straightforward. But for most real problems that involve real data, there are many possible algorithms, and a little more ingenuity is required to find the best one.

To illustrate, let's consider a slightly more involved problem that can be formulated in terms of the following "real world" situation. Suppose you organized a petition drive in your community, and have now collected all of the lists of signatures from your volunteers. Before you can submit your petition to the governor, you need to remove duplicate signatures from the combined list. Rather than try to do this by hand, you decide to scan all of the names into a file, and design an algorithm to remove the duplicates for you. Because the signatures are numbered, you want the final list of unique names to be in their original order.

As we design an algorithm for this problem, we will keep in mind the four steps outlined in Section 7.1:

1. Understand the problem.

2. Design an algorithm.

3. Implement your algorithm as a program.

4. Analyze your algorithm for clarity, correctness, and efficiency.

Reflection 8.17 *Before you read any further, make sure you understand this problem. If you were given this list of names on paper, what algorithm would you use to remove the duplicates?*

The input to our problem is a list of items, and the output is a new list of unique items in the same order they appeared in the original list. We will start with an intuitive algorithm and work through a process of refinement to get progressively better solutions. In this process, we will see how a critical look at the algorithms we write can lead to significant improvements.

A first algorithm

There are several different approaches we could use to solve this problem. We will start with an algorithm that iterates over the items and, for each one, marks any duplicates found further down the list. Once all of the duplicates are marked, we can remove them from a copy of the original list. The following example illustrates this approach with a list containing four unique names, abbreviated A, B, C, and D. The algorithm starts at the beginning of the list, which contains the name A. Then we search down the list and record the index of the duplicate, marked in red.

$$\text{A} \quad \text{B} \quad \text{C} \quad \text{B} \quad \text{D} \quad \text{A} \quad \text{D} \quad \text{B}$$
$$\uparrow \quad \underbrace{\phantom{\text{B} \quad \text{C} \quad \text{B} \quad \text{D} \quad \text{A} \quad \text{D} \quad \text{B}}}_{\text{search for A}}$$

Next we move on to the second item, B, and mark its duplicates.

$$\text{A} \quad \text{B} \quad \text{C} \quad \text{B} \quad \text{D} \quad \text{A} \quad \text{D} \quad \text{B}$$
$$\phantom{\text{A}} \uparrow \quad \underbrace{\phantom{\text{C} \quad \text{B} \quad \text{D} \quad \text{A} \quad \text{D} \quad \text{B}}}_{\text{search for B}}$$

Some items, like the next item, C, do not have any duplicates.

$$\text{A} \quad \text{B} \quad \text{C} \quad \text{B} \quad \text{D} \quad \text{A} \quad \text{D} \quad \text{B}$$
$$\phantom{\text{A} \quad \text{B}} \uparrow \quad \underbrace{\phantom{\text{B} \quad \text{D} \quad \text{A} \quad \text{D} \quad \text{B}}}_{\text{search for C}}$$

The item after C is already marked as a duplicate, so we skip the search.

$$\text{A B C B D A D B}$$
$$\phantom{\text{A B C }}\uparrow$$

The next item, D, is not marked so we search for its duplicates down the list.

$$\text{A B C B D A D B}$$
$$\phantom{\text{A B C B }}\uparrow \underbrace{\phantom{\text{A D B}}}_{\text{search for D}}$$

Finally, we finish iterating over the list, but find that the remaining items are already marked as duplicates.

Once we know where all the duplicates are, we can make a copy of the original list and remove the duplicates from the copy. This algorithm, partially written in Python, is shown below. We keep track of the "marked" items with a list of their indices. The portions in red need to be replaced with calls to appropriate functions, which we will develop soon.

```python
def removeDuplicates1(data):
    """Return a list containing only the unique items in data.

    Parameter:
        data: a list

    Return value: a new list of the unique values in data,
                  in their original order
    """

    duplicateIndices = [ ]        # indices of duplicate items
    for index in range(len(data)):
        if index is not in duplicateIndices:
            positions = indices of later duplicates of data[index]
            duplicateIndices.extend(positions)

    unique = data.copy()
    for index in duplicateIndices:
        remove data[index] from unique

    return unique
```

To implement the red portion of the `if` statement on line 13, we need to search for `index` in the list `duplicateIndices`. We could use the Python `in` operator (`if index in duplicateIndices:`), but let's instead revisit how to write a search function from scratch. Recall that, in Section 6.5, we developed a linear search algorithm (named `find`) that returns the index of the first occurrence of a substring in a string. We can use this algorithm as a starting point for an algorithm to find an item in a list.

Reflection 8.18 *Look back at the* `find` *function on Page 283. Modify the function so that it returns the index of the first occurrence of a target item in a list.*

The linear search function, based on the `find` function but applied to a list, follows.

```
def linearSearch(data, target):
    """Find the index of the first occurrence of target in data.

    Parameters:
        data: a list object to search in
        target: an object to search for

    Return value: the index of the first occurrence of target in data
    """

    for index in range(len(data)):
        if data[index] == target:
            return index
    return -1
```

The function iterates over the indices of the list named `data`, checking whether each item equals the target value. If a match is found, the function returns the index of the match. Otherwise, if no match is ever found, the function returns −1. We can call this function in the `if` statement on line 13 of `removeDuplicates1` to determine whether the current value of `index` is already in `duplicateIndices`.

Reflection 8.19 *What would this* `if` *statement look like?*

We want the condition in the `if` statement to be true if `index` is not in the list, i.e., if `linearSearch` returns −1. So the `if` statement should look like:

```
if linearSearch(duplicateIndices, index) == -1:
```

In the red portion on line 14, we need to find *all* of the indices of items equal to `data[index]` that occur *later* in the list. A function to do this will be very similar to `linearSearch`, but differ in two ways. First, it must return a list of indices instead of a single index. Second, it will require a third parameter that specifies where the search should begin.

Reflection 8.20 *Write a new version of* `linearSearch` *that differs in the two ways outlined above.*

The resulting function could look like this:

```
def linearSearchAll(data, target, start):
    """Find the indices of all occurrences of target in data.

    Parameters:
        data: a list object to search in
        target: an object to search for
        start: the index in data to start searching from

    Return value: a list of indices of all occurrences of
                  target in data
    """

    found = [ ]
    for index in range(start, len(data)):
        if data[index] == target:
            found.append(index)
    return found
```

With these two new functions, we can fill in the first two missing parts of our `removeDuplicates1` function:

```
def removeDuplicates1(data):
    """ (docstring omitted) """

    duplicateIndices = [ ]           # indices of duplicate items
    for index in range(len(data)):
        if linearSearch(duplicateIndices, index) == -1:
            positions = linearSearchAll(data, data[index], index + 1)
            duplicateIndices.extend(positions)

    unique = data.copy()
    for index in duplicateIndices:
        remove data[index] from unique

    return unique
```

Once we have the list of duplicates, we need to remove them from the list. The algorithm above suggests using the `pop` method to do this, since `pop` takes an index as a parameter:

```
unique.pop(index)
```

However, this is likely to cause a problem. To see why, suppose we have the list [1, 2, 3, 2, 3] and we want to remove the duplicates at indices 3 and 4:

```
>>> data = [1, 2, 3, 2, 3]
>>> data.pop(3)
>>> data.pop(4)
```

Reflection 8.21 *What is the problem above?*

The problem is that, after we delete the item at index 3, the list looks like [1, 2, 3, 3], so the next item we want to delete is at index 3 instead of index 4. In fact, there is no index 4!

An alternative approach would be to use a list accumulator to build the **unique** list up from an empty list. To do this, we will need to iterate over the original list and append items to **unique** if their indices are not in **duplicateIndices**. The following function replaces the last loop with a new list accumulator loop that uses this approach instead.

```
def removeDuplicates1(data):
    """ (docstring omitted) """

    duplicateIndices = [ ]          # indices of duplicate items
    for index in range(len(data)):
        if linearSearch(duplicateIndices, index) == -1:
            positions = linearSearchAll(data, data[index], index + 1)
            duplicateIndices.extend(positions)

    unique = [ ]
    for index in range(len(data)):
        if linearSearch(duplicateIndices, index) == -1:
            unique.append(data[index])

    return unique
```

We now finally have a working function! (Try it out.) However, this does not mean that we should immediately leave this problem behind. It is important that we take some time to critically *reflect* on our solution. Is it correct? (If you covered Section 7.3, this would be a good time to develop some unit tests.) Can the function be simplified or made more efficient? What is the algorithm's time complexity?

Reflection 8.22 *The function above can be simplified a bit. Look at the similarity between the two* **for** *loops. Can they be combined?*

The two **for** loops can, in fact, be combined. If the condition in the first **if** statement is true, then this must be the first time we have seen this particular list item. Therefore, we can append it to the **unique** list right then. The resulting function is a bit more streamlined:

```
1  def removeDuplicates1(data):
2      """ (docstring omitted) """
3
4      duplicateIndices = [ ]
5      unique = [ ]
6      for index in range(len(data)):
7          if linearSearch(duplicateIndices, index) == -1:
8              positions = linearSearchAll(data, data[index], index + 1)
9              duplicateIndices.extend(positions)
10             unique.append(data[index])
11     return unique
```

Now let's analyze the asymptotic time complexity of this algorithm in the worst case. The input to the algorithm is the parameter `data`. As usual, we will call the length of this list n (i.e., `len(data)` is n). The statements on lines 4, 5, and 11 each count as one elementary step. The rest of the algorithm is contained in the `for` loop, which iterates n times. So the `if` statement on line 7 is executed n times and, in the worst case, the statements on lines 8–10 are each executed n times as well. But how many elementary steps are hidden in each of these statements?

Reflection 8.23 *How many elementary steps are required in the worst case by the call to* `linearSearch` *on line 7? (You might want to refer back to Page 283, where we talked about the time complexity of a linear search.)*

We saw in Section 6.5 that linear search is a linear-time algorithm (hence the name). So the number of elementary steps required by line 7 is proportional to the length of the list that is passed in as a parameter. The length of the list that we pass into `linearSearch` in this case, `duplicateIndices`, will be zero in the first iteration of the loop, but may contain as many as $n - 1$ indices in the second iteration. So the total number of elementary steps in each of these later iterations is at most $n - 1$, which is just n asymptotically.

Reflection 8.24 *Why could* `duplicateIndices` *have length* $n - 1$ *after the first iteration of the loop? What value of* `data` *would make this happen?*

So the total number of elementary steps executed by line 7 is at most $1 + (n-1)(n)$, one in the first iteration of the loop, then at most n for each iteration thereafter. This simplifies to $n^2 - n + 1$ which, asymptotically, is just n^2.

The `linearSearchAll` function, called on line 8, is also a linear search, but this function only iterates from `start` to the end of the list, so it requires about $n -$ `start` elementary steps instead of n, as long as the call to `append` that is in the body of the loop qualifies as an elementary step.

Reflection 8.25 *Do you think* `append` *qualifies as an elementary step? In other words, does it take the same amount of time regardless of the list or the item being appended?*

> **Box 8.3: Amortization**
>
> Consider a typical call to the `append` method, like `data.append(item)`. How many elementary steps are involved in this call? We have seen that some Python operations, like assignment statements, require a constant number of elementary steps while others, like the `count` method, can require a linear number of elementary steps. The `append` method is an interesting case because the number of elementary steps depends on how much space is available in the chunk of memory that has been allocated to hold the list. If there is room available at the end of this chunk of memory, then a reference to the appended item can be placed at the end of the list in constant time. However, if the list is already occupying all of the memory that has been allocated to it, the `append` has to allocate a larger chunk of memory, and then copy the entire list to this larger chunk. The amount of time to do this is proportional to the current length of the list. To reconcile these different cases, we can use the average number of elementary steps over a long sequence of `append` calls, a technique known as *amortization*.
>
> When a new chunk of memory needs to be allocated, Python allocates more than it actually needs so that the next few `append` calls will be quicker. The amount of extra memory allocated is cleverly proportional to the length of the list, so the extra memory grows a little more each time. Therefore, a sequence of `append` calls on the same list will consist of sequences of several quick operations interspersed with increasingly rare slow operations. If you take the average number of elementary steps over all of these appends, it turns out to be a constant number (or very close to it). Thus an `append` can be considered one elementary step!

This is actually a tricky question to answer, as explained in Box 8.3. In a nutshell, the *average* time required over a sequence of `append` calls is constant (or close enough to it), so it is safe to characterize an `append` as an elementary step. So it is safe to say that a call to `linearSearchAll` requires about n − `start` elementary steps. When the `linearSearchAll` function is called on line 8, the value of `index + 1` is passed in for `start`. So when `index` has the value 0, the `linearSearchAll` function requires about n − `start` = n − (`index` + 1) = $n - 1$ elementary steps. When `index` has the value 1, it requires about n − (`index` + 1) = $n - 2$ elementary steps. And so on. So the total number of elementary steps in line 8 is

$$(n-1) + (n-2) + \cdots + 1.$$

We have seen this sum before (see Box 4.2); it is equal to the triangular number

$$\frac{n(n-1)}{2} = \frac{n^2}{2} - \frac{n}{2}$$

which is the same as n^2 asymptotically.

Finally, lines 9 and 10 involve a total of n elementary steps asymptotically. This is easier to see with line 10 because it involves at most one `append` per iteration. The `extend` method called on line 9 effectively appends all of the values in `positions`

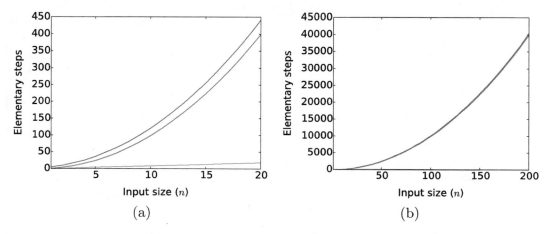

Figure 8.3 Two views of the time complexities $n^2 + 2n + 2$ (blue), n^2 (green), and n (red). (This is the same as Figure 6.4.)

to the end of duplicateIndices. Although one call to extend may append more than one index to duplicateIndices at a time, overall, each index of data can be appended at most once, so the total number of elementary steps over all of the calls to extend must be proportional to n. In summary, we have determined the following numbers of elementary steps for each line.

Line	Elementary steps
4	1
5	1
6	n
7	n^2
8	n^2
9	n
10	n
11	1

Since the maximum number of elementary steps for any line is n^2, this is the asymptotic time complexity of the algorithm. Algorithms with asymptotic time complexity n^2 are called *quadratic-time algorithms*.

How do quadratic-time algorithms compare to the linear-time algorithms we have encountered in previous chapters? The answer is communicated graphically by Figure 8.3: quadratic-time algorithms are *a lot* slower than linear-time algorithms. However, they still tend to finish in a reasonable amount of time, unless n is extremely large.

Another visual way to understand the difference between a linear-time algorithm and a quadratic-time algorithm is shown in Figure 8.4. Suppose each square represents one elementary step. On the left is a representation of the work required in a linear-time algorithm. If the size of the input to the linear-time algorithm

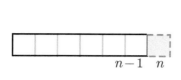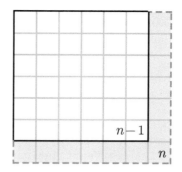

Figure 8.4 Linear vs. quadratic growth.

increases from $n-1$ to n ($n = 7$ in the pictures), then the algorithm must execute one additional elementary step (in gray). On the right is a representation of the work involved in a quadratic-time algorithm. If the size of the input to the quadratic-time algorithm increases from $n-1$ to n, then the algorithm gains $2n-1$ additional steps. So we can see that the amount of work that a quadratic-time algorithm must do grows much more quickly than the work required by a linear-time algorithm!

A more elegant algorithm

In our current algorithm, we collect a list of indices of duplicate items, and search this list before deciding whether to append a new item to the growing list of unique items.

Reflection 8.26 *In our current algorithm, how else could we tell if the current item,* data[index], *is a duplicate, without referring to* duplicateIndices?

Since we are now constructing the list of unique items in the main `for` loop, we could decide whether the current item (data[index]) is a duplicate by searching for it in unique, instead of searching for index in duplicateIndices. This change eliminates the need for the duplicateIndices list altogether and greatly simplifies the algorithm, as illustrated below:

```
def removeDuplicates2(data):
    """ (docstring omitted) """

    duplicateIndices = [ ]
    unique = [ ]
    for index in range(len(data)):
        if linearSearch(unique, data[index]) == -1:
            positions = linearSearchAll(data, data[index], index + 1)
            duplicateIndices.extend(positions)
            unique.append(data[index])
    return unique
```

In addition, since we are not storing indices any longer, we can iterate over the items in the list instead of the indices, giving a much cleaner look to the algorithm:

```
1  def removeDuplicates2(data):
2      """ (docstring omitted) """
3
4      unique = [ ]
5      for item in data:
6          if linearSearch(unique, item) == -1:
7              unique.append(item)
8      return unique
```

Reflection 8.27 *This revised algorithm is certainly more elegant. But is it more efficient?*

To answer this question, let's revisit our time complexity analysis. The `for` loop still iterates n times and, in the worst case, both the call to `linearSearch` and the `append` are still executed in every iteration of the loop. We saw above that the number of elementary steps executed by the `linearSearch` function depends on the length of the list that is passed in. In this case, `unique` can be no longer than the number of previous iterations, since at most one item is appended to it in each iteration. So the length of `unique` can grow by at most one in each iteration, meaning that the number of elementary steps executed by `linearSearch` can also grow by at most one in each iteration and the total number of elementary steps executed by all of the calls to `linearSearch` is at most

$$1 + 2 + 3 + \cdots + (n-1) = \frac{n(n-1)}{2}$$

or, once again, n^2 asymptotically. In summary, the numbers of elementary steps for each line are now:

Line	Elementary steps
4	1
5	n
6	n^2
7	n
8	1

Like our previous algorithm, the maximum value in the table is n^2, so our new algorithm is also a quadratic-time algorithm.

A more efficient algorithm

Reflection 8.28 *Do you think it is possible to design a more efficient algorithm for this problem? If so, what part(s) of the algorithm would need to be made more efficient? Are there parts of the algorithm that are absolutely necessary and therefore cannot be made more efficient?*

To find all of the duplicates in a list, it seems obvious that we need to look at every one of the n items in the list (using a `for` loop or some other kind of loop). Therefore, the time complexity of any algorithm for this problem must be at least n, or linear-time. But is a linear-time algorithm actually possible? Apart from the loop, the only significant component of the algorithm remaining is the linear search in line 6. Can the efficiency of this step be improved from linear time to something better?

Searching for data efficiently, as we do in a linear search, is a fundamental topic that is covered in depth in later computer science courses. There are a wide variety of innovative ways to store data to facilitate fast access, most of which are beyond the scope of an introductory book. However, we have already seen one alternative. Recall from Box 8.2 that a dictionary is cleverly implemented so that access can be considered a constant-time operation. So if we can store information about duplicates in a dictionary, we should be able to perform the search in line 6 in constant time!

The trick is to store the items that we have already seen as keys in a dictionary. In particular, when we append a value of `item` to the list of unique items, we also make `item` a new key in the dictionary. Then we can test whether we want to append each new value of `item` by checking whether `item` is already a key in `seen`. The new function with these changes follows.

```
def removeDuplicates3(data):
    """ (docstring omitted) """

    seen = { }
    unique = [ ]
    for item in data:
        if item not in seen:
            unique.append(item)
            seen[item] = True
    return unique
```

In our new function, we associate the value `True` with each key, but this is an arbitrary choice because we never actually use these values.

Since every statement in the body of the `for` loop is now one elementary step, the `removeDuplicates3` function is a linear-time algorithm. As we saw in Figure 8.3, this makes a significant difference! Exercise 8.5.2 asks you to investigate this difference for yourself.

Exercises

8.5.1. Show how to modify each of the three functions we wrote so that they each instead return a list of only those items in `data` that have duplicates. For example, if `data` were [1, 2, 3, 1, 3, 1], the function should return the list [1, 3].

8.5.2. In this exercise, you will write a program that tests whether the linear-time `removeDuplicates3` function really is faster than the first two versions that we wrote. First, write a function that creates a list of n random integers between 0 and $n-1$ using a list accumulator and the `random.randrange` function. Then write another function that calls each of the three functions with such a list as the argument. Time how long each call takes using the `time.time` function, which returns the current time in elapsed seconds since a fixed "epoch" time (usually midnight on January 1, 1970). By calling `time.time` before and after each call, you can find the number of seconds that elapsed.

Repeat your experiment with $n = 100, 1000, 10,000$, and $100,000$ (this will take a long time). Describe your results.

8.5.3. In a round-robin tournament, every player faces every other player exactly once, and the player with the most head-to-head wins is deemed the champion. The following partially written algorithm simulates a round-robin tournament. Assume that each of the steps expressed as a comment is one elementary step.

```
def roundRobin(players):
    # initialize all players' win counts to zero
    for player1 in players:
        for player2 in players:
            if player2 != player1:
                # player1 challenges player2
                # increment the win count of the winner
    # return the player with the most wins (or tie)
```

What is the asymptotic time complexity of this algorithm?

8.5.4. The nested loop in the previous function actually generates far too many contests. Not only does it create situations in which the two players are the same person, necessitating the `if` statement, it also considers every head-to-head contest twice. The illustration on the left below (a) shows all of the contests created by the nested `for` loop in the previous function in a four-player tournament.

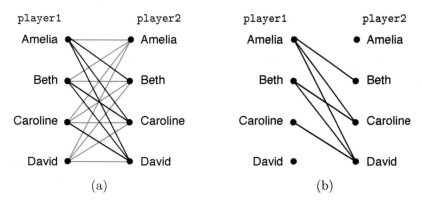

A line between two players represents the player on the left challenging the player on the right. Notice, for example, that Amelia challenges Caroline *and* Caroline challenges Amelia. Obviously, these are redundant, so we would like to avoid them. The red lines represent all of the unnecessary contests. The remaining contests that we need to consider are illustrated on the right side above (b).

(a) Let's think about how we can design an algorithm to create just the contests we need. Imagine that we are iterating from top to bottom over the `players` list on the left side of (b) above. Notice that each value of `player1` on the left only challenges values of `player2` on the right that come after it in the list. First, Amelia needs to challenge Beth, Caroline, and David. Then Beth only needs to challenge Caroline and David because Amelia challenged her in the previous round. Then Caroline only needs to challenge David because both Amelia and Beth already challenged her in previous rounds. Finally, when we get to David, everyone has already challenged him, so nothing more needs to be done.

Modify the nested `for` loop in the previous exercise so that it implements this algorithm instead.

(b) What is the asymptotic time complexity of your algorithm?

8.5.5. Suppose you have a list of projected daily stock prices, and you wish to find the best days to buy and sell the stock to maximize your profit. For example, if the list of daily stock prices was [3, 2, 1, 5, 3, 9, 2], you would want to buy on day 2 and sell on day 5, for a profit of $8 per share. Similar to the previous exercise, you need to check all possible pairs of days, such that the sell day is after the buy day. Write a function

```
profit(prices)
```

that returns a tuple containing the most profitable buy and sell days for the given list of prices. For example, for the list of daily prices above, your function should return (2, 5).

*8.6 LINEAR REGRESSION

Suppose you work in college admissions, and would like to determine how well various admissions data predict success in college. For example, if an applicant earned a 3.1 GPA in high school, is that a predictor of cumulative college GPA? Are SAT scores better or worse predictors?

Analyses such as these, looking for relationships between two (or more) variables, are called *regression analyses*. In a regression analysis, we would like to find a function or formula that accurately predicts the value of a *dependent variable* based on the value of an *independent variable*. In the college admissions problem, the independent variables are high school GPA and SAT scores, and the dependent variable is cumulative GPA in college. A regression analysis would choose one independent variable (e.g., high school GPA) and try to find a function that accurately predicts the value of one dependent variable (e.g., college GPA). Regression is a fundamental technique of *data mining*, which seeks to extract patterns and other meaningful information from large data sets.

The input to a regression problem is a data set consisting of n pairs of values. The first value in each pair is a value of the independent variable and the second value is the corresponding value of the dependent variable. For example, consider the following very small data set.

High school GPA	College GPA
3.1	2.8
3.8	3.7
3.4	3.2

We can also think of each row in the table, which represents one student, as an (x, y) point where x is the value of the independent variable and y is the value of the dependent variable.

The most common type of regression analysis is called *linear regression*. A linear regression finds a straight line that most closely approximates the relationship between the two variables. The most commonly used linear regression technique, called the *least squares* method, finds the line that minimizes the sum of the squares of the vertical distances between the line and our data points. This is represented graphically below.

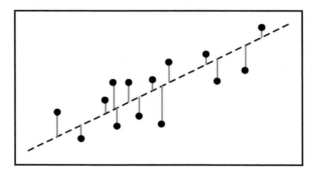

The red line segments in the figure represent the vertical distances between the points and the dashed line. This dashed line represents the line that results in the minimum total squared vertical distance for these points.

Mathematically, we are trying to find the line $y = mx + b$ (with slope m and y-intercept b) for which

$$\sum (y - (mx + b))^2$$

is the minimum. The x and y in this notation represent any one of the points (x, y) in our data set; $(y - (mx + b))$ represents the vertical distance between the height of (x, y) (given by y) and the height of the line at (x, y) (given by $mx + b$). The upper case Greek letter sigma (Σ) means that we are taking the sum over all such points (x, y) in our data set.

To find the least squares line for a data set, we could test all of the possible lines, and choose the one with the minimum total squared distance. However, since there are an infinite number of such lines, this "brute force" approach would take a very long time. Fortunately, the least squares line can be found exactly using calculus. The slope m of this line is given by

$$m = \frac{n \cdot \sum(xy) - \sum x \cdot \sum y}{n \cdot \sum(x^2) - (\sum x)^2}$$

and the y-intercept b is given by
$$b = \frac{\sum y - m \sum x}{n}.$$
Although the notation is these formulas may look imposing, the quantities are really quite simple:

- n is the number of points
- $\sum x$ is the sum of the x coordinates of all of the points (x, y)
- $\sum y$ is the sum of the y coordinates of all of the points (x, y)
- $\sum (xy)$ is the sum of x times y for all of the points (x, y)
- $\sum (x^2)$ is the sum of the squares of the x coordinates of all of the points (x, y)

For example, suppose we had only three points: $(5, 4)$, $(3, 2)$, and $(8, 3)$. Then

- $\sum x = 5 + 3 + 8 = 16$
- $\sum y = 4 + 2 + 3 = 9$
- $\sum (xy) = (5 \cdot 4) + (3 \cdot 2) + (8 \cdot 3) = 20 + 6 + 24 = 50$
- $\sum (x^2) = 5^2 + 3^2 + 8^2 = 25 + 9 + 64 = 98$

Therefore,
$$m = \frac{n \cdot \sum(xy) - \sum x \cdot \sum y}{n \cdot \sum(x^2) - (\sum x)^2} = \frac{3 \cdot 50 - 16 \cdot 9}{3 \cdot 98 - 16^2} = \frac{3}{19}$$
and
$$b = \frac{\sum y - m \sum x}{n} = \frac{9 - (3/19) \cdot 16}{3} = \frac{41}{19}.$$
Plugging in these values, we find that the formula for the least squares line is
$$y = \left(\frac{3}{19}\right) x + \frac{41}{19},$$
which is plotted below.

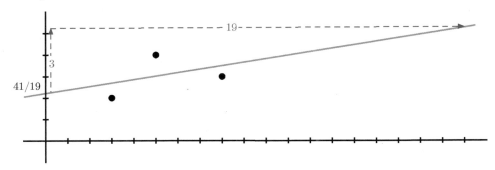

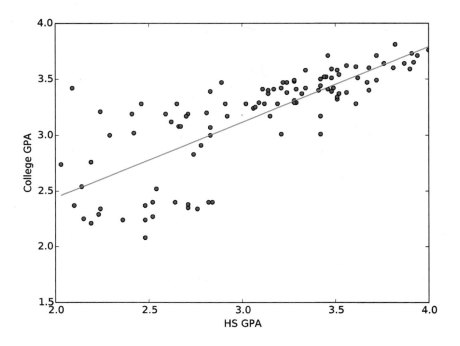

Figure 8.5 Scatter plot showing high school GPA and corresponding cumulative college GPA with regression line.

Given these formulas, it is fairly easy to write a function `linearRegression` to find the least squares regression line. Suppose the function takes as parameters a list of x coordinates named `x` and a list of y coordinates named `y` (just like the `matplotlib plot` function). The x coordinates are values of the independent variable and the y coordinates are the values of the dependent variable. Then we can use four accumulator variables to store the four sums above. For example, $\sum x$ can be computed with

```
n = len(x)              # number of points
sumx = 0                # sum of x coordinates
for index in range(n):
    sumx = sumx + x[index]
```

Once all of the sums are computed, m and b can be computed and returned with

```
return m, b
```

Exercise 8.6.1 asks you to complete the implementation of this function.

Once we have a function that performs linear regression, it is fairly simple to plot a set of points with the regression line:

```
def plotRegression(x, y, xLabel, yLabel):
    """Plot points in x and y with a linear regression line.

    Parameters:
        x: a list of x values (independent variable)
        y: a list of y values (dependent variable)
        xLabel: a string to label the x axis
        yLabel: a string to label the y axis

    Return value: None
    """

    pyplot.scatter(x, y)
    m, b = linearRegression(x, y)
    minX = min(x)
    maxX = max(x)
    pyplot.plot([minX, maxX], [m * minX + b, m * maxX + b], color = 'red')
    pyplot.xlabel(xLabel)
    pyplot.ylabel(yLabel)
    pyplot.show()
```

Returning to our college admissions problem, suppose the high school GPA values are in a list named `hsGPA` and the college GPA values are in a list named `collegeGPA`. Then we can get our regression line by calling

```
plotRegression(hsGPA, collegeGPA, 'HS GPA', 'College GPA')
```

An example plot based on real data is shown in Figure 8.5.

Reflection 8.29 *What can you discern from this plot? Does high school GPA do a good job of predicting college GPA?*

In the exercises below, and in Project 8.4, you will have the opportunity to investigate this problem in more detail. Projects 8.3 and 8.5 also use linear regression to approximate the demand curve for an economics problem and predict flood levels on the Snake River, respectively.

Exercises

8.6.1. Complete the function

```
linearRegression(x, y)
```

The function should return the slope and y-intercept of the least squares regression line for the points whose x and y coordinates are stored in the lists x and y, respectively. (Your function should use only one loop.)

8.6.2. The table below lists the homework and exam scores for a class (one row per student). Write a program that uses the completed `linearRegression` function from the previous exercise and the `plotRegression` function above to plot a linear regression line for this data.

HW	Exam
63	73
91	99
81	98
67	82
100	97
87	99
91	96
74	77
26	33
100	98
78	100
59	81
85	38
69	74

8.6.3. On the book web site, you will find a CSV data file named `sat.csv` that contains GPA and SAT data for 105 students. Write a function

   `readData(filename)`

that reads the data from this file and returns a tuple of two lists containing the data in the first and fourth columns of the file (high school GPAs and college GPAs). Then use the `plotRegression` function (which will call your completed `linearRegression` function from the previous exercise) to plot this data with a linear regression line to determine whether there is a correlation between high school GPA and college GPA. (Your plot should look like Figure 8.5.)

8.6.4. A standard way to measure how well a regression line fits a set of data is to compute the coefficient of determination, often called the R^2 coefficient, of the regression line. R^2 is defined to be

$$R^2 = 1 - \frac{S}{T}$$

where S and T are defined as follows:

$$S = \sum (y - (mx + b))^2$$

$T = \sum (y - \bar{y})^2$, where $\bar{y}$ is the mean y value

For example, the example with three points in the text: $(5, 4)$, $(3, 2)$, and $(8, 3)$. We saw that these points have regression line

$$y = \left(\frac{3}{14}\right)x + \frac{13}{7}.$$

(So $m = 3/14$ and $b = 13/7$.) Then

- $\bar{y} = (4 + 2 + 3)/3 = 3$
- $T = \sum(y - \bar{y})^2 = (4 - 3)^2 + (2 - 3)^2 + (3 - 3)^2 = 2$

- $S = \sum(y - (mx+b))^2 = (4 - 41/14)^2 + (2 - 5/2)^2 + (3 - 25/7)^2 = 169/98$
- $R^2 = 1 - (169/98)/2 = 1 - 169/196 = 27/196 \approx 0.137755$

The R^2 coefficient is always between 0 and 1, with values closer to 1 indicating a better fit.

Write a function

```
rSquared(x, y, m, b)
```

that returns the R^2 coefficient for the set of points whose x and y coordinates are stored in the lists x and y, respectively. The third and fourth parameters are the slope and y-intercept of the regression line for the set of points. For example, to apply your function to the example above, we would call

```
rSquared([5, 3, 8], [4, 2, 3], 3/14, 13/7)
```

8.6.5. An alternative linear regression method, called a *Deming regression* finds the line that minimizes the squares of the perpendicular distances between the points and the line rather than the vertical distances. While the traditional least squares method accounts only for errors in the y values, this technique accounts for errors in both x and y. The slope and y-intercept of the line found by this method are[2]

$$m = \frac{s_{yy} - s_{xx} + \sqrt{(s_{yy} - s_{xx})^2 + 4(s_{xy})^2}}{2s_{xy}}$$

and

$$b = \bar{y} - m\bar{x}$$

where

- $\bar{x} = (1/n) \sum x$, the mean of the x coordinates of all of the points (x, y)
- $\bar{y} = (1/n) \sum y$, the mean of the y coordinates of all of the points (x, y)
- $s_{xx} = (1/(n-1)) \sum (x - \bar{x})^2$
- $s_{yy} = (1/(n-1)) \sum (y - \bar{y})^2$
- $s_{xy} = (1/(n-1)) \sum (x - \bar{x})(y - \bar{y})$

Write a function

```
linearRegressionDeming(x, y)
```

that computes these values of m and b. Test your function by using it in the `plotRegression` function above.

*8.7 DATA CLUSTERING

In some large data sets, we are interested in discovering groups of items that are similar in some way. For example, suppose we are trying to determine whether a suite of test results is indicative of a malignant tumor. If we have these test results for a large number of patients, together with information about whether each patient has

---
[2]These formulas assume that the variances of the x and y errors are equal.

been diagnosed with a malignant tumor, then we can test whether clusters of similar test results correspond uniformly to the same diagnosis. If they do, the clustering is evidence that the tests can be used to test for malignancy, and the clusters give insights into the range of results that correspond to that diagnosis. Because data clustering can result in deep insights like this, it is another fundamental technique used in *data mining*.

Algorithms that cluster items according to their similarity are easiest to understand initially with two-dimensional data (e.g., longitude and latitude) that can be represented visually. Therefore, before you tackle the tumor diagnosis problem (as an exercise), we will look at a data set containing the geographic locations of vehicular collisions in New York City.[3]

The clustering problem turns out to be very difficult, and there are no known efficient algorithms that solve it exactly. Instead, people use *heuristics*. A heuristic is an algorithm that does not necessarily give the best answer, but tends to work well in practice. Colloquially, you might think of a heuristic as a "good rule of thumb." We will discuss a common clustering heuristic known as *k-means clustering*. In *k*-means clustering, the data is partitioned into *k* clusters, and each cluster has an associated *centroid*, defined to be the mean of the points in the cluster. The *k*-means clustering heuristic consists of a number of iterations in which each point is (re)assigned to the cluster of the closest centroid, and then the centroids of each cluster are recomputed based on their potentially new membership.

Defining similarity

Similarity among items in a data set can be defined in a variety of ways; here we will define similarity simply in terms of normal Euclidean distance. If each item in our data set has m numerical attributes, then we can think of these attributes as a point in m-dimensional space. Given two m-dimensional points p and q, we can find the distance between them using the formula

$$\text{distance}(p, q) = \sqrt{(p[0] - q[0])^2 + (p[1] - q[1])^2 + \cdots + (p[m-1] - q[m-1])^2}.$$

In Python, we can represent each point as a list of m numbers. For example, each of the following three lists might represent six test results ($m = 6$) for one patient:

```
p1 = [85, 92, 45, 27, 31, 0.0]
p2 = [85, 64, 59, 32, 23, 0.0]
p3 = [86, 54, 33, 16, 54, 0.0]
```

But since we are not going to need to change any element of a point once we read it from a data file, a better alternative is to store each point in a tuple, as follows:

---
[3] http://nypd.openscrape.com

Box 8.4: Privacy in the Age of Big Data

Companies collect (and buy) a lot of data about their customers, including demographics (age, address, marital status), education, financial and credit history, buying habits, and web browsing behavior. They then mine this data, using techniques like clustering, to learn more about customers so they can target customers with advertising that is more likely to lead to sales. But when does this practice lead to unacceptable breaches of privacy? For example, a recent article explained how a major retailer is able to figure out when a woman is pregnant before her family does.

When companies store this data online, it also becomes vulnerable to unauthorized access by hackers. In recent years, there have been several high-profile incidents of retail, government, and financial data breaches. As our medical records also begin to migrate online, more people are taking notice of the risks involved in storing "big data."

So as you continue to work with data, remember to always balance the reward with the inherent risk. Just because we *can* do something doesn't mean that we *should* do it.

```
p1 = (85, 92, 45, 27, 31, 0.0)
p2 = (85, 64, 59, 32, 23, 0.0)
p3 = (86, 54, 33, 16, 54, 0.0)
```

To find the distance between tuples p1 and p2, we can compute

$$\sqrt{(85-85)^2 + (92-64)^2 + (45-59)^2 + (27-32)^2 + (31-23)^2 + (0-0)^2} \approx 32.7$$

and to find the distance between tuples p1 and p3, we can compute

$$\sqrt{(85-86)^2 + (92-54)^2 + (45-33)^2 + (27-16)^2 + (31-54)^2 + (0-0)^2} \approx 47.3.$$

Because the distance between p1 and p2 is less than the distance between p1 and p3, we consider the results for patient p1 to be more similar to the results for patient p2 than to the results for patient p3.

Reflection 8.30 *What is the distance between tuples* p2 *and* p3*? Which patient's results are more similar to each of* p2 *and* p3*?*

A k-means clustering example

To illustrate in more detail how the heuristic works, consider the following small set of six points, and suppose we want to group them into $k = 2$ clusters. The lower left corner of the rectangle has coordinates $(0, 0)$ and each square is one unit on a side. So our six points have coordinates $(0.5, 5)$, $(1.75, 5.75)$, $(3.5, 2)$, $(4, 3.5)$, $(5, 2)$, and $(7, 1)$.

412 ■ Data analysis

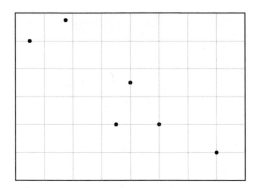

We begin by randomly choosing $k = 2$ of our points to be the initial centroids. In the figure below, we chose one centroid to be the red point at $(0.5, 5)$ and the other centroid to be the blue point at $(7, 1)$.

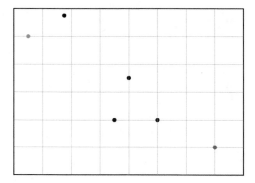

In the first iteration of the heuristic, for every point, we compute the distance between the point and each of the two centroids, represented by the dashed line segments. We assign each point to a cluster associated with the centroid to which it is closest. In this example, the three points circled in red below are closest to the red centroid and the three points circled in blue below are closest to the blue centroid.

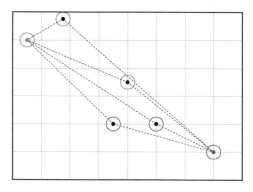

Once every point is assigned to a cluster, we compute a new centroid for each cluster, defined to be the mean of all the points in the cluster. In our example, the x and y coordinates of the new centroid of the red cluster are $(0.5 + 1.75 + 4)/3 = 25/12 \approx 2.08$

and $(5 + 5.75 + 3.5)/3 = 4.75$. The x and y coordinates of the new centroid of the blue cluster are $(3.5 + 5 + 7)/3 = 31/6 \approx 5.17$ and $(2 + 2 + 1)/3 = 5/3 \approx 1.67$. These are the new red and blue points shown below.

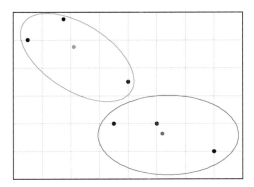

In the second iteration of the heuristic, we once again compute the closest centroid for each point. As illustrated below, the point at $(4, 3.5)$ is now grouped with the lower right points because it is closer to the blue centroid than to the red centroid (distance$((2.08, 4.75), (4, 3.5)) \approx 2.29$ and distance$((5.17, 1.67), (4, 3.5)) \approx 2.17$).

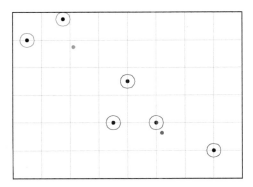

Next, we once again compute new centroids. The x and y coordinates of the new centroid of the red cluster are $(0.5 + 1.75)/2 = 1.125$ and $(5 + 5.75)/2 = 5.375$. The x and y coordinates of the new centroid of the blue cluster are $(3.5 + 4 + 5 + 7)/4 = 4.875$ and $(1 + 2 + 2 + 3.5)/4 = 2.125$. These new centroids are shown below.

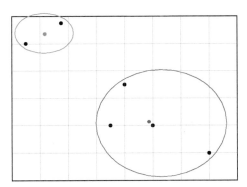

In the third iteration, when we find the closest centroid for each point, we find that nothing changes. Therefore, these clusters are the final ones chosen by the heuristic.

Implementing k-means clustering

To implement this heuristic in Python, we define the function named **kmeans** below with three parameters: a list of tuples named **data**, the number of clusters **k**, and **iterations**, the number of iterations in the heuristic. If we were calling the function with the points in the previous example, then **data** would be assigned the list

> [(0.5, 5), (1.75, 5.75), (3.5, 2), (4, 3.5), (5, 2), (7, 1)]

```
def kmeans(data, k, iterations):
    """Cluster data into k clusters using the given number of
        iterations of the k-means clustering algorithm.

    Parameters:
        data: a list of points
        k: the number of desired clusters
        iterations: the number of iterations of the algorithm

    Return value:
        a tuple containing the list of clusters and the list
        of centroids; each cluster is represented by a list
        of indices of the points assigned to that cluster
    """

    n = len(data)
    centroids = random.sample(data, k)   # k initial centroids

    for step in range(iterations):
```

The function begins by using the **sample** function from the **random** module to choose **k** random tuples from **data** to be the initial centroids, assigned to the variable named **centroids**. The remainder of the function consists of **iterations** passes over the list of points controlled by the **for** loop above.

Next, within the loop, we first create a list of **k** empty lists, one for each cluster. For example, if **k** was 4 then, after the following three lines, **clusters** would be assigned the list [[ ], [ ], [ ], [ ]].

```
        clusters = []
        for i in range(k):
            clusters.append([])
```

The first empty list, **clusters[0]**, will be used to hold the indices of the points in the cluster associated with **centroids[0]**; the second empty list, **clusters[1]**, will be used to hold the indices of the points in the cluster associated with `centroids[1]`; and so on. Some care has to be taken to create this list of lists; `clusters = [[]] * k` will not work. Exercise 8.7.6 asks you to investigate this further.

Next, we iterate over the indices of the points in `data`. In each iteration, we need to find the centroid that is closest to the point `data[dataIndex]`.

```
for dataIndex in range(n):
    minIndex = 0
    for clustIndex in range(1, k):
        dist = distance(centroids[clustIndex], data[dataIndex])
        if dist < distance(centroids[minIndex], data[dataIndex]):
            minIndex = clustIndex
    clusters[minIndex].append(dataIndex)
```

The inner `for` loop above is essentially the same algorithm that we used in Section 8.1 to find the minimum value in a list. The difference here is that we need to find the *index* of the centroid with the minimum distance to `data[dataIndex]`. (We leave writing this `distance` function as Exercise 8.7.1.) Therefore, we maintain the index of the closest centroid in the variable named `minIndex`. Once we have found the final value of `minIndex`, we append `dataIndex` to the list of indices assigned to the cluster with index `minIndex`.

After we have assigned all of the points to a cluster, we compute each new centroid with a function named `centroid`. We leave writing this function as Exercise 8.7.2.

```
for clustIndex in range(k):
    centroids[clustIndex] = centroid(clusters[clustIndex], data)
```

Finally, once the outer `for` loop finishes, we return a tuple containing the list of clusters and the list of centroids.

```
return (clusters, centroids)
```

The complete function is reproduced below:

```
def kmeans(data, k, iterations):
    """ (docstring omitted) """
    
    n = len(data)
    centroids = random.sample(data, k)
    
    for step in range(iterations):
        clusters = []
        for i in range(k):
            clusters.append([])
```

```
    for dataIndex in range(n):
        minIndex = 0
        for clustIndex in range(1, k):
            dist = distance(centroids[clustIndex], data[dataIndex])
            if dist < distance(centroids[minIndex], data[dataIndex]):
                minIndex = clustIndex
        clusters[minIndex].append(dataIndex)

    for clustIndex in range(k):
        centroids[clustIndex] = centroid(clusters[clustIndex], data)

return (clusters, centroids)
```

Locating bicycle safety programs

Suppose that, in response to an increase in the number of vehicular accidents in New York City involving cyclists, we want to find the best locations to establish a limited number of bicycle safety programs. To do the most good, we would like to centrally locate these programs in areas containing the most cycling injuries. In other words, we want to locate safety programs at the centroids of clusters of accidents.

We can use the k-means clustering heuristic to find these centroids. First, we need to read the necessary attributes from a data file containing the accident information. This data file,[4] which is available on the book web site, is tab-separated and contains a row for each of 7,642 collisions involving bicycles between August, 2011 and August, 2013 in all five boroughs of New York City. Associated with each collision are 68 attributes that include information about the location, injuries, and whether pedestrians or cyclists were involved. The first few rows of the data file look like this (with triangles representing tab characters):

```
borocode▷precinct▷year▷month▷lon▷lat▷street1▷street2▷collisions▷...
3▷60▷2011▷8▷-73.987424▷40.58539▷CROPSEY AVENUE▷SHORE PARKWAY▷4▷...
3▷60▷2011▷8▷-73.98584▷40.576503▷MERMAID AVENUE▷WEST 19 STREET▷2▷...
3▷60▷2011▷8▷-73.983088▷40.579161▷NEPTUNE AVENUE▷WEST 15 STREET▷2▷...
3▷60▷2011▷8▷-73.986609▷40.574894▷SURF AVENUE▷WEST 20 STREET▷1▷...
3▷61▷2011▷8▷-73.955477▷40.598939▷AVENUE U▷EAST 16 STREET▷2▷...
```

For our clustering algorithm, we will need a list of accident locations as (longitude, latitude) tuples. To limit the data in our analysis, we will extract the locations of only those accidents that occurred in Manhattan (borocode 1) during 2012. The function to accomplish this is very similar to what we have done previously.

---

[4]The data file contains the rows of the file at http://nypd.openscrape.com/#/collisions.csv.gz (accessed October 1, 2013) that involved a bicyclist.

```
def readFile(filename):
    """Read locations of 2012 bicycle accidents in Manhattan
        into a list of tuples.

    Parameter:
        filename: the name of the data file

    Return value: a list of (longitude, latitude) tuples
    """
    inputFile = open(filename, 'r', encoding = 'utf-8')
    header = inputFile.readline()
    data = [ ]
    for line in inputFile:
        row = line.split('\t')
        borough = int(row[0])
        year = int(row[2])
        if (borough == 1) and (year == 2012):
            longitude = float(row[4])
            latitude = float(row[5])
            data.append( (longitude, latitude) )
    inputFile.close()
    return data
```

Once we have the data as a list of tuples, we can call the `kmeans` function to find the clusters and centroids. If we had enough funding for six bicycle safety programs, we would call the function with $k = 6$ as follows:

```
data = readFile('collisions_cyclists.txt')
(clusters, centroids) = kmeans(data, 6, 100)
```

To visualize the clusters, we can plot the points with `matplotlib`, with each cluster in a different color and centroids represented by stars. The following `plotClusters` function does just this. The result is shown in Figure 8.6.

```
import matplotlib.pyplot as pyplot

def plotClusters(clusters, data, centroids):
    """Plot clusters and centroids in unique colors.

    Parameters:
        clusters: a list of k lists of data indices
        data: a list of points
        centroids: a list of k centroid points

    Return value: None
    """
```

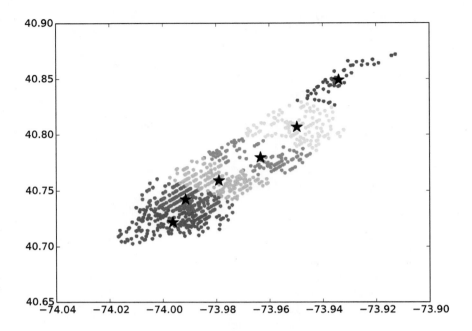

Figure 8.6 Six clusters of bicycle accidents in Manhattan with the centroid for each shown as a star.

```
colors = ['blue', 'red', 'yellow', 'green', 'purple', 'orange']
for clustIndex in range(len(clusters)):   # plot clusters of points
    x = []
    y = []
    for dataIndex in clusters[clustIndex]:
        x.append(data[dataIndex][0])
        y.append(data[dataIndex][1])
    pyplot.scatter(x, y, 10, color = colors[clustIndex])

x = []
y = []
for centroid in centroids:      # plot centroids
    x.append(centroid[0])
    y.append(centroid[1])
pyplot.scatter(x, y, 200, marker = '*', color = 'black')
pyplot.show()
```

In Exercise 8.7.3, we ask you to combine the `kmeans`, `readFile`, and `plotClusters` functions into a complete program that produces the plot in Figure 8.6.

Reflection 8.31 *If you look at the "map" in Figure 8.6, you will notice that the orientation of Manhattan appears to be more east-west than it is in reality. Do you know why?*

In addition to completing this program, Exercise 8.7.4 asks you to apply k-means clustering to the tumor diagnosis problem that we discussed at the beginning of the chapter.

Exercises

8.7.1. Write a function

```
distance(p, q)
```

that returns the Euclidean distance between k-dimensional points p and q, each of which is represented by a tuple with length k.

8.7.2. Write the function

```
centroid(cluster, data)
```

that is needed by the kmeans function. The function should compute the centroid of the given cluster, which is a list of indices of points in data. Remember that the points in data can be of any length, and the centroid that the function returns must be a tuple of the same length. It will probably be most convenient to build up the centroid in a list and then convert it to a tuple using the tuple function. If the cluster is empty, then return a randomly chosen point from data with random.choice(data).

8.7.3. Combine the distance and centroid functions that you wrote in the previous two exercises with the kmeans, readFile, and plotClusters functions to create a complete program (with a main function) that produces the plot in Figure 8.6. The data file, collisions_cyclists.txt, is available on the book web site.

8.7.4. In this exercise, you will apply k-means clustering to the tumor diagnosis problem from the beginning of this section, using a data set of breast cancer biopsy results. This data set, available on the book web site, contains nine test results for each of 683 individuals.[5] The first few lines of the comma-separated data set look like:

```
1000025,5,1,1,1,2,1,3,1,1,2
1002945,5,4,4,5,7,10,3,2,1,2
1015425,3,1,1,1,2,2,3,1,1,2
1016277,6,8,8,1,3,4,3,7,1,2
1017023,4,1,1,3,2,1,3,1,1,2
1017122,8,10,10,8,7,10,9,7,1,4
    ⋮
```

The first number in each column is a unique sample number and the last number in each column is either 2 or 4, where 2 means "benign" and 4 means "malignant."

---

[5]The data set on the book's web site is the original file from http://archive.ics.uci.edu/ml/datasets/Breast+Cancer+Wisconsin+(Original) with 16 rows containing missing attributes removed.

Of the 683 samples in the data set, 444 were diagnosed as benign and 239 were diagnosed as malignant. The nine numbers on each line between the sample number and the diagnosis are test result values that have been normalized to a 1–10 scale. (There is additional information accompanying the data on the book's web site.)

The first step is to write a function

```
readFile(filename)
```

that reads this data and returns two lists, one of test results and the other of diagnoses. The test results will be in a list of 9-element tuples and the diagnosis values will be a list of integers that are in the same order as the test results.

Next, write a program, with the following `main` function, that clusters the data into $k = 2$ groups. If one of these two groups contains predominantly benign samples and the other contains predominantly malignant samples, then we can conclude that these test results can effectively discriminate between malignant and benign tumors.

```
def main():
    data, diagnosis = readFile('breast-cancer-wisconsin.csv')
    clusters, centroids = kmeans(data, 2, 10)
    for clustIndex in range(2):
        count = {2: 0, 4: 0}
        for index in clusters[clustIndex]:
            count[diagnosis[index]] = count[diagnosis[index]] + 1
        print('Cluster', clustIndex)
        print('  benign:', count[2], 'malignant:', count[4])
```

To complete the program, you will need to incorporate the `distance` and `centroid` functions from Exercises 8.7.1 and 8.7.2, the `kmeans` function, and your `readFile` function. The data file, `breast-cancer-wisconsin.csv`, is available on the book web site. Describe the results.

8.7.5. Given data on 100 customers' torso length and chest measurements, use k-means clustering to cluster the customers into three groups: S, M, and L. (Later, you can use this information to design appropriately sized t-shirts.) You can find a list of measurements on the book web site.

8.7.6. In the `kmeans` function, we needed to create a new list of empty clusters at the beginning of each iteration. To create this list, we did the following:

```
clusters = []
for i in range(k):
    clusters.append([])
```

Alternatively, we could have created a list of lists using the * repetition operator like this:

```
clusters = [[]] * k
```

However, this will not work correctly, as evidenced by the following:

```
>>> clusters = [[]] * 5
>>> clusters
[[], [], [], [], []]
>>> clusters[2].append(4)
>>> clusters
[[4], [4], [4], [4], [4]]
```

What is going on here?

8.7.7. Modify the `kmeans` function so that it does not need to take the number of iterations as a parameter. Instead, the function should iterate until the list of clusters stops changing. In other words, the loop should iterate while the list of clusters generated in an iteration is not the same as the list of clusters from the previous iteration.

8.8 SUMMARY

Those who know how to manipulate and extract meaning from data will be well positioned as the decision makers of the future. Algorithmically, the simplest way to store data is in a list. We developed algorithms to summarize the contents of a list with various *descriptive statistics*, *modify* the contents of lists, and create *histograms* describing the frequency of values in a list. The beauty of these techniques is that they can be used with a wide variety of data types and applications. But before any of them can be used on real data, the data must be read from its source and wrangled into a usable form. To this end, we also discussed basic methods for *reading* and *formatting* tabular data both from local files and from the web.

In later sections, we went beyond simply describing data sets to *data mining* techniques that can make predictions from them. *Linear regression* seeks a linear pattern in data and then uses this pattern to predict missing data points. The *k-means clustering* algorithm partitions data into clusters of like elements to elicit hidden relationships.

Algorithms that manipulate lists can quickly become much more complicated than what we have seen previously, and therefore paying attention to their time complexity is important. To illustrate, we worked through a sequence of three increasingly more elegant and more efficient algorithms for removing duplicates from a list. In the end, we saw that the additional time taken to think through a problem carefully and reduce its time complexity can pay dividends.

8.9 FURTHER DISCOVERY

Sherlock Holmes was an early data scientist, always insisting on fact-based theories (and not vice versa). The chapter epigraph is from Sir Arthur Conan Doyle's short story, *The Adventure of the Copper Beeches* [11].

A good resource for current data-related news is the "data journalism" web site *FiveThirtyEight* at `http://fivethirtyeight.com`. If you are interested in learning more about the emerging field of data science, one resource is *Doing Data Science* by Rachel Schutt and Cathy O'Neil [49].

The article referenced in Box 8.4 is from *The New York Times* [12]. The non-profit Electronic Frontier Foundation (EFF), founded in 1990, works at the forefront of issues of digital privacy and free speech. To learn more about contemporary privacy issues, visit its website at http://www.eff.org. For more about ethical issues in computing in general, we recommend *Computer Ethics* by Deborah Johnson and Keith Miller [21].

8.10 PROJECTS

Project 8.1 Climate change

The causes and consequences of global warming are of intense interest. The consensus of the global scientific community is that the primary cause of global warming is an increased concentration of "greenhouse gasses," primarily carbon dioxide (CO_2) in the atmosphere. In addition, it is widely believed that human activity is the cause of this increase, and that it will continue into the future if we do not limit what we emit into the atmosphere.

To understand the causes of global warming, and to determine whether the increase in CO_2 is natural or human-induced, scientists have reconstructed ancient climate patterns by analyzing the composition of deep ocean sediment core samples. In a core sample, the older sediment is lower and the younger sediment is higher. These core samples contain the remains of ancient bottom-dwelling organisms called *foraminifera* that grew shells made of calcium carbonate ($CaCO_3$).

Virtually all of the oxygen in these calcium carbonate shells exists as one of two stable isotopes: oxygen-16 (^{16}O) and oxygen-18 (^{18}O). Oxygen-16 is, by far, the most common oxygen isotope in our atmosphere and seawater at about 99.76% and oxygen-18 is the second most common at about 0.2%. The fraction of oxygen-18 incorporated into the calcium carbonate shells of marine animals depends upon the temperature of the seawater. Given the same seawater composition, colder temperatures will result in a higher concentration of oxygen-18 being incorporated into the shells. Therefore, by analyzing the ratio of oxygen-18 to oxygen-16 in an ancient shell, scientists can deduce the temperature of the water at the time the shell formed. This ratio is denoted $\delta^{18}O$; higher values of $\delta^{18}O$ represent lower temperatures.

Similarly, it is possible to measure the relative amounts of carbon isotopes in calcium carbonate. Carbon can exist as one of two stable isotopes: carbon-12 (^{12}C) and carbon-13 (^{13}C). (Recall from Section 4.4 that carbon-14 (^{14}C) is radioactive and used for radiocarbon dating.) $\delta^{13}C$ is a measure of the ratio of carbon-13 to carbon-12. The value of $\delta^{13}C$ can decrease, for example, if there were a sudden injection of ^{12}C-rich (i.e., ^{13}C-depleted) carbon. Such an event would likely cause an increase in warming due to the increase in greenhouse gasses.

In this project, we will examine a large data set [60] containing over 17,000 $\delta^{18}O$ and $\delta^{13}C$ measurements from deep ocean sediment core samples, representing conditions over a period of about 65 million years. From this data, we will be able to visualize deep-sea temperature patterns over time (based on the δ^{18}O measurements) and the accompanying values of δ^{13}C. We can try to answer two questions with this data. First, is there a correspondence between changes in carbon output and changes in temperature? Second, are recent levels of carbon in the atmosphere and ocean "natural," based on what has happened in the past?

Part 1: Read and plot the data

First, download the file named `2008CompilationData.csv` from the book web site.[6] Examine the file and write a function that creates four parallel lists containing information about δ^{18}O and δ^{13}C readings: one list containing the δ^{18}O measurements, one list containing the δ^{13}C measurements, one list containing the sites of the measurements, and a fourth list containing the ages of the measurements. (Note that the `Age (ma)` column represents the ages of the samples in millions of years.) If a row is missing either the δ^{18}O value or the δ^{13}C value, do not include that row in your lists.

Using `matplotlib`, create scatter plots of the δ^{18}O and δ^{13}C measurements separately with respect to time (we will use the list of sites later). Be sure to appropriately label your axes. Because there is a lot of data here, your plots will be clearer if you make the dots small by passing `s = 1` into the `scatter` function. To create two plots in separate windows, precede the first plot with `pyplot.figure(1)` and precede the second plot with `pyplot.figure(2)`:

```
pyplot.figure(1)

# plot d18O data here

pyplot.figure(2)

# plot d13C data here

pyplot.show()   # after all the figures
```

Part 2: Smooth and plot the data

Inconsistencies in sampling, and local environmental variation, result in relatively noisy plots, but this can be fixed with a smoothing function that computes means in a window moving over the data. This is precisely the algorithm we developed back in Section 1.3 (and Exercise 8.2.14). Write a function that implements this smoothing algorithm, and use it to create three new lists — for δ^{18}O, δ^{13}C, and the ages — that are smoothed over a window of length 5.

To compare the oxygen and carbon isotope readings, it will be convenient to create two plots, one over the other, in the same window. We can do this with the `matplotlib subplot` function:

---

[6] This is a slightly "cleaned" version of the original data file from `http://www.es.ucsc.edu/~jzachos/Data/2008CompilationData.xls`

```
pyplot.figure(3)
pyplot.subplot(2, 1, 1) # arguments are (rows, columns, subplot #)

# plot d18O here

pyplot.subplot(2, 1, 2)

# plot d13C here
```

Part 3: The PETM

Looking at your smoothed data, you should now clearly see a spike in temperature (i.e., rapid decrease in $\delta^{18}O$) around 55 ma. This event is known as the Palaeocene-Eocene Thermal Maximum (PETM), during which deep-sea temperatures rose 5° to 6° Celsius in fewer than 10,000 years!

To investigate further what might have happened during this time, let's "zoom in" on this period using data from only a few sites. Create three new lists, one of ages and the other two of measurements, that contain data only for ages between 53 and 57 ma and only for sites 527, 690, and 865. Sites 527 and 690 are in the South Atlantic Ocean, and site 865 is in the Western Pacific Ocean. (Try using list comprehensions to create the three lists.)

Plot this data in the same format as above, using two subplots.

Question 8.1.1 *What do you notice? What do your plots imply about the relationship between carbon and temperature?*

Part 4: Recent history

To gain insight into what "natural" CO_2 levels have looked like in our more recent history, we can consult another data set, this time measuring the CO_2 concentrations in air bubbles trapped inside ancient Antarctic ice [39]. Download this tab-separated data file from the book web site, and write code to extract the data into two parallel lists — a list of ages and a list of corresponding CO_2 concentrations. The CO_2 concentrations are measured in parts per million by volume (ppmv) and the ages are in years.

Next, plot your data. You should notice four distinct, stable cycles, each representing a glacial period followed by an interglacial period. To view the temperature patterns during the same period, we can plot the most recent 420,000 years of our $\delta^{18}O$ readings. To do so, create two new parallel lists — one containing the $\delta^{18}O$ readings from sites 607, 659, and 849 for the last 420,000 years, and the other containing the corresponding ages. Arrange a plot of this data and your plot of CO_2 concentrations in two subplots, as before.

Question 8.1.2 *What do you notice? What is the maximum CO_2 concentration during this period?*

Part 5: Very recent history

In 1958, geochemist Charles David Keeling began measuring atmospheric CO_2 concentrations at the Mauna Loa Observatory (MLO) on the big island of Hawaii. These measurements have continued for the past 50 years, and are now known as the *Keeling curve*. Available on the book web site (and at `http://scrippsco2.ucsd.edu/data/mlo.html`) is a CSV data file containing weekly CO_2 concentration readings (in ppmv) since 1958. Read this data into two parallel lists, one containing the CO_2 concentration readings and the other containing the dates of the readings. Notice that the file contains a long header in which each line starts with a quotation mark. To read past this header, you can use the following `while` loop:

```
line = inputFile.readline()   # inputFile is your file object name
while line[0] == '"':
    line = inputFile.readline()
```

Each of the dates in the file is a string in `YYYY-MM-DD` format. To convert each of these date strings to a fractional year (e.g., `2013-07-01` would be `2013.5`), we can use the following formula:

```
year = float(date[:4]) + (float(date[5:7]) + float(date[8:10])/31) / 12
```

Next, plot the data.

Question 8.1.3 *How do these levels compare with the maximum level from the previous 420,000 years? Is your plot consistent with the pattern of "natural" CO_2 concentrations from the previous 420,000 years? Based on these results, what conclusions can we draw about the human impact on atmospheric CO_2 concentrations?*

Project 8.2 Does education influence unemployment?

Conventional wisdom dictates that those with more education have an easier time finding a job. In this project, we will investigate this claim by analyzing 2012 data from the U.S. Census Bureau.[7]

Part 1: Get the data

Download the data file from the book web site and look at it in a text editor. You will notice that it is a tab-separated file with one line per U.S. metropolitan area. In each line are 60 columns containing population data broken down by educational attainment and employment status. There are four educational attainment categories: less than high school graduate (i.e., no high school diploma), high school graduate, some college or associate's degree, and college graduate.

Economists typically define unemployment in terms of the "labor force," people

---

[7] *Educational Attainment by Employment Status for the Population 25 to 64 Years* (Table B23006), U.S. Census Bureau, 2012 American Community Survey, `http://factfinder2.census.gov/faces/tableservices/jsf/pages/productview.xhtml?pid=ACS_12_1YR_B23006`

who are available for employment at any given time, whether or not they are actually employed. So we will define the unemployment rate to be the fraction of the (civilian) labor force that is unemployed. To compute the unemployment rate for each category of educational attainment, we will need data from the following ten columns of the file:

Column index	Contents
2	Name of metropolitan area
3	Total population of metropolitan area
11	No high school diploma, total in civilian labor force
15	No high school diploma, unemployed in civilian labor force
25	High school graduate, total in civilian labor force
29	High school graduate, unemployed in civilian labor force
39	Some college, total in civilian labor force
43	Some college, unemployed in civilian labor force
53	College graduate, total in civilian labor force
57	College graduate, unemployed in civilian labor force

Read this data from the file and store it in six lists that contain the names of the metropolitan areas, the total populations, and the unemployment rates for each of the four educational attainment categories. Each unemployment rate should be stored as a floating point value between 0 and 100, rounded to one decimal point. For example, 0.123456 should be stored as 12.3.

Next, use four calls to the `matplotlib plot` function to plot the unemployment rate data in a single figure with the metropolitan areas on the x-axis, and the unemployment rates for each educational attainment category on the y-axis. Be sure to label each plot and display a legend. You can place the names of the metropolitan areas on the x-axis with the `xticks` function:

`pyplot.xticks(range(len(names)), names, rotation = 270, fontsize = 'small')`

The first argument is the locations of the ticks on the x-axis, the second argument is a list of labels to place at those ticks, and the third and fourth arguments optionally rotate the text and change its size.

Part 2: Filter the data

Your plot should show data for 366 metropolitan areas. To make trends easier to discern, let's narrow the data down to the most populous areas. Create six new lists that contain the same data as the original six lists, but for only the thirty most populous metropolitan areas. Be sure to maintain the correct correspondence between values in the six lists. Generate the same plot as above, but for your six shorter lists.

Part 3: Analysis

Write a program to answer each of the following questions.

Question 8.2.1 *In which of the 30 metropolitan areas is the unemployment rate higher for HS graduates than for those without a HS diploma?*

Question 8.2.2 *Which of the 30 metropolitan areas have the highest and lowest unemployment rates for each of the four categories of educational attainment? Use a single loop to compute all of the answers, and do not use the built-in* `min` *and* `max` *functions.*

Question 8.2.3 *Which of the 30 metropolitan areas has the largest difference between the unemployment rates of college graduates and those with only a high school diploma? What is this difference?*

Question 8.2.4 *Print a formatted table that ranks the 30 metropolitan areas by the unemployment rates of their college graduates. (Hint: use a dictionary.)*

Project 8.3 Maximizing profit

Part 4 of this project assumes that you have read Section 8.6.

Suppose you are opening a new coffee bar, and need to decide how to price each shot of espresso. If you price your espresso too high, no customers will buy from you. On the other hand, if you price your espresso too low, you will not make enough money to sustain your business.

Part 1: Poll customers

To determine the most profitable price for espresso, you poll 1,000 potential daily customers, asking for the maximum price they would be willing to pay for a shot of espresso at your coffee bar. To simulate these potential customers, write a function

    randCustomers(n)

that returns a list of `n` normally distributed prices with mean $4.00 and standard deviation $1.50. Use this function to generate a list of maximum prices for your 1,000 potential customers, and display of histogram of these maximum prices.

Part 2: Compute sales

Next, based on this information, you want to know how many customers would buy espresso from you at any given price. Write a function

    sales(customers, price)

that returns the number of customers willing to buy espresso if it were priced at the given `price`. The first parameter `customers` is a list containing the maximum price that each customer is willing to pay. Then write another function

    plotDemand(customers, lowPrice, highPrice, step)

that uses your `sales` function to plot a *demand curve*. A demand curve has price on the x-axis and the quantity of sales on the y-axis. The prices on the x-axis should run from `lowPrice` to `highPrice` in increments of `step`. Use this function to draw a demand curve for prices from free to $8.00, in increments of a quarter.

Part 3: Compute profits

Suppose one pound (454 g) of roasted coffee beans costs you $10.00 and you use 8 g of coffee per shot of espresso. Each "to go" cup costs you $0.05 and you estimate that half of your customers will need "to go" cups. You also estimate that you have about $500 of fixed costs (wages, utilities, etc.) for each day you are open. Write a function

```
profits(customers, lowPrice, highPrice, step, perCost, fixedCost)
```

that plots your profit at each price. Your function should return the maximum profit, the price at which the maximum profit is attained, and the number of customers who buy espresso at that price. Do not use the built-in `min` and `max` functions.

Question 8.3.1 *How should you price a shot of espresso to maximize your profit? At this price, how much profit do you expect each day? How many customers should you expect each day?*

Part 4: Find the demand function

If you have not already done so, implement the linear regression function discussed in Section 8.6. (See Exercise 8.6.1.) Then use linear regression on the data in your demand plot in Part 2 to find the linear function that best approximates the demand. This is called the *demand* function. Linear demand functions usually have the form

$$Q = b - m \cdot P,$$

where Q is the quantity sold and P is the price. Modify your `plotDemand` function so that it computes the regression line and plots it with the demand curve.

Question 8.3.2 *What is the linear demand function that best approximates your demand curve?*

Project 8.4 Admissions

This project is a continuation of the problem begun in Section 8.6, and assumes that you have read that section.

Suppose you work in a college admissions office and would like to study how well admissions data (high school GPA and SAT scores) predict success in college.

Part 1: Get the data

We will work with a limited data source consisting of data for 105 students, available on the book web site.[8] The file is named `sat.csv`.

Write a function

```
readData(fileName)
```

that returns four lists containing the data from `sat.csv`. The four lists will contain high school GPAs, math SAT scores, verbal (critical reading) SAT scores, and cumulative college GPAs. (This is an expanded version of the function in Exercise 8.6.3.)

Then a write another function

```
plotData(hsGPA, mathSAT, crSAT, collegeGPA)
```

that plots all of this data in one figure. We can do this with the `matplotlib subplot` function:

```
pyplot.figure(1)
pyplot.subplot(4, 1, 1) # arguments are (rows, columns, subplot #)

# plot HS GPA data here

pyplot.subplot(4, 1, 2)

# plot SAT math here

pyplot.subplot(4, 1, 3)

# plot SAT verbal here

pyplot.subplot(4, 1, 4)

# plot college GPA here
```

Reflection 8.32 *Can you glean any useful information from these plots?*

Part 2: Linear regression

As we discussed in Section 8.6, a linear regression is used to analyze how well an independent variable predicts a dependent variable. In this case, the independent variables are high school GPA and the two SAT scores. If you have not already, implement the linear regression function discussed in Section 8.6. (See Exercise 8.6.1.) Then use the `plotRegression` function from Section 8.6 to individually plot each independent variable, plus combined (math plus verbal) SAT scores against college GPA, with a regression line. (You will need four separate plots.)

---

[8]Adapted from data available at http://onlinestatbook.com/2/case_studies/sat.html

Question 8.4.1 *Judging from the plots, how well do you think each independent variable predicts college GPA? Is there one variable that is a better predictor than the others?*

Part 3: Measuring fit

As explained in Exercise 8.6.4, the coefficient of determination (or R^2 coefficient) is a mathematical measure of how well a regression line fits a set of data. Implement the function described in Exercise 8.6.4 and modify the `plotRegression` function so that it returns the R^2 value for the plot.

Question 8.4.2 *Based on the R^2 values, how well does each independent variable predict college GPA? Which is the best predictor?*

Part 4: Additional analyses

Choose two of the independent variables and perform a regression analysis.

Question 8.4.3 *Explain your findings.*

*Project 8.5 Preparing for a 100-year flood

This project assumes that you have read Section 8.6.

Suppose we are undertaking a review of the flooding contingency plan for a community on the Snake River, just south of Jackson, Wyoming. To properly prepare for future flooding, we would like to know the river's likely height in a 100-year flood, the height that we should expect to see only once in a century. This 100-year designation is called the flood's *recurrence interval*, the amount of time that typically elapses between two instances of the river reaching that height. Put another way, there is a 1/100, or 1%, chance that a 100-year flood happens in any particular year.

River heights are measured by stream gauges maintained by the U.S. Geological Survey (USGS)[9]. A snippet of the data from the closest Snake River gauge, which can be downloaded from the USGS[10] or the book's web site, is shown below.

---

[9] http://nwis.waterdata.usgs.gov/nwis
[10] http://nwis.waterdata.usgs.gov/nwis/peak?site_no=13018750\&agency_cd=USGS\&format=rdb

```
#
# U.S. Geological Survey
# National Water Information System
  ⋮
#
agency_cd ▷ site_no ▷ peak_dt ▷ peak_tm ▷ peak_va ▷ peak_cd ▷ gage_ht ▷ ...
5s ▷ 15s ▷ 10d ▷ 6s ▷ 8s ▷ 27s ▷ 8s ▷ ...
USGS ▷ 13018750 ▷ 1976-06-04 ▷ ▷ 15800 ▷ 6 ▷ 7.80 ▷ ...
USGS ▷ 13018750 ▷ 1977-06-09 ▷ ▷ 11000 ▷ 6 ▷ 6.42 ▷ ...
USGS ▷ 13018750 ▷ 1978-06-10 ▷ ▷ 19000 ▷ 6 ▷ 8.64 ▷ ...
  ⋮
USGS ▷ 13018750 ▷ 2011-07-01 ▷ ▷ 19900 ▷ 6 ▷ 8.75 ▷ ...
USGS ▷ 13018750 ▷ 2012-06-06 ▷ ▷ 16500 ▷ 6 ▷ 7.87 ▷ ...
```

The file begins with several comment lines preceded by the hash (#) symbol. The next two lines are header rows; the first contains the column names and the second contains codes that describe the content of each column, e.g., 5s represents a string of length 5 and 10d represents a date of length 10. Each column is separated by a tab character, represented above by a right-facing triangle (▷). The header rows are followed by the data, one row per year, representing the peak event of that year. For example, in the first row we have:

- agency_cd (agency code) is USGS
- site_no (site number) is 13018750 (same for all rows)
- peak_dt (peak date) is 1976-06-04
- peak_tm (peak time) is omitted
- peak_va (peak streamflow) is 15800 cubic feet per second
- peak_cd (peak code) is 6 (we will ignore this)
- gage_ht (gauge height) is 7.80 feet

So for each year, we essentially have two gauge values: the peak streamflow in cubic feet per second and the maximum gauge height in feet.

If we had 100 years of gauge height data in this file, we could approximate the water level of a 100-year flood with the maximum gauge height value. However, our data set only covers 37 years (1976 to 2012) and, for 7 of those years, the gauge height value is missing. Therefore, we will need to estimate the 100-year flood level from the limited data we are given.

Part 1: Read the data

Write a function

```
readData(filename)
```

that returns lists of the peak streamflow and gauge height data (as floating point numbers) from the Snake River data file above. Your function will need to first read past the comment section and header lines to get to the data. Because we do not know how many comment lines there might be, you will need to use a `while` loop containing a call to the `readline` function to read past the comment lines.

Notice that some rows in the data file are missing gauge height information. If this information is missing for a particular line, use a value of 0 in the list instead.

Your function should return two lists, one containing the peak streamflow rates and one containing the peak gauge heights. A function can return two values by simply separating them with a comma, .e.g.,

```
return flows, heights
```

Then, when calling the function, we need to assign the function call to two variable names to capture these two lists:

```
flows, heights = readData('snake_peak.txt')
```

Part 2: Recurrence intervals

To associate the 100-year recurrence interval with an estimated gauge height, we can associate each of our known gauge heights with a recurrence interval, plot this data, and then use regression to extrapolate out to 100 years. A flood's recurrence interval is computed by dividing $(n+1)$, where n is the number of years on record, by the rank of the flood. For example, suppose we had only three gauge heights on record. Then the recurrence interval of the maximum (rank 1) height is $(3+1)/1 = 4$ years, the recurrence interval of the second largest height is $(3+1)/2 = 2$ years, and the recurrence interval of the smallest height is $(3+1)/3 = 4/3$ years. (However, these inferences are unlikely to be at all accurate because there is so little data!)

Write a function

```
getRecurrenceIntervals(n)
```

that returns a list of recurrence intervals for **n** floods, in order of lowest to highest. For example if **n** is 3, the function should return the list `[1.33, 2.0, 4.0]`.

After you have written this function, write another function

```
plotRecurrenceIntervals(heights)
```

that plots recurrence intervals and corresponding gauge heights (also sorted from smallest to largest). Omit any missing gauge heights (with value zero). Your resulting plot should look like Figure 8.7.

Part 3: Find the river height in a 100-year flood

To estimate the gauge height corresponding to a 100-year recurrence interval, we need to extend the "shape" of this curve out to 100 years. Mathematically speaking, this means that we need to find a mathematical function that predicts the peak

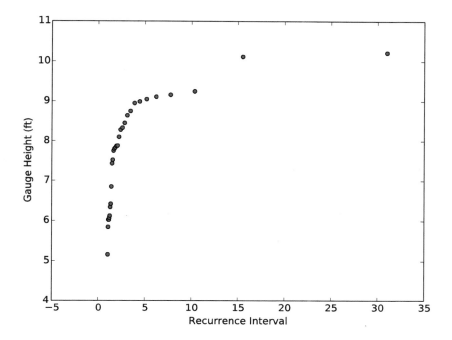

Figure 8.7 The peak gauge height for each recurrence interval.

gauge height for each recurrence interval. Once we have this function, we can plug in 100 to find the gauge height for a 100-year flood.

What we need is a regression analysis, as we discussed in Section 8.6. But linear regression only works properly if the data exhibits a linear relationship, i.e., we can draw a straight line that closely approximates the data points.

Reflection 8.33 *Do you think we can use linear regression on the data in Figure 8.7?*

This data in Figure 8.7 clearly do not have a linear relationship, so a linear regression will not produce a good approximation. The problem is that the x coordinates (recurrence intervals) are increasing multiplicatively rather than additively; the recurrence interval for the flood with rank $r + 1$ is $(r + 1)/r$ times the recurrence interval for the flood with rank r. However, we will share a trick that allows us to use linear regression anyway. To illustrate the trick we can use to turn this non-linear curve into a "more linear" one, consider the plot on the left in Figure 8.8, representing points $(2^0, 0), (2^1, 1), (2^2, 2), \ldots, (2^{10}, 10)$. Like the plot in Figure 8.7, the x coordinates are increasing multiplicatively; each x coordinate is twice the one before it. The plot on the right in Figure 8.8 contains the points that result from taking the logarithm base 2 of each x coordinate ($\log_2 x$), giving

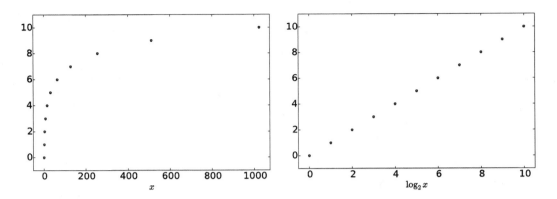

Figure 8.8 On the left is a plot of the points $(2^0, 0), (2^1, 1), (2^2, 2), \ldots, (2^{10}, 10)$, and on the right is a plot of the points that result from taking the logarithm base 2 of the x coordinate of each of these points.

$(0,0), (1,1), (2,2), \ldots, (10,10)$. Notice that this has turned an exponential plot into a linear one.

We can apply this same technique to the `plotRecurrenceIntervals` function you wrote above to make the curve approximately linear. Write a new function

   `plotLogRecurrenceIntervals(heights)`

that modifies the `plotRecurrenceIntervals` function so that it makes a new list of logarithmic recurrence intervals, and then computes the linear regression line based on these x values. Use logarithms with base 10 for convenience. Then plot the data and the regression line using the `linearRegression` function from Exercise 8.6.1. To find the 100-year flood gauge height, we want the regression line to extend out to 100 years. Since we are using logarithms with base 10, we want the x coordinates to run from $\log_{10} 1 = 0$ to $\log_{10} 100 = 2$.

Question 8.5.1 *Based on Figure 8.8, what is the estimated river level for a 100-year flood? How can you find this value exactly in your program? What is the exact value?*

Part 4: An alternative method

As noted earlier in this section, there are seven gauge height values missing from the data file. But all of the peak streamflow values are available. If there is a linear correlation between peak streamflow and gauge height, then it might be more accurate to find the 100-year peak streamflow value instead, and then use the linear regression between peak streamflow and gauge height to find the gauge height that corresponds to the 100-year peak streamflow value.

First, write a function

   `plotFlowsHeights(flows, heights)`

that produces a scatter plot with peak streamflow on the x-axis and the same year's gauge height on the y axis. Do not plot data for which the gauge height is missing. Then also plot the least squares linear regression for this data.

Next, write a function

```
plotLogRecurrenceIntervals2(flows)
```

that modifies the `plotLogRecurrenceIntervals` function from Part 3 so that it find the 100-year peak streamflow value instead.

Question 8.5.2 *What is the 100-year peak streamflow rate?*

Once you have found the 100-year peak streamflow rate, use the linear regression formula to find the corresponding 100-year gauge height.

Question 8.5.3 *What is the gauge height that corresponds to the 100-year peak streamflow rate?*

Question 8.5.4 *Compare the two results. Which one do you think is more accurate? Why?*

Project 8.6 Voting methods

Although most of us are used to simple plurality-based elections in which the candidate with the most votes wins, there are other voting methods that have been proven to be fairer according to various criteria. In this project, we will investigate two other such voting methods. For all of the parts of the project, we will assume that there are four candidates named Amelia, Beth, Caroline, and David (abbreviated A, B, C, and D).

Part 1: Get the data

The voting results for an election are stored in a data file containing one ballot per line. Each ballot consists of one voter's ranking of the candidates. For example, a small file might look like:

```
B A D C
D B A C
B A C D
```

To begin, write a function

```
readVotes(fileName)
```

that returns these voting results as a list containing one list for each ballot. For example, the file above should be stored in a list that looks like this:

```
[['B', 'A', 'D', 'C'], ['D', 'B', 'A', 'C'], ['A', 'B', 'C', 'D']]
```

There are three sample voting result files on the book web site. Also feel free to create your own.

Part 2: Plurality voting

First, we will implement basic plurality voting. Write a function

   `plurality(ballots)`

that prints the winner (or winners if there is a tie) of the election based on a plurality count. The parameter of the function is a list of ballots like that returned by the `readVotes` function. Your function should first iterate over all of the ballots and count the number of first-place votes won by each candidate. Store these votes in a dictionary containing one entry for each candidate. To break the problem into more manageable pieces, write a "helper function"

   `printWinners(points)`

that determines the winner (or winners if there is a tie) based on this dictionary (named `points`), and then prints the outcome. Call this function from your `plurality` function.

Part 3: Borda count

Next, we will implement a vote counting system known as a Borda count, named after Jean-Charles de Borda, a mathematician and political scientist who lived in 18th century France. For each ballot in a Borda count, a candidate receives a number of points equal to the number of lower-ranked candidates on the ballot. In other words, with four candidates, the first place candidate on each ballot is awarded three points, the second place candidate is awarded two points, the third place candidate is awarded one point, and the fourth place candidate receives no points. In the example ballot above, candidate B is the winner because candidate A receives $2 + 1 + 2 = 5$ points, candidate B receives $3 + 2 + 3 = 8$ points, candidate C receives $0 + 0 + 1 = 1$ points, and candidate D receives $1 + 3 + 0 = 4$ points. Note that, like a plurality count, it is possible to have a tie with a Borda count.

Write a function

   `borda(ballots)`

that prints the winner (or winners if there is a tie) of the election based on a Borda count. Like the `plurality` function, this function should first iterate over all of the ballots and count the number of points won by each candidate. To make this more manageable, write another "helper function" to call from within your loop named

   `processBallot(points, ballot)`

that processes each individual ballot and adds the appropriate points to the dictionary of accumulated points named `points`. Once all of the points have been accumulated, call the `printWinners` above to determine the winner(s) and print the outcome.

Part 4: Condorcet voting

Marie Jean Antoine Nicolas de Caritat, Marquis de Condorcet was another mathematician and political scientist who lived about the same time as Borda. Condorcet proposed a voting method that he considered to be superior to the Borda count. In this method, every candidate participates in a virtual head-to-head election between herself and every other candidate. For each ballot, the candidate who is ranked higher wins. If a candidate wins every one of these head-to-head contests, she is determined to be the Condorcet winner. Although this method favors the candidate who is most highly rated by the majority of voters, it is also possible for there to be no winner.

Write a function

```
condorcet(ballots)
```

that prints the Condorcet winner of the election or indicates that there is none. (If there is a winner, there can only be one.)

Suppose that the list of candidates is assigned to `candidates`. (Think about how you can get this list.) To simulate all head-to-head contests between one candidate named `candidate1` and all of the rest, we can use the following `for` loop:

```
for candidate2 in candidates:
    if candidate2 != candidate1:
        # head-to-head between candidate1 and candidate2 here
```

This loop iterates over all of the candidates and sets up a head-to-head contest between each one and `candidate1`, as long as they are not the same candidate.

Reflection 8.34 *How can we now use this loop to generate contests between all pairs of candidates?*

To generate all of the contests with all possible values of `candidate1`, we can nest this `for` loop in the body of another `for` loop that also iterates over all of the candidates, but assigns them to `candidate1` instead:

```
for candidate1 in candidates:
    for candidate2 in candidates:
        if candidate2 != candidate1:
            # head-to-head between candidate1 and candidate2 here
```

Reflection 8.35 *This nested `for` loop actually generates too many pairs of candidates. Can you see why?*

To simplify the body of the nested loop (where the comment is currently), write another "helper function"

```
head2head(ballots, candidate1, candidate2)
```

that returns the candidate that wins in a head-to-head vote between `candidate1` and `candidate2`, or `None` is there is a tie. Your `condorcet` function should call this function for every pair of different candidates. For each candidate, keep track of the number of head-to-head wins in a dictionary with one entry per candidate.

Reflection 8.36 *The most straightforward algorithm to decide whether* `candidate1` *beats* `candidate2` *on a particular ballot iterates over all of the candidates on the ballot. Can you think of a way to reorganize the ballot data before calling* `head2head` *so that the* `head2head` *function can decide who wins each ballot in only one step instead?*

Part 5: Compare the three methods

Execute your three functions on each of the three data files on the book web site and compare the results.

Project 8.7 Heuristics for traveling salespeople

Imagine that you drive a delivery truck for one of the major package delivery companies. Each day, you are presented with a list of addresses to which packages must be delivered. Before you can call it a day, you must drive from the distribution center to each package address, and back to the distribution center. This cycle is known as a *tour*. Naturally, you wish to minimize the total distance that you need to drive to deliver all the packages, i.e., the total length of the tour. For example, Figures 8.9 and 8.10 show two different tours.

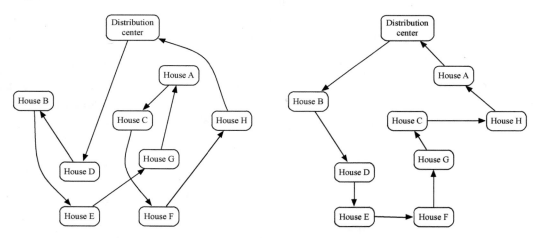

Figure 8.9 An inefficient tour.   Figure 8.10 A more efficient tour.

This is known as the *traveling salesperson problem (TSP)*, and it is notoriously difficult. In fact, as far as anyone knows, the only way to come up with a guaranteed correct solution is to essentially enumerate all possible tours and choose the best one.

But since, for n locations, there are $n!$ (n factorial) different tours, this is practically impossible.

Unfortunately, the TSP has many important applications, several of which seem at first glance to have nothing at all to do with traveling or salespeople, including circuit board drilling, controlling robots, designing networks, x-ray crystallography, scheduling computer time, and assembling genomes. In these cases, a heuristic must be used. A heuristic does not necessarily give the best answer, but it tends to work well in practice.

For this project, you will design your own heuristic, and then work with a genetic algorithm, a type of heuristic that mimics the process of evolution to iteratively improve problem solutions.

Part 1: Write some utility functions

Each point on your itinerary will be represented by (x, y) coordinates, and the input to the problem is a list of these points. A tour will also be represented by a list of points; the order of the points indicates the order in which they are visited. We will store a list of points as a list of tuples.

The following function reads in points from a file and returns the points as a list of tuples. We assume that the file contains one point per line, with the x and y coordinates separated by a space.

```
def readPoints(filename):
    inputFile = open(filename, 'r')
    points = []
    for line in inputFile:
        values = line.split()
        points.append((float(values[0]), float(values[1])))
    return points
```

To begin, write the following three functions. The first two will be needed by your heuristics, and the third will allow you to visualize the tours that you create. To test your functions, and the heuristics that you will develop below, use the example file containing the coordinates of 96 African cities (`africa.tsp`) on the book web site.

1. `distance(p, q)` returns the distance between points p and q, each of which is stored as a two-element tuple.

2. `tourLength(tour)` returns the total length of a tour, stored as a list of tuples. Remember to include the distance from the last point back to the first point.

3. `drawTour(tour)` draws a tour using turtle graphics. Use the `setworldcoordinates` method to make the coordinates in your drawing window more closely match the coordinates in the data files you use. For example, for the `africa.tsp` data file, the following will work well:

   ```
   screen.setworldcoordinates(-40, -25, 40, 60)
   ```

Part 2: Design a heuristic

Now design your own heuristic to find a good TSP tour. There are many possible ways to go about this. Be creative. Think about what points should be next to each other. What kinds of situations should be fixed? Use the `drawTour` function to visualize your tours and help you design your heuristic.

Part 3: A genetic algorithm

A *genetic algorithm* attempts to solve a hard problem by emulating the process of evolution. The basic idea is to start with a population of feasible solutions to the problem, called individuals, and then iteratively try to improve the fitness of this population through the evolutionary operations of recombination and mutation. In genetics, recombination is when chromosomes in a pair exchange portions of their DNA during meiosis. The illustration below shows how a crossover would affect the bases in a particular pair of (single stranded) DNA molecules.

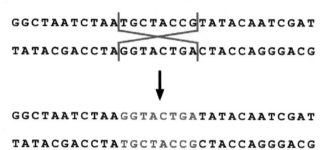

Mutation occurs when a base in a DNA molecule is replaced with a different base or when bases are inserted into or deleted from a sequence. Most mutation is the result of DNA replication errors but environmental factors can also lead to mutations in DNA.

To apply this technique to the traveling salesperson problem, we first need to define what we mean by an individual in a population. In genetics, an individual is represented by its DNA, and an individual's fitness, for the purposes of evolution, is some measure of how well it will thrive in its environment. In the TSP, we will have a population of tours, so an individual is one particular tour — a list of cities. The most natural fitness function for an individual is the length of the tour; a shorter tour is more fit than a longer tour.

Recombination and mutation on tours are a bit trickier conceptually than they are for DNA. Unlike with DNA, swapping two subsequences of cities between two tours is not likely to produce two new valid tours. For example, suppose we have two tours [a, b, c, d] and [b, a, d, c], where each letter represents a point. Swapping the two middle items between the tours will produce the offspring [a, a, d, d] and [b, b, c, c], neither of which are permutations of the cities. One way around this is to delete from the first tour the cities in the portion to be swapped from the second tour, and then insert this portion from the second tour. In the above example, we would delete points a and d from the first tour, leaving [b, c], before inserting [a, d] in the middle. Doing likewise for the second tour gives us children [b, a, d, c] and [a, b, c, d]. But we are not limited in genetic programming to recombination that more or less mimics that found in nature. A recombination operation can be anything that creates new offspring by somehow combining two parents. A large part of this project involves brainstorming about and experimenting with different techniques.

We must also rethink mutation since we cannot simply replace an arbitrary city with another city and end up with a valid tour. One idea might be to swap the positions of two randomly selected cities instead. But there are other possibilities as well.

Your mission is to improve upon a baseline genetic algorithm for TSP. Be creative! You may change anything you wish as long as the result can still be considered a genetic algorithm. To get started, download the baseline program from the book web site. Try running it with the 96-point instance on the book web site. Take some time to understand how the program works. Ask questions. You may want to refer to the Python documentation if you don't recall how a particular function works. Most of the work is performed by the following four functions:

- `makePopulation(cities)`: creates an initial population (of random tours)

- `crossover(mom, pop)`: performs a recombination operation on two tours and returns the two offspring

- `mutate(individual)`: mutates an individual tour

- `newGeneration(population)`: update the population by performing a crossover and mutating the offspring

Write the function

`histogram(population)`

that is called from the `report` function. (Use a Python dictionary.) This function should print a frequency chart (based on tour length) that gives you a snapshot of the diversity in your population. Your histogram function should print something like this:

```
Population diversity
   1993.2714596455853 : ****
   2013.1798076309087 : **
   2015.1395212505120 : ****
   2017.1005248468230 : ********************************
   2020.6881282400334 : *
   2022.9044855489917 : *
   2030.9623523675089 : *
   2031.4773010231959 : *
   2038.0257926528227 : *
   2040.7438913120230 : *
   2042.8148398732630 : *
   2050.1916058477627 : *
```

This will be very helpful as you strive to improve the algorithm: recombination in a homogeneous population is not likely to get you very far.

Brainstorm ways to **improve the algorithm**. Try lots of different things, ranging from tweaking parameters to completely rewriting any of the four functions described above. You are free to change anything, as long as the result still resembles a genetic algorithm. Keep careful records of what works and what doesn't to include in your submission.

On the book web site is a link to a very good reference [42] that will help you think of new things to try. Take some time to skim the introductory sections, as they will give you a broader sense of the work that has been done on this problem. Sections 2 and 3 contain information on genetic algorithms; Section 5 contains information on various recombination/crossover operations; and Section 7 contains information on possible mutation operations. As you will see, this problem has been well studied by researchers over the past few decades! (To learn more about this history, we recommend *In Pursuit of the Traveling Salesman* by William Cook [8].)

CHAPTER 9

Flatland

> Suffice it that I am the completion of your incomplete self. You are a Line, but I am a Line of Lines, called in my country a Square: and even I, infinitely superior though I am to you, am of little account among the great nobles of Flatland, whence I have come to visit you, in the hope of enlightening your ignorance.
>
> Edwin A. Abbott
> *Flatland: A Romance of Many Dimensions (1884)*

THE title of this chapter is a reference to the highly-recommended book of the same name written by Edwin A. Abbott in 1884 [1]. In it, the narrator, a square who lives in the two-dimensional world of Flatland, discovers and grapples with comprehending the three-dimensional world of Spaceland, while simultaneously recognizing the profound advantages he has over those living in the zero- and one-dimensional worlds of Pointland and Lineland.

Analogously, we have discovered the advantages of one-dimensional data (strings and lists) over zero-dimensional numbers and characters. In this chapter, we will discover the further possibilities afforded us by understanding how to work with two-dimensional data. We will begin by looking at how we can create a two-dimensional table of data read in from a file. Then we will explore a powerful two-dimensional simulation technique called *cellular automata*. At the end of the chapter are several projects that illustrate how simulations similar to cellular automata can be used to model a wide variety of problems. We also spend some time exploring how two-dimensional images are stored and manipulated.

9.1 TWO-DIMENSIONAL DATA

There are a few different ways to store two-dimensional data. To illustrate the most straightforward technique, let's revisit the tabular data sets with which we worked in Chapter 8. One of the simplest was the CSV file containing monthly extreme temperature readings in Madison, Wisconsin from Exercise 8.4.6 (`madison_temp.csv`):

```
STATION,STATION_NAME,DATE,EMXT,EMNT
GHCND:USW00014837,MADISON DANE CO REGIONAL AIRPORT WI US,19700101,33,-294
GHCND:USW00014837,MADISON DANE CO REGIONAL AIRPORT WI US,19700201,83,-261
GHCND:USW00014837,MADISON DANE CO REGIONAL AIRPORT WI US,19700301,122,-139
⋮
```

Because all of the data in this file is based on conditions at the same site, the first two columns are identical in every row. The third column contains the dates of the first of each month in which data was collected. The fourth and fifth columns contain the maximum and minimum monthly temperatures, respectively, which are in tenths of a degree Celsius (i.e., 33 represents 3.3° C). Previously, we would have extracted this data into three parallel lists containing dates, maximum temperatures, and minimum temperatures, like this:

```
def readData():
    """Read monthly extreme temperature data into a table.

    Parameters: none

    Return value: a list of lists containing monthly extreme
                  temperature data
    """

    dataFile = open('madison_temp.csv', 'r')
    header = dataFile.readline()
    dates = []
    maxTemps = []
    minTemps = []
    for line in dataFile:
        row = line.split(',')
        dates.append(row[2])
        maxTemps.append(int(row[3]))
        minTemps.append(int(row[4]))
    dataFile.close()
    return dates, maxTemps, minTemps
```

Alternatively, we may wish to extract tabular data into a single table. This is especially convenient if the data contains many columns. For example, the three columns above could be stored in a unified tabular structure like the following:

DATE	EMXT	EMNT
19700101	33	-294
19700201	83	-261
19700301	122	-139
⋮	⋮	⋮

We can represent this structure as a list of rows, where each row is a list of values in that row. In other words, the table above can be stored like this:

[['19700101', 33, -294], ['19700201', 83, -261], ['19700301', 122, -139], ...]

To better visualize this list as a table, we can reformat its presentation a bit:

```
[
   [ '19700101',   33, -294 ],
   [ '19700201',   83, -261 ],
   [ '19700301',  122, -139 ],
   ⋮
]
```

Notice that, in the function above, we already have each of these row lists contained in the list named `row`. Therefore, to create this structure, we can simply append each value of `row`, with the temperature values converted to integers and the first two redundant columns removed, to a growing list of rows named `table`:

```
def readData():
    """ (docstring omitted) """

    dataFile = open('madison_temp.csv', 'r')
    header = dataFile.readline()
    table = []
    for line in dataFile:
        row = line.split(',')
        row[3] = int(row[3])
        row[4] = int(row[4])
        table.append(row[2:])    # add a new row to the table
    dataFile.close()
    return table
```

Since each element of `table` is a list containing a row of this table, the first row is assigned to `table[0]`, as illustrated below. Similarly, the second row is assigned to `table[1]` and the third row is assigned to `table[2]`.

$$\underbrace{[\ ['19700101', 33, -294]}_{\text{table}[0]},\ \underbrace{['19700201', 83, \overbrace{-261}^{\text{table}[1][2]}]}_{\text{table}[1]},\ \underbrace{['19700301', \overbrace{122}^{\text{table}[2][1]}, -139]}_{\text{table}[2]}, ...\ ]$$

Reflection 9.1 *How would you access the minimum temperature in February, 1970 (date value '19700201') from this list?*

The minimum temperature in February, 1970 is the third value in the list `table[1]`. Since `table[1]` is a list, we can use indexing to access individual items contained in it. Therefore, the third value in `table[1]` is `table[1][2]`, which equals -261 (i.e., -26.1° C), as indicated above. Likewise, `table[2][1]` is the maximum temperature in March, 1970: 122, or 12.2° C.

Reflection 9.2 *In general, how can you access the value in row* `r` *and column* `c`*?*

Notice that, for a particular value `table[r][c]`, `r` is the index of the row and `c` is the index of the column. So if we know the row and column of any desired value, it is easy to retrieve that value with this convenient notation.

Now suppose we want to search this table for the minimum temperature in a particular month, given the corresponding date string. To access this value in the table, we will need both its row and column indices. We already know that the column index must be 2, since the minimum temperatures are in the third column. To find the correct row index, we need to search all of the values in the first column until we find the row that contains the desired string. Once we have the correct row index `r`, we can simply return the value of `table[r][2]`. The following function does exactly this.

```
def getMinTemp(table, date):
    """Return the minimum temperature for the month
       with the given date string.
    
    Parameters:
        table: a table containing extreme temperature data
        date: a date string
    
    Return value:
        the minimum temperature for the given date
        or None if the date does not exist
    """
    
    numRows = len(table)
    for r in range(numRows):
        if table[r][0] == date:
            return table[r][2]
    return None
```

The `for` loop iterates over the the indices of the rows in the table. For each row with index `r`, we check if the first value in that row, `table[r][0]`, is equal to the date we are looking for. If it is, we return the value in column 2 of that row. On the other hand, if we get all the way through the loop without returning a value, the desired date must not exist, so we return `None`.

Exercises

From this point on, we will generally not specify what the name and parameters of a function should be. Instead, we would like you to design the function(s).

9.1.1. Show how the following table can be stored in a list named `scores`.

Student ID	SAT M	SAT CR
10305	700	610
11304	680	590
10254	710	730
12007	650	690
10089	780	760

9.1.2. In the list you created above, how do you refer to each of the following?

(a) the SAT M value for student 10089

(b) the SAT CR value for student 11304

(c) the SAT M value for student 10305

(d) the SAT CR value for student 12007

9.1.3. Alternatively, a table could be stored as a list of columns.

(a) Show how to store the table in Exercise 9.1.1 in this way.

(b) Redo Exercise 9.1.2 using this new list.

(c) Why is this method less convenient when reading data in from a file?

9.1.4. Write a program that calls the `readData` function to read temperatures into a table, and then repeatedly asks for a date string to search for in the table. For each date string entered, your function should call the `getMinTemp` function to get the corresponding minimum temperature. For example, your function should print something like the following:

```
Minimum temperature for which date (q to quit)? 20050501
The minimum temperature for 20050501 was -2.2 degrees Celsius.

Minimum temperature for which date (q to quit)? 20050801
The minimum temperature for 20050801 was 8.3 degrees Celsius.

Minimum temperature for which date (q to quit)? q
```

9.1.5. Write a function that does the same thing as the `getMinTemp` function above, but returns the maximum temperature for a particular date instead.

9.1.6. If a data set contains a unique key value that is frequently searched, we can alternatively store the data in a dictionary. Each row in the table can be associated with its particular key value, which makes searching for a key value very efficient. For example, the temperature table

```
[
    [ '19700101',   33, -294 ],
    [ '19700201',   83, -261 ],
    [ '19700301',  122, -139 ]
]
```

could be stored in a dictionary as

```
{ '19700101': [33, -294],
  '19700201': [83, -261],
  '19700301': [122, -139] }
```

Rewrite the `readData` and `getMinTemp` functions so that the temperature data is stored in this way instead. Then incorporate these new functions into your program from Exercise 9.1.4.

9.1.7. Write a function that reads the earthquake data from the CSV file at http://earthquake.usgs.gov/earthquakes/feed/v1.0/summary/2.5_month.csv into a table with four columns containing the latitude, longitude, depth, and magnitude of each earthquake. All four values should be stored as floating point numbers. (This is the same URL that we used in Section 8.4.)

9.1.8. Write a function that takes as a parameter a table returned by your function from Exercise 9.1.7 and prints a formatted table of the data. Your table should look similar to this:

```
Latitude  Longitude  Depth  Magnitude
--------  ---------  -----  ---------
   33.49    -116.46   19.8        1.1
   33.14    -115.65    2.1        1.8
   -2.52     146.12   10.0        4.6
     ⋮
```

9.1.9. Write a function that takes as a parameter a table returned by your function from Exercise 9.1.7, and repeatedly prompts for a minimum earthquake magnitude. The function should then create a new table containing the rows corresponding to earthquakes with at least that magnitude, and then print this table using your function from Exercise 9.1.8. The output from your function should look similar to this:

```
Minimum magnitude (q to quit)? 6.2

Latitude  Longitude  Depth  Magnitude
--------  ---------  -----  ---------
  -46.36      33.77   10.0        6.2
  -37.68     179.69   22.0        6.7
    1.93     126.55   35.0        7.1
   -6.04     148.21   43.2        6.6

Minimum magnitude (q to quit)? 7

Latitude  Longitude  Depth  Magnitude
--------  ---------  -----  ---------
    1.93     126.55   35.0        7.1

Minimum magnitude (q to quit)? 8
There were no earthquakes with magnitude at least 8.0.

Minimum magnitude (q to quit)? q
```

9.2 THE GAME OF LIFE

A *cellular automaton* is a rectangular grid of discrete cells, each of which has an associated state or value. Each cell represents an individual entity, such as an organism or a particle. At each time step, every cell can simultaneously change its state according to some *rule* that depends only on the states of its neighbors. Depending upon on the rules used, cellular automata may evolve *global, emergent* behaviors based only on these *local* interactions.

The most famous example of a cellular automaton is the *Game of Life*, invented by mathematician John Conway in 1970. In the Game of Life, each cell can be in one of two states: alive or dead. At each time step, the makeup of a cell's neighborhood dictates whether or not it will pass into the next generation alive (or be reborn if it is dead). Depending upon the initial configuration of cells (which cells are initially alive and which are dead), the Game of Life can produce amazing patterns.

We consider each cell in the Game of Life to have eight neighbors, as illustrated below:

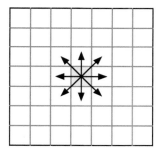

In each step, each of the cells simultaneously observes the states of its neighbors, and may change its state according to the following rules:

1. If a cell is alive and has fewer than two live neighbors, it dies from loneliness.

2. If a cell is alive and has two or three live neighbors, it remains alive.

3. If a cell is alive and has more than three live neighbors, it dies from overcrowding.

4. If a cell is dead and has exactly three live neighbors, it is reborn.

To see how these rules affect the cells in the Game of Life, consider the initial configuration in the top left of Figure 9.1. Dead cells are represented by white squares and live cells are represented by black squares. To apply rule 1 to the initial configuration, we need to check whether there are any live cells with fewer than two live neighbors. As illustrated below, there are two such cells, each marked with D.

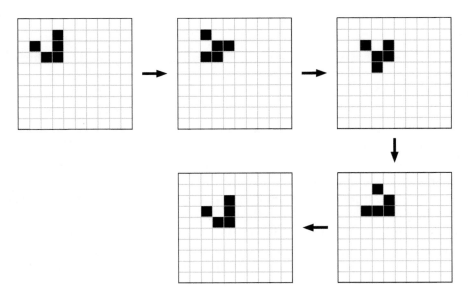

Figure 9.1 The first five generations of a "glider" cellular automaton.

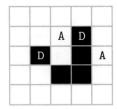

According to rule 1, these two cells will die in the next generation. To apply rule 2, we need to check whether there are any live cells that have two or three live neighbors. Since this rule applies to the other three live cells, they will remain alive into the next generation. There are no cells that satisfy rule 3, so we move on to rule 4. There are two dead cells with exactly three live neighbors, marked with **A**. According to rule 4, these two cells will come alive in the next generation.

Reflection 9.3 *Show what the second generation looks like, after applying these rules.*

The figure in the top center of Figure 9.1 shows the resulting second generation, followed by generations three, four, and five. After five generations, as illustrated in the bottom center of Figure 9.1, the grid has returned to its initial state, but it has moved one cell down and to the right of its initial position. If we continued computing generations, we would find that it would continue in this way indefinitely, or until it collides with a border. For this reason, this initial configuration generates what is known as a "glider."

The game of life ■ 451

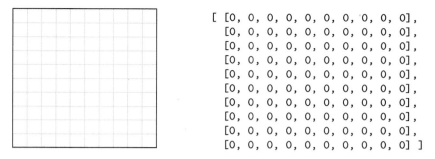
```
[ [0, 0, 0, 0, 0, 0, 0, 0, 0, 0],
  [0, 0, 0, 0, 0, 0, 0, 0, 0, 0],
  [0, 0, 0, 0, 0, 0, 0, 0, 0, 0],
  [0, 0, 0, 0, 0, 0, 0, 0, 0, 0],
  [0, 0, 0, 0, 0, 0, 0, 0, 0, 0],
  [0, 0, 0, 0, 0, 0, 0, 0, 0, 0],
  [0, 0, 0, 0, 0, 0, 0, 0, 0, 0],
  [0, 0, 0, 0, 0, 0, 0, 0, 0, 0],
  [0, 0, 0, 0, 0, 0, 0, 0, 0, 0],
  [0, 0, 0, 0, 0, 0, 0, 0, 0, 0] ]
```

Figure 9.2 Views of the "empty" cellular automaton as a grid and as a list.

Creating a grid

To implement this cellular automaton, we first need to create an empty grid of cells. For simplicity, we will keep it relatively small, with 10 rows and 10 columns. We can represent each cell by a single integer value which equals 1 if the cell is alive or 0 if it is dead. For clarity, it is best to assign these values to meaningful names:

```
ALIVE = 1
DEAD = 0
```

An initially empty grid, like the one on the left side of Figure 9.2, will be represented by a list of row lists, each of which contains a zero for every column. For example, the list on the right side of Figure 9.2 represents the grid to its left.

Reflection 9.4 *How can we easily create a list of many zeros?*

If the number of columns in the grid is assigned to **columns**, then each of these rows can be created with a **for** loop:

```
row = []
for c in range(columns):
    row.append(DEAD)
```

We can then create the entire grid by simply appending a copy of **row** for every row in the grid to a list named **grid**:

```
def makeGrid(rows, columns):
    """Create a rows x columns grid of zeros.

    Parameters:
        rows: the number of rows in the grid
        columns: the number of columns in the grid

    Return value: a list of ROWS lists of COLUMNS zeros
    """

    grid = []
```

```
        for r in range(rows):
            row = []
            for c in range(columns):
                row.append(DEAD)
            grid.append(row)
        return grid
```

This nested loop can actually be simplified somewhat by substituting

```
    row = [DEAD] * columns
```

for the three statements that construct `row` because all of the values in each row are the same.

Initial configurations

The cellular automaton will evolve differently, depending upon the initial configuration of alive and dead cells. (There are several examples of this in the exercises at the end of this section.) We will assume that all cells are dead initially, except for those we explicitly specify. Each cell can be conveniently represented by a (row, column) tuple. The coordinates of the initially live cells can be stored in a list of tuples, and passed into the following function to initialize the grid. (Recall that a tuple is like a list, except that it is immutable and enclosed in parentheses instead of square brackets.)

```
    def initialize(grid, coordinates):
        """Set a given list of coordinates to 1 in the grid.

        Parameters:
            grid: a grid of values for a cellular automaton
            coordinates: a list of coordinates

        Return value: None
        """

        for (r, c) in coordinates:
            grid[r][c] = ALIVE
```

The function iterates over the list of tuples and sets the cell at each position to be alive. For example, to match the initial configuration in the upper left of Figure 9.1, we would pass in the list

```
    [(1, 3), (2, 3), (3, 3), (3, 2), (2, 1)]
```

Notice that by using a generic tuple as the index variable, we can conveniently assign the two values in each tuple in the list to `r` and `c`.

The game of life ■ 453

> **Box 9.1: NumPy arrays in two dimensions**
>
> Two-dimensional (and higher) data can also be represented with a NumPy `array` object. (See Box 8.1.) As in one dimension, we can initialize an `array` by either passing in a list, or passing a size to one of several functions that fill the array with particular values. Here are some examples.
>
> ```
> >>> import numpy
> >>> temps = numpy.array([[3.3, -29.4], [8.3, -26.1], [12.2, -13.9]])
> >>> temps
> array([[  3.3, -29.4],
>        [  8.3, -26.1],
>        [ 12.2, -13.9]])
> >>> grid = numpy.zeros((2, 4))   # zero-filled with 2 rows, 4 cols
> >>> grid
> array([[ 0.,  0.,  0.,  0.],
>        [ 0.,  0.,  0.,  0.]])
> ```
>
> In the second case, the tuple `(2, 4)` specifies the "shape" of the array: three rows and four columns. We can modify individual `array` elements with indexing by simply specifying the comma-separated row and column in a single pair of square braces:
>
> ```
> >>> grid[1, 3] = 1
> >>> grid
> array([[ 0.,  0.,  0.,  0.],
>        [ 0.,  0.,  0.,  1.]])
> ```
>
> As we saw in Box 8.1, the real power of NumPy arrays lies in the ability to change every element in a single statement. For example, the following statement adds one to every element of the `temps` array.
>
> ```
> >>> temps = temps + 1
> >>> temps
> array([[  4.3, -28.4],
>        [  9.3, -25.1],
>        [ 13.2, -12.9]])
> ```
>
> For more details, see `http://numpy.org`.

Surveying the neighborhood

To algorithmically carry out the rules in the Game of Life, we will need a function that returns the number of live neighbors of any particular cell.

Reflection 9.5 *Consider a cell at position (r, c). What are the coordinates of the eight neighbors of this cell?*

The coordinates of the eight neighbors are visualized in the following grid with coordinates (r, c) in the center.

$(r-1, c-1)$	$(r-1, c)$	$(r-1, c+1)$
$(r, c-1)$	(r, c)	$(r, c+1)$
$(r+1, c-1)$	$(r+1, c)$	$(r+1, c+1)$

We could use an eight-part `if/elif/else` statement to check whether each neighbor is alive. However, an easier approach is illustrated below.

```
def neighborhood(grid, row, column):
    """Finds the number of live neighbors of the cell in the
       given row and column.

    Parameters:
        grid: a two-dimensional grid of cells
        row: the row index of a cell
        column: the column index of a cell

    Return value:
        the number of live neighbors of the cell at (row, column)
    """

    offsets = [ (-1, -1), (-1, 0), (-1, 1),
                (0, -1),           (0, 1),
                (1, -1),  (1, 0),  (1, 1)  ]
    rows = len(grid)
    columns = len(grid[0])
    count = 0
    for offset in offsets:
        r = row + offset[0]
        c = column + offset[1]
        if (r >= 0 and r < rows) and (c >= 0 and c < columns):
            if grid[r][c] == ALIVE:
                count = count + 1
    return count
```

The list `offsets` contains tuples with the offsets of all eight neighbors. We iterate over these offsets, adding each one to the given row and column to get the coordinates of each neighbor. Then, if the neighbor is on the grid and is alive, we increment a counter.

Performing one pass

Once we can count the number of live neighbors, we can simulate one generation by iterating over all of the cells and updating them appropriately. But before we look at how to iterate over every cell in the grid, let's consider how we can iterate over just the first row.

Reflection 9.6 *What is the name of the first row of* `grid`*?*

The first row of `grid` is named `grid[0]`. Since `grid[0]` is a list, we already know how to iterate over it, either by value or by index.

Reflection 9.7 *Should we iterate over the indices of* `grid[0]` *or over its values? Does it matter?*

As we saw above, we need to access a cell's neighbors by specifying their relative row and column indices. So we are going to need to know the indices of the row and column of each cell as we iterate over them. This means that we need to iterate over the indices of `grid[0]`, which are also its column indices, rather than over its values. Therefore, the `for` loop looks like this, assuming the number of columns is assigned to the variable name `columns`:

```
for c in range(columns):
    # update grid[0][c] here
```

Notice that, in this loop, the row number stays the same while the column number (`c`) increases. We can generalize this idea to iterate over any row with index `r` by simply replacing the row index with `r`:

```
for c in range(columns):
    # update grid[r][c] here
```

Now, to iterate over the entire grid, we need to repeat the loop above with values of `r` ranging from 0 to `rows-1`, where `rows` is assigned the number of rows in the grid. We can do this by nesting the loop above in the body of another `for` loop that iterates over the rows:

```
for r in range(rows):
    for c in range(columns):
        # update grid[r][c] here
```

Reflection 9.8 *In what order will the cells of the grid be visited in this nested loop? In other words, what sequence of* `r,c` *values does the nested loop generate?*

The value of `r` is initially set to 0. While `r` is 0, the inner `for` loop iterates over values of `c` from 0 to `COLUMNS - 1`. So the first cells that will be visited are

grid[0][0], grid[0][1], grid[0][2], ..., grid[0][9]

Once the inner `for` loop finishes, we go back up to the top of the outer `for` loop. The value of `r` is incremented to 1, and the inner `for` loop executes again. So the next cells that will be visited are grid[1][0], grid[1][1], grid[1][2], ..., grid[1][9]

This process repeats with r incremented to 2, 3, ..., 9, until finally the cells in the last row are visited:

grid[9][0], grid[9][1], grid[9][2], ..., grid[9][9]

Therefore, visually, the cells in the grid are being visited row by row:

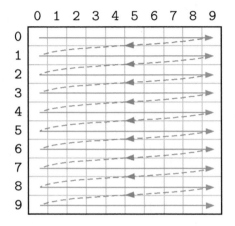

Reflection 9.9 *How would we change the nested loop so that the cells in the grid are visited column by column instead?*

To visit the cells column by column, we can simply swap the positions of the loops:

```
for c in range(columns):
    for r in range(rows):
        # update grid[r][c] here
```

Reflection 9.10 *In what order are the cells of the grid visited in this new nested loop?*

In this new nested loop, for each value of c, the inner for loop iterates all of the values of r, visiting all of the cells in that column. So the first cells that will be visited are

grid[0][0], grid[1][0], grid[2][0], ..., grid[9][0]

Then the value of c is incremented to 1 in the outer for loop, and the inner for loop executes again. So the next cells that will be visited are

grid[0][1], grid[1][1], grid[2][1], ..., grid[9][1]

This process repeats with consecutive values of c, until finally the cells in the last column are visited:

grid[0][9], grid[1][9], grid[2][9], ..., grid[9][9]

Therefore, in this case, the cells in the grid are being visited column by column instead:

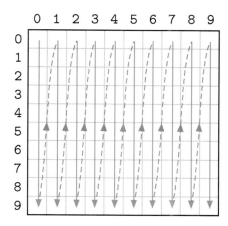

Reflection 9.11 *Do you think the order in which the cells are updated in a cellular automaton matters?*

In most cases, the order does not matter. We will update the cells row by row.

Updating the grid

Now we are ready to implement one generation of the Game of Life by iterating over all of the cells and applying the rules to each one. For each cell in position (r, c), we first need to find the number of neighbors by calling the **neighbors** function that we wrote above. Then we set the cell's new value, if it changes, according to the four rules, as follows.

```
for r in range(rows):
    for c in range(columns):
        neighbors = neighborhood(grid, r, c)
        if grid[r][c] == ALIVE and neighbors < 2:       # rule 1
            grid[r][c] = DEAD
        elif grid[r][c] == ALIVE and neighbors > 3:     # rule 3
            grid[r][c] = DEAD
        elif grid[r][c] == DEAD and neighbors == 3:     # rule 4
            grid[r][c] = ALIVE
```

Reflection 9.12 *Why is rule 2 not represented in the code above?*

Since rule 2 does not change the state of any cells, there is no reason to check for it.

Reflection 9.13 *There is one problem with the algorithm we have developed to update cells. What is it? (Think about the values referenced by the* **neighborhood** *function when it is applied to neighboring cells. Are the values from the previous generation or the current one?)*

To see the subtle problem, suppose that we change cell (r, c) from alive to dead. Then, when the live neighbors of the next cell in position $(r+1, c)$ are being counted, the cell at (r, c) will not be counted. But it should have been because it was alive in the previous generation. To fix this problem, we cannot modify the grid directly while we are updating it. Instead, we need to make a copy of the grid before each generation. When we count live neighbors, we will look at the original grid, but make modifications in the copy. Then, after we have looked at all of the cells, we can update the grid by assigning the updated copy to the main grid. These changes are shown below in red.

```
newGrid = copy.deepcopy(grid)
for r in range(rows):
    for c in range(columns):
        neighbors = neighborhood(grid, r, c)
        if grid[r][c] == ALIVE and neighbors < 2:      # rule 1
            newGrid[r][c] = DEAD
        elif grid[r][c] == ALIVE and neighbors > 3:    # rule 3
            newGrid[r][c] = DEAD
        elif grid[r][c] == DEAD and neighbors == 3:    # rule 4
            newGrid[r][c] = ALIVE
grid = newGrid
```

The `deepcopy` function from the `copy` module creates a completely independent copy of the grid.

Now that we can simulate one generation, we simply have to repeat this process to simulate many generations. The complete function is shown below. The grid is initialized with our `makeGrid` and `initialize` functions, then the nested loop that updates the grid is further nested in a loop that iterates for a specified number of generations.

```
def life(rows, columns, generations, initialCells):
    """Simulates the Game of Life for the given number of
    generations, starting with the given live cells.

    Parameters:
        rows: the number of rows in the grid
        columns: the number of columns in the grid
        generations: the number of generations to simulate
        initialCells: a list of (row, column) tuples indicating
                      the positions of the initially alive cells

    Return value:
        the final configuration of cells in a grid
    """

    grid = makeGrid(rows, columns)
    initialize(grid, initialCells)
    for g in range(generations):
```

```
            newGrid = copy.deepcopy(grid)
            for r in range(rows):
                for c in range(columns):
                    neighbors = neighborhood(grid, r, c)
                    if grid[r][c] == ALIVE and neighbors < 2:     # rule 1
                        newGrid[r][c] = DEAD
                    elif grid[r][c] == ALIVE and neighbors > 3:   # rule 3
                        newGrid[r][c] = DEAD
                    elif grid[r][c] == DEAD and neighbors == 3:   # rule 4
                        newGrid[r][c] = ALIVE
            grid = newGrid
    return grid
```

On the book web site, you can find an enhanced version of this function that uses turtle graphics to display the evolution of the system with a variety of initial configurations.

Exercises

9.2.1. Download the enhanced Game of Life program from the book web site and run it with each of the following lists of coordinates set to be alive in the initial configuration. Use at least a 50 × 50 grid. Describe what happens in each case.

(a) [(1, 3), (2, 3), (3, 3), (3, 2), (2, 1)]

(b) [(9, 10), (10, 10), (11, 10)]

(c) [(10, c + 1), (10, c + 4), (11, c), (12, c),
    (12, c + 4), (13, c), (13, c + 1), (13, c + 2),
    (13, c + 3)] with c = COLUMNS - 5

(d) [(r + 1, c + 2), (r + 2, c + 4), (r + 3, c + 1),
    (r + 3, c + 2), (r + 3, c + 5), (r + 3, c + 6),
    (r + 3, c + 7)] with r = ROWS // 2 and c = COLUMNS // 2

9.2.2. Modify the **neighborhood** function so that it treats the grid as if all four sides "wrap around." For example, in the 7 × 7 grid below on the left, the neighbors of (4, 6) include (3, 0), (4, 0), and (5, 0). In the grid on the right, the neighbors of the corner cell (6, 6) include (0, 0), (0, 5), (0, 6), (5, 0), and (6, 0).

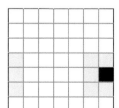 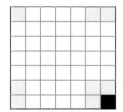

9.2.3. Write a function that prints the contents of a parameter named **grid**, which is a list of lists. The contents of each row should be printed on one line with spaces in between. Each row should be printed on a separate line.

9.2.4. Write a function that takes an integer n as a parameter, and returns an $n \times n$ multiplication table as a list of lists. The value in row r and column c should be the product of r and c.

9.2.5. Write a function that takes an integer n as a parameter, and returns an $n \times n$ grid (list of lists) in which all cells on the main diagonal contain a 1 and the rest of the cells contain a 0. For example, if $n = 5$, your function should return the grid

```
[[1, 0, 0, 0, 0],
 [0, 1, 0, 0, 0],
 [0, 0, 1, 0, 0],
 [0, 0, 0, 1, 0],
 [0, 0, 0, 0, 1]]
```

9.2.6. Write a function that takes an integer n as a parameter, and returns an $n \times n$ grid (list of lists) in which all cells on *and below* the main diagonal contain a 1 and the rest of the cells contain a 0. For example, if $n = 5$, your function should return

```
[[1, 0, 0, 0, 0],
 [1, 1, 0, 0, 0],
 [1, 1, 1, 0, 0],
 [1, 1, 1, 1, 0],
 [1, 1, 1, 1, 1]]
```

9.2.7. Write a function that takes as a parameter a two-dimensional grid (list of lists) of numbers, and prints (a) the sum of each row, (b) the sum of each column, and (c) the sum of all the entries.

9.2.8. Write a function that takes as parameters a two-dimensional grid (list of lists) and a value to search for, and returns the (row, column) where the value first appears, if it appears anywhere in the grid, and $(-1, -1)$ otherwise.

9.2.9. Write a function that returns a two-dimensional grid (list of lists) representation of an 8×8 checkerboard in which the squares in both directions alternate between the values 'B' and 'W'.

9.2.10. A magic square is a grid of numbers for which the sum of all the columns and all the rows is the same. For example, in the magic square below, all the rows and columns add up to 15.

5	7	3
1	6	8
9	2	4

The following algorithm generates magic squares with odd-length sides, using the consecutive numbers $1, 2, 3, \ldots$.

(a) Randomly put 1 in some position in your square.

(b) Look in the square diagonally to the lower right of the previous square, wrapping around to the if you go off the edge to the right or bottom.

  i. If this square is unoccupied, put the next number there.

  ii. Otherwise, put the next number directly above the previous number (again wrapping to the bottom if you are on the top row).

(c) Continue step (b) until all the positions are filled.

Write a function that takes an odd integer n as a parameter and returns an $n \times n$ magic square.

9.2.11. A two-dimensional grid can also be stored as a dictionary in which the keys are tuples representing grid positions. For example, the small grid

```
[ [0, 1],
  [1, 1] ]
```

would be stored as the following dictionary:

```
{ (0, 0): 0,
  (0, 1): 1,
  (1, 0): 1,
  (1, 1): 1  }
```

Rewrite the Game of Life program on the book web site so that it stores the grid in this way instead. The following four functions will need to change: `emptyGrid`, `initialize`, `neighborhood`, and `life`.

9.3 DIGITAL IMAGES

Digital photographs and other images are also "flat" two-dimensional objects that can be manipulated with the same techniques that we discussed in previous sections.

Colors

A digital image is a two-dimensional grid (sometimes called a *bitmap*) in which each cell, called a *pixel* (short for "picture element"), contains a value representing its color. In a grayscale image, the colors are limited to shades of gray. These shades are more commonly referred to as levels of *brightness* or *luminance*, and in theory are represented by values between 0 and 1, 0 being black and 1 being white. As we briefly explained in Box 3.2, each pixel in a color image can be represented by a (red, green, blue), or *RGB*, tuple. Each component, or *channel*, of the tuple represents the brightness of the respective color. The value $(0,0,0)$ is black and $(1,1,1)$ is white. Values between these can represent any color in the spectrum. For example, $(0, 0.5, 0)$ is a medium green and $(1, 0.5, 0)$ is orange. In practice, each channel is represented by eight bits (one byte) or, equivalently, a value between 0 and 255. So black is represented by $(0,0,0)$ or

00000000 00000000 00000000,

white is $(255, 255, 255)$ or

11111111 11111111 11111111,

and orange is $(255, 127, 0)$ or

11111111 01111111 00000000.

> **Box 9.2: Additive vs. subtractive color models**
>
> Colors in digital images are usually represented in one of two ways. In the text, we focus on the RGB color model because it is the most common, especially for digital cameras and displays. RGB is called an *additive* color model because the default "no color" $(0, 0, 0)$ is black, and adding all of the colors, represented by $(255, 255, 255)$, is white. In contrast, printers use a *subtractive* color model that "subtracts" from the brightness of a white paper background by applying color. The most common subtractive color model is CMY (short for "Cyan Magenta Yellow"). In CMY, the "no color" $(0, 0, 0)$ is white, and $(255, 255, 255)$ combines cyan, magenta, and yellow at full intensities to produce black. In practice, a black channel is added to the CMY color model because black is so common in print, and combining cyan, magenta, and yellow to produce black tends to be both imperfect and expensive in practice. The resulting four-color model is called CMYK where K stands for "Key" or "blacK," depending on who you ask.
>
> Some color models also allow an *alpha channel* to specify transparency. An alpha value of 0 means the color is completely transparent (i.e., invisible), 255 means it is opaque, and values in between correspond to degrees of translucency. Translucency effects are implemented by combining the translucent color in the foreground with the background color to an extent specified by the alpha channel. In other words,
>
> $$\text{displayed color} = \alpha \cdot \text{foreground color} + (1 - \alpha) \cdot \text{background color}.$$

RGB is most commonly used for images produced by digital cameras and viewed on a screen. Another encoding, called *CMYK* is used for print. See Box 9.2 for more details.

Reflection 9.14 *If we use eight bits to represent the intensity of each channel, can we still represent any color in the spectrum? If not, how many different colors can we represent?*

Using eight bits per channel, we cannot represent the continuous range of values between 0 and 1 that would be necessary to represent any color in the spectrum. In effect, we are only able to represent 254 values between 0 and 1: 1/255, 2/255, ..., 254/255. This is another example of how some objects represented in a computer are limited versions of those existing in nature. Looking at it another way, by using 8 bits per channel, or 24 bits total, we can represent $2^{24} = 16,777,216$ distinct colors. The good news is that, while this does not include all the colors in the spectrum, it is greater than the number of colors distinguishable by the human eye.

Reflection 9.15 *Assuming eight bits are used for each channel, what RGB tuple represents pure blue? What tuple represents purple? What color is $(0, 128, 128)$?*

Bright blue is $(0, 0, 255)$ and any tuple with equal parts red and blue, for example $(128, 0, 128)$ is a shade of purple. The tuple $(0, 128, 128)$, equal parts medium green and medium blue, is teal.

> **Box 9.3: Image storage and compression**
>
> Digital images are stored in a variety of file formats. Three of the most common are BMP, GIF, and JPEG. The technical details of each format can be quite complex, so we will just highlight the key differences, advantages, and disadvantages.
>
> All image files begin with a short header that contains information about the dimensions of the image, the number of bits that are used to encode the color of each pixel (usually 24), and other format-specific characteristics. The header is followed by information about the actual pixels. In a BMP (short for "BitMaP") file, the pixels are simply stored row by row, starting from the bottom left corner (upside down). Assuming 24-bit color, the size of a BMP file is roughly three bytes per pixel. For example, the 300×200 pixel color image in Figure 9.3 requires about $300 \cdot 200 \cdot 3 = 180,000$ bytes ≈ 180 KB in BMP format.
>
> GIF (short for "Graphics Interchange Format") files try to cut down on the amount of memory required to store an image in two ways. First, rather than store the actual color of each pixel individually, GIF files encode each pixel with an 8-bit index into a table of $2^8 = 256$ image-dependent colors. Second, GIF files compress the resulting pixel data using the Lempel-Ziv-Welch (LZW) data compression algorithm (see Box 6.1). This compression algorithm is *lossless*, meaning that the original data can be completely recovered from the compressed data, resulting in no loss of image quality. So the size of a GIF file is typically less than one byte per pixel. The color image in Figure 9.3 requires about 46 KB in GIF format.
>
> JPEG (short for "Joint Photographic Experts Group," the name of the group that created it) files use a *lossy compression* algorithm to further cut down on their size. "Lossy" means that information is lost from the original image. However, the lossy compression algorithm used in JPEG files selectively removes characteristics that are less noticeable to the naked eye, resulting in very little noticeable difference in quality. The color image in Figure 9.3 requires about 36 KB in JPEG format.

The digital images produced by digital cameras can be quite large. For example, a high quality camera can produce an image that is about 5120 pixels wide and 3413 pixels high, and therefore contains a total of $5120 \times 3413 = 17,474,560$ pixels. At one byte per pixel, a grayscale image of this size would require about 17 MB of storage. A 5120 by 3413 color image requires $5120 \times 3413 \times 3 = 52,423,680$ bytes, or about 50 MB, of storage, since it requires 24 bits, or three bytes, per pixel. In practice, color image files are *compressed* to take up much less space. (See Box 9.3.)

Image filters

To illustrate some basic image processing techniques, let's consider how we can produce a grayscale version of a color image. An operation such as this is known as an *image filter* algorithm. Photo-editing software typically includes several different image filters for enhancing digital photographs.

To change an image to grayscale, we need to convert every color pixel (an

RGB tuple) to a gray pixel with similar brightness. A white pixel (RGB color $(255, 255, 255)$) is the brightest, so we would map this to a grayscale brightness of 255 while a black pixel (RGB color $(0, 0, 0)$) is the least bright, so we would map this to a grayscale brightness of 0.

Reflection 9.16 *How can we compute the brightness of a color pixel in general?*

Consider the RGB color $(250, 50, 200)$. The red and blue channels of this color contribute a lot of brightness to the color while the green channel does not. So, to estimate the overall brightness, we can simply average the three values. In this case, $(250 + 50 + 200)/3 \approx 167$. In RGB, any tuple with equal parts red, green, and blue will be a shade of gray. Therefore, we can encode this shade of gray in RGB with the tuple $(167, 167, 167)$. A function to perform this conversion is straightforward:

```
def color2gray(color):
    """Convert a color to a shade of gray.

    Parameter:
        color: a tuple representing an RGB color

    Return value: a tuple representing an equivalent gray
    """

    brightness = (color[0] + color[1] + color[2]) // 3
    return (brightness, brightness, brightness)
```

The parameter `color` is a three-element tuple of integers between 0 and 255. The function computes the average of the three channels and returns a tuple representing a shade of gray with that brightness.

To apply this transformation to an entire image, we need to iterate over the positions of all of the pixels. Since an image is a two-dimensional object, we can process its pixels row by row as we did in the previous section:

```
for r in range(rows):
    for c in range(columns):
        # process the pixel at position (r, c)
```

To be consistent with the language typically used in image processing, we will use different names for the variables, however. Rather than referring to the size of an image in terms of rows and columns, we will use height and width. And we will use x and y (with $(0,0)$ in the top left corner) to denote the horizontal and vertical positions of a pixel instead of the row and column numbers. So the following is equivalent to the nested loop above:

```
for y in range(height):
    for x in range(width):
        # process the pixel at coordinates (x, y)
```

The standard Python module for displaying images (and creating graphical interface elements like windows and buttons) is called `tkinter` (This name is short for "Tk interface." Tk is a widely used graphical programming package that predates Python; `tkinter` provides an "interface" to Tk.) Because simple image manipulation in `tkinter` is slightly more complicated than we would like, we will interact with `tkinter` indirectly through a simple class named `Image`. The `Image` class is available in the module `image.py` on the book web site. Download this file and copy it into the same folder as your programs for this section.

The following program illustrates how to use the `Image` class to read a digital image file, iterate over its pixels, and produce a new image that is a grayscale version of the original. Each of the methods and functions below is described in Appendix B.8.

```
import image

def grayscale(photo):
    """Convert a color image to grayscale.

    Parameter:
        photo: an Image object

    Return value: a new grayscale Image object
    """

    width = photo.width()
    height = photo.height()
    newPhoto = image.Image(width, height, title = 'Grayscale image')
    for y in range(height):
        for x in range(width):
            color = photo.get(x, y)
            newPhoto.set(x, y, color2gray(color))
    return newPhoto

def main():
    penguin = image.Image(file = 'penguin.gif', title = 'Penguin')
    penguinGray = grayscale(penguin)
    penguin.show()
    penguinGray.show()
    image.mainloop()

main()
```

Let's look at the `grayscale` function first. The lone parameter named `photo` is the `Image` object that we want to turn to grayscale. The first two statements in the function call the `width` and `height` methods of `photo` to get the image's dimensions. Then the third statement creates a new, empty `Image` object with the same dimensions. This will be our grayscale image. Next, we iterate over all of the

Figure 9.3 The original image of a penguin and the grayscale version.

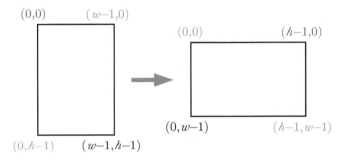

Figure 9.4 Rotating an image 90 degrees clockwise. After rotation, the corners with the same colors should line up. The width and height of the image are represented by `w` and `h`, respectively.

pixels in `photo`. Inside the nested loop, we call the `get` method to get the color of the pixel at each position `(x,y)` in `photo`. The color is returned as a three-element tuple of integers between 0 and 255. Next, we `set` the pixel at the same position in `newPhoto` to the color returned by the `color2gray` function that we wrote above. Once the nested loop has finished, we return the grayscale photo.

In the `main` function, we create an `Image` object named `penguin` from a GIF file named `penguin.gif` that can be found on the book web site. (GIF is a common image file format; see Box 9.3 for more about image files.) We then call the `grayscale` function with `penguin`, and assign the resulting grayscale image to `penguinGray`. Finally, we display both images in their own windows by calling the `show` method of each one. The `mainloop` function at the end causes the program to wait until all of the windows have been closed before it quits the program. The results are shown in Figure 9.3. This simple filter is just the beginning; we leave several other fun image filters as exercises.

Transforming images

There are, of course, many other ways we might want to transform an image. For example, we commonly need to rotate landscape images 90 degrees clockwise. This is illustrated in Figure 9.4. From the figure, we notice that the pixel in the corner at $(0,0)$ in the original image needs to be in position $(h-1,0)$ after rotation. Similarly, the pixel in the corner at $(w-1,0)$ needs to be in position $(h-1, w-1)$ after rotation. The transformations for all four corners are shown below.

Before		After
$(0,0)$	$\Rightarrow$	$(h-1,0)$
$(w-1,0)$	$\Rightarrow$	$(h-1, w-1)$
$(w-1, h-1)$	$\Rightarrow$	$(0, w-1)$
$(0, h-1)$	$\Rightarrow$	$(0,0)$

Reflection 9.17 *Do you see a pattern in these transformations? Use this pattern to infer a general rule about where each pixel at coordinates (x, y) should be in the rotated image.*

The first thing to notice is that the width and height of the image are swapped, so the x and y coordinates in the original image need to be swapped in the rotated image. However, just swapping the coordinates leads to the mirror image of what we want. Notice from the table above that the y coordinate of each rotated corner is the same as the x coordinate of the corresponding original corner. But the x coordinate of each rotated corner is the $h-1$ minus the y coordinate of the corresponding corner in the original image. So we want to draw each pixel at (x, y) in the original image at position $(h-1-y, x)$ in the rotated image. The following function does this. Notice that it is identical to the **grayscale** function, with the exceptions of parts of two statements in red.

```
def rotate90(photo):
    """Rotate an image 90 degrees clockwise.

    Parameter:
        photo: an Image object

    Return value: a new rotated Image object
    """

    width = photo.width()
    height = photo.height()
    newPhoto = image.Image(height, width, title = 'Rotated image')
    for y in range(height):
        for x in range(width):
            color = photo.get(x, y)
            newPhoto.set(height - y - 1, x, color)
    return newPhoto
```

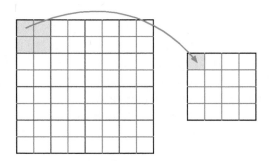

Figure 9.5 Reducing an image by one quarter.

Let's look at one more example, and then we will leave several more as exercises. Suppose we want to reduce the size of an image to one quarter of its original size. In other words, we want to reduce both the width and height by half. In the process, we are obviously going to lose three quarters of the pixels. Which ones do we throw away? One option would be to group the pixels of the original image into 2 × 2 blocks and choose the color of one of these four pixels for the corresponding pixel in the reduced image, as illustrated in Figure 9.5. This is accomplished by the following function. Again, it is very similar to the previous functions.

```
def reduce(photo):
    """Reduce an image to one quarter of its size.

    Parameter:
        photo: an Image object

    Return value: a new reduced Image object
    """
    width = photo.width()
    height = photo.height()
    newPhoto = image.Image(width // 2, height // 2,
                           title = 'Reduced image')
    for y in range(0, height, 2):
        for x in range(0, width, 2):
            color = photo.get(x, y)
            newPhoto.set(x // 2, y // 2, color)
    return newPhoto
```

Although this works, a better option would be to average the three channels of the four pixels in the block, and use this average color in the reduced image. This is left as an exercise.

Once we have filters like this, we can combine them in any way we like. For example, we can create an image of a small, upside down, grayscale penguin:

```
def main():
    penguin = image.Image(file = 'penguin.gif', title = 'Penguin')
    penguinSmall = reduce(penguin)
    penguinGray = grayscale(penguinSmall)
    penguinRotate1 = rotate90(penguinGray)
    penguinRotate2 = rotate90(penguinRotate1)
    penguinRotate2.show()
    image.mainloop()
```

By implementing some of the additional filters in the exercises below, you can devise many more fun creations.

Exercises

9.3.1. Real grayscale filters take into account how different colors are perceived by the human eye. Human sight is most sensitive to green and least sensitive to blue. Therefore, for a grayscale filter to look more realistic, the intensity of the green channel should contribute the most to the grayscale luminance and the intensity of the blue channel should contribute the least. The following formula is a common way to weigh these intensities:

$$\text{luminance} = 0.2126 \cdot \text{red} + 0.7152 \cdot \text{green} + 0.0722 \cdot \text{blue}$$

Modify the `color2gray` function in the text so that it uses this formula instead.

9.3.2. The colors in an image can be made "warmer" by increasing the yellow tone. In the RGB color model, this is accomplished by increasing the intensities of both the red and green channels. Write a function that returns an **Image** object that is warmer than the original by some factor that is passed as a parameter. If the factor is positive, the image should be made warmer; if the factor is negative, it should be made less warm.

9.3.3. The colors in an image can be made "cooler" by increasing the intensity of the blue channel. Write a function that returns an **Image** object that is cooler than the original by some factor that is passed as a parameter. If the factor is positive, the image should be made cooler; if the factor is negative, it should be made less cool.

9.3.4. The overall brightness in an image can be adjusted by increasing the intensity of all three channels. Write a function that returns an **Image** object that is brighter than the original by some factor that is passed as a parameter. If the factor is positive, the image should be made brighter; if the factor is negative, it should be made less bright.

9.3.5. A negative image is one in which the colors are the opposite of the original. In other words, the intensity of each channel is 255 minus the original intensity. Write a function that returns an **Image** object that is the negative of the original.

9.3.6. Write a function that returns an **Image** object that is a horizontally flipped version of the original. Put another way, the image should be reflected along an imaginary vertical line drawn down the center. See the example on the left of Figure 9.6.

 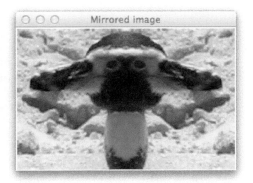

Figure 9.6 Horizontally flipped and mirrored versions of the original penguin image from Figure 9.3.

9.3.7. Write a function that returns an `Image` object with left half the same as the original but with right half that is a mirror image of the original. (Imagine placing a mirror along a vertical line down the center of an image, facing the left side.) See the example on the right of Figure 9.6.

9.3.8. In the text, we wrote a function that reduced the size of an image to one quarter of its original size by replacing each 2 × 2 block of pixels with the pixel in the top left corner of the block. Now write a function that reduces an image by the same amount by instead replacing each 2 × 2 block with a pixel that has the average color of the pixels in the block.

9.3.9. An image can be blurred by replacing each pixel with the average color of its eight neighbors. Write a function that returns a blurred version of the original.

9.3.10. An item can be further blurred by repeatedly applying the blur filter you wrote above. Write a function that returns a version of the original that has been blurred any number of times.

9.3.11. Write a function that returns an image that is a cropped version of the original. The portion of the original image to return will be specified by a rectangle, as illustrated below.

The function should take in four additional parameters that specify the (x, y) coordinates of the top left and bottom right corners (shown above) of the crop rectangle.

9.4 SUMMARY

As we saw in the previous chapter, a lot of data is naturally stored in two-dimensional tables. So it makes sense that we would also want to store this data in a two-dimensional structure in a program. We discussed two ways to do this. First, we can store the data in a *list of lists* in which each inner list contains one row of data. Second, in Exercises 9.1.6 and 9.2.11, we looked at how two-dimensional data can be stored in a *dictionary*. The latter representation has the advantage that it can be searched efficiently, if the data has an appropriate key value.

Aside from storing data, two-dimensional structures have many other applications. Two-dimensional *cellular automata* are widely used to model a great variety of phenomena. The *Game of Life* is probably the most famous, but cellular automata can also be used to model actual cellular systems, pigmentation patterns on sea shells, climate change, racial segregation (Project 9.1), ferromagnetism (Project 9.2), and to generate pseudorandom numbers. Digital images are also stored as two-dimensional structures, and *image filters* are simply algorithms that manipulate these structures.

9.5 FURTHER DISCOVERY

The chapter epigraph is from *Flatland: A Romance of Many Dimensions*, written by Edwin A. Abbott in 1884 [1].

If you are interested in learning more about cellular automata, or emergent systems in general, we recommend *Turtles, Termites, and Traffic Jams* by Mitchell Resnick [44] and *Emergence* by Steven Johnson [22] as good places to start. Also, we recommend *Agent-Based Models* by Nigel Gilbert [15] if you would like to learn more about using cellular automata in the social sciences. *Biological Computation* by Ehud Lamm and Ron Unger [27] is about using biologically-inspired computational techniques, such as cellular automata, genetic algorithms and artificial neural networks, to solve hard problems.

9.6 PROJECTS

Project 9.1 Modeling segregation

In 1971, Thomas Schelling (who in 2005 was co-recipient of the Nobel Prize in Economics) proposed a theoretical model for how racial segregation occurs in urban areas [47]. In the *Schelling model*, as it is now called, individuals belonging to one of two groups live in houses arranged in a grid. Let's call the two groups the Plain-Belly Sneetches and the Star-Belly Sneetches [50]. Each cell in the grid contains a house that is either vacant or inhabited by a Plain-Belly or a Star-Belly. Because each cell represents an individual with its own independent attribute(s), simulations such as these are known as *agent-based simulations*. Contrast this approach with the population models in Chapter 4 in which there were no discernible individuals.

Instead, we were concerned there only with aggregate sizes of populations of identical individuals.

In an instance of the Schelling model, the grid is initialized to contain some proportion of Plain-Bellies, Star-Bellies, and unoccupied cells (say 0.45, 0.45, 0.10, respectively) with their locations chosen at random. At each step, a Sneetch looks at each of its eight neighbors. (If a neighbor is off the side of the grid, wrap around to the other side.) If the fraction of a cell's neighbors that are different from itself exceeds some "tolerance threshold," the Sneetch moves to a randomly chosen unoccupied cell. Otherwise, the Sneetch stays put. For example, if the tolerance threshold is 3/8, then a Sneetch will move if more than three of its neighbors are different. We would like to answer the following question.

Question 9.1.1 *Are there particular tolerance thresholds at which the two groups always segregate themselves over time?*

Create a simulation of the Schelling model to answer this question. Visualize its behavior using the turtle graphics functions provided for the Game of Life. Experiment with different tolerance thresholds, and then answer the following questions, in addition to the one above.

Question 9.1.2 *Are there tolerance thresholds at which segregation happens only some of the time? Or does it occur all of the time for some tolerance thresholds and never for others?*

Question 9.1.3 *In the patterns you observed when answering the previous questions, was there a "tipping point" or "phase transition" in the tolerance threshold? In other words, is there a value of the tolerance threshold that satisfies the following property: if the tolerance threshold is below this value, then one thing is certain to happen and if the tolerance threshold is above this value then another thing is certain to happen?*

Question 9.1.4 *If the cells become segregated, are there "typical" patterns of segregation or is segregation different every time?*

Question 9.1.5 *The Schelling model demonstrates how a "macro" (i.e., global) property like segregation can evolve in an unpredictable way out of a sequence of entirely "micro" (i.e., local) events. (Indeed, Schelling even wrote a book titled* Micromotives and Macrobehavior *[48].) Such properties are called* emergent. *Can you think of other examples of emergent phenomena?*

Question 9.1.6 *Based on the outcome of this model, can you conclude that segregation happens because individuals are racist? Or is it possible that something else is happening?*

Project 9.2 Modeling ferromagnetism

The *Ising model* is a model of a ferromagnet at a microscopic scale. Each atom in the magnet has an individual polarity, known as spin up (+1) or spin down (−1). When the spins are random, the material is not magnetic on its own; rather, it is paramagnetic, only becoming magnetic in the presence of an external magnetic field. On the other hand, when the spins all point in the same direction, the material becomes ferromagnetic, with its own magnetic field. Each spin is influenced by neighboring spins and the ambient temperature. At lower temperatures, the spins are more likely to "organize," pointing in the same direction, causing the material to become ferromagnetic.

This situation can be modeled on a grid in which each cell has the value 1 or −1. At each time step, a cell may change its spin, depending on the spins of its four neighbors and the temperature. (To avoid inconvenient boundary conditions, treat your grid as if it wraps around side to side and bottom to top. In other words, use modular arithmetic when determining neighbors.)

Over time, the material will seek the lowest energy state. Therefore, at each step, we will want the spin of an atom to flip if doing so puts the system in a lower energy state. However, we will sometimes also flip the spin even if doing so results in a higher energy state; the probability of doing so will depend on the energy and the ambient temperature. We will model the energy associated with each spin as the number of neighbors with opposing spins. So if the spin of a particular atom is +1 and its neighbors have spins −1, −1, +1, and −1, then the energy is 3. Obviously the lowest energy state results when an atom and all of its neighbors have the same spin.

The most common technique used to decide whether a spin should flip is called the *Metropolis algorithm*. Here is how it works. For a particular particle, let E_{old} denote the energy with the particle's current spin and let E_{new} denote the energy that would result if the particle flipped its spin. If $E_{\text{new}} < E_{\text{old}}$, then we flip the spin. Otherwise, we flip the spin with probability

$$e^{-(E_{\text{new}} - E_{\text{old}})/T}$$

where e is Euler's number, the base of the natural logarithm, and T is the temperature.

Implement this simulation and visualize it using the turtle graphics functions provided for the Game of Life. Initialize the system with particles pointing in random directions. Once you have it working, try varying the temperature T between 0.1 and 10.0. (The temperature here is in energy units, not Celsius or Kelvin.)

Question 9.2.1 *At what temperature (roughly) does the system reach equilibrium, i.e., settle into a consistent state that no longer changes frequently?*

Question 9.2.2 *As you vary the temperature, does the system change its behavior gradually? Or is the change sudden? In other words, do you observe a "phase transition" or "tipping point?"*

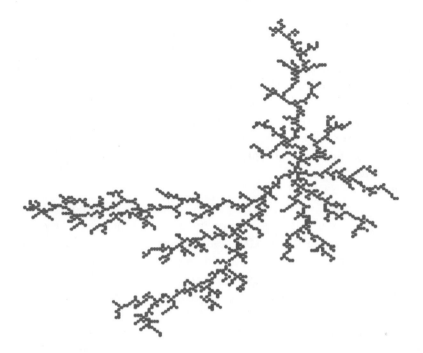

Figure 9.7 A two-dimensional Brownian tree.

Question 9.2.3 *In the Ising model, there is no centralized agent controlling which atoms change their polarity and when. Rather, all of the changes occur entirely based on an atom's local environment. Therefore, the global property of ferromagnetism occurs as a result of many local changes. Can you think of other examples of this so-called* emergent *phenomenon?*

Project 9.3 Growing dendrites

In this project, you will simulate a phenomenon known as *diffusion-limited aggregation* (DLA). DLA models the growth of clusters of particles such as snowflakes, soot, and some mineral deposits known as dendrites. In this type of system, we have some particles (or molecules) of a substance in a low concentration in a fluid (liquid or gas). The particles move according to Brownian motion (i.e., a random walk) and "stick" to each other when they come into contact. In two dimensions, this process creates patterns like the one in Figure 9.7. These structures are known as *Brownian trees* and they are *fractals* in a mathematical sense.

To simplify matters, we will model the fluid as a two dimensional grid with integer coordinates. The lower left-hand corner will have coordinates $(0,0)$ and the upper right-hand corner will have coordinates $(199, 199)$. We will place an initial seed particle in the middle at $(100, 100)$, which we will refer to as the origin. We will only let particles occupy discrete positions in this grid.

In the simulation, we will introduce new particles into the system, one at a time, some distance from the growing cluster. Each of these new particles will follow a random walk on the grid (as in Section 5.1), either eventually sticking to a fixed particle in the cluster or walking far enough away that we abandon it and move on to the next particle. If the particle sticks, it will remain in that position for the remainder of the simulation.

To keep track of the position of each fixed particle in the growing cluster, we will need a 200 × 200 grid. Each element in the grid will be either zero or one, depending upon whether there is a particle fixed in that location. To visualize the growing cluster, you will also set up a turtle graphics window with world coordinates matching the dimensions of your grid. When a particle sticks, place a dot in that location. Over time, you should see the cluster emerge.

The simulation can be described in five steps:

1. Initialize your turtle graphics window and your grid.

2. Place a seed particle at the origin (both in your grid and graphically), and set the radius R of the cluster to be 0. As your cluster grows, R will keep track of the maximum distance of a fixed particle from the origin.

3. Release a new particle in a random position at a distance of $R + 1$ from the origin. (You will need to choose a random angle relative to the origin and then use trigonometry to find the coordinates of the random particle position.)

4. Let the new particle perform a random walk on the grid. If, at any step, the particle comes into contact with an existing particle, it "sticks." Update your grid, draw a dot at this location, and update R if this particle is further from the origin than the previous radius. If the particle does not stick within 200 moves, abandon it and create a new particle. (In a more realistic simulation, you would only abandon the particle when it wanders outside of a circle with a radius that is some function of R.)

5. Repeat the previous two steps for `particles` particles.

Write a program that implements this simulation.

Additional challenges

There are several ways in which this model can be enhanced and/or explored further. Here are some ideas:

- Implement a three-dimensional DLA simulation. A video illustrating this can be found on the book web site. Most aspects of your program will need to be modified to work in three dimensions on a three-dimensional grid. To visualize the cluster in three dimensions, you can use a module called `vpython`.

  For instructions on how to install VPython on your computer, visit `http://vpython.org` and click on one of the "Download" links on the left side of the page.

As of this writing, the latest version of VPython only works with Python 2.7, so installing this version of Python will be part of the process. To force your code to behave in most respects like Python 3, even while using Python 2.7, add the following as the first line of code in your program files, above any other `import` statements:

```
from __future__ import division, print_function
```

Once you have `vpython` installed, you can import it with

```
import visual as vp
```

Then simply place a sphere at each particle location like this:

```
vp.sphere(pos = (0, 0, 0), radius = 1, color = vp.color.blue)
```

This particular call draws a blue sphere with radius 1 at the origin. No prior setup is required in `vpython`. At any time, you can zoom the image by dragging while holding down the left and right buttons (or the Option key on a Mac). You can rotate the image by dragging while holding down the right mouse button (or the Command key on a Mac). There is very good documentation for `vpython` available online at `vpython.org`.

- Investigate what happens if the particles are "stickier" (i.e., can stick from further away).

- What if you start with a line of seed particles across the bottom instead of a single particle? Or a circle of seed particles and grow inward? (Your definition of distance or radius will need to change, of course, in these cases.)

CHAPTER 10

Self-similarity and recursion

> Clouds are not spheres, mountains are not cones, coastlines are not circles, and bark is not smooth, nor does lightning travel in a straight line.
>
> Benoît Mandelbrot
> *The Fractal Geometry of Nature (1983)*

> Though this be madness, yet there is method in't.
>
> William Shakespeare
> *Hamlet (Act II, Scene II)*

HAVE you ever noticed, while laying on your back under a tree, that each branch of the tree resembles the tree itself? If you could take any branch, from the smallest to the largest, and place it upright in the ground, it could probably be mistaken for a smaller tree. This phenomenon, called *self-similarity*, is widespread in nature. There are also computational problems that, in a more abstract way, exhibit self-similarity. In this chapter, we will discuss a computational technique, called *recursion*, that we can use to elegantly solve problems that exhibit this property. As the second quotation above suggests, although recursion may seem somewhat foreign at first, it really is quite natural and just takes some practice to master.

10.1 FRACTALS

Nature is not geometric, at least not in a traditional sense. Instead, natural structures are complex and not easily described. But many natural phenomena do share a common characteristic: if you zoom in on any part, that part resembles the whole. For example, consider the images in Figure 10.1. In the bottom two images, we can see that if we zoom in on parts of the rock face and tree, these parts resemble the whole. (In nature, the resemblance is not always exact, of course.) These kinds of

478 ■ Self-similarity and recursion

Figure 10.1 Fractal patterns in nature. Clockwise from top left: a nautilus shell [61], the coastline of Norway [62], a closeup of a leaf [63], branches of a tree, a rock outcropping, and lightning [64]. The insets in the bottom two images show how smaller parts resemble the whole.

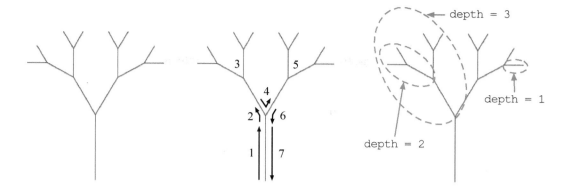

Figure 10.2 A tree produced by tree(george, 100, 4). The center figure illustrates what is drawn by each numbered step of the function. The figure on the right illustrates the self-similarity in the tree.

structures are called *fractals*, a term coined by mathematician Benoît Mandelbrot, who developed the first theory of fractal geometry.

A fractal tree

An algorithm for creating a fractal shape is *recursive*, meaning that it invokes itself on smaller and smaller scales. Let's consider the example of the simple tree shown on the left side of Figure 10.2. Notice that each of the two main branches of the tree is a smaller tree with the same structure as the whole. As illustrated in the center of Figure 10.2, to create this fractal tree, we first draw a trunk and then, for each branch, we draw two smaller trees at 30-degree angles using the same algorithm. Each of these smaller trees is composed of a trunk and two yet smaller trees, again drawn with the same tree-drawing algorithm. This process could theoretically continue forever, producing a tree with infinite complexity. In reality, however, the process eventually stops by invoking a non-recursive *base case*. The base case in Figure 10.2 is a "tree" that consists of only a single line segment.

This recursive structure is shown more precisely on the right side of Figure 10.2. The *depth* of the tree is a measure of its distance from the base case. The overall tree in Figure 10.2 has depth 4 and each of its two main branches is a tree with depth 3. Each of the two depth 3 trees is composed of two depth 2 trees. Finally, each of the four depth 2 trees is composed of two depth 1 trees, each of which is only a line segment.

The following tree function uses turtle graphics to draw this tree.

```
import turtle

def tree(tortoise, length, depth):
    """Recursively draw a tree.

    Parameters:
        tortoise: a Turtle object
        length: the length of the trunk
        depth: the desired depth of recursion

    Return value: None
    """

    if depth <= 1:                          # the base case
        tortoise.forward(length)
        tortoise.backward(length)
    else:                                   # the recursive case
1       tortoise.forward(length)
2       tortoise.left(30)
3       tree(tortoise, length * (2 / 3), depth - 1)
4       tortoise.right(60)
5       tree(tortoise, length * (2 / 3), depth - 1)
6       tortoise.left(30)
7       tortoise.backward(length)

def main():
    george = turtle.Turtle()
    george.left(90)
    tree(george, 100, 4)
    screen = george.getscreen()
    screen.exitonclick()

main()
```

Let's look at what happens when we run this program. The initial statements in the **main** function initialize the turtle and orient it to the north. Then the **tree** function is called with **tree(george, 100, 4)**. On the lines numbered 1–2, the turtle moves forward **length** units to draw the trunk, and then turns 30 degrees to the left. This is illustrated in the center of Figure 10.2; the numbers correspond to the line numbers in the function. Next, to draw the smaller tree, we call the **tree** function *recursively* on line 3 with two-thirds of the length, and a value of **depth** that is one less than what was passed in. The **depth** parameter controls how long we continue to draw smaller trees recursively. After the call to **tree** returns, the turtle turns 60 degrees to the right on line 4 to orient itself to draw the right tree. On line 5, we recursively call the **tree** function again with arguments that are identical to those on line 3. When that call returns, the turtle retraces its steps in lines 6–7 to return to the origin.

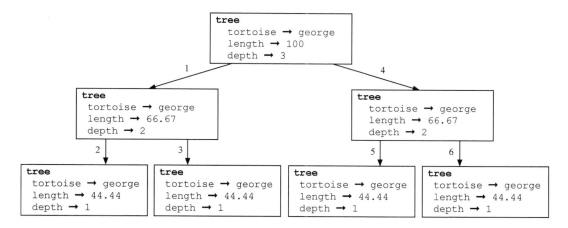

Figure 10.3 An illustration of the recursive calls in `tree(george, 100, 4)`.

The case when `depth` is at most 1 is called the *base case* because it does not make a recursive call to the function; this is where the recursion stops.

Reflection 10.1 *Try running the tree-growing function with a variety of parameters. Also, try changing the turn angles and the amount `length` is shortened. Do you understand the results you observe?*

Figure 10.3 illustrates the recursive function calls that are made by the `tree` function when `length` is 100 and `depth` is 3. The top box represents a call to the `tree` function with parameters `tortoise = george`, `length = 100`, and `depth = 3`. This function calls two instances of `tree` with `length = 100 * (2 / 3) = 66.67` and `depth = 2`. Then, each of these instances of `tree` calls two instances of `tree` with `length = 66.67 * (2 / 3) = 44.44` and `depth = 1`. Because `depth` is 1, each of these instances of `tree` simply draws a line segment and returns.

Reflection 10.2 *The numbers on the lines in Figure 10.3 represent the order in which the recursive calls are made. Can you see why that is?*

Reflection 10.3 *What would happen if we removed the base case from the algorithm by deleting the first four statements, so that the line numbered 1 was always the first statement executed?*

The base case is extremely important to the correctness of this algorithm. Without the base case, the function would continue to make recursive calls forever!

A fractal snowflake

One of the most famous fractal shapes is the *Koch curve*, named after Swedish mathematician Helge von Koch. A Koch curve begins with a single line segment with length ℓ. Then that line segment is divided into three equal parts, each with

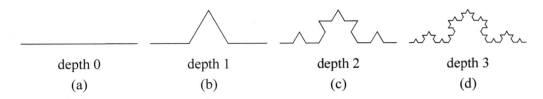

depth 0   depth 1   depth 2   depth 3
(a)   (b)   (c)   (d)

Figure 10.4   The Koch curve at depths 0, 1, 2, and 3.

length $\ell/3$. The middle part of this divided segment is replaced by two sides of an equilateral triangle (with side length $\ell/3$), as depicted in Figure 10.4(b). Next, each of the four line segments of length $\ell/3$ is divided in the same way, with the middle segment again replaced by two sides of an equilateral triangle with side length $\ell/9$, etc., as shown in Figure 10.4(c)–(d). As with the tree above, this process could theoretically go on forever, producing an infinitely intricate pattern.

Notice that, like the tree, this shape exhibits self-similarity; each "side" of the Koch curve is itself a Koch curve with smaller depth. Consider first the Koch curve in Figure 10.4 with depth 1. It consists of four smaller Koch curves with depth 0 and length $\ell/3$. Likewise, the Koch curve with depth 2 consists of four smaller Koch curves with depth 1 and the Koch curve with depth 3 consists of four smaller Koch curves with depth 2.

We can use our understanding of this self-similarity to write an algorithm to produce a Koch curve. To draw a Koch curve with depth d ($d > 0$) and overall horizontal length ℓ, we do the following:

1. Draw a Koch curve with depth $d - 1$ and overall length $\ell/3$.

2. Turn left 60 degrees.

3. Draw another Koch curve with depth $d - 1$ and overall length $\ell/3$.

4. Turn right 120 degrees.

5. Draw another Koch curve with depth $d - 1$ and overall length $\ell/3$.

6. Turn left 60 degrees.

7. Draw another Koch curve with depth $d - 1$ and overall length $\ell/3$.

The base case occurs when $d = 0$. In this case, we simply draw a line with length ℓ.

Reflection 10.4 *Follow the algorithm above to draw (on paper) a Koch curve with depth 1. Then follow the algorithm again to draw one with depth 2.*

This algorithm can be directly translated into Python:

```
def koch(tortoise, length, depth):
    """Recursively draw a Koch curve.

    Parameters:
        tortoise: a Turtle object
        length: the length of a line segment
        depth: the desired depth of recursion

    Return value: None
    """

    if depth == 0:                                          # base case
        tortoise.forward(length)
    else:                                                   # recursive case
        koch(tortoise, length / 3, depth - 1)
        tortoise.left(60)
        koch(tortoise, length / 3, depth - 1)
        tortoise.right(120)
        koch(tortoise, length / 3, depth - 1)
        tortoise.left(60)
        koch(tortoise, length / 3, depth - 1)

def main():
    george = turtle.Turtle()
    koch(george, 400, 3)
    screen = george.getscreen()
    screen.exitonclick()

main()
```

Reflection 10.5 *Run this program and experiment by calling* koch *with different values of* length *and* depth.

We can attach three Koch curves at 120-degree angles to produce an intricate snowflake shape like that in Figure 10.5.

Reflection 10.6 *Look carefully at Figure 10.5. Can you see where the three individual Koch curves are connected?*

The following function draws this Koch snowflake.

484 ■ Self-similarity and recursion

Figure 10.5 A Koch snowflake.

```
def kochSnowFlake(tortoise, length, depth):
    """Recursively draw a Koch snowflake.

    Parameters:
        tortoise: a Turtle object
        length: the length of a line segment
        depth: the desired depth of recursion

    Return value: None
    """

    for side in range(3):
        koch(tortoise, length, depth)
        tortoise.right(120)
```

Reflection 10.7 *Insert this function into the previous program and call it from* **main**. *Try making different snowflakes by increasing the number of sides (and decreasing the right turn angle).*

Imagine a Koch snowflake made from *infinitely* recursive Koch curves. Paradoxically, while the area inside any Koch snowflake is clearly finite (because it is bounded), the length of its border is infinite! In fact, the distance between any two points on its border is infinite! To see this, notice that, at every stage in its construction, each line segment is replaced with four line segments that are one-third the length of the original. Therefore, the total length of that "side" increases by one-third. Since this happens infinitely often, the perimeter of the snowflake continues to grow forever.

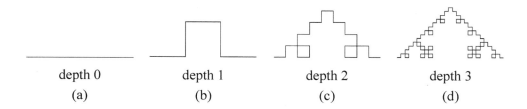

Figure 10.6 Depths 0, 1, 2, and 3 of a quadratic Koch curve.

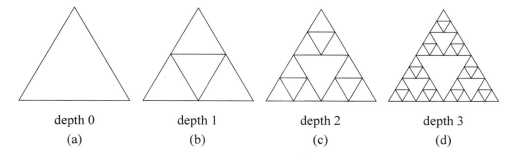

Figure 10.7 Depths 0, 1, 2, and 3 of a Sierpinski triangle.

Exercises

10.1.1. Modify the recursive tree-growing function so that it branches at random angles between 10 and 60 degrees (instead of 30 degrees) and it shrinks the trunk/branch length by a random fraction between 0.5 and 0.75. Do your new trees now look more "natural"?

10.1.2. The *quadratic Koch curve* is similar to the Koch curve, but replaces the middle segment of each side with three sides of a square instead, as illustrated in Figure 10.6. Write a recursive function

```
quadkoch(tortoise, length, depth)
```

that draws the quadratic Koch curve with the given segment length and depth.

10.1.3. Each of the following activities is recursive in the sense that each step can be considered a smaller version of the original activity. Describe how this is the case and how the "input" gets smaller each time. What is the base case of each operation below?

(a) Evaluating an arithmetic expression like $7 + (15 - 3)/4$.

(b) The chain rule in calculus (if you have taken calculus).

(c) One hole of golf.

(d) Driving directions to some destination.

10.1.4. Generalize the Koch snowflake function with an additional parameter so that it can be used to draw a snowflake with any number of sides.

486 ■ Self-similarity and recursion

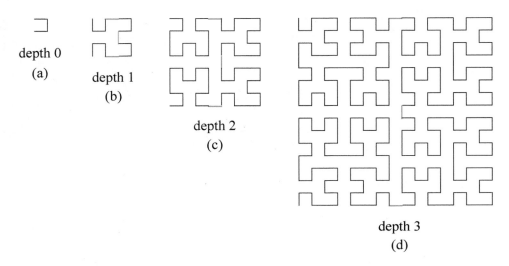

Figure 10.8   Hilbert space-filling curves with depths 0, 1, 2, and 3.

10.1.5. The Sierpinski triangle, depicted in Figure 10.7, is another famous fractal. The fractal at depth 0 is simply an equilateral triangle. The triangle at depth 1 is composed of three smaller triangles, as shown in Figure 10.7(b). (The larger outer triangle and the inner "upside down" triangle are indirect effects of the positions of these three triangles.) At depth 2, each of these three triangles is replaced by three smaller triangles. And so on. Write a recursive function

```
sierpinski(tortoise, p1, p2, p3, depth)
```

that draws a Sierpinski triangle with the given `depth`. The triangle's three corners should be at coordinates `p1`, `p2`, and `p3` (all tuples). It will be helpful to also write two smaller functions that you can call from `sierpinski`: one to draw a simple triangle, given the coordinates of its three corners, and one to compute the midpoint of a line segment.

10.1.6. The Hilbert space-filling curve, shown in Figure 10.8, is a fractal path that visits all of the cells in a square grid in such a way that close cells are visited close together in time. For example, the figure below shows how a depth 2 Hilbert curve visits the cells in an 8 × 8 grid.

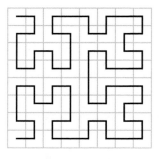

Assume the turtle is initially pointing north (up). Then the following algorithm draws a Hilbert curve with depth $d >= 0$. The algorithm can be in one of two

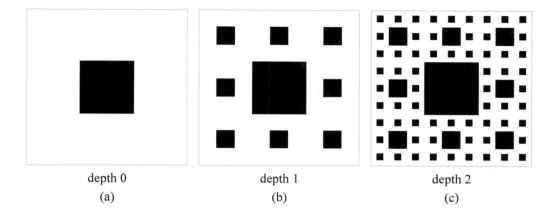

Figure 10.9 Sierpinski carpets with depths 0, 1, and 2. (The gray bounding box shows the extent of the drawing area; it is not actually part of the fractal.)

different modes. In the first mode, steps 1 and 11 turn right, and steps 4 and 8 turn left. In the other mode, these directions are reversed (indicated in square brackets below). Steps 2 and 10 make recursive calls that switch this mode.

1. Turn 90 degrees to the right [left].
2. Draw a depth $d-1$ Hilbert curve with left/right swapped.
3. Draw a line segment.
4. Turn 90 degrees to the left [right].
5. Draw a depth $d-1$ Hilbert curve.
6. Draw a line segment.
7. Draw a depth $d-1$ Hilbert curve.
8. Turn 90 degrees to the left [right].
9. Draw a line segment.
10. Draw a depth $d-1$ Hilbert curve with left and right swapped.
11. Turn 90 degrees to the right [left].

The base case of this algorithm ($d < 0$) draws nothing. Write a recursive function

`hilbert(tortoise, reverse, depth)`

that draws a Hilbert space-filling curve with the given `depth`. The Boolean parameter `reverse` indicates which mode the algorithm should draw in. (Think about how you can accommodate both drawing modes by changing the angle of the turns.)

10.1.7. A fractal pattern called the Sierpinski carpet is shown in Figure 10.9. At depth 0, it is simply a filled square one-third the width of the overall square space containing the fractal. At depth 1, this center square is surrounded by eight one-third size Sierpinski carpets with depth 0. At depth 2, the center square is surrounded by eight one-third size Sierpinski carpets with depth 1. Write a function

`carpet(tortoise, upperLeft, width, depth)`

488 ■ Self-similarity and recursion

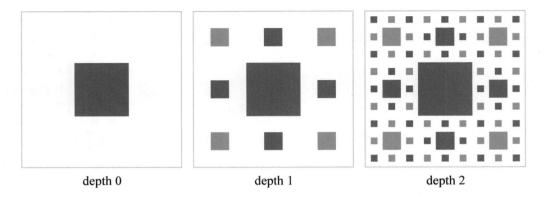

Figure 10.10 Colorful Sierpinski carpets with depths 0, 1, and 2.

that draws a Sierpinski carpet with the given `depth`. The parameter `upperLeft` refers to the coordinates of the upper left corner of the fractal and `width` refers to the overall width of the fractal.

10.1.8. Modify your Sierpinski carpet function from the last exercise so that it displays the color pattern shown in Figure 10.10.

10.2 RECURSION AND ITERATION

We typically only solve problems recursively when they obviously exhibit self-similarity or seem "naturally recursive," as with fractals. But recursion is not some obscure problem-solving technique. Although recursive algorithms may seem quite different from iterative algorithms, recursion and iteration are actually just two sides of the same coin. Every iterative algorithm can be written recursively, and vice versa. This realization may help take some of the mystery out of this seemingly foreign technique.

Consider the problem of summing the numbers in a list. Of course, this is easily achieved iteratively with a `for` loop:

```
def sumList(data):
    """Compute the sum of the values in a list.

    Parameter:
        data: a list of numbers

    Return value: the sum of the values in the list
    """

    sum = 0
    for value in data:
        sum = sum + value
    return sum
```

Let's think about how we could achieve the same thing recursively. To solve a problem recursively, we need to think about how we could solve it if we had a solution to a smaller *subproblem*. A subproblem is the same as the original problem, but with only part of the original input.

In the case of summing the numbers in a list named `data`, a subproblem would be summing the numbers in a *slice* of `data`. Consider the following example:

$$\text{data} \longrightarrow [1, \underbrace{7, 3, 6}_{\text{data[1:]}}]$$

If we had the sum of the numbers in `data[1:]`, i.e., `sumList(data[1:])`, then we could compute the value of `sumList(data)` by simply adding this sum to `data[0]`. In other words, `sumList(data)` is equal to `data[0]` plus the solution to the subproblem `sumList(data[1:])`. In terms of the example above, if we knew that `sumList([7, 3, 6])` returned `16`, then we could easily find that `sumList([1, 7, 3, 6])` is 1 + 16 = 17.

Reflection 10.8 *Will this work if `data` is empty?*

Since there is no `data[0]` or `data[1:]` when `data` is empty, the method above will not work in this case. But we can easily check for this case and simply return 0; this is the base case of the function. Putting these two parts together, we have the following function:

```
def sumList(data):
    """ (docstring omitted) """

    if len(data) == 0:                      # base case
        return 0
    return data[0] + sumList(data[1:])      # recursive case
```

But does this actually work? Yes, it does. To see why, let's look at Figure 10.11. At the top of the diagram, in box (a), is a representation of a `main` function that has called `sumList` with the argument `[1, 7, 3, 6]`. Calling `sumList` creates a new instance of the `sumList` function, represented in box (b), with `data` assigned the values `[1, 7, 3, 6]`. Since `data` is not empty, the function will return `1 + sumList([7, 3, 6])`, the value enclosed in the gray box. To evaluate this return value, we must call `sumList` again with the argument `[7, 3, 6]`, resulting in another instance of the `sumList` function, represented in box (c). The instance of `sumList` in box (b) must wait to return its value until the instance in box (c) returns. Again, since `data` is not empty, the instance of `sumList` in box (c) will return `7 + sumList([3, 6])`. Evaluating this value requires that we call `sumList` again, resulting in the instance of `sumList` in box (d). This process continues until the instance of `sumList` in box (e) calls `sumList([])`, creating the instance of `sumList` in box (f). Since this value of the `data` parameter is empty, the instance of

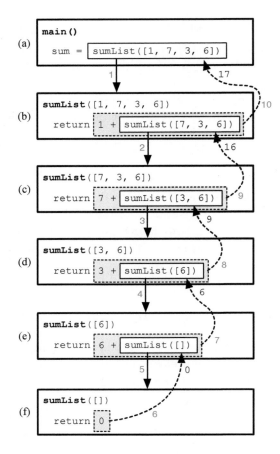

Figure 10.11 A representation of the function calls in the recursive sumList function. The red numbers indicate the order in which the events occur. The black numbers next to the arrows are return values.

sumList in box (f) immediately returns 0 to the instance of sumList that called it, in box (e). Now that the instance of sumList in box (e) has a value for sumList([]), it can return 6 + 0 = 6 to the instance of sumList in box (d). Since the instance of sumList in box (d) now has a value for sumList([6]), it can return 3 + 6 = 9 to the instance of sumList in box (c). The process continues up the chain until the value 17 is finally returned to main.

Notice that the sequence of function calls, moving down the figure from (a) to (f), only ended because we eventually reached the *base case* in which data is empty, which resulted in the function returning without making another recursive call. Every recursive function must have a non-recursive base case, and each recursive call must get one step closer to the base case. This may sound familiar; it is very similar to the way we must think about while loops. Each iteration of a while loop must move one step closer to the loop condition becoming false.

Solving a problem recursively

The following five questions generalize the process we followed to design the recursive `sumList` function. We will illustrate each question by summarizing how we answered it in the design of the `sumList`.

1. *What does a subproblem look like?*

   A subproblem is to compute the sum of a slice of the list.

2. *Suppose we could ask an all-knowing oracle for the solution to any subproblem (but not for the solution to the problem itself). Which subproblem solution would be the most useful for solving the original problem?*

   The most useful solution would be the solution for a sublist that contains all but one element of the original list, e.g., `sumList(data[1:])`.

3. *How do we find the solution to the original problem using this subproblem solution? Implement this as the recursive step of our recursive function.*

   The solution to `sumList(data)` is `data[0] + sumList(data[1:])`. Therefore, the recursive step in our function should be

   ```
   return data[0] + sumList(data[1:])
   ```

4. *What are the simplest subproblems that we can solve non-recursively, and what are their solutions? Implement your answer as the base case of the recursive function.*

   The simplest subproblem would be to compute the sum of an empty list, which is 0, of course. So the base case should be

   ```
   if len(data) == 0:
       return 0
   ```

5. *For any possible parameter value, will the recursive calls eventually reach the base case?*

   Yes, since an empty list will obviously reach the base case and any other list will result in a sequence of recursive calls that each involve a list that is one element shorter.

Reflection 10.9 *How could we have answered question 2 differently? What is another subproblem that involves all but one element of the original list? Using this subproblem instead, answer the rest of the questions to write an alternative recursive* `sumList` *function.*

An alternative subproblem that involves all but one element would be `sumList(data[:-1])` (all but the last element). Then the solution to the original problem would be `sumList(data[:-1]) + data[-1]`. The base case is the same, so the complete function is

```
def sumList(data):
    """ (docstring omitted) """

    if len(data) == 0:                      # base case
        return 0
    return sumList(data[:-1]) + data[-1]    # recursive case
```

Palindromes

Let's look at another example. A *palindrome* is any sequence of characters that reads the same forward and backward. For example, radar, star rats, and now I won are all palindromes. An iterative function that determines whether a string is a palindrome is shown below.

```
def palindrome(s):
    """Determine whether a string is a palindrome.

    Parameter: a string s

    Return value: a Boolean value indicating whether s is a palindrome
    """

    for index in range(len(s) // 2):
        if s[index] != s[-(index + 1)]:
            return False
    return True
```

Let's answer the five questions above to develop an equivalent recursive algorithm for this problem.

Reflection 10.10 *First, what does a subproblem look like?*

A subproblem would be to determine whether a slice of the string (i.e., a substring) is a palindrome.

Reflection 10.11 *Second, if you could know whether any slice is a palindrome, which would be the most helpful?*

It is often helpful to look at an example. Consider the following:

$$s \longrightarrow \text{'n}\underbrace{\text{ow I wo}}_{s[1:-1]}\text{n'}$$

If we begin by looking at the first and last characters and determine that they are *not* the same, then we know that the string is not a palindrome. But if they *are* the same, then the question of whether the string is a palindrome is decided by whether the slice that omits the first and last characters, i.e., `s[1:-1]`, is a palindrome. So it would be helpful to know the result of `palindrome(s[1:-1])`.

Reflection 10.12 *Third, how could we use this information to determine whether the whole string is a palindrome?*

If the first and last characters are the same and `s[1:-1]` is a palindrome, then `s` is a palindrome. Otherwise, `s` is not a palindrome. In other words, our desired return value is the answer to the following Boolean expression.

```
return s[0] == s[-1] and palindrome(s[1:-1])
```

If the first part is true, then the answer depends on whether the slice is a palindrome (`palindrome(s[1:-1])`). Otherwise, if the first part is false, then the entire **and** expression is false. Furthermore, due to the short circuit evaluation of the **and** operator, the recursive call to `palindrome` will be skipped.

Reflection 10.13 *What are the simplest subproblems that we can solve non-recursively, and what are their solutions? Implement your answer as the base case.*

The simplest string is, of course, the empty string, which we can consider a palindrome. But strings containing a single character are also palindromes, since they read the same forward and backward. So we know that any string with length at most one is a palindrome. But we also need to think about strings that are obviously *not* palindromes. Our discussion above already touched on this; when the first and last characters are different, we know that the string cannot be a palindrome. Since this situation is already handled by the Boolean expression above, we do not need a separate base case for it.

Putting this all together, we have the following elegant recursive function:

```
def palindrome(s):
    """ (docstring omitted) """

    if len(s) <= 1:     # base case
        return True
    return s[0] == s[-1] and palindrome(s[1:-1])
```

Let's look more closely at how this recursive function works. On the left side of Figure 10.12 is a representation of the recursive calls that are made when the `palindrome` function is called with the argument `'now I won'`. From the `main` function in box (a), `palindrome` is called with `'now I won'`, creating the instance of `palindrome` in box (b). Since the first and last characters of the parameter are equal (the `s[0] == s[-1]` part of the return statement is not shown to make the pictures less cluttered), the function will return the value of `palindrome('ow I wo')`. But,

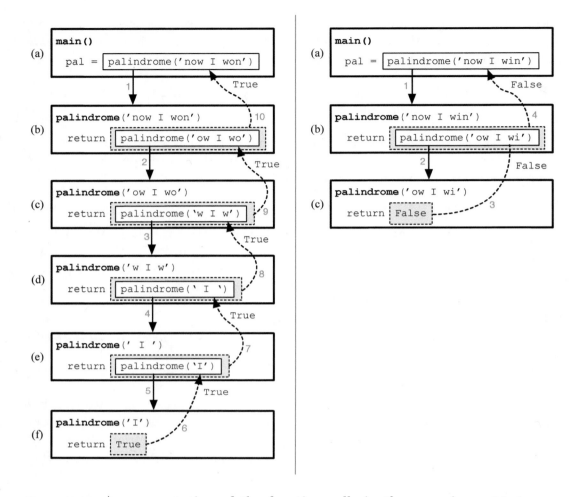

Figure 10.12 A representation of the function calls in the recursive **palindrome** function. The red numbers indicate the order in which the events happen. On the left is an instance in which the function reaches the base case and returns **True**. On the right is an instance in which the function returns **False**.

in order to get this value, it needs to call **palindrome** again, creating the instance in box (c). These recursive calls continue until we reach the base case in box (f), where the length of the parameter is one. The instance of **palindrome** in box (f) returns **True** to the previous instance of **palindrome** in box (e). Now the instance in box (e) returns to (d) the value **True** that it just received from (f). The value of **True** is propagated in this way all the way up the chain until it eventually reaches **main**, where it is assigned to the variable named **pal**.

To see how the function returns **False**, let's consider the example on the right side of Figure 10.12. In this example, the recursive **palindrome** function is called from **main** in box (a) with the non-palindromic argument **'now I win'**, which creates the instance of **palindrome** in box (b). As before, since the first and last characters of the parameter are equal, the function will return the value of `palindrome('ow I wi')`. Calling `palindrome` with this parameter creates the instance in box (c). But now, since the first and last characters of the parameter are not equal, Boolean expression returned by the function is `False`, so the instance of `palindrome` in box (c) returns `False`, and this value is propagated up to the `main` function.

Guessing passwords

One technique that hackers use to compromise computer systems is to rapidly try all possible passwords up to some given length.

Reflection 10.14 *How many possible passwords are there with length n, if there are c possible characters to choose from?*

The number of different passwords with length n is

$$\underbrace{c \cdot c \cdot c \cdots c}_{n \text{ times}} = c^n.$$

For example, there are

$$26^8 = 208,827,064,576 \approx 208 \text{ billion}$$

different eight-character passwords that use only lower-case letters. But there are

$$67^{12} = 8,182,718,904,632,857,144,561 \approx 8 \text{ sextillion}$$

different twelve-character passwords that draw from the lower and upper-case letters, digits, and the five special characters $, &, #, ?, and !, which is why web sites prompt you to use long passwords with all of these types of characters! When you use a long enough password and enough different characters, this "guess and check" method is useless.

Let's think about how we could generate a list of possible passwords by first considering the simpler problem of generating all binary strings (or "bit strings") of a given length. This is the same problem, but using only two characters: '0' and '1'. For example, the list of all binary strings with length three is ['000', '001', '010', '011', '100', '101', '110', '111'].

Thinking about this problem iteratively can be daunting. However, it becomes easier if we think about the problem's relationship to smaller versions of itself (i.e., self-similarity). As shown below, a list of binary strings with a particular length can be created easily if we already have a list of binary strings that are one bit shorter. We simply make two copies of the list of shorter binary strings, and then precede all of the strings in the first copy with a 0 and all of the strings in the second copy with a 1.

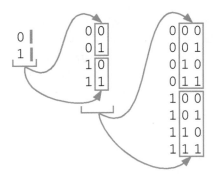

In the illustration above, the list of binary strings with length 2 is created from two copies of the list of binary strings with length 1. Then the list of binary strings with length 3 is created from two copies of the list of binary strings with length 2. In general, the list of all binary strings with a given length is the concatenation of

(a) the list of all binary strings that are one bit shorter and preceded by zero and

(b) the list of all binary strings that are one bit shorter and preceded by one.

Reflection 10.15 *What is the base case of this algorithm?*

The base case occurs when the length is 0, and there are no binary strings. However, the problem says that the return value should be a list of strings, so we will return a list containing an empty string in this case. The following function implements our recursive algorithm:

```
def binary(length):
    """Return a list of all binary strings with the given length.

    Parameter:
        length: the length of the binary strings

    Return value:
        a list of all binary strings with the given length
    """

    if length == 0:                             # base case
        return ['']

    shorter = binary(length - 1)    # recursively get shorter strings

    bitStrings = []                             #  create a list
    for shortString in shorter:                 #    of bit strings
        bitStrings.append('0' + shortString)    #    with prefix 0
    for shortString in shorter:                 #  append bit strings
        bitStrings.append('1' + shortString)    #    with prefix 1

    return bitStrings                           #  return all bit strings
```

In the recursive step, we assign a list of all bit strings with length that is one shorter to `shorter`, and then create two lists of bit strings with the desired `length`. The first, named `bitStrings0`, is a list of bit strings consisting of each bit string in `shorter`, preceded by '0'. Likewise, the second list, named `bitStrings1`, contains the shorter bit strings preceded by '1'. The return value is the list consisting of the concatenation of `bitStrings0` and `bitStrings1`.

Reflection 10.16 *Why will this algorithm not work if we return an empty list (`[ ]`) in the base case? What will be returned?*

If we return an empty list in the base case instead of a list containing an empty string, the function will return an empty list. To see why, consider what happens when we call `binary(1)`. Then `shorter` will be assigned the empty list, which means that there is nothing to iterate over in the two `for` loops, and the function returns the empty list. Since `binary(2)` calls `binary(1)`, this means that `binary(2)` will also return the empty list, and so on for any value of `length`!

Reflection 10.17 *Reflecting the algorithm we developed, the function above contains two nearly identical `for` loops, one for the '0' prefix and one for the '1' prefix. How can we combine these two loops?*

We can combine the two loops by repeating a more generic version of the loop for each of the characters '0' and '1':

```
def binary(length):
    """ (docstring omitted) """

    if length == 0:
        return ['']

    shorter = binary(length - 1)

    bitStrings = []
    for character in ['0', '1']:
        for shortString in shorter:
            bitStrings.append(character + shortString)

    return bitStrings
```

We can use a very similar algorithm to generate a list of possible passwords. The only difference is that, instead of preceding each shorter string with 0 and 1, we need to precede each shorter string with every character in the set of allowable characters. The following function, with a string of allowable characters assigned to an additional parameter, is a simple generalization of our `binary` function.

```
def passwords(length, characters):
    """Return a list of all possible passwords with the given length,
       using the given characters.

    Parameters:
        length: the length of the passwords
        characters: a string containing the characters to use

    Return value:
        a list of all possible passwords with the given length,
        using the given characters
    """

    if length == 0:
        return ['']

    shorter = passwords(length - 1, characters)

    passwordList = []
    for character in characters:
        for shorterPassword in shorter:
            passwordList.append(character + shorterPassword)

    return passwordList
```

Reflection 10.18 *How would we call the* `passwords` *function to generate a list of all bit strings with length 5? What about all passwords with length 4 containing the characters* `'abc123'`*? What about all passwords with length 8 containing lower case letters? (Do not actually try this last one!)*

Exercises

*Write a **recursive** function for each of the following problems.*

10.2.1. Write a recursive function

```
sum(n)
```

that returns the sum of the integers from 1 to `n`.

10.2.2. Write a recursive function

```
factorial(n)
```

that returns the value of $n! = 1 \cdot 2 \cdot 3 \cdots n$.

10.2.3. Write a recursive function

```
power(a, n)
```

that returns the value of a^n without using the `**` operator.

10.2.4. Write a recursive function

> `minList(data)`

that returns the minimum of the items in the list of numbers named `data`. You may use the built-in `min` function for finding the minimum of two numbers (only).

10.2.5. Write a recursive function

> `length(data)`

that returns the length of the list `data` without using the `len` function.

10.2.6. Euclid's algorithm for finding the greatest common divisor (GCD) of two integers uses the fact that the GCD of m and n is the same as the GCD of n and m mod n (the remainder after dividing m by n). Write a function

> `gcd(m, n)`

that recursively implements Euclid's GCD algorithm. For the base case, use the fact that `gcd(m, 0)` is m.

10.2.7. Write a recursive function

> `reverse(text)`

that returns the reverse of the string named `text`.

10.2.8. Write a recursive function

> `int2string(n)`

that converts an integer value `n` to its string equivalent, *without* using the `str` function. For example, `int2string(1234)` should return the string '1234'.

10.2.9. Write a recursive function

> `countUpper(s)`

that returns the number of uppercase letters in the string `s`.

10.2.10. Write a recursive function

> `equal(list1, list2)`

that returns a Boolean value indicating whether the two lists are equal without testing whether `list1 == list2`. You may only compare the lengths of the lists and test whether individual list elements are equal.

10.2.11. Write a recursive function

> `powerSet(n)`

that returns a list of all subsets of the integers $1, 2, \ldots, n$. A subset is a list of zero or more unique items from a set. The set of all subsets of a set is also called the *power set*. For example, `subsets(n)` should return the list `[[], [1], [2], [2, 1], [3], [3, 1], [3, 2], [3, 2, 1]]`. (Hint: this is similar to the `binary` function.)

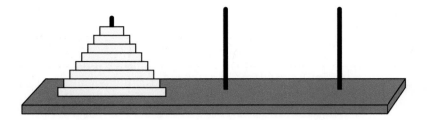

Figure 10.13 Tower of Hanoi with eight disks.

10.2.12. Suppose you work for a state in which all vehicle license plates consist of a string of letters followed by a string of numbers, such as `'ABC 123'`. Write a recursive function

```
licensePlates(length, letters, numbers)
```

that returns a list of strings representing all possible license plates of this form, with `length` letters and numbers chosen from the given strings. For example, `licensePlates(2, 'XY', '12')` should return the following list of 16 possible license plates consisting of two letters drawn from `'XY'` followed by two digits drawn from `'12'`:

```
['XX 11', 'XX 21', 'XY 11', 'XY 21', 'XX 12', 'XX 22',
 'XY 12', 'XY 22', 'YX 11', 'YX 21', 'YY 11', 'YY 21',
 'YX 12', 'YX 22', 'YY 12', 'YY 22']
```

(Hint: this is similar to the `passwords` function.)

10.3 THE MYTHICAL TOWER OF HANOI

The *Tower of Hanoi* is a game that was first marketed in 1883 by French mathematician Édouard Lucas. As illustrated in Figure 10.13, the game is played on a board with three pegs. One peg holds some number of disks with unique diameters, ordered smallest to largest. The objective of the game is to move this "tower" of disks from their original peg to another peg, one at a time, without ever placing a larger disk on top of a smaller one. The game was purported to have originated in an ancient legend. In part, the game's instruction sheet reads:

> According to an old Indian legend, the Brahmins have been following each other for a very long time on the steps of the altar in the Temple of Benares, carrying out the moving of the Sacred Tower of Brahma with sixty-four levels in fine gold, trimmed with diamonds from Golconde. When all is finished, the Tower and the Brahmins will fall, and that will be the end of the world![1]

This game is interesting because it is naturally solved using the following recursive insight. To move n disks from the first peg to the third peg, we must first be able to

---

[1] http://www.cs.wm.edu/~pkstoc/toh.html

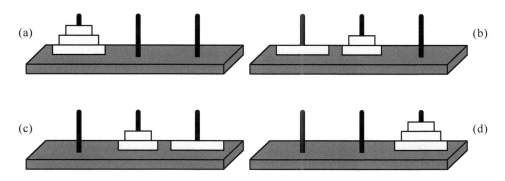

Figure 10.14 Illustration of the recursive algorithm for Tower of Hanoi with three disks.

move the bottom (largest) disk on the first peg to the bottom position on the third peg. The only way to do this is to somehow move the top $n-1$ disks from the first peg to the second peg, to get them out of the way, as illustrated in Figure 10.14(a)–(b). But notice that moving $n-1$ disks is a subproblem of moving n disks because it is the same problem but with only part of the input. The source and destination pegs are different in the original problem and the subproblem, but this can be handled by making the source, destination, and intermediate pegs additional inputs to the problem. Because this step is a subproblem, we can perform it recursively! Once this is accomplished, we are free to move the largest disk from the first peg to the third peg, as in Figure 10.14(c). Finally, we can once again recursively move the $n-1$ disks from the second peg to the third peg, shown in Figure 10.14(d). In summary, we have the following recursive algorithm:

1. Recursively move the top $n-1$ disks from the source peg to the intermediate peg, as in Figure 10.14(b).

2. Move one disk from the source peg to the destination peg, as in Figure 10.14(c).

3. Recursively move the $n-1$ disks from the intermediate peg to the destination peg, as in Figure 10.14(d).

Reflection 10.19 *What is the base case in this recursive algorithm? In other words, what is the simplest subproblem that will be reached by these recursive calls?*

The simplest base case would be if there were no disks at all! In this case, we simply do nothing.

We cannot actually write a Python function to move the disks for us, but we can write a function that gives us instructions on how to do so. The following function accomplishes this, following exactly the algorithm described above.

```
def hanoi(n, source, destination, intermediate):
    """Print instructions for solving the Tower of Hanoi puzzle.

    Parameters:
        n: the number of disks
        source: the source peg
        destination: the destination peg
        intermediate: the other peg

    Return value: None
    """

    if n >= 1:
        hanoi(n - 1, source, intermediate, destination)
        print('Move a disk from peg', source, 'to peg', destination)
        hanoi(n - 1, intermediate, destination, source)
```

The parameters `n`, `source`, `destination`, and `intermediate` represent the number of disks, the source peg, the destination peg, and the remaining peg that can be used as a temporary resting place for disks that we need to get out of the way. In our examples, peg 1 was the source, peg 3 was the destination, and peg 2 was the intermediate peg. Notice that the base case is implicit: if n is less than one, the function simply returns.

When we execute this function, we can name our pegs anything we want. For example, if we name our pegs A, B, and C, then

```
hanoi(8, 'A', 'C', 'B')
```

will print instructions for moving eight disks from peg A to peg C, using peg B as the intermediate.

Reflection 10.20 *Execute the function with three disks. Does it work? How many steps are necessary? What about with four and five disks? Do you see a pattern?*

*Is the end of the world nigh?

The original game's instruction sheet claimed that the world would end when the monks finished moving 64 disks. So how long does this take? To derive a general answer, let's start by looking at small numbers of disks. When there is one disk, only one move is necessary: move the disk from the source peg to the destination peg. When there are two disks, the algorithm moves the smaller disk to the intermediate peg, then the larger disk to the destination peg, and finally the smaller disk to the destination peg, for a total of three moves.

When n is three, we need to first move the top two disks to the intermediate peg which, we just deduced, requires three moves. Then we move the largest disk to the destination peg, for a total of four moves so far. Finally, we move the two disks from the intermediate peg to the destination peg, which requires another three moves, for a total of seven moves.

In general, notice that the number of moves required for n disks is the number of moves required for $n-1$ disks, plus one move for the bottom disk, plus the number of moves required for $n-1$ disks again. In other words, if the function $M(n)$ represents the number of moves required for n disks, then

$$M(n) = M(n-1) + 1 + M(n-1) = 2M(n-1) + 1.$$

Does this look familiar? This is a *difference equation*, just like those in Chapter 4. In this context, a function that is defined in terms of itself is also called a *recurrence relation*. The pattern produced by this recurrence relation is illustrated by the following table.

n	$M(n)$
1	1
2	3
3	7
4	15
5	31
⋮	⋮

Reflection 10.21 *Do you see the pattern in the table? What is the formula for $M(n)$ in terms of n?*

$M(n)$ is always one less than 2^n. In other words, the algorithm requires

$$M(n) = 2^n - 1$$

moves to solve the the game when there are n disks. This expression is called a *closed form* for the recurrence relation because it is defined only in terms of n, not $M(n-1)$. According to our formula, moving 64 disks would require

$$2^{64} - 1 = 18,446,744,073,709,551,615$$

moves. I guess the end of the world is not coming any time soon!

10.4 RECURSIVE LINEAR SEARCH

In Sections 6.5 and 8.5, we developed (and used) the fundamental linear search algorithm to find the index of a target item in a list. In this section, we will develop a recursive version of this function and then show that it has the same time complexity as the iterative version. Later, in Section 11.1, we will develop a more efficient algorithm for searching in a sorted list.

Reflection 10.22 *What does a subproblem look like in the search problem? What would be the most useful subproblem to have an answer for?*

In the search problem, a subproblem is to search for the target item in a smaller list. Since we can only "look at" one item at a time, the most useful subproblem will be to search for the target item in a sublist that contains all but one item. This way, we can break the original problem into two parts: determine if the target is this one item and determine if the target is in the rest of the list. It will be convenient to have this one item be the first item, so we can then solve the subproblem with the slice starting at index 1. Of course, if the first item is the one we are searching for, we can avoid the recursive call altogether. Otherwise, we return the index that is returned by the search of the smaller list.

Reflection 10.23 *What is the base case for this problem?*

We have already discussed one base case: if the target item is the first item in the list, we simply return its index. Another base case would be when the list is empty. In this case, the item for which we are searching is not in the list, so we return -1. The following function (almost) implements the recursive algorithm we have described.

```
def linearSearch(data, target):
    """Recursively find the index of the first occurrence of
       target in data.

    Parameters:
        data: a list object to search in
        target: an object to search for

    Return value: index of the first occurrence of target in data
    """
    if len(data) == 0:                      # base case 1: not found
        return -1
    if target == data[0]:                   # base case 2: found
        return ??
    return linearSearch(data[1:], target)   # recursive case
```

However, as indicated by the red question marks above, we have a problem.

Reflection 10.24 *What is the problem indicated by the red question marks? Why can we not just return 0 in that case?*

When we find the target at the beginning of the list and are ready to return its index in the original list, we do not know what it was! We know that it is at index 0 in the current sublist being searched, but we have no way of knowing where this sublist starts in the original list. Therefore, we need to add a third parameter to the function that keeps track of the original index of the first item in the sublist `data`. In each recursive call, we add one to this argument since the index of the new front item in the list will be one more than that of the current front item.

```
def linearSearch(data, target, first):
    """ (docstring omitted) """

    if len(data) == 0:                                  # base case 1: not found
        return -1
    if target == data[0]:                               # base case 2: found
        return first
    return linearSearch(data[1:], target, first + 1)    # recursive case
```

With this third parameter, we need to now initially call the function like

```
position = linearSearch(data, target, 0)
```

We can now also make this function more efficient by eliminating the need to take slices of the list in each recursive call. Since we now have the index of the first item in the sublist under consideration, we can pass the entire list as a parameter each time and use the value of `first` in the second base case.

```
def linearSearch(data, target, first):
    """ (docstring omitted) """

    if (first < 0) or (first >= len(data)):             # base case 1: not found
        return -1
    if target == data[first]:                           # base case 2: found
        return first
    return linearSearch(data, target, first + 1)        # recursive case
```

As shown above, this change also necessitates a change in our first base case because the length of the list is no longer decreasing to zero. Since the intent of the function is to search in the list between indices `first` and `len(data) - 1`, we will consider the list under consideration to be empty if the value of `first` is greater than the last index in the list. Just to be safe, we also make sure that `first` is at least zero.

Efficiency of recursive linear search

Like the iterative linear search, this algorithm has linear time complexity. But we cannot derive this result in the same way we did for the iterative version. Instead, we need to use a recurrence relation, as we did to find the number of moves required in the Tower of Hanoi algorithm.

Let $T(n)$ represent the maximum (worst case) number of comparisons required by a linear search when the length of the list is n. When we look at the algorithm above, we see that there are two comparisons that the algorithm must make before reaching a recursive function call. But since it only matters asymptotically that this number is a constant, we will simply represent the number of comparisons before the recursive call as a constant value c. Therefore, the number of comparisons necessary in the base case in which the list is empty ($n = 0$) is $T(0) = c$. In recursive cases, there are additional comparisons to be made in the recursive call.

Reflection 10.25 *How many more comparisons are made in the recursive call to* `linearSearch`*?*

The size of the sublist yet to be considered in each recursive call is $n - 1$, one less than in the current instance of the function. Therefore, the number of comparisons in each recursive call must be the number of comparisons required by a linear search on a list with length $n - 1$, which is $T(n - 1)$. So the total number of comparisons is

$$T(n) = T(n - 1) + c.$$

But this is not very helpful in determining what the time complexity of the linear search is. To get this recurrence relation into a more useful form, we can think about the recurrence relation as saying that the value of $T(n)$ is equal to, or can be replaced by, the value of $T(n - 1) + c$, as illustrated below:

Reflection 10.26 *Now what can we replace $T(n - 1)$ with?*

$T(n - 1)$ is just $T(n)$ with $n - 1$ substituted for n. Therefore, using the definition of $T(n)$ above,

$$T(n - 1) = T(n - 1 - 1) + c = T(n - 2) + c.$$

So we can also substitute $T(n - 2) + c$ for $T(n - 1)$. Similarly,

$$T(n - 2) = T(n - 2 - 1) + c = T(n - 3) + c.$$

and

$$T(n - 3) = T(n - 3 - 1) + c = T(n - 4) + c.$$

This sequence of substitutions is illustrated in Figure 10.15. The right side of the figure illustrates the accumulation of c's as we proceed downward. Since c is the number of comparisons made before each recursive call, these values on the right represent the number of comparisons made so far. Notice that the number subtracted from n in the argument of T at each step is equal to the multiplier in front of the accumulated c's at that step. In other words, to the right of each $T(n - i)$, the accumulated value of c's is $i \cdot c$. When we finally reach $T(0)$, which is the same as $T(n - n)$, the total on the right must be nc. Finally, we showed above that $T(0) = c$, so the total number of comparisons is $(n + 1)c$. This expression is called the *closed*

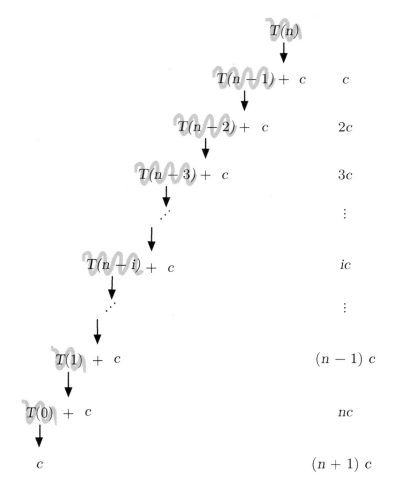

Figure 10.15 An illustration of how to derive a closed form for the recurrence relation $T(n) = T(n-1) + c$.

form of the recurrence relation because it does not involve any values of $T(n)$. Since $(n+1)c$ is proportional to n asymptotically, recursive linear search is a linear-time algorithm, just like the iterative linear search. Intuitively, this should make sense because the two algorithms essentially do the same thing: they both look at every item in the list until the target is found.

Exercises

10.4.1. Our first version of the `linearSearch` function, without the `first` parameter, can work if we only need to know whether the target is contained in the list or not. Write a working version of this function that returns `True` if the target item is contained in the list and `False` otherwise.

10.4.2. Unlike our final version of the `linearSearch` function, the function you wrote in the previous exercise uses slicing. Is this still a linear-time algorithm?

10.4.3. Write a new version of recursive linear search that looks at the last item in the list, and recursively calls the function with the sublist not containing the last item instead.

10.4.4. Write a new version of recursive linear search that only looks at every other item in the list for the target value. For example, `linearSearch([1, 2, 3, 4, 2], 2, 0)` should return the index 4 because it would not find the target, 2, at index 1.

10.4.5. Write a recursive function

> `sumSearch(data, total, first)`

that returns the first index in `data`, greater than or equal to `first`, for which the sum of the values in `data[:index + 1]` is greater than or equal to `total`. If the sum of all of the values in the list is less than `total`, the function should return −1. For example, `sumSearch([2, 1, 4, 3], 4)` returns index 2 because $2 + 1 + 4 \geq 4$ but $2 + 1 < 4$.

10.5 DIVIDE AND CONQUER

The algorithm for the Tower of Hanoi game elegantly used recursion to divide the problem into three simpler subproblems: recursively move $n - 1$ disks, move one disk, and then recursively move $n - 1$ disks again. Such algorithms came to be called *divide and conquer algorithms* because they "divide" a hard problem into two or more subproblems, and then "conquer" each subproblem recursively with the same algorithm. The divide and conquer technique has been found to yield similarly elegant, and often quite efficient, algorithms for a wide variety of problems.

It is actually useful to think of divide and conquer algorithms as comprising three steps instead of two:

1. *Divide* the problem into two or more subproblems.

2. *Conquer* each subproblem recursively.

3. *Combine* the solutions to the subproblems into a solution for the original problem.

In the Tower of Hanoi algorithm, the "combine" step was essentially free. Once the subproblems had been solved, we were done. But other problems do require this step at the end.

In this section, we will look at three more relatively simple examples of problems that are solvable by divide and conquer algorithms.

Buy low, sell high

Suppose that you have created a model to predict the future daily closing prices of a particular stock. With a list of these daily stock prices, you would like to determine when to buy and sell the stock to maximize your profit. For example, when should you buy and sell if your model predicts that the stock will be priced as follows?

Day	1	2	3	4	5	6	7	8	9	10
Price	3.90	3.60	3.65	3.71	3.78	4.95	3.21	4.50	3.18	3.53

It is tempting to look for the minimum price ($3.18) and then look for the maximum price after that day. But this clearly does not work with this example. Even choosing the second smallest price ($3.21) does not give the optimal answer. The most profitable choice is to buy on day 2 at $3.60 and sell on day 6 at $4.95, for a profit of $1.35 per share.

One way to find this answer is to look at all possible pairs of buy and sell dates, and pick the pair with the maximum profit. (See Exercise 8.5.5.) Since there are $n(n-1)/2$ such pairs, this yields an algorithm with time complexity n^2. However, there is a more efficient way. Consider dividing the list of prices into two equal-size lists (or as close as possible). Then the optimal pair of dates must either reside in the left half of the list, the right half of the list, or straddle the two halves, with the buy date in the left half and sell date in the right half. This observation can be used to design the following divide and conquer algorithm:

1. *Divide* the problem into two subproblems: (a) finding the optimal buy and sell dates in the left half of the list and (b) finding the optimal buy and sell dates in the right half of the list.

2. *Conquer* the two subproblems by executing the algorithm recursively on these two smaller lists of prices.

3. *Combine* the solutions by choosing the best buy and sell dates from the left half, the right half, and from those that straddle the two halves.

Reflection 10.27 *Is there an easy way to find the best buy and sell dates that straddle the two halves, with the buy date in the left half and sell date in the right half?*

At first glance, it might look like the "combine" step would require another recursive call to the algorithm. But finding the optimal buy and sell dates with this particular restriction is actually quite easy. The best buy date in the left half must be the one with the minimum price, and the best sell date in the right half must be the one with the maximum price. So finding these buy and sell dates simply amounts to finding the minimum price in the left half of the list and the maximum price in the right half, which we already know how to do.

Before we write a function to implement this algorithm, let's apply it to the list of prices above:

[3.90, 3.60, 3.65, 3.71, 3.78, 4.95, 3.21, 4.50, 3.18, 3.53]

The algorithm divides the list into two halves, and recursively finds the maximum profit in the left list [3.90, 3.60, 3.65, 3.71, 3.78] and the maximum profit in the right list [4.95, 3.21, 4.50, 3.18, 3.53]. In each recursive call, the algorithm is executed again but, for now, let's assume that we magically get these two maximum profits: 3.78 − 3.60 = 0.18 in the left half and 4.50 − 3.21 = 1.29 in the right half. Next, we find the maximum profit possible by holding the stock from the first half to the second half. Since the minimum price in the first half is 3.60 and the maximum price in the second half is 4.95, this profit is 1.35. Finally, we return the maximum of these three profits, which is 1.35.

Reflection 10.28 *Since this is a recursive algorithm, we also need a base case. What is the simplest list in which to find the optimal buy and sell dates?*

A simple case would be a list with only two prices. Then obviously the optimal buy and sell dates are days 1 and 2, as long as the price on day 2 is higher than the price on day 1. Otherwise, we are better off not buying at all (or equivalently, buying and selling on the same day). An even easier case would be a list with less than two prices; then once again we either never buy at all, or buy and sell on the same day.

The following function implements this divide and conquer algorithm, but just finds the optimal profit, not the actual buy and sell days. Finding these days requires just a little more work, which we leave for you to think about as an exercise.

```
 1  def profit(prices):
 2      """Find the maximum achievable profit from a list of daily
 3         stock prices.
 4
 5      Parameter:
 6          prices: a list of daily stock prices
 7
 8      Return value: the maximum profit
 9      """
10
11      if len(prices) <= 1:                        # base case
12          return 0
13
14      midIndex = len(prices) // 2                 # divide in half
15      leftProfit = profit(prices[:midIndex])      # conquer 2 halves
16      rightProfit = profit(prices[midIndex:])
17
18      buy = min(prices[:midIndex])                # min price on left
19      sell = max(prices[midIndex:])               # max price on right
20      midProfit = sell - buy
21      return max(leftProfit, rightProfit, midProfit) # combine 3 cases
```

The base case covers situations in which the list contains either one price or no prices. In these cases, the profit is 0. In the divide step, we find the index of the middle (or close to middle) price, which is the boundary between the left and right halves of the list. We then call the function recursively for each half. From each of these recursive calls, we get the maximum possible profits in the two halves. Finally,

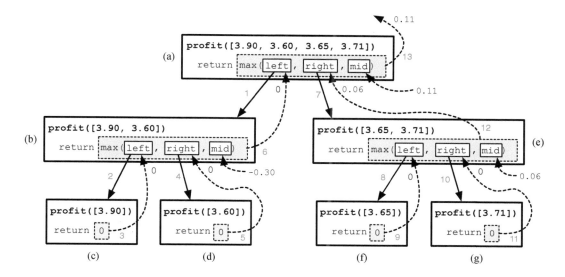

Figure 10.16 A representation of the function calls in the recursive `profit` function. The red numbers indicate the order in which the events happen.

for the combine step, we find the minimum price in the left half of the list and the maximum price in the right half of the list. The final return value is then the maximum of the maximum profit from the left half of the list, the maximum profit from the right half of the list, and the maximum profit from holding the stock from a day in the first half to a day in the second half.

Reflection 10.29 *Call this function with the list of prices that we used in the example above.*

That this algorithm actually works may seem like magic at this point. But, rest assured, like all recursive algorithms, there is a perfectly good reason why it works. The process is sketched out in Figure 10.16 for a small list containing just the first four prices in our example. As in Figures 10.11 and 10.12, each bold rectangle represents an instance of a function call. Each box contains the function's name and argument, and a representation of how the return value is computed. In all but the base cases, this is the maximum of `leftProfit` (represented by `left`), `rightProfit` (represented by `right`), and `midProfit` (represented by `mid`). At the top of Figure 10.16, the `profit` function is called with a list of four prices. This results in calling `profit` recursively with the first two prices and the last two prices. The first recursive call (on line 15) is represented by box (b). This call to `profit` results in two more recursive calls, labeled (c) and (d), each of which is a base case. These two calls both return the value 0, which is separately assigned to `leftProfit` and `rightProfit` (`left` and `right`) in box (b). The profit from holding the stock in the combine step in box (b), -0.30, is assigned to `midProfit` (`mid`). And then the maximum value, which in this case is 0, is returned back to box (a) and assigned to leftProfit. Once this recursive call returns, the second recursive call (on line 16) is made from box (a), resulting in a similar sequence of function calls, as illustrated in boxes (e)–(g). This second recursive call from (a) results in 0.06 being assigned to rightProfit. Finally, the maximum of leftProfit, rightProfit, and midProfit, which is 3.71 - 3.60 = 0.11, is returned by the original function call. The red numbers on the arrows in the figure indicate the complete ordering of events.

Navigating a maze

Suppose we want to design an algorithm to navigate a robotic rover through an unknown, obstacle-filled terrain. For simplicity, we will assume that the landscape is laid out on a grid and the rover is only able to "see" and move to the four grid cells to its east, south, west, and north in each step, as long as they do not contain obstacles that the rover cannot move through.

To navigate the rover to its destination on the grid (or determine that the destination cannot be reached), we can use a technique called *depth-first search*. The depth-first search algorithm explores a grid by first exploring in one direction as far as it can from the source. Then it *backtracks* to follow paths that branch off in each of the other three directions. Put another way, a depth-first search divides the problem of searching for a path to the destination into four subproblems: search for a path starting from the cell to the east, search for a path starting from the cell to the south, search for a path starting from the cell to the west, and search for a path starting from the cell to the north. To solve each of these subproblems, the algorithm follows this identical procedure again, just from a different starting point. In terms of the three divide and conquer steps, the depth-first search algorithm looks like this:

1. *Divide* the problem into four subproblems. Each subproblem searches for a path that starts from one of the four neighboring cells.

2. *Conquer* the subproblems by recursively executing the algorithm from each of the neighboring cells.

3. *Combine* the solutions to the subproblems by returning success if any of the subproblems were successful. Otherwise, return failure.

To illustrate, consider the grid in Figure 10.17(a). In this example, we are attempting to find a path from the green cell in position $(1,1)$ to the red cell in position $(3,0)$. The black cells represent obstacles that we cannot move through. The depth-first search algorithm will visit neighboring cells in clockwise order: east, south, west, north. In the first step, the algorithm starts at cell $(1,1)$ and looks east to $(1,2)$, but cannot move in that direction due to an obstacle. Therefore, it next explores the cell to the south in position $(2,1)$, colored blue in Figure 10.17(b). From this cell, it recursively executes the same algorithm, first looking east to $(2,2)$. Since this cell is not blocked, it is the next one visited, as represented in Figure 10.17(c). The depth-first search algorithm is recursively called again from cell $(2,2)$, but the

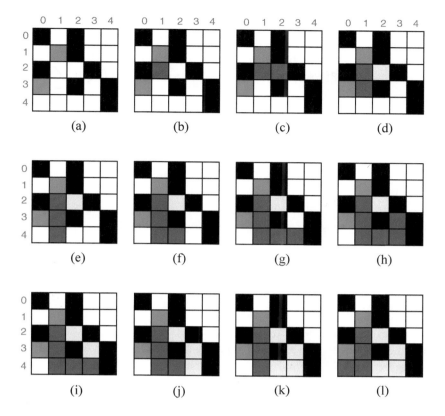

Figure 10.17 An illustration of a depth-first search on a grid.

cells to the east, south, and north are blocked; and the cell to the west has already been visited. Therefore, the depth-first search algorithm returns failure to the cell at $(2, 1)$. We color cell $(2, 2)$ light blue to indicate that it has been visited, but is no longer on the path to the destination. From cell $(2, 1)$, the algorithm has already looked east, so it now moves south to $(3, 1)$, as shown in Figure 10.17(d). In the next step, shown in Figure 10.17(e), the algorithm moves south again to $(4, 1)$ because the cell to the east is blocked.

Reflection 10.30 *When it is at cell $(3, 1)$, why does the algorithm not "see" the destination in cell $(3, 0)$?*

It does not yet "see" the destination because it looks east and south before it looks west. Since there is an open cell to the south, the algorithm will follow that possibility first. In the next steps, shown in Figure 10.17(f)–(g), the algorithm is able to move east, and in Figure 10.17(h), it is only able to move north. At this point, the algorithm backtracks to $(4, 1)$ over three steps, as shown in Figure 10.17(i)–(k), because all possible directions have already been attempted from cells $(3, 3)$, $(4, 3)$, and $(4, 2)$. From cell $(4, 1)$, the algorithm next moves west to cell $(4, 0)$, as shown in Figure 10.17(l), because it has already moved east and there is no cell to the south.

Finally, from cell (4,0), it cannot move east, south, or west; so it moves north where it finally finds the destination.

The final path shown in blue illustrates a path from the source to destination. Of course, this is not the path that the algorithm followed, but this path can now be remembered for future trips.

Reflection 10.31 *Did the depth-first search algorithm find the shortest path?*

A depth-first search is not guaranteed to find the shortest, or even a short, path. But it will find *a* path if one exists. Another algorithm, called a *breadth-first search*, can be used to find the shortest path.

Reflection 10.32 *What is the base case of the depth-first search algorithm? For what types of source cells can the algorithm finish without making any recursive calls?*

There are two kinds of base cases in depth-first search, corresponding to the two possible outcomes. One base case occurs when the source cell is not a "legal" cell from which to start a new search. These source cells are outside the grid, blocked, or already visited. In these cases, we simply return failure. The other base case occurs when the source cell is the same as the destination cell. In this case, we return success.

The depth-first search algorithm is implemented by the following function. The function returns `True` (success) if the destination was reached by a path and `False` (failure) if the destination could not be found.

```
BLOCKED = 0    # site is blocked
OPEN = 1       # site is open and not visited
VISITED = 2    # site is open and already visited

def dfs(grid, source, dest):
    """Perform a depth-first search on a grid to determine if there
       is a path between a source and destination.

    Parameters:
        grid: a 2D grid (list of lists)
        source: a (row, column) tuple to start from
        dest: a (row, column) tuple to reach

    Return value: Boolean indicating whether destination was reached
    """

    (row, col) = source
    rows = len(grid)
    columns = len(grid[0])
```

```
        if (row < 0) or (row >= rows) \
          or (col < 0) or (col >= columns) \
          or (grid[row][col] == BLOCKED) \
          or (grid[row][col] == VISITED):    # dead end (base case)
            return False                     #   so return False

        if source == dest:                   # dest found (base case)
            return True                      #   so return True

        grid[row][col] = VISITED             # visit this cell

        if dfs(grid, (row, col + 1), dest):  # search east
            return True                      #   and return if dest found
        if dfs(grid, (row + 1, col), dest):  # else search south
            return True                      #   and return if dest found
        if dfs(grid, (row, col - 1), dest):  # else search west
            return True                      #   and return if dest found
        if dfs(grid, (row - 1, col), dest):  # else search north
            return True                      #   and return if dest found

        return False                         # destination was not found
```

The variable names BLOCKED, VISITED, and OPEN represent the possible status of each cell. For example, the grid in Figure 10.17 is represented with

```
grid = [[BLOCKED, OPEN, BLOCKED, OPEN, OPEN],
        [OPEN, OPEN, BLOCKED, OPEN, OPEN],
        [BLOCKED, OPEN, OPEN, BLOCKED, OPEN],
        [OPEN, OPEN, BLOCKED, OPEN, BLOCKED],
        [OPEN, OPEN, OPEN, OPEN, BLOCKED]]
```

When a cell is visited, its value is changed from OPEN to VISITED by the dfs function. There is a program available on the book web site that includes additional turtle graphics code to visualize how the cells are visited in this depth-first search. Download this program and run it on several random grids.

Reflection 10.33 *Our dfs function returns a Boolean value indicating whether the destination was reached, but it does not actually give us the path (as marked in blue in Figure 10.17). How can we modify the function to do this?*

This modification is actually quite simple, although it may take some time to understand how it works. The idea is to add another parameter, a list named path, to which we append each cell after we mark it as visited. The values in this list contain the sequence of cells visited in the recursive calls. However, we remove the cell from path if we get to the end of the function where we return False because getting this far means that this cell is not part of a successful path after all. In our example in Figure 10.17, initially coloring a cell blue is analogous to appending that cell to the path, while recoloring a cell light blue when backtracking is analogous

Exercises

*Write a **recursive** divide and conquer function for each of the following problems. Each of your functions should contain at least two recursive calls.*

10.5.1. Write a recursive divide and conquer function

    `power(a, n)`

that returns the value of a^n, utilizing the fact that $a^n = (a^{n/2})^2$ when n is even and $a^n = (a^{(n-1)/2})^2 \cdot a$ when n is odd. Assume that n is a non-negative integer.

10.5.2. The Fibonacci sequence is a sequence of integers in which each integer is the sum of the previous two. The first two Fibonacci numbers are $1, 1$, so the sequence begins $1, 1, 2, 3, 5, 8, 13, \ldots$. Write a function

    `fibonacci(n)`

that returns the nth Fibonacci number.

10.5.3. In the `profit` function, we defined the left half as ending at index `midIndex - 1` and the right half starting at index `midIndex`. Would it also work to have the left half end at index `midIndex` and the right half start at index `midIndex + 1`? Why or why not?

10.5.4. The `profit` function in the text takes a single list as the parameter and calls the function recursively with slices of this list. In this exercise, you will write a more efficient version of this function

    `profit(prices, first, last)`

that does not use slicing in the arguments to the recursive calls. Instead, the function will pass in the entire list in each recursive call, with the two additional parameters assigned the first and last indices of the sublist that we want to consider. In the divide step, the function will need to assign `midIndex` the index that is midway between `first` and `last` (which is usually not `last // 2`). To find the maximum profit achievable with a list of prices, the function must initially be called with `profit(prices, 0, len(prices) - 1)`.

10.5.5. Modify the version of the function that you wrote in Exercise 10.5.4 so that it returns the most profitable buy and sell days instead of the maximum profit.

10.5.6. Write a divide and conquer version of the recursive linear search from Section 10.4 that checks if the middle item is equal to the target in each recursive call and then recursively calls the function with the first half and second half of the list, as needed. Your function should return the index in the list where the target value was found, or -1 if it was not found. If there are multiple instances of the target in the list, your function will not necessarily return the minimum index at which the target can be found. (This function is quite similar to Exercise 10.5.4.)

10.5.7. Write a new version of the depth-first search function

   dfs(grid, source, dest, path)

in which the parameter path contains the sequence of cell coordinates that comprise a path from source to dest in the grid when the function returns. The initial value of path will be an empty list. In other words, to find the path in Figure 10.17, your function will be called like this:

```
grid = [[BLOCKED, OPEN, BLOCKED, OPEN, OPEN],
        [OPEN, OPEN, BLOCKED, OPEN, OPEN],
        [BLOCKED, OPEN, OPEN, BLOCKED, OPEN],
        [OPEN, OPEN, BLOCKED, OPEN, BLOCKED],
        [OPEN, OPEN, OPEN, OPEN, BLOCKED]]
path = []
success = dfs(grid, (1, 1), (3, 0), path)
if success:
    print('A path was found!')
    print(path)
else:
    print('A path was not found.')
```

In this example, the final value of path should be

   [(1, 1), (2, 1), (3, 1), (4, 1), (4, 0)]

10.5.8. Write a recursive function

   numPaths(n, row, column)

that returns the number of distinct paths in an $n \times n$ grid from the cell in the given row and column to the cell in position $(n-1, n-1)$. For example, if $n = 3$, then the number of paths from $(0,0)$ to $(2,2)$ is six, as illustrated below.

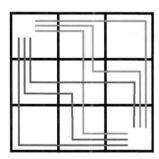

10.5.9. In Section 10.2, we developed a recursive function named binary that returned a list of all binary strings with a given length. We can design an alternative divide and conquer algorithm for the same problem by using the following insight. The list of n-bit binary strings with the common prefix p (with length less than n) is the concatenation of

   (a) the list of n-bit binary strings with the common prefix p + '0' and
   (b) the list of n-bit binary strings with the common prefix p + '1'.

For example, the list of all 4-bit binary strings with the common prefix 01 is the list of 4-bit binary strings with the common prefix 010 (namely, 0100 and 0101) plus the list of 4-bit binary strings with the common prefix 011 (namely, 0110 and 0111).

Write a recursive divide and conquer function

```
binary(prefix, n)
```

that uses this insight to return a list of all n-bit binary strings with the given prefix. To compute the list of 4-bit binary strings, you would call the function initially with binary('', 4).

*10.6 LINDENMAYER SYSTEMS

Aristid Lindenmayer was a Hungarian biologist who, in 1968, invented an elegant mathematical system for describing the growth of plants and other multicellular organisms. This type of system, now called a *Lindenmayer system*, is a particular type of *formal grammar*. In this section, we will explore basic Lindenmayer systems. In Project 10.1, you will have the opportunity to augment what we do here to create Lindenmayer systems that can produce realistic two-dimensional images of plants, trees, and bushes.

Formal grammars

A formal grammar defines a set of *productions* (or *rules*) for constructing strings of characters. For example, the following very simple grammar defines three productions that allow for the construction of a handful of English sentences.

$S \to N\ V$

$N \to$ our dog | the school bus | my foot

$V \to$ ate my homework | swallowed a fly | barked

The first production, $S \to N\ V$ says that the symbol S (a special *start symbol*) may be replaced by the string $N\ V$. The second production states that the symbol N (short for "noun phrase") may be replaced by one of three strings: our dog, the school bus, or my foot (the vertical bar (|) means "or"). The third production states that the symbol V (short for "verb phrase") may be replaced by one of three other strings. The following sequence represents one way to use these productions to derive a sentence.

$S \Rightarrow N\ V \Rightarrow$ my foot $V \Rightarrow$ my foot swallowed a fly

The derivation starts with the start symbol S. Using the first production, S is replaced with the string $N\ V$. Then, using the second production, N is replaced with the string "my foot". Finally, using the third production, V is replaced with "swallowed a fly."

Formal grammars were invented by linguist Noam Chomsky in the 1950's as a model for understanding the common characteristics of human language. Formal grammars are used extensively in computer science to both describe the syntax of programming languages and as formal models of computation. As a model of computation, a grammar's productions represent the kinds of operations that are possible, and the resulting strings, formally called *words*, represent the range of possible outputs. The most general formal grammars, called *unrestricted grammars* are computationally equivalent to Turing machines. Putting restrictions on the types of productions that are allowed in grammars affects their computational power. The standard hierarchy of grammar types is now called the *Chomsky hierarchy*.

A Lindenmayer system (or L-system) is a special type of grammar in which

(a) all applicable productions are applied in parallel at each step, and

(b) some of the symbols represent turtle graphics drawing commands.

The parallelism is meant to mimic the parallel nature of cellular division in the plants and multicellular organisms that Lindenmayer studied. The turtle graphics commands represented by the symbols in a derived string can be used to draw the growing organism.

Instead of a start symbol, an L-system specifies an *axiom* where all derivations begin. For example, the following grammar is a simple L-system:

Axiom: F
Production: F → F-F++F-F

We can see from the following derivation that parallel application of the single production very quickly leads to very long strings:

F ⇒ F-F++F-F
  ⇒ F-F++F-F-F-F++F-F++F-F++F-F-F-F++F-F
  ⇒ F-F++F-F-F-F++F-F++F-F++F-F-F-F++F-F-F-F++F-F-F-F++F-F++F-F++F-F-F-F
    ++F-F++F-F++F-F-F-F++F-F++F-F++F-F-F-F++F-F-F-F++F-F-F-F++F-F++F-F++
    F-F-F-F++F-F
  ⇒ F-F++F-F-F-F++F-F++F-F++F-F-F-F++F-F-F-F++F-F-F-F++F-F++F-F++F-F-F-F
    ++F-F++F-F++F-F-F-F++F-F++F-F++F-F-F-F++F-F-F-F++F-F-F-F++F-F++F-F++
    F-F-F-F++F-F-F-F++F-F-F-F++F-F++F-F++F-F-F-F++F-F-F-F++F-F-F-F++F-F+
    +F-F++F-F-F-F++F-F++F-F++F-F-F-F++F-F++F-F++F-F-F-F++F-F-F-F++F-F-F-
    F++F-F++F-F++F-F-F-F++F-F++F-F++F-F-F-F++F-F++F-F++F-F-F-F++F-F-F-F+
    +F-F-F-F++F-F++F-F++F-F-F-F++F-F++F-F++F-F-F-F++F-F++F-F++F-F-F-F++F
    -F-F-F++F-F-F-F++F-F++F-F++F-F-F-F++F-F-F-F++F-F++F-F++F-F++F-F++F-F-
    F-F++F-F-F-F++F-F-F-F++F-F++F-F++F-F-F-F++F-F++F-F++F-F-F-F++F-F++F-
    F++F-F-F-F++F-F-F-F++F-F++F-F++F-F-F-F++F-F
  ⇒ ...

In the first step of the derivation, the production is applied to replace `F` with
`F-F++F-F`. In the second step, all four instances of `F` are replaced with `F-F++F-F`.
The same process occurs in the third step, and the resulting string grows very
quickly. The number of strings generated from the axiom in a derivation is called
the *depth* of the derivation. If we stopped the above derivation after the last string
shown, then its depth would be four because four strings were generated from the
axiom.

Reflection 10.34 *How can repeated applications of the production(s) be framed as a recursive algorithm?*

Like a recursive solution, we are applying the same algorithm ("apply all applicable productions") at each stage of a derivation. At each stage, the inputs are growing larger, and we are getting closer to our desired depth. In other words, we can think of each step in the derivation as applying the algorithm to a longer string, but with the depth decreased by one. For example, in the derivation above, applying a derivation with depth four to the axiom `F` is the same as applying a derivation with depth three to `F-F++F-F`, which is the same as applying a derivation of length two to the next string, etc. In general, applying a derivation with depth d to a string is the same as applying a derivation with depth $d-1$ to that string after the productions have been applied one time. As a recursive algorithm in pseudocode, this process looks like this:

```
def derive(string, productions, depth):
    if depth <= 0:           # base case
        return string

    # newString = result of productions applied once to string

    return derive(newString, productions, depth - 1)
```

We will later look more closely at how to represent the productions and implement the part of the function represented by the comment.

As mentioned above, each symbol in an L-system represents a turtle graphics command:

- `F` means "move forward"

- `-` means "turn left"

- `+` means "turn right"

Interpreted in this way, every derived string represents a sequence of instructions for a turtle to follow. The distance moved for an `F` symbol can be chosen when the string is drawn. But the angle that the turtle turns when it encounters a `-` or `+`

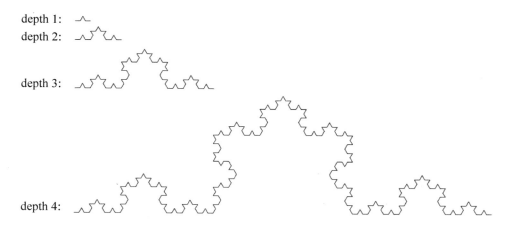

Figure 10.18 Koch curves resulting from a Lindenmayer system.

symbol must be specified by the L-system. For the L-system above, we will specify an angle of 60 degrees:

    Axiom: F
    Production: F → F-F++F-F
    Angle: 60 degrees

Reflection 10.35 *Carefully follow the turtle graphics instructions (on graph paper) in each of the first two strings derived from this L-system (*F-F++F-F *and* F-F++F-F-F-F++F-F++F-F++F-F-F-F++F-F*). Do the pictures look familiar?*

An annotated sketch of the shorter string is shown below.

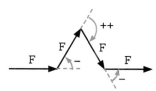

Starting on the left, we first move forward. Then we turn left 60 degrees and move forward again. Next, we turn right twice, a total of 120 degrees. Finally, we move forward, turn left again 60 degrees, and move forward one last time. Does this look familiar? As shown in Figure 10.18, the strings derived from this L-system produce Koch curves. Indeed, Lindenmayer systems produce fractals!

Here is another example:

    Axiom: FX
    Productions: X → X-YF
                 Y → FX+Y
    Angle: 90 degrees

This L-system produces a well-known fractal known as a dragon curve, shown in Figure 10.19.

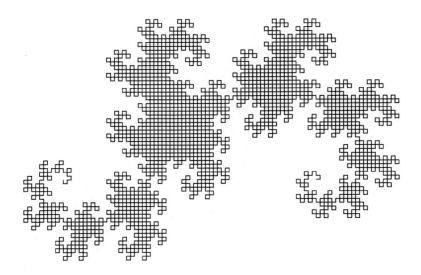

Figure 10.19 A dragon curve resulting from a Lindenmayer system.

Implementing L-systems

To implement Lindenmayer systems in Python, we need to answer three questions:

1. How do we represent a Linenmayer system in a program?

2. How do we apply productions to generate strings?

3. How do we draw the sequence of turtle graphics commands in an L-system string?

Clearly, the axiom and subsequent strings generated by an L-system can be stored as Python string objects. The productions can conveniently be stored in a dictionary. For each production, we create an entry in the dictionary with key equal to the symbol on the lefthand side and value equal to the string on the righthand side. For example, the productions for the dragon curve L-system would be stored as the following dictionary:

```
{'X': 'X-YF', 'Y':'FX+Y'}
```

Once we have the productions in a dictionary, applying them to a string is relatively easy. We iterate over the string one character at a time. For each character, we check if that character is a key in the production dictionary. If it is, we apply the associated production by appending the value associated with that key to the end of a new string. If the character is not in the dictionary, then we simply append the same character to the end of the new string. The following code accomplishes this:

```
newString = ''
for symbol in string:
    if symbol in productions:
        newString = newString + productions[symbol]
    else:
        newString = newString + symbol
```

To apply the productions again, we want to repeat this process on the new string. This can be accomplished recursively by calling the same function on `newString`. To control how many times we apply the productions, we will need a parameter named `depth` that we decrement with each recursive call, precisely the way we did with fractals in Section 10.1. When `depth` is 0, we do not want to apply the productions at all, so we return the string untouched.

```
def derive(string, productions, depth):
    """Recursively apply productions to axiom 'depth' times.

    Parameters:
        string: a string of L-system symbols
        productions: a dictionary containing L-system productions
        depth: the number of times the productions are applied

    Return value:
        the new string reflecting the application of productions
    """

    if depth <= 0:          # base case
        return string

    newString = ''          # beginning of recursive case
    for symbol in string:
        if symbol in productions:
            newString = newString + productions[symbol]
        else:
            newString = newString + symbol
    return derive(newString, productions, depth - 1)

def main():
    kochProductions = {'F': 'F-F++F-F'}
    result = derive('F', kochProductions, 3)
    print(result)

main()
```

The `main` function above derives a string for the Koch curve with depth 3.

Reflection 10.36 *Run the program above. Then modify the* `main` *function so that it derives the depth 4 string for the dragon curve.*

524 ■ Self-similarity and recursion

Of course, Lindenmayer systems are much more satisfying when you can draw them. We will leave that to you as an exercise. In Project 10.1, we explore how to augment L-systems so they can produce branching shapes that closely resemble real plants.

Exercises

10.6.1. Write a function

```
drawLSystem(tortoise, string, angle, distance)
```

that draws the picture described by the given L-system **string**. Your function should correctly handle the special symbols we discussed in this section (F, +, -). Any other symbols should be ignored. The parameters **angle** and **distance** give the angle the turtle turns in response to a + or - command, and the distance the turtle draws in response to an F command, respectively. For example, the following program should draw the smallest Koch curve.

```
def main():
    george = turtle.Turtle()
    screen = george.getscreen()
    george.hideturtle()

    drawLSystem(george, 'F-F++F-F', 60, 10)

    screen.update()
    screen.exitonclick()

main()
```

10.6.2. Apply your **drawLSystem** function from Exercise 10.6.1 to each of the following strings:

(a) F-F++F-F++F-F++F-F++F-F++F-F (angle = 60 degrees, distance = 20)

(b) FX-YF-FX+YF-FX-YF+FX+YF-FX-YF-FX+YF+FX-YF+FX+YF (angle = 90 degrees, distance = 20)

10.6.3. Write a function

```
lsystem(axiom, productions, depth, angle, distance, position, heading)
```

that calls the **derive** function with the first three parameters, and then calls your **drawLSystem** function from Exercise 10.6.1 with the new string and the values of **angle** and **distance**. The last two parameters specify the initial **position** and **heading** of the turtle, before **drawLSystem** is called. This function combines all of your previous work into a single L-system generator.

10.6.4. Call your **lsystem** function from Exercise 10.6.3 on each the following L-systems:

(a)
  Axiom: F
  Production: F → F-F++F-F
  Angle: 60 degrees

  distance = 10, position = (−400, 0), heading = 0, depth = 4

(b)

   Axiom: FX
   Productions: X → X-YF
                 Y → FX+Y
   Angle: 90 degrees

   distance = 5, position = (0,0), heading = 0, depth = 12

(c)

   Axiom: F-F-F-F
   Production: F → F-F+F+FF-F-F+F
   Angle: 90 degrees

   distance = 3, position = (−100,−100), heading = 0, depth = 3

(d)

   Axiom: F-F-F-F
   Production: F → FF-F-F-F-F-F+F
   Angle: 90 degrees

   distance = 5, position = (0,−200), heading = 0, depth = 3

10.6.5. By simply changing the axiom, we can turn the L-system for the Koch curve discussed in the text an L-system for a Koch snowflake composed of three Koch curves. Show what the axiom needs to be. Use your `lsystem` function from Exercise 10.6.3 to work out and test your answer.

10.7 SUMMARY

Some problems, like many natural objects, "naturally" exhibit self-similarity. In other words, a problem solution is simply stated in terms of solutions to smaller versions of itself. It is often easier to see how to solve such problems recursively than it is iteratively. An algorithm that utilizes this technique is called *recursive*. We suggested answering five questions to solve a problem recursively:

1. *What does a subproblem look like?*

2. *Which subproblem solution would be the most useful for solving the original problem?*

3. *How do we find the solution to the original problem using this subproblem solution? Implement this as the recursive step of our recursive function.*

4. *What are the simplest subproblems that we can solve non-recursively, and what are their solutions? Implement your answer as the base case of the recursive function.*

5. *For any possible parameter value, will the recursive calls eventually reach the base case?*

We designed recursive algorithms for about a dozen different problems in this chapter to illustrate how widely recursion can be applied. But learning how to solve problems recursively definitely takes time and practice. The more problems you solve, the more comfortable you will become!

526 ■ Self-similarity and recursion

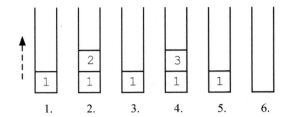

Figure 10.20 The results of a sequence of stack operations.

10.8 FURTHER DISCOVERY

The first epigraph at the beginning of this chapter is from the first page of Benoît Mandelbrot's *The Fractal Geometry of Nature* [31]. The second is from Shakespeare's *Hamlet*, Act II, Scene II.

Aristid Lindenmayer's work was described in a now freely available book titled, *The Algorithmic Beauty of Plants* [43], published in 1990.

10.9 PROJECTS

*Project 10.1 Lindenmayer's beautiful plants

For this project, we assume you have read Section 10.6.

Aristid Lindenmayer was specifically interested in modeling the branching behavior of plants. To accomplish this, we need to introduce two more symbols: [and]. For example, consider the following L-system:

    Axiom: X
    Productions: X → F[-X]+X
                   F → FF
    Angle: 30 degrees

These two new symbols involve the use of a simple data structure called a *stack*. A stack is simply a list in which we only append to and delete from one end. The append operation is called a *push* and the delete operation is called a *pop* (hence the name of the list `pop` method). For example, consider the following sequence of operations on a hypothetical stack object named `stack`, visualized in Figure 10.20.

1. stack.push(1)
2. stack.push(2)
3. x = stack.pop()
4. stack.push(3)
5. y = stack.pop()
6. z = stack.pop()

Although we implement a stack as a restricted list, it is usually visualized as a vertical stack of items in which we always push and pop items from the top. The result of the first push operation above results in the leftmost picture in Figure 10.20. After the second push operation, we have two numbers on the stack, with the second number on top of the first, as in the second picture in Figure 10.20. The third operation, a pop, removes the top item, which is then assigned to the variable name x. The fourth operation pushes the value 3 on the top of the stack. The fifth operation pops this value and assigns it to the variable name y. Finally, the bottom value is popped and assigned to the variable name z. The final values of x, y, and z are 2, 3, and 1, respectively.

In Python, we can represent the stack as an initially empty list, implement the push operation as an **append** and implement the pop operation as a **pop** with no arguments (which defaults to deleting the last item in the list). So the equivalent sequence of Python statements is:

```
stack = [ ]          # empty stack; stack is now [ ]
stack.append(1)      # push 1;      stack is now [1]
stack.append(2)      # push 2;      stack is now [1, 2]
x = stack.pop()      # x is now 2;  stack is now [1]
stack.append(3)      # push 3;      stack is now [1, 3]
y = stack.pop()      # y is now 3;  stack is now [1]
z = stack.pop()      # z is now 1;  stack is now [ ]
```

In a Lindenmayer system, the [symbol represents a push operation and the] symbol represents a pop operation. More specifically,

- [means "push the turtle's current position and heading on a stack," and

-] means "pop a position and heading from the stack and set the turtle's current position and heading to these values."

Let's now return to the Lindenmayer system above. Applying the productions of this Lindenmayer system twice results in the following string.

$$X \Rightarrow F[-X]+X \Rightarrow FF[-F[-X]+X]+F[-X]+X$$

The X symbols are used only in the derivation process and do not have any meaning for turtle graphics, so we simply skip them when we are drawing. So the string FF[-F[-X]+X]+F[-X]+X represents the simple "tree" below. On the left is a drawing of the tree; on the right is a schematic we will use to explain how it was drawn.

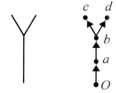

528 ■ Self-similarity and recursion

The turtle starts at the origin, marked O, with a heading of 90 degrees (north). The first two F symbols move the turtle forward from the origin to point a and then point b. The next symbol, [, means that we push the current position and heading (b, 90 degrees) on the stack.

$$\boxed{\begin{array}{c} \\ (b, 90 \text{ degrees}) \end{array}}$$

The next two symbols, -F, turn the turtle left 30 degrees (to a heading of 120 degrees) and move it forward, to point c. The next symbol is another [, which pushes the current position and heading, (c, 120 degrees), on the stack. So now the stack contains two items—(b, 90 degrees) and (c, 120 degrees)—with the last item on top.

$$\boxed{\begin{array}{c} (c, 120 \text{ degrees}) \\ (b, 90 \text{ degrees}) \end{array}}$$

The next three symbols, -X], turn the turtle left another 30 degrees (to a heading of 150 degrees), but then restore its heading to 120 degrees by popping (c, 120 degrees) from the stack.

$$\boxed{\begin{array}{c} \\ (b, 90 \text{ degrees}) \end{array}}$$

The next three symbols, +X], turn the turtle 30 degrees to the right (to a heading of 90 degrees), but then pop (b, 90 degrees) from the stack, moving the turtle back to point b, heading north.

(So, in effect, the previous six symbols, [-X]+X did nothing.) The next two symbols, +F, turn the turtle 30 degrees to the right (to a heading of 60 degrees) and move it forward to point d. Similar to before, the last six symbols, [-X]+X, while pushing states onto the stack, have no visible effect.

Continued applications of the productions in the L-system above will produce strings that draw the same sequence of trees that we created in Section 10.1. More involved L-systems will produce much more interesting trees. For example, the following two L-systems produce the trees in Figure 10.21.

Axiom:	X	Axiom:	F
Productions:	X → F-[[X]+X]+F[+FX]-X	Production:	F → FF-[-F+F+F]+[+F-F-F]
	F → FF	Angle:	22.5 degrees
Angle:	25 degrees		

Figure 10.21 Two trees from *The Algorithmic Beauty of Plants* ([43], p. 25).

Question 10.1.1 *Using an angle of 20 degrees, draw the figure corresponding to the string*

```
FF-[-F+F+F]+[+F-F-F]
```

(Graph paper might make this easier.)

Question 10.1.2 *Using an angle of 30 degrees, draw the figure corresponding to the string*

```
FF-[[FF+F]+FF+F]+FF[+FFFF+F]-FF+F
```

Part 1: Draw L-systems with a stack

If you have not completed Exercises 10.6.1 and 10.6.3, do that first. Then incorporate these functions, with the `derive` from Section 10.6, into a complete program. Your `main` function should call the `lsystem` function to draw a particular L-system.

Next, augment the `drawLSystem` function from Exercise 10.6.1 so that it correctly draws L-system strings containing the [and] characters. Do this by incorporating a single stack into your function, as we described above. Test your function with the three tree-like L-systems above.

Part 2: Draw L-systems recursively

The `drawLSystem` function can be implemented without an explicit stack by using recursion. Think of the `drawLSystem` function as drawing the figure corresponding to a string situated inside matching square brackets. We will pass the index of the first character after the left square bracket ([) as an additional parameter:

drawLSystem(tortoise, string, startIndex, angle, distance)

The function will return the index of the matching right square bracket (]). (We can pretend that there are imaginary square brackets around the entire string for the initial call of the function, so we initially pass in 0 for startIndex.) The recursive function will iterate over the indices of the characters in string, starting at startIndex. (Use a while loop, for reasons we will see shortly.) When it encounters a non-bracket character, it should do the same thing it did earlier. When the function encounters a left bracket, it will save the turtle's current position and heading, and then recursively call the function with startIndex assigned to the index of the character after the left bracket. When this recursive call returns, the current index should be set to the index returned by the recursive call, and the function should reset the turtle's position and heading to the saved values. When it encounters a right bracket, the function will return the index of the right bracket.

For example, the string below would be processed left to right but when the first left bracket is encountered, the function would be called recursively with index 5 passed in for startIndex.

```
   0    5                          26
[FFFF [ -FF [-F [-X] +X] +F [-X] +X ] +FF [ -F [-X] +X ] +F [ -X ] +X ]
            drawLSystem(..., 5, ...)
```

This recursive call will return 26, the index of the corresponding right bracket, and the + symbol at index 27 would be the next character processed in the loop. The function will later make two more recursive calls, marked with the two additional braces above.

Using this description, rewrite drawLSystem as a recursive function that does not use an explicit stack. Test your recursive function with the same tree-like L-systems, as above.

Question 10.1.3 *Why can the stack used in Part 1 be replaced by recursion in Part 2? Referring back to Figures 10.11 and 10.12, how are recursive function calls similar to pushing and popping from a stack?*

Question 10.1.4 *Use your program to draw the following additional Lindenmayer systems. For each one, set distance = 5, position = (0, −300), heading = 90, and depth = 6.*

    Axiom: X
    Productions: X → F[+X]F[-X]+X
                     F → FF
    Angle: 30 degrees

Axiom:	H
Productions:	H → HFX[+H][-H]
	X → X[-FFF][+FFF]FX
Angle:	25.7 degrees

Project 10.2 Gerrymandering

This project assumes that you have read Section 9.3 and the part of Section 10.5 on depth-first search.

U.S. states are divided into electoral districts that each elect one person to the U.S. House of Representatives. Each district is supposed to occupy a contiguous area and represent an approximately equal number of residents. Ideally, it is also *compact*, i.e., not spread out unnecessarily. More precisely, it should have a relatively small perimeter relative to the area that it covers or, equivalently, enclose a relatively large area for a shape with its perimeter length. A perfect circle is the most compact shape, while an elongated shape, like the one on the right below, is not compact.

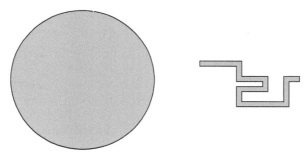

These two shapes actually have the same perimeter, but the shape on the right contains only about seven percent of the area of the circle on the left.

In some states, the majority political party has control over periodic redistricting. Often, the majority exploits this power by drawing district lines that favor their chances for re-election, a practice that has come to be known as *gerrymandering*. These districts often take on bizarre, non-compact shapes.

Several researchers have developed algorithms that redistrict states objectively to optimize various measures of compactness. For example, the image below on the left shows a recent district map for the state of Ohio. The image on the right shows a more compact district map.[2]

---

[2]These figures were produced by an algorithm developed by Brian Olson and retrieved from http://bdistricting.com

Ohio congressional districts   More compact Ohio districts

The districts on the right certainly appear to be more compact (less gerrymandered), but how much better are they? In this project, we will write a program that answers this question by determining the compactness of the districts in images like these.

Part 1: Measuring compactness

The *compactness* of a region can be measured in several ways. We will consider three possibilities:

1. First, we can measure the mean of the distance between each voter and the *centroid* of the district. The centroid is the "average point," computed by averaging all of the x and y values inside the district. We might expect a gerrymandered district to have a higher mean distance than a more compact district. Since we will not actually have information fine enough to compute this value for individual voters, we will compute the average distance between the centroid and each pixel in the image of the district.

2. Second, we can measure the standard deviation of the distance between each pixel and the centroid of the district. The standard deviation measures the degree of variability from the average. Similar to above, we might expect a gerrymandered district to have higher variability in this distance. The standard deviation of a list of values (in this case, distances) is the square root of the variance. (See Exercise 8.1.10.)

3. Third, we can compare the area of the district to the area of a (perfectly compact) circle with the same perimeter. In other words, we can define

$$\text{compactness} = \frac{\text{area of district with perimeter } p}{\text{area of circle with perimeter } p}.$$

Intuitively, a circle with a given perimeter encloses the maximum area possible for that perimeter and hence has compactness 1. A gerrymandered shape with the same perimeter encloses far less area, as we saw in the illustration above, and hence has compactness less than 1.

Suppose that we measure a particular district and find that it has area A and perimeter p. To find the value for the denominator of our formula, we need to

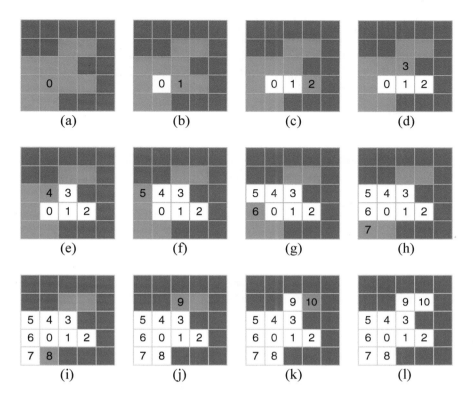

Figure 10.22 An example of the recursive flood fill algorithm.

express the area of a circle, which is normally expressed in terms of the radius r (i.e., πr^2), in terms of p instead. To do this, recall that the formula for the perimeter of a circle is $p = 2\pi r$. Therefore, $r = p/(2\pi)$. Substituting this into the standard formula, we find that the area of a circle with perimeter p is

$$\pi r^2 = \pi \left(\frac{p}{2\pi}\right)^2 = \frac{\pi p^2}{4\pi^2} = \frac{p^2}{4\pi}.$$

Finally, incorporating this into the formula above, we have

$$\text{compactness} = \frac{A}{\frac{p^2}{4\pi}} = \frac{4\pi A}{p^2}.$$

To compute values for the first and second compactness measures, we need a list of the coordinates of all of the pixels in each district. With this list, we can find the centroid of the district, and then compute the mean and standard deviation of the distances from this centroid to the list of coordinates. To compute the third metric, we need to be able to determine the perimeter and area of each district.

Part 2: Measure the districts

We can accomplish all of this by using a variant of depth-first search called a *flood fill* algorithm. The idea is to start somewhere inside a district and then use DFS to explore the pixels in that district until a pixel with a different color is reached. This is illustrated in Figure 10.22. In this small example, the red and blue squares represent two different districts on a very small map. To explore the red district, we start at some square inside the district, in this case the square marked 0. We then explore outward using a depth-first search. As we did in Section 10.5, we will explore in clockwise order: east, south, west, north. In Figure 10.22(b), we mark the first square as visited by coloring it white and then recursively visit the square to the east, marked 1. After marking square 1 as visited (colored white), the algorithm explores square 2 to the east recursively, as shown in Figure 10.22(c). After marking square 2 as visited, the algorithm backtracks to square 1 because all four neighbors of square 2 are either a different color or have already been visited. From square 1, the algorithm next explores square 3 to the north, as shown in Figure 10.22(d). This process continues until the entire red area has been visited. The numbers indicate the order in which the squares are first visited. As each square is visited for the first time, the algorithm also appends its coordinates to a list (as discussed at the end of Section 10.5). When the algorithm finishes, this list contains the coordinates of all of the squares in the red region.

Based on the `dfs` function from Section 10.5, implement this flood fill algorithm in the function

```
measureDistrict(map, x, y, color, points)
```

The five parameters have the following meanings:

- `map` is the name of an `Image` object (see Section 9.3) containing the district map. The flood fill algorithm will be performed on the pixels in this object rather than on a separate two-dimensional grid. There are several state maps available on the book web site, all of which look similar to the maps of Ohio above.

- `x` and `y` are the coordinates of the pixel from which to begin the depth-first search.

- `color` is the color of the district being measured. This is used to tell whether the current pixel is in the desired region. You may notice that the colors of the pixels in each district are not entirely uniform across the district. (The different shades represent different population densities.) Therefore, the algorithm cannot simply check whether the color of the current pixel is equal to `color`. Rather, it needs to check whether the color of the current pixel is *close to* `color`. Since colors are represented as three-element tuples, we can treat them as three-dimensional points and use the traditional Euclidean distance formula to determine "closeness:"

$$\text{distance}((x_1, y_1, z_1), (x_2, y_2, z_2)) = \sqrt{(x_1 - x_2)^2 + (y_1 - y_2)^2 + (z_1 - z_2)^2}$$

Start with a distance threshold for closeness of 100, and adjust as needed.

- **points** will be a list of coordinates of the pixels that the algorithm visited. When you call the function, initially pass in an empty list. When the function returns, this list should be populated with the coordinates in the district.

Your function should return a tuple containing the total perimeter and total area obtained from a DFS starting at (x, y). The perimeter can be obtained by counting the number of times the algorithm reaches a pixel that is outside of the region (think base case), and the area is the total number of pixels that are visited inside the region. For example, as shown below, the region from Figure 10.22 has perimeter 18 and area 11. The red numbers indicate the order in which the flood fill algorithm will count each border.

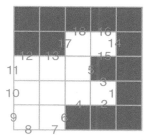

Given these measurements, the compactness of this region is

$$\frac{4\pi \cdot 11}{18^2} \approx 0.4266.$$

The value of **points** after calling the function on this example would be

```
[(3, 1), (3, 2), (3, 3), (2, 2), (2, 1), (2, 0), (3, 0),
 (4, 0), (4, 1), (1, 2), (1, 3)]
```

The centroid of these points, derived by the averaging the x and y coordinates, is $(28/11, 15/11) \approx (2.54, 1.36)$. Then the mean distance to the centroid is approximately 1.35 and the standard deviation is approximately 0.54.

Part 3: Compare district maps

To compute the average compactness metrics for a particular map, write a function

    `compactness(imageName, districts)`

that computes the three compactness measurements that we discussed above for the district map with file name **imageName**. You can find maps for several states on the book web site. The second parameter **districts** will contain a list of starting coordinates (two-element tuples) for the districts on the map. These are also available on the book web site. Your function should iterate over this list of tuples, and call your **measureDistrict** function with **x** and **y** set to the coordinates in each one.

536 ■ Self-similarity and recursion

The function should return a three-element tuple containing the average value, over all of the districts, for each of the three metrics. To make sure your flood fill is working properly, it will also be helpful to display the map (using the `show` method of the `Image` class) and update it (using the `update` method of the `Image` class) in each iteration. You should see the districts colored white, one by one.

For at least three states, compare the existing district map and the more compact district map, using the three compactness measures. What do you notice?

To drive your program, write a `main` function that calls the `compactness` function with a particular map, and then reports the results. As always, think carefully about the design of your program and what additional functions might be helpful.

Technical notes

1. The images supplied on the book web site have low resolution to keep the depth of the recursive calls in check. As a result, your compactness results will only be a rough approximation. Also, shrinking the image sizes caused some of the boundaries between districts to become "fuzzy." As a result, you will see some unvisited pixels along these boundaries when the flood fill algorithm is complete.

2. Depending on your particular Python installation, the depth of recursion necessary to analyze some of these maps may exceed the maximum allowed. To increase the allowed recursion depth, you can call the `sys.setrecursionlimit` function at the top of your program. For example,

   ```
   import sys
   sys.setrecursionlimit(10000)
   ```

   However, set this value carefully. Use the smallest value that works. Setting the maximum recursion depth too high may crash Python on your computer! If you cannot find a value that works on your computer, try shrinking the image file instead (and scaling the starting coordinates appropriately).

Project 10.3 Percolation

This project assumes that you are familiar with two-dimensional grids from Section 9.2 and have read the part of Section 10.5 on depth-first search.

Suppose we have a grid of squares called *sites*. A site can either be open (white) or blocked (black).

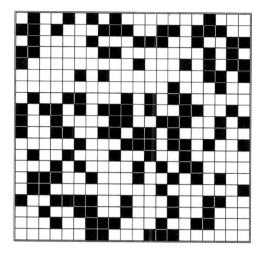

Now imagine that we pour a liquid uniformly over the top of the grid. The liquid will fill the open sites at the top and percolate to connected open sites until the liquid fills all of the open sites that are reachable from an open site at the top. We say that the grid *percolates* if at least one open site in the bottom row is full at the end. For example, the grid below on the left percolates, while the grid on the right does not.

This system can be used to model a variety of naturally occurring phenomena. Most obviously, it can model a porous rock being saturated with water. Similarly, it can model an oil (or natural gas) field; in this case, the top of the grid represents the oil underground and percolation represents the oil reaching the surface. Percolation systems can also be used to model the flow of current through a network of transistors, whether a material conducts electricity, the spread of disease or forest fires, and even evolution.

We can represent how "porous" a grid is by a *vacancy probability*, the probability that any particular site is open. For a variety of applications, scientists are interested in knowing the probability that a grid with a particular vacancy probability will percolate. In other words, we would like to know the percolation probability for any vacancy probability p. Despite decades of research, there is no known mathematical solution to this problem. Therefore, we will use a Monte Carlo simulation to estimate it.

Recall from Chapter 5 that a Monte Carlo simulation flips coins (metaphorically speaking) at each step in a computation. For example, to estimate the distance traveled by a random walk, we performed many random walks and took the average final distance from the origin. In this problem, for any given vacancy probability p, we will create many random grids and then test whether they percolate. By computing the number that do percolate divided by the total number of trials, we will estimate the percolation probability for vacancy probability p.

Part 1: Does it percolate?

To decide whether a grid percolates, we can use a depth-first search. Recall from Section 10.5 that the depth-first search algorithm completely explores a space by first exploring as far away as possible from the source. Then it *backtracks* to follow paths that branch off. To decide whether a given grid percolates, we must do a depth-first search from each of the open sites in the top row. Once this is done, we simply look at whether any site in the bottom row has been visited. If so, the system percolates. Write a function

```
percolates(grid, draw)
```

that decides whether a given grid percolates. The second parameter is a Boolean that indicates whether the grid should also be drawn using turtle graphics. There is a skeleton program on the book web site in which the drawing code has already been written. Notice that some of the functions include a Boolean parameter that indicates whether the percolation should be visualized.

Part 2: Find the percolation probability

For any particular vacancy probability p, we can design a Monte Carlo simulation to estimate the percolation probability:

1. Create a random grid in which, with vacancy probability p, any particular site is open.

2. Test to see if this grid percolates.

3. Repeat steps 1 and 2 a large number of times (say, 10,000), keeping track of the number of grids that percolate.

4. Divide the number that percolate by the total number of trials. This is your estimated percolation probability.

Implement this algorithm with a function

```
percMonteCarlo(rows, columns, p, trials)
```

Part 3: When is a grid likely to percolate?

For what vacancy probability does the percolation probability reach at least 1/2? In other words, what must the vacancy probability be for a system to be more likely than not to percolate?

To answer this question, also write a function

```
percPlot(minP, maxP, stepP, trials)
```

that plots vacancy probability on the x axis and percolation probability on the y axis for vacancy probabilities minP, minP + stepP, ..., maxP. Each percolation probability should be derived from a Monte Carlo simulation with the given number of trials.

You should discover a *phase transition*: if the vacancy probability is less than a particular *threshold* value, the system almost certainly does not percolate; but if the vacancy probability is greater than this threshold value, the system almost certainly *does* percolate. What is this threshold value?

CHAPTER 11

Organizing data

Search is an unsolved problem. We have a good 90 to 95% of the solution, but there is a lot to go in the remaining 10%.

Marissa Mayer, President and CEO of Yahoo!
Los Angeles Times interview (2008)

IN this age of "big data," we take search algorithms for granted. Without web search sites that are able to sift through billions of pages in a fraction of a second, the web would be practically useless. Similarly, large data repositories, such as those maintained by the U.S. Geological Survey (USGS) and the National Institutes of Health (NIH), would be useless without the ability to search for specific information. Even the operating systems on our personal computers now supply integrated search capabilities to help us navigate our increasingly large collections of files.

To enable fast access to these sets of data, it must be organized in some way. A method for organizing data is known as a *data structure*. Hidden data structures in the implementations of the list and dictionary abstract data types enable their methods to access and modify their contents quickly. (The data structure behind a dictionary was briefly explained in Box 8.2.) In this chapter, we will explore one of the simplest ways to organize data—maintaining it in a sorted list—and the benefits this can provide. We will begin by developing a significantly faster search algorithm that can take advantage of knowing that the data is sorted. Then we will design three algorithms to sort data in a list, effectively creating a sorted list data structure. If you continue to study computer science, you can look forward to seeing many more sophisticated data structures in the future that enable a wide variety of efficient algorithms.

11.1 BINARY SEARCH

The spelling checkers that are built into most word processing programs work by searching through a list of English words, seeking a match. If the word is found, it is considered to be spelled correctly. Otherwise, it is assumed to be a misspelling. These word lists usually contain about a quarter million entries. If the words in the list are in no particular order, then we have no choice but to search through it one item at a time from the beginning, until either we happen to find the word we seek or we reach the end. We previously encountered this algorithm, called a *linear search* (or *sequential search*) because it searches in a linear fashion from beginning to end, and has linear time complexity.

Now let's consider the improvements we can make if the word list has been sorted in alphabetical order, as they always are. At the very least, if we use a linear search on a sorted list, we know that we can abandon the search if we reach a word that is alphabetically after the word we seek. But we can do even better. Think about how we would search a physical, bound dictionary for the word "espresso." Since "E" is in the first half of the alphabet, we might begin by opening the book to a point about 1/4 of the way through. Suppose that, upon doing so, we find ourselves on a page containing words beginning with the letter "G." We would then flip backwards several pages, perhaps finding ourselves on a page on which the last word is "eagle." Next, we would flip a few pages forward, and so on, continuing to hone in on "espresso" until we find it.

Reflection 11.1 *Can we apply this idea to searching in a sorted list of values?*

We can search a list in a similar way, except that we usually do not know much about the distribution of the list's contents, so it is hard to make that first guess about where to start. In this case, the best strategy is to start in the middle. After comparing the item we seek to the middle item, we continue on the half of the list that must contain the item. Because we are effectively dividing the list into two halves each time, this algorithm is called *binary search*.

For example, suppose we wanted to search for the number 70 in the following sorted list of numbers. (We will use numbers instead of words in our example to save space.)

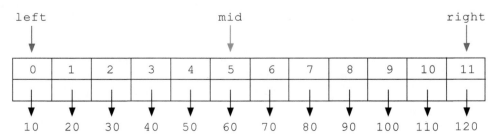

As we hone in on our target, we will update two variables named `left` and `right` to keep track of the first and last indices of the sublist that we are still considering.

> **Box 11.1: Databases**
>
> A database is a structured file (or set of files) that contains a large amount of searchable data. The most common type of database, called a *relational database*, consists of some number of tables. Each row in a table represents one piece of data. Every row has a unique *key* that can be used to search for that row. For example, the tables below represent a small portion of the earthquake data that we worked with in Section 8.4.
>
> Earthquakes
>
> | QuakeID | Latitude | Longitude | Mag | NetID |
> |---|---|---|---|---|
> | nc72076126 | 40.1333 | -123.863 | 1.8 | NC |
> | ak10812068 | 59.8905 | -151.2392 | 2.5 | AK |
> | nc72076101 | 37.3242 | -122.1015 | 1.8 | NC |
> | ci11369570 | 34.3278 | -116.4663 | 1.2 | CI |
> | ci11369562 | 35.0418 | -118.3227 | 1.4 | CI |
> | ci11369546 | 32.0487 | -115.0075 | 3.2 | CI |
>
> Networks
>
> | NetID | NetName |
> |---|---|
> | AK | Alaska Regional |
> | CI | Southern California |
> | NC | Northern California |
> | US | US National |
> | UW | Pacific Northwest |
>
> The table on the left contains information about individual earthquakes, each of which is identified with a `QuakeID` that acts as its key. The last column contains a two-letter network code that identifies the preferred source of information about that earthquake. The table on the right contains the names associated with each two-letter code. The two-letter codes also act as the keys for this table.
>
> Relational databases are queried using a programming language called SQL. A simple SQL query looks like this:
>
> > `select Mag from Earthquakes where QuakeID = 'nc72076101'`
>
> This query is asking for the magnitude (`Mag`), from the `Earthquakes` table, of the earthquake with `QuakeID nc72076101`. The response to this query would be the value `1.8`. Searching a table quickly for a particular key value is facilitated by an *index*. An index is data structure that maps key values to rows in a table (similar to a Python dictionary). The key values in the index can be maintained in sorted order so that any key value, and hence any row, can be found quickly using a binary search. (Database indices are more commonly maintained in practice in a hash table or a specialized data structure called a B-tree.)

In addition, we will maintain a variable named `mid` that is assigned to the index of the middle value of this sublist. (When there are two middle values, we choose the leftmost one.) In each step, we will compare the target item to the item at index `mid`. If the target is equal to this middle item, we return `mid`. Otherwise, we either set `right` to be `mid - 1` (to hone in on the left sublist) or we set `left` to be `mid + 1` (to hone in on the right sublist).

In the list above, we start by comparing the item at index `mid` (60) to our target item (70). Then, because 70 > 60, we decide to narrow our search to the second half of the list. To do this, we assign `left` to `mid + 1`, which is the index of the item immediately after the middle item. In this case, we assign `left` to 5 + 1 = 6, as shown below.

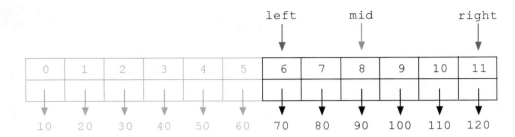

Then we update `mid` to be the index of the middle item in this sublist between `left` and `right`, in this case, 8. Next, since 70 is less than the new middle value, 90, we discard the second half of the sublist by assigning `right` to `mid - 1`, in this case, 8 - 1 = 7, as shown below.

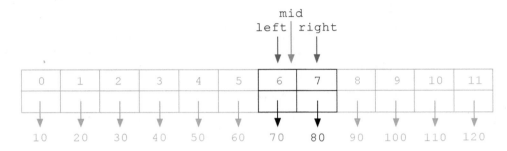

Then we update `mid` to be 6, the index of the "middle" item in this short sublist. Finally, since the item at index `mid` is the one we seek, we return the value of `mid`.

Reflection 11.2 *What would have happened if we were looking for a non-existent number like 72 instead?*

If we were looking for 72 instead of 70, all of the steps up to this point would have been the same, except that when we looked at the middle item in the last step, it would not have been equal to our target. Therefore, picking up from where we left off, we would notice that 72 is greater than our middle item 70, so we update `left` to be the index after `mid`, as shown below.

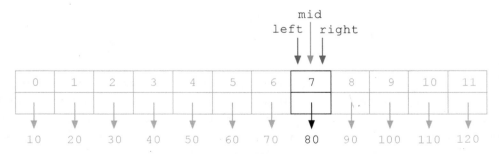

Now, since `left` and `right` are both equal to 7, `mid` must be assigned to 7 as well. Then, since 72 is less than the middle item, 80, we continue to blindly follow the algorithm by assigning `right` to be one less than `mid`.

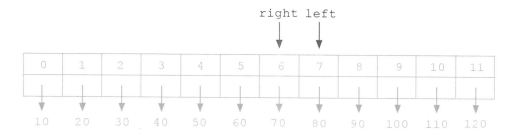

At this point, since `right` is to the left of `left` (i.e., `left > right`), the sublist framed by `left` and `right` is empty! Therefore, 72 must not be in the list, and we return −1.

This description of the binary search algorithm can be translated into a Python function in a very straightforward way:

```
def binarySearch(keys, target):
    """Find the index of target in a sorted list of keys.

    Parameters:
        keys: a list of key values
        target: a value for which to search

    Return value:
        the index of an occurrence of target in keys
    """

    n = len(keys)
    left = 0
    right = n - 1
    while left <= right:
        mid = (left + right) // 2
        if target < keys[mid]:
            right = mid - 1
        elif target > keys[mid]:
            left = mid + 1
        else:
            return mid
    return -1
```

Notice that we have named our list parameter `keys` (instead of the usual `data`) because, in real database applications (see Box 11.1), we typically try to match a target value to a particular feature of a data item, rather than to the entire data item. This particular feature is known as a *key* value. For example, in searching a phone directory, if we enter "Cumberbatch" in the search field, we are not looking for a directory entry (the data item) in which the entire contents contain only the word "Cumberbatch." Instead, we are looking for a directory entry in which just the last name (the key value) matches Cumberbatch. When the search term is found, we return the entire directory entry that corresponds to this key value. In our function, we return the index at which the key value was found which, if we had data associated with the key, might provide us with enough information to find it in an associated data structure. We will look at an example of this in Section 11.2.

The `binarySearch` function begins by initializing `left` and `right` to the first and last indices in the list. Then, while `left <= right`, we compute the new value of `mid` and compare `keys[mid]` to the target we seek. If they are equal, we simply return the value of `mid`. Otherwise, we adjust `left` or `right` to hone in on `target`. If we get to the point where `left > right`, the loop ends and we return −1, having not found `target`.

Reflection 11.3 *Write a* `main` *function that calls* `binarySearch` *with the list that we used in our example. Search for* 70 *and* 72.

Reflection 11.4 *Insert a statement in the* `binarySearch` *function, after* `mid` *is assigned its value, that prints the values of* `left`, `right`, *and* `mid`. *Then search for more target values. Do you see why* `mid` *is assigned to the printed values?*

Efficiency of iterative binary search

How much better is binary search than linear search? When we analyzed the linear search in Section 6.7, we counted the worst case number of comparisons between the target and a list item, so let's perform the same analysis for binary search. Since the binary search contains a `while` loop, we will need to think more carefully this time about when the worst case happens.

Reflection 11.5 *Under what circumstances will the binary search algorithm perform the most comparisons between the target and a list item?*

In the worst case, the `while` loop will iterate all the way until `left > right`. Therefore, the worst case number of item comparisons will be necessary when the item we seek is not found in the list. Let's start by thinking about the worst case number of item comparisons for some small lists. First, suppose we have a list with length $n = 4$. In the worst case, we first look at the item in the middle of this list, and then are faced with searching a sublist with length 2. Next, we look at the middle item of this sublist and, upon not finding the item, search a sublist of length 1. After one final comparison to this single item, the algorithm will return −1. So we needed a total of 3 comparisons for a list of length 4.

Reflection 11.6 *Now what happens if we double the size of the list to $n = 8$?*

After we compare the middle item in a list with length $n = 8$ to our target, we are left with a sublist with length 4. We already know that a list with length 4 requires 3 comparisons in the worst case, so a list with length 8 must require $3 + 1 = 4$ comparisons in the worst case. Similarly, a list with length 16 must require only one

List length n	Worst case comparisons c
1	1
2	2
4	3
8	4
16	5
$\vdots$	$\vdots$
$2^{10} = 1,024$	11
$\vdots$	$\vdots$
$2^{20} \approx 1$ million	21
$\vdots$	$\vdots$
$2^{30} \approx 1$ billion	31

Table 11.1 The worst case number of comparisons for a binary search on lists with increasing lengths.

more comparison than a list with length 8, for a total of 5. And so on. This pattern is summarized in Table 11.1. Notice that a list with over a billion items requires at most 31 comparisons!

Reflection 11.7 *Do you see the pattern in Table 11.1? For list of length n, how many comparisons are necessary in the worst case?*

In each row of the table, the length of the list (n) is 2 raised to the power of 1 less than the number of comparisons (c), or

$$n = 2^{c-1}.$$

Therefore, for a list of size n, the binary search requires

$$c = \log_2 n + 1$$

comparisons in the worst case. So binary search is a *logarithmic-time algorithm*. Since the time complexity of linear search is proportional to n, this means that linear search is *exponentially* slower than binary search.

This is a degree of speed-up that is *only* possible through algorithmic refinement; a faster computer simply cannot have this kind of impact. Figure 11.1 shows a comparison of actual running times of both search algorithms on some small lists. The time required by binary search is barely discernible as the red line parallel to the x-axis. But the real power of binary search becomes evident on very long lists. As suggested by Table 11.1, a binary search takes almost no time at all, even on huge lists, whereas a linear search, which must potentially examine every item, can take a very long time.

548 ■ Organizing data

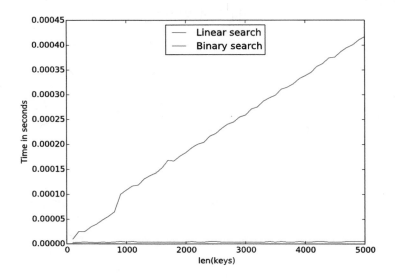

Figure 11.1 A comparison of the execution times of linear search and binary search on small sorted lists.

A spelling checker

Now let's apply our binary search to the spelling checker problem with which we started this section. We can write a program that reads an alphabetized word list, and then allows someone to repeatedly enter a word to see if it is spelled correctly (i.e., is present in the list of words). A list of English words can be found on computers running Mac OS X or Linux in the file /usr/share/dict/words, or one can be downloaded from the book web site. This list is already sorted if you consider an upper case letter to be equivalent to its lower case counterpart. (For example, "academy" directly precedes "Acadia" in this file.) However, as we saw in Chapter 6, Python considers upper case letters to come before lower case letters, so we actually still need to sort the list to have it match Python's definition of "sorted." For now, we can use the sort method; in the coming sections, we will develop our own sorting algorithms. The following function implements our spelling checker.

```
def spellcheck():
    """Repeatedly ask for a word to spell-check and print the result.

    Parameters: none

    Return value: None
    """

    dictFile = open('/usr/share/dict/words', 'r', encoding = 'utf-8')
    wordList = [ ]
```

```
        for word in dictFile:
            wordList.append(word[:-1])
        dictFile.close()
        wordList.sort()

        word = input('Enter a word to spell-check (q to quit): ')
        while word != 'q':
            index = binarySearch(wordList, word)
            if index != -1:
                print(word, 'is spelled correctly.')
            else:
                print(word, 'is not spelled correctly.')
            print()
            word = input('Enter a word to spell-check (q to quit): ')
```

The function begins by opening the word list file and reading each word (one word per line) into a list. Since each line ends with a newline character, we slice it off before adding the word to the list. After all of the words have been read, we sort the list. The following `while` loop repeatedly prompts for a word until the letter `q` is entered. Notice that we ask for a word before the `while` loop to initialize the value of `word`, and then again at the bottom of the loop to set up for the next iteration. In each iteration, we call the binary search function to check if the word is contained in the list. If the word was found (`index != -1`), we state that it is spelled correctly. Otherwise, we state that it is not spelled correctly.

Reflection 11.8 *Combine this function with the* `binarySearch` *function together in a program. Run the program to try it out.*

Recursive binary search

You may have noticed that the binary search algorithm displays a high degree of self-similarity. In each step, the problem is reduced to solving a subproblem involving half of the original list. In particular, the problem of searching for an item between indices `left` and `right` is reduced to the subproblem of searching between `left` and `mid - 1`, or the subproblem of searching between `mid + 1` and `right`. Therefore, binary search is a natural candidate for a recursive algorithm.

Like the recursive linear search from Section 10.4, this function will need two base cases. In the first base case, when the list is empty (`left > right`), we return `-1`. In the second base case, `keys[mid] == item`, so we return `mid`. If neither of these cases holds, we solve one of the two subproblems recursively. If the item we are searching for is less than the middle item, we recursively call the binary search with the `right` argument set to `mid - 1`. Or, if the item we are searching for is greater than the middle item, we recursively call the binary search with the `left` argument set to `mid + 1`.

```
def binarySearch(keys, target, left, right):
    """Recursively find the index of target in a sorted list of keys.

    Parameters:
        keys: a list of key values
        target: a value for which to search

    Return value:
        the index of an occurrence of target in keys
    """

    if left > right:               # base case 1: not found
        return -1
    mid = (left + right) // 2
    if target == keys[mid]:        # base case 2: found
        return mid
    if target < keys[mid]:         # recursive cases
        return binarySearch(keys, target, left, mid - 1)    # 1st half
    return binarySearch(keys, target, mid + 1, right)       # 2nd half
```

Reflection 11.9 *Repeat Reflections 11.3 and 11.4 with the recursive binary search function. Does the recursive version "look at" the same values of* `mid`*?*

Efficiency of recursive binary search

Like the iterative binary search, this algorithm also has logarithmic time complexity. But, as with the recursive linear search, we have to derive this result differently using a recurrence relation. Again, let the function $T(n)$ denote the maximum (worst case) number of comparisons between the target and a list item in a binary search when the length of the list is n. In the function above, there are two such comparisons before reaching a recursive function call. As we did in Section 10.4, we will simply represent the number of comparisons before each recursive call with the constant c. When $n = 0$, we reach a base case with no recursive calls, so $T(0) = c$.

Reflection 11.10 *How many more comparisons are there in a recursive call to* `binarySearch`*?*

Since each recursive call divides the size of the list under consideration by (about) half, the size of the list we are passing into each recursive call is (about) $n/2$. Therefore, the number of comparisons in each recursive call must be $T(n/2)$. The total number of comparisons is then

$$T(n) = T(n/2) + c.$$

Now we can use the same substitution method that we used with recursive linear search to arrive at a closed form expression in terms of n. First, since $T(n) = T(n/2) + c$, we can substitute $T(n)$ with $T(n/2) + c$:

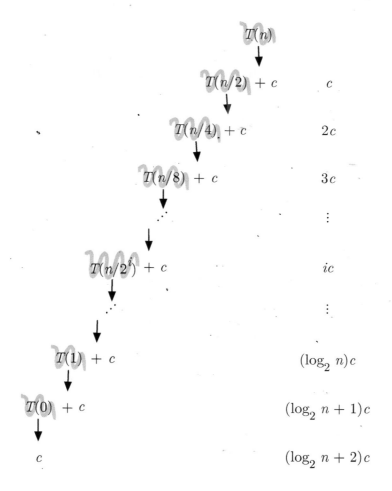

Figure 11.2 An illustration of how to derive a closed form for the recurrence relation $T(n) = T(n/2) + c$.

$$T(n)$$
$$\downarrow$$
$$T(n/2) + c$$

Now we need to replace $T(n/2)$ with something. Notice that $T(n/2)$ is just $T(n)$ with $n/2$ substituted for n. Therefore, using the definition of $T(n)$ above,

$$T(n/2) = T(n/2/2) + c = T(n/4) + c.$$

Similarly,

$$T(n/4) = T(n/4/2) + c = T(n/8) + c$$

and

$$T(n/8) = T(n/8/2) + c = T(n/16) + c.$$

This sequence of substitutions is illustrated in Figure 11.2. Notice that the denominator under the n at each step is a power of 2 whose exponent is the multiplier in front of the accumulated c's at that step. In other words, for each denominator 2^i, the accumulated value on the right is $i \cdot c$. When we finally reach $T(1) = T(n/n)$, the denominator has become $n = 2^{\log_2 n}$, so $i = \log_2 n$ and the total on the right must be $(\log_2 n)c$. Finally, we know that $T(0) = c$, so the total number of comparisons is

$$T(n) = (\log_2 n + 2)\,c.$$

Therefore, as expected, the recursive binary search is a logarithmic-time algorithm.

Exercises

11.1.1. How would you modify each of the binary search functions in this section so that, when a target is not found, the function also prints the values in `keys` that would be on either side of the target if it were in the list?

11.1.2. Write a function that takes a file name as a parameter and returns the number of misspelled words in the file. To check the spelling of each word, use binary search to locate it in a list of words, as we did above. (Hint: use the `strip` string method to remove extraneous punctuation.)

11.1.3. Write a function that takes three parameters—minLength, maxLength, and step—and produces a plot like Figure 11.1 comparing the worst case running times of binary search and linear search on lists with length minLength, minLength + step, minLength + 2 * step, ..., maxLength. Use a slice of the list derived from list(range(maxLength)) as the sorted list for each length. To produce the worst case behavior of each algorithm, search for an item that is not in the list (e.g., −1).

11.1.4. The function below plays a guessing game against the pseudorandom number generator. What is the worst case number of guesses necessary for the function to win the game for any value of `n`, where `n` is a power of 2? Explain your answer.

```
import random

def guessingGame(n):
    secret = random.randrange(1, n + 1)
    left = 1
    right = n
    guessCount = 1
    guess = (left + right) // 2
    while guess != secret:
        if guess > secret:
            right = guess - 1
        else:
            left = guess + 1
        guessCount = guessCount + 1
        guess = (left + right) // 2
    return guessCount
```

11.2 SELECTION SORT

Sorting is a well-studied problem, and a wide variety of sorting algorithms have been designed, including the one used by the familiar `sort` method of the `list` class. We will develop and compare three other common algorithms, named *selection sort*, *insertion sort*, and *merge sort*.

Reflection 11.11 *Before you read further, think about how you would sort a list of data items (names, numbers, books, socks, etc.) in some desired order. Write down your algorithm informally.*

The *selection sort* algorithm is so called because, in each step, it *selects* the next smallest value in the list and places it in its proper sorted position, by swapping it with whatever is currently there. For example, consider the list of numbers [50, 30, 40, 20, 10, 70, 60]. To sort this list in ascending order, the selection sort algorithm first finds the smallest number, 10. We want to place 10 in the first position in the list, so we swap it with the number that is currently in that position, 50, resulting in the modified list

   [10, 30, 40, 20, 50, 70, 60]

Next, we find the second smallest number, 20, and swap it with the number in the second position, 30:

   [10, 20, 40, 30, 50, 70, 60]

Then we find the third smallest number, 30, and swap it with the number in the third position, 40:

   [10, 20, 30, 40, 50, 70, 60]

Next, we find the fourth smallest number, 40. But since 40 is already in the fourth position, no swap is necessary. This process continues until we reach the end of the list.

Reflection 11.12 *Work through the remaining steps in the selection sort algorithm. What numbers are swapped in each step?*

Implementing selection sort

Now let's look at how we can implement this algorithm, using a more detailed representation of this list which, as usual, we will name `data`:

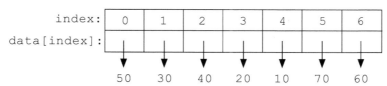

To begin, we want to search for the smallest value in the list, and swap it with the value at index 0. We have already seen how to accomplish key parts of this step. First, writing a function

```
swap(data, i, j)
```

to swap two values in a list was the objective of Exercise 8.2.5. In this `swap` function, `i` and `j` are the indices of the two values in the list `data` that we want to swap. And on Page 354, we developed a function to find the minimum value in a list. The `min` function used the following algorithm:

```
minimum = data[0]
for item in data[1:]:
    if item < minimum:
        minimum = item
```

The variable named `minimum` is initialized to the first value in the list, and updated with smaller values as they are encountered in the list.

Reflection 11.13 *Once we have the final value of* `minimum`, *can we implement the first step of the selection sort algorithm by swapping* `minimum` *with the value currently at index 0?*

Unfortunately, it is not quite that easy. To swap the positions of two values in `data`, we need to know their *indices* rather than their *values*. Therefore, in our selection sort algorithm, we will need to substitute `minimum` with a reference to the index of the minimum value. Let's call this new variable `minIndex`. This substitution results in the following alternative algorithm.

```
n = len(data)
minIndex = 0
for index in range(1, n):
    if data[index] < data[minIndex]:
        minIndex = index
```

Notice that this algorithm is really the same as the previous algorithm; we are just now referring to the minimum value indirectly through its index (`data[minIndex]`) instead of directly through the variable `minimum`. Once we have the index of the minimum value in `minIndex`, we can swap the minimum value with the value at index 0 by calling

```
if minIndex != 0:
    swap(data, 0, minIndex)
```

Reflection 11.14 *Why do we check if* `minIndex != 0` *before calling the* `swap` *function?*

In our example list, these steps will find the smallest value, 10, at index 4, and then call `swap(data, 0, 4)`. We first check if `minIndex != 0` so we do not needlessly swap the value in position 0 with itself. This swap results in the following modified list:

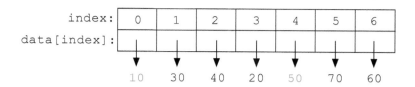

In the next step, we need to do almost exactly the same thing, but for the second smallest value.

Reflection 11.15 *How we do we find the second smallest value in the list?*

Notice that, now that the smallest value is "out of the way" at the front of the list, the second smallest value in `data` must be the smallest value in `data[1:]`. Therefore, we can use exactly the same process as above, but on `data[1:]` instead. This requires only four small changes in the code, marked in red below.

```
minIndex = 1
for index in range(2, n):
    if data[index] < data[minIndex]:
        minIndex = index
if minIndex != 1:
    swap(data, 1, minIndex)
```

Instead of initializing `minIndex` to 0 and starting the `for` loop at 1, we initialize `minIndex` to 1 and start the `for` loop at 2. Then we swap the smallest value into position 1 instead of 0. In our example list, this will find the smallest value in `data[1:]`, 20, at index 3. Then it will call `swap(data, 1, 3)`, resulting in the following modified list:

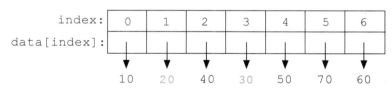

Similarly, the next step is to find the index of the smallest value starting at index 2, and then swap it with the value in index 2:

```
minIndex = 2
for index in range(3, n):
    if data[index] < data[minIndex]:
        minIndex = index
if minIndex != 2:
    swap(data, 2, minIndex)
```

In our example list, this will find the smallest value in `data[2:]`, 30, at index 3. Then it will call `swap(data, 2, 3)`, resulting in:

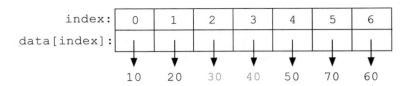

We continue by repeating this sequence of steps, with increasing values of the numbers in red, until we reach the end of the list.

To implement this algorithm in a function that works for any list, we need to situate the repeated sequence of steps in a loop that iterates over the increasing values in red. We can do this by replacing the initial value of `minIndex` in red with a variable named `start`:

```
minIndex = start
for index in range(start + 1, n):
    if data[index] < data[minIndex]:
        minIndex = index
if minIndex != start:
    swap(data, start, minIndex)
```

Then we place these steps inside a `for` loop that has `start` take on all of the integers from 0 to n − 2:

```
 1  def selectionSort(data):
 2      """Sort a list of values in ascending order using the
 3         selection sort algorithm.
 4
 5      Parameter:
 6          data: a list of values
 7
 8      Return value: None
 9      """
10
11      n = len(data)
12      for start in range(n - 1):
13          minIndex = start
14          for index in range(start + 1, n):
15              if data[index] < data[minIndex]:
16                  minIndex = index
17          if minIndex != start:
18              swap(data, start, minIndex)
```

Reflection 11.16 *In the outer* `for` *loop of the* `selectionSort` *function, why is the last value of* `start` *n − 2 instead of n − 1? Think about what steps would be executed if* `start` *were assigned the value n − 1.*

Reflection 11.17 *What would happen if we called* `selectionSort` *with the list* `['dog', 'cat', 'monkey', 'zebra', 'platypus', 'armadillo']`*? Would it work? If so, in what order would the words be sorted?*

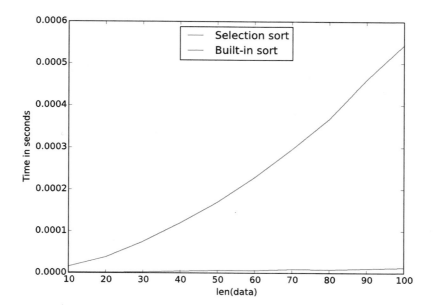

Figure 11.3 A comparison of the execution times of selection sort and the list `sort` method on small randomly shuffled lists.

Because the comparison operators are defined for both numbers and strings, we can use our `selectionSort` function to sort either kind of data. For example, call `selectionSort` on each of the following lists, and then print the results. (Remember to incorporate the `swap` function from Exercise 8.2.5.)

```
numbers = [50, 30, 40, 20, 10, 70, 60]
animals = ['dog', 'cat', 'monkey', 'zebra', 'platypus', 'armadillo']
heights = [7.80, 6.42, 8.64, 7.83, 7.75, 8.99, 9.25, 8.95]
```

Efficiency of selection sort

Next, let's look at the time complexity of the selection sort algorithm. We can derive the asymptotic time complexity by counting the number of times the most frequently executed elementary step executes. The most frequently executed step in the `selectionSort` function is the comparison in the `if` statement on line 15. Since line 15 is in the body of the inner `for` loop (on line 14) and this inner `for` loop iterates a different number of times for each value of `start` in the outer `for` loop (on line 12), we must ask how many times the inner `for` loop iterates for each value of `start`. When `start` is 0, the inner `for` loop on line 14 runs from 1 to $n-1$, for a total of $n-1$ iterations. So line 15 is also executed $n-1$ times. Next, when `start` is 1, the inner `for` loop runs from 2 to $n-1$, a total of $n-2$ iterations, so line 15 is executed $n-2$ times. With each new value of `start`, there is one less iteration of the inner `for` loop. Therefore, the total number of times that line 15 is executed is

$$(n-1) + (n-2) + (n-3) + \cdots$$

Where does this sum stop? To find out, we look at the last iteration of the outer `for` loop, when `start` is `n - 2`. In this case, the inner `for` loop runs from `n - 1` to `n - 1`, for only one iteration. So the total number of steps is

$$(n-1) + (n-2) + (n-3) + \cdots + 3 + 2 + 1.$$

What is this sum equal to? You may recall that we have encountered this sum a few times before (see Box 4.2):

$$(n-1) + (n-2) + (n-3) + \cdots + 3 + 2 + 1 = \frac{n(n-1)}{2} = \frac{1}{2}n^2 - \frac{1}{2}n.$$

Ignoring the constant $1/2$ in front of n^2 and the low order term $(1/2)n$, we find that this expression is asymptotically proportional to n^2; hence selection sort has quadratic time complexity.

Figure 11.3 shows the results of an experiment comparing the running time of selection sort to the `sort` method of the list class. (Exercise 11.3.4 asks you to replicate this experiment.) The parabolic blue curve in Figure 11.3 represents the quadratic time complexity of the selection sort algorithm. The red curve at the bottom of Figure 11.3 represents the running time of the `sort` method. Although this plot compares the algorithms on very small lists on which both algorithms are very fast, we see a marked difference in the growth rates of the execution times. We will see why the `sort` method is so much faster in Section 11.4.

Querying data

Suppose we want to write a program that allows someone to query the USGS earthquake data that we worked with in Section 8.4. Although we did not use IDs then, each earthquake was identified by a unique ID such as `ak10811825`. The first two characters identify the monitoring network (`ak` represents the Alaska Regional Network) and the last eight digits represent a unique code assigned by the network. So our program needs to search for a given ID, and then return the attributes associated with that ID, say the earthquake's latitude, longitude, magnitude, and depth. The earthquake IDs will act as *keys*, on which we will perform our searches. This associated data is sometimes called *satellite data* because it revolves around the key.

To use the efficient binary search algorithm in our program, we need to first sort the data by its key values. When we read this data into memory, we can either read it into parallel lists, as we did in Section 8.4, or we can read it into a table (i.e., a list of lists), as we did in Section 9.1. In this section, we will modify our selection sort algorithm to handle the first option. We will leave the second option as an exercise.

First, we will read the data into two lists: a list of keys (i.e., earthquake IDs) and a list of tuples containing the satellite data. For example, the satellite data for one earthquake that occurred at 19.5223 latitude and -155.5753 longitude with magnitude 1.1 and depth 13.6 km will be stored in the tuple (19.5223, -155.5753, 1.1, 13.6). Let's call these two lists `ids` and `data`, respectively. (We will leave writing the function to read the IDs and data into their respective lists as an exercise.) By design, these two lists are *parallel* in the sense that the satellite data in `data[index]` belongs to the earthquake with ID in `ids[index]`. When we sort the earthquakes by ID, we will need to make sure that the associations with the satellite data are maintained. In other words, if, during the sort of the list `ids`, we swap the values in `ids[9]` and `ids[4]`, we also need to swap `data[9]` and `data[4]`.

To do this with our existing selection sort algorithm is actually quite simple. We pass both lists into the selection sort function, but make all of our sorting decisions based entirely on a list of `keys`. Then, when we swap two values in `keys`, we also swap the values in the same positions in `data`. The modified function looks like this (with changes in red):

```
def selectionSort(keys, data):
    """Sort parallel lists of keys and data values in ascending
       order using the selection sort algorithm.

    Parameters:
        keys: a list of keys
        data: a list of data values corresponding to the keys

    Return value: None
    """

    n = len(keys)
    for start in range(n - 1):
        minIndex = start
        for index in range(start + 1, n):
            if keys[index] < keys[minIndex]:
                minIndex = index
        swap(keys, start, minIndex)
        swap(data, start, minIndex)
```

Once we have the sorted parallel lists `ids` and `data`, we can use binary search to retrieve the index of a particular ID in the list `ids`, and then use that index to retrieve the corresponding satellite data from the list `data`. The following function does this repeatedly with inputted earthquake IDs.

```
def queryQuakes(ids, data):
    key = input('Earthquake ID (q to quit): ')
    while key != 'q':
        index = binarySearch(ids, key, 0, len(ids) - 1)
        if index >= 0:
            print('Location: ' + str(data[index][:2]) + '\n' +
                  'Magnitude: ' + str(data[index][3]) + '\n' +
                  'Depth: ' + str(data[index][2]) + '\n')
        else:
            print('An earthquake with that ID was not found.')
        key = input('Earthquake ID (q to quit): ')
```

A `main` function that ties all three pieces together looks like this:

```
def main():
    ids, data = readQuakes()      # left as an exercise
    selectionSort(ids, data)
    queryQuakes(ids, data)
```

Exercises

11.2.1. Can you find a list of length 5 that requires more comparisons (on line 15) than another list of length 5? In general, with lists of length n, is there a worst case list and a best case list with respect to comparisons? How many comparisons do the best case and worst case lists require?

11.2.2. Now consider the number of swaps. Can you find a list of length 5 that requires more swaps (on line 18) than another list of length 5? In general, with lists of length n, is there a worst case list and a best case list with respect to swaps? How many swaps do the best case and worst case lists require?

11.2.3. The inner `for` loop of the selection sort function can be eliminated by using two built-in Python functions instead:

```
def selectionSort2(data):
    n = len(data)
    for start in range(n - 1):
        minimum = min(data[start:])
        minIndex = start + data[start:].index(minimum)
        if minIndex != start:
            swap(data, start, minIndex)
```

Is this function more or less efficient than the `selectionSort` function we developed above? Explain.

11.2.4. Suppose we already have a list that is sorted in ascending order, and want to insert new values into it. Write a function that inserts an item into a sorted list, maintaining the sorted order, without re-sorting the list.

11.2.5. Write a function that reads earthquake IDs and earthquake satellite data, consisting of latitude, longitude, depth and magnitude, from the data file on the web at

http://earthquake.usgs.gov/earthquakes/feed/v1.0/summary/2.5_month.csv

and returns two parallel lists, as described on Page 559. The satellite data for each earthquake should be stored as a tuple of floating point values. Then use this function to complete a working version of the program whose `main` function is shown on Page 560. (Remember to incorporate the recursive binary search and the `swap` function from Exercise 8.2.5.) Look at the above URL in a web browser to find some earthquake IDs to search for or do the next exercise to have your program print a list of all of them.

11.2.6. Add to the `queryQuakes` function on Page 560 the option to print an alphabetical list of all earthquakes, in response to typing `list` for the earthquake ID. For example, the output should look something like this:

```
Earthquake ID (q to quit): ci37281696
Location: (33.4436667, -116.6743333)
Magnitude: 0.54
Depth: 13.69

Earthquake ID (q to quit): list
    ID              Location            Magnitude   Depth
----------    ----------------------    ---------   -----
ak11406701    (63.2397, -151.4564)         5.5       1.3
ak11406705    (58.9801, -152.9252)        69.2       2.3
ak11406708    (59.7555, -152.6543)        80.0       1.9
    ...

uw60913561    (41.8655, -119.6957)         0.2       2.4
uw60913616    (44.2917, -122.6705)         0.0       1.3

Earthquake ID (q to quit):
```

11.2.7. An alternative to storing the earthquake data in two parallel lists is to store it in one table (a list of lists or a list of tuples). For example, the beginning of a table containing the earthquakes shown in the previous exercise would look like this:

```
[['ak11406701', 63.2397, -151.4564, 5.5, 1.3],
 ['ak11406705', 58.9801, -152.9252, 69.2, 2.3],
 ...
]
```

Rewrite the `readQuakes`, `selectionSort`, `binarySearch`, and `queryQuakes` functions so that they work with the earthquake data stored in this way instead. Your functions should assume that the key value for each earthquake is in column 0. Combine your functions into a working program that is driven by the following `main` function:

```
def main():
    quakes = readQuakes()
    selectionSort(quakes)
    queryQuakes(quakes)
```

11.2.8. The Sieve of Eratosthenes is a simple algorithm for generating prime numbers that has a structure that is similar to the nested loop structure of selection sort. In this algorithm, we begin by initializing a list of n Boolean values named `prime` as follows. (In this case, $n = 12$.)

```
prime = | F | F | T | T | T | T | T | T | T | T | T  | T  |
          0   1   2   3   4   5   6   7   8   9   10   11
```

At the end of the algorithm, we want `prime[index]` to be `False` if `index` is not prime and `True` if `index` is prime. The algorithm continues by initializing a loop `index` variable to 2 (indicated by the arrow below) and then setting the list value of every multiple of 2 to be `False`.

```
| F | F | T | T | F | T | F | T | F | T | F  | T  |
  0   1   2   3   4   5   6   7   8   9   10   11
          ↑
```

Next, the loop index variable is incremented to 3 and, since `prime[3]` is `True`, the list value of every multiple of 3 is set to be `False`.

```
| F | F | T | T | F | T | F | T | F | F | F  | T  |
  0   1   2   3   4   5   6   7   8   9   10   11
              ↑
```

Next, the loop index variable is incremented to 4. Since `prime[4]` is `False`, we do not need to set any of its multiples to `False`, so we do not do anything.

```
| F | F | T | T | F | T | F | T | F | F | F  | T  |
  0   1   2   3   4   5   6   7   8   9   10   11
                  ↑
```

This process continues with the loop index variable set to 5:

```
| F | F | T | T | F | T | F | T | F | F | F  | T  |
  0   1   2   3   4   5   6   7   8   9   10   11
                      ↑
```

And so on. How far must we continue to increment `index` before we know we are done? Once we are done filling in the list, we can iterate over it one more time to build the list of prime numbers, in this case, [2, 3, 5, 7, 11]. Write a function that implements this algorithm to return a list of all prime numbers less than or equal to a parameter n.

11.3 INSERTION SORT

Our second sorting algorithm, named *insertion sort*, is familiar to anyone who has sorted a hand of playing cards. Working left to right through our hand, the insertion sort algorithm inserts each card into its proper place with respect to the previously arranged cards. For example, consider our previous list, arranged as a hand of cards:

We start with the second card to the left, 30, and decide whether it should stay where it is or be inserted to the left of the first card. In this case, it should be inserted to the left of 50, resulting in the following slightly modified ordering:

Then we consider the third card from the left, 40. We see that 40 should be inserted between 30 and 50, resulting in the following order.

Next, we consider 20, and see that it should be inserted all the way to the left, before 30.

564 ■ Organizing data

This process continues with 10, 70, and 60, at which time the hand is sorted.

Implementing insertion sort

To implement this algorithm, we need to repeatedly find the correct location to insert an item among the items to the left, assuming that the items to the left are already sorted. Let's name the index of the item that we wish to insert `insertIndex` and the item itself `itemToInsert`. In other words, we assign

```
itemToInsert = data[insertIndex]
```

To illustrate, suppose that `insertIndex` is 4 (and `itemToInsert` is 10), as shown below:

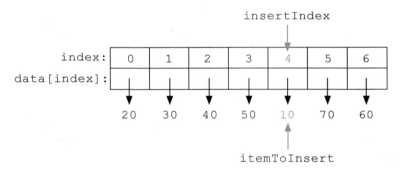

We need to compare `itemToInsert` to each of the items to the left, first at index `insertIndex - 1`, then at `insertIndex - 2`, `insertIndex - 3`, etc. When we come to an item that is less than or equal to `itemToInsert` or we reach the beginning of the list, we know that we have found the proper location for the item. This process can be expressed with a `while` loop:

```
itemToInsert = data[insertIndex]
index = insertIndex - 1
while index >= 0 and data[index] > itemToInsert:
    index = index - 1
```

The variable named `index` tracks which item we are currently comparing to `itemToInsert`. The value of `index` is decremented while it is still at least zero and the item at position `index` is still greater than `itemToInsert`. At the end of the loop, because the `while` condition has become false, either `index` has reached `-1` or `data[index] <= itemToInsert`. In either case, we want to insert `itemToInsert` into position `index + 1`. In the example above, we would reach the beginning of the list, so we want to insert `itemToInsert` into position `index + 1 = 0`.

To actually insert `itemToInsert` in its correct position, we need to delete `itemToInsert` from its current position, and insert it into position `index + 1`.

```
data.pop(insertIndex)
data.insert(index + 1, itemToInsert)
```

In the insertion sort algorithm, we want to repeat this process for each value of `insertIndex`, starting at 1, so we enclose these steps in a `for` loop:

```
def insertionSort(data):
    """Sort a list of values in ascending order using the
       insertion sort algorithm.

    Parameter:
        data: a list of values

    Return value: None
    """

    n = len(data)
    for insertIndex in range(1, n):
        itemToInsert = data[insertIndex]
        index = insertIndex - 1
        while index >= 0 and data[index] > itemToInsert:
            index = index - 1
        data.pop(insertIndex)
        data.insert(index + 1, itemToInsert)
```

Although this function is correct, it performs more work than necessary. To see why, think about how the `pop` and `insert` methods must work, based on the picture of the list on Page 564. First, to delete (`pop`) `itemToInsert`, which is at position `insertIndex`, all of the items to the right, from position `insertIndex + 1` to position `n - 1`, must be shifted one position to the left. Then, to insert `itemToInsert` into position `index + 1`, all of the items to the right, from position `index + 2` to `n - 1`, must be shifted one position to the right. So the items from position `insertIndex + 1` to position `n - 1` are shifted twice, only to end up back where they started.

A more efficient technique would be to only shift those items that need to be shifted, and do so while we are already iterating over them. The following modified algorithm does just that.

```
 1  def insertionSort(data):
 2      """ (docstring omitted) """
 3  
 4      n = len(data)
 5      for insertIndex in range(1, n):
 6          itemToInsert = data[insertIndex]
 7          index = insertIndex - 1
 8          while index >= 0 and data[index] > itemToInsert:
 9              data[index + 1] = data[index]
10              index = index - 1
11          data[index + 1] = itemToInsert
```

The red assignment statement in the `for` loop copies each item at position `index` one position to the right. Therefore, when we get to the end of the loop, position `index + 1` is available to store `itemToInsert`.

Reflection 11.18 *To get a better sense of how this works, carefully work through the steps with the three remaining items to be inserted in the illustration on Page 564.*

Reflection 11.19 *Write a `main` function that calls the `insertionSort` function to sort the list from the beginning of this section:* [50, 30, 40, 20, 10, 70, 60].

Efficiency of insertion sort

Is the insertion sort algorithm any more efficient than selection sort? To discover its time complexity, we first need to identify the most frequently executed elementary step(s). In this case, these appear to be the two assignment statements on lines 9–10 in the body of the `while` loop. However, the most frequently executed step is *actually* the test of the `while` loop condition on line 8. To see why, notice that the condition of a `while` loop is always tested when the loop is first reached, and again after each iteration. Therefore, a `while` loop condition is always tested once more than the number of times the `while` loop body is executed. As we did with selection sort, we will count the number of times the `while` loop condition is tested for each value of the index in the outer `for` loop.

Unlike with selection sort, we need to think about the algorithm's best case and worst case behavior since the number of iterations of the `while` loop depends on the values in the list being sorted.

Reflection 11.20 *What are the minimum and maximum number of iterations executed by the `while` loop for a particular value of `insertIndex`?*

In the best case, it is possible that the item immediately to the left of `itemToInsert` is less than `itemToInsert`, and therefore the condition is tested only once. Therefore, since there are $n - 1$ iterations of the outer `for` loop, there are only $n - 1$ steps total for the entire algorithm. So, in the best case, insertion sort is a linear-time algorithm.

On the other hand, in the worst case, `itemToInsert` may be the smallest item in the list and the `while` loop executes until `index` is less than 0. Since `index` is initialized to `insertIndex - 1`, the condition on line 8 is tested with `index` equal to `insertIndex - 1`, then `insertIndex - 2`, then `insertIndex - 3`, etc., until `index` is -1, at which time the `while` loop ends. In all, this is `insertIndex + 1` iterations of the `while` loop. Therefore, when `insertIndex` is 1 in the outer `for` loop, there are two iterations of the `while` loop; when `insertIndex` is 2, there are three iterations of the `while` loop; when `insertIndex` is 3, there are four iterations of the `while` loop; etc. In total, there are $2 + 3 + 4 + \cdots$ iterations of the `while` loop. As we did with selection sort, we need to figure out when this pattern ends. Since the last value of `insertIndex` is `n - 1`, the `while` loop condition is tested at most `n` times in the last iteration. So the total number of iterations of the `while` loop is

$$2 + 3 + 4 + \cdots + (n-1) + n.$$

Using the same trick we used with selection sort, we find that the total number of steps is

$$2 + 3 + 4 + \cdots + (n-1) + n = \frac{n(n+1)}{2} - 1 = \frac{1}{2}n^2 + \frac{1}{2}n - 1.$$

Ignoring the constants and lower order terms, this means that insertion sort is also a quadratic-time algorithm in the worst case.

So which case is more representative of the efficiency of insertion sort, the best case or the worst case? Computer scientists are virtually always interested in the worst case over the best case because the best case scenario for an algorithm is usually fairly specific and very unlikely to happen in practice. On the other hand, the worst case gives a robust upper limit on the performance of the algorithm. Although more challenging, it is also possible to find the average case complexity of some algorithms, assuming that all possible cases are equally likely to occur.

We can see from Figure 11.4 that the running time of insertion sort is almost identical to that of selection sort in practice. Since the best case and worst case of selection sort are the same, it appears that the worst case analysis of insertion sort is an accurate measure of performance on random lists. Both algorithms are still significantly slower than the built-in `sort` method, and this difference is even more apparent with longer lists. We will see why in the next section.

Exercises

11.3.1. Give a particular 10-element list that requires the worst case number of comparisons in an insertion sort. How many comparisons are necessary for this list?

11.3.2. Give a particular 10-element list that requires the best case number of comparisons in an insertion sort. How many comparisons are necessary for this list?

11.3.3. Write a function that compares the time required to sort a long list of English words using insertion sort to the time required by the `sort` method of the list class.

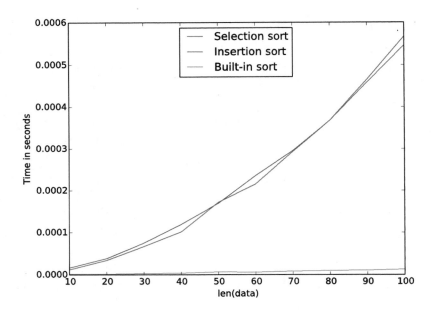

Figure 11.4 A comparison of the execution times of selection sort, insertion sort, and the list `sort` method on small randomly shuffled lists.

We can use the function `time` in the module of the same name to record the time required to execute each function. The `time.time()` function returns the number of seconds that have elapsed since January 1, 1970; so we can time a function by getting the current time before and after we call the function, and finding the difference.

As we saw in Section 11.1, a list of English words can be found on computers running Mac OS X or Linux in the file `/usr/share/dict/words`, or one can be downloaded from the book web site. This list is already sorted if you consider an upper case letter to be equivalent to its lower case counterpart. However, since Python considers upper case letters to come before lower case letters, the list is not sorted for our purposes.

Notes:

- When you read in the list of words, remember that each line will end with a newline character. Be sure to remove that newline character before adding the word to your list.
- Make a separate copy of the original list for each sorting algorithm. If you pass a list that has already been sorted by the `sort` method to your insertion sort, you will not get a realistic time measurement because a sorted list is a best case scenario for insertion sort.

How many seconds did each sort require? (Be patient; insertion sort will take several minutes!) If you can be *really* patient, try timing selection sort as well.

11.3.4. Write a function

   sortPlot(minLength, maxLength, step)

   that produces a plot like Figure 11.4 comparing the running times of insertion sort, selection sort, and the sort method of the list class on shuffled lists with length minLength, minLength + step, minLength + 2 * step, ..., maxLength. At the beginning of your function, produce a shuffled list with length maxLength with

   data = list(range(maxLength))
   random.shuffle(data)

   Then time each function for each list length using a new, unsorted slice of this list.

11.3.5. A sorting algorithm is called *stable* if two items with the same value always appear in the sorted list in the same order as they appeared in the original list. Are selection sort and insertion sort stable sorts? Explain your answer in each case.

11.3.6. A third simple quadratic-time sort is called *bubble sort* because it repeatedly "bubbles" large items toward the end of the list by swapping each item repeatedly with its neighbor to the right if it is larger than this neighbor.

   For example, consider the short list [3, 2, 4, 1]. In the first pass over the list, we compare pairs of items, starting from the left, and swap them if they are out of order. The illustration below depicts in red the items that are compared in the first pass, and the arrows depict which of those pairs are swapped because they are out of order.

   3 2 4 1   2 3 4 1   2 3 4 1   2 3 1 4

   At the end of the first pass, we know that the largest item (in blue) is in its correct location at the end of the list. We now repeat the process above, but stop before the last item.

   2 3 1 4   2 3 1 4   2 1 3 4

   After the second pass, we know that the two largest items (in blue) are in their correct locations. On this short list, we make just one more pass.

   2 1 3 4   1 2 3 4

   After $n - 1$ passes, we know that the last $n - 1$ items are in their correct locations. Therefore, the first item must be also, and we are done.

   1 2 3 4

   Write a function that implements this algorithm.

11.3.7. In the bubble sort algorithm, if no items are swapped during a pass over the list, the list must be in sorted order. So the bubble sort algorithm can be made somewhat more efficient by detecting when this happens, and returning early if it does. Write a function that implements this modified bubble sort algorithm. (Hint: replace the outer `for` loop with a `while` loop and introduce a Boolean variable that controls the `while` loop.)

11.3.8. Write a modified version of the insertion sort function that sorts two parallel lists named `keys` and `data`, based on the values in `keys`, like the parallel list version of selection sort on Page 559.

570 ■ Organizing data

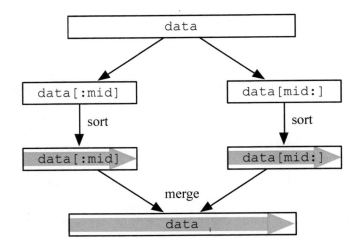

Figure 11.5 An illustration of the merge sort algorithm.

11.4 EFFICIENT SORTING

In the preceding sections, we developed two sorting algorithms, but discovered that they were both significantly less efficient than the built-in **sort** method. The algorithm behind the **sort** method is based on a recursive sorting algorithm called *merge sort*.[1] The merge sort algorithm, like all recursive algorithms, constructs a solution from solutions to subproblems, in this case, sorts of smaller lists. As illustrated in Figure 11.5, the merge sort algorithm divides the original list into two halves, and then recursively sorts each of these halves. Once these halves are sorted, it merges them into the final sorted list.

Merge sort is a *divide and conquer* algorithm, like those you may recall from Section 10.5. Divide and conquer algorithms generally consist of three steps:

1. *Divide* the problem into two or more subproblems.

2. *Conquer* each subproblem recursively.

3. *Combine* the solutions to the subproblems into a solution for the original problem.

Reflection 11.21 *Based on Figure 11.5, what are the divide, conquer, and combine steps in the merge sort algorithm?*

The *divide* step of merge sort is very simple: just divide the list in half. The *conquer* step recursively calls the merge sort algorithm on the two halves. The *combine* step merges the two sorted halves into the final sorted list. This elegant algorithm can be implemented by the following function:

---

[1] The Python sorting algorithm, called *Timsort*, has elements of both merge sort and insertion sort. If you would like to learn more, go to http://bugs.python.org/file4451/timsort.txt

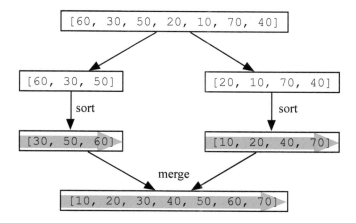

Figure 11.6 An illustration of the merge sort algorithm with the list [60, 30, 50, 20, 10, 70, 40].

```
def mergeSort(data):
    """Recursively sort a list in place in ascending order,
       using the merge sort algorithm.

    Parameter:
        data: a list of values to sort

    Return value: None
    """

    n = len(data)
    if n > 1:
        mid = n // 2                # divide list in half
        left = data[:mid]
        right = data[mid:]
        mergeSort(left)             # recursively sort first half
        mergeSort(right)            # recursively sort second half
        merge(left, right, data)    # merge sorted halves into data
```

Reflection 11.22 *Where is the base case in this function?*

The base case in this function is implicit; when n <= 1, the function just returns because a list containing zero or one values is, of course, already sorted.

All that is left to flesh out mergeSort is to implement the merge function. Suppose we want to sort the list [60, 30, 50, 20, 10, 70, 40]. As illustrated in Figure 11.6, the merge sort algorithm first divides this list into the two sublists left = [60, 30, 50] and right = [20, 10, 70, 40]. After recursively sorting each of these lists, we have left = [30, 50, 60] and right = [10, 20, 40, 70]. Now we want to efficiently merge these two sorted lists into one final sorted list. We

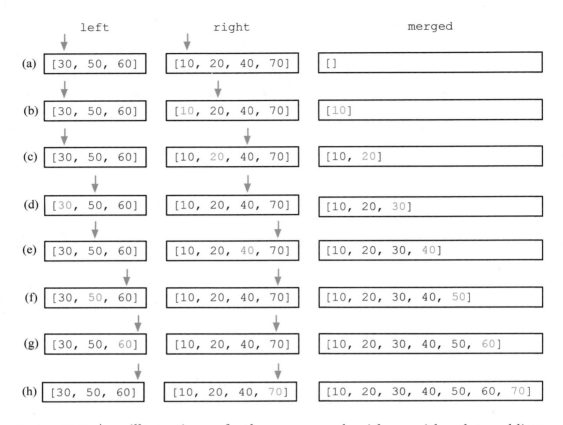

Figure 11.7 An illustration of the merge algorithm with the sublists `left = [60, 30, 50]` and `right = [20, 10, 70, 40]`.

could, of course, concatenate the two lists and then call merge sort with them. But that would be far too much work; because the individual lists are sorted, we can do much better!

Since `left` and `right` are sorted, the first item in the merged list must be the minimum of the first item in `left` and the first item in `right`. So we place this minimum item into the first position in the merged list, and remove it from `left` or `right`. Then the next item in the merged list must again be one of the items at the front of `left` or `right`. This process continues until we run out of items in one of the lists.

This algorithm is illustrated in Figure 11.7. Rather than delete items from `left` and `right` as we append them to the merged list, we will simply maintain an index for each list to remember the next item to consider. The red arrows in Figure 11.7 represent these indices which, as shown in part (a), start at the left side of each list. In parts (a)–(b), we compare the two front items in `left` and `right`, append the minimum (`10` from `right`) to the merged list, and advance the right index. In parts (b)–(c), we compare the first item in `left` to the second item in `right`, again append the minimum (`20` from `right`) to the merged list, and advance the right index. This process continues until, after step (g), when the index in `left` exceeds the length of the list. At this point, we simply extend the merged list with whatever is left over in `right`, as shown in part (h).

Reflection 11.23 *Work through steps (a) through (h) on your own to make sure you understand how the merge algorithm works.*

This merge algorithm is implemented by the following function.

```
1  def merge(left, right, merged):
2      """Merge two sorted lists, named left and right, into
3         one sorted list named merged.
4
5      Parameters:
6          left: a sorted list
7          right: another sorted list
8          merged: the merged sorted list
9
10         Return value: None
11     """
12
13     merged.clear()      # clear contents of merged
14     leftIndex = 0       # index in left
15     rightIndex = 0      # index in right
16
17     while leftIndex < len(left) and rightIndex < len(right):
18         if left[leftIndex] <= right[rightIndex]:
19             merged.append(left[leftIndex])    # left value is smaller
20             leftIndex = leftIndex + 1
21         else:
22             merged.append(right[rightIndex])  # right value is smaller
23             rightIndex = rightIndex + 1
24
25     if leftIndex >= len(left):                # items remaining in right
26         merged.extend(right[rightIndex:])
27     else:                                     # items remaining in left
28         merged.extend(left[leftIndex:])
```

The `merge` function begins by clearing out the contents of the merged list and initializing the indices for the left and right lists to zero. The `while` loop starting at line 17 constitutes the main part of the algorithm. While both indices still refer to items in their respective lists, we compare the items at the two indices and append the smallest to `merged`. When this loop finishes, we know that either `leftIndex >= len(left)` or `rightIndex >= len(right)`. In the first case (lines 25–26), there are still items remaining in `right` to append to `merged`. In the second case (lines 27–28), there are still items remaining in `left` to append to `merged`.

Reflection 11.24 *Write a program that uses the merge sort algorithm to sort the list in Figure 11.6.*

Internal vs. external sorting

We have been assuming that the data that we want to sort is small enough to fit in a list in a computer's memory. The selection and insertion sort algorithms must have the entire list in memory at once because they potentially pass over the entire list in each iteration of their outer loops. For this reason, they are called *internal sorting algorithms*.

But what if the data is larger than the few gigabytes that can fit in memory all at once? This is routinely the situation with real databases. In these cases, we need an *external sorting algorithm*, one that can sort a data set in secondary storage by just bringing pieces of it into memory at a time. The merge sort algorithm is just such an algorithm. In the `merge` function, each of the sorted halves could reside in a file on disk and just the current front item in each half can be read into memory at once. When the algorithm runs out of items in one of the sorted halves, it can simply read in some more. The merged list can also reside in a file on disk; when a new item is added to end of the merged result, it just needs to be written to the merged file after the previous item. Exercise 11.4.7 asks you to write a version of the `merge` function that merges two sorted files in this way.

Efficiency of merge sort

To discover the time complexity of merge sort, we need to set up a recurrence relation like the one we developed for the recursive binary search. We can again let $T(n)$ represent the number of steps required by the algorithm with a list of length n. The `mergeSort` function consists of two recursive calls to `mergeSort`, each one with a list that is about half as long as the original. Like binary search, each of these recursive calls must require $T(n/2)$ comparisons. But unlike binary search, there are two recursive calls instead of one. Therefore,

$$T(n) = 2T(n/2) + ?$$

where the question mark represents the number of steps in addition to the recursive calls, in particular to split `data` into `left` and `right` and to call the `merge` function.

Reflection 11.25 *How many steps must be required to split the list into `left` and `right`?*

Since every item in `data` is copied through slicing to one of the two lists, this must require about n steps.

Reflection 11.26 *How many steps does the `merge` function require?*

Efficient sorting ■ 575

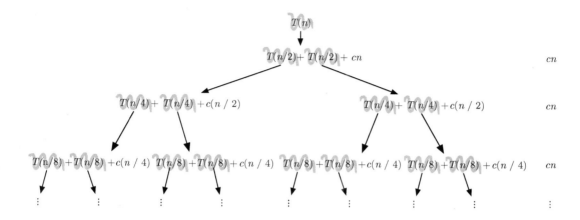

Figure 11.8 An illustration of how to derive a closed form for the recurrence relation $T(n) = 2T(n/2) + cn$.

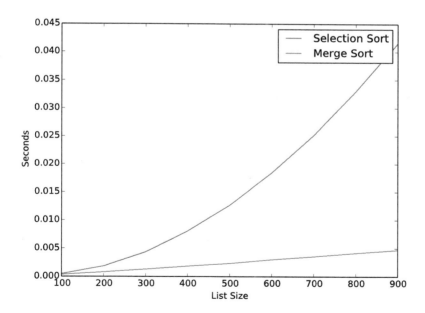

Figure 11.9 A comparison of the execution times of selection sort and merge sort on small randomly shuffled lists.

Because either `leftIndex` or `rightIndex` is incremented in every iteration, the `merge` function iterates at most once for each item in the sorted list. So the total number of steps must be proportional to n. Therefore, the additional steps in the recurrence relation must also be proportional to n. Let's call this number cn, where c is some constant. Then the final recurrence relation describing the number of steps in merge sort is

$$T(n) = 2T(n/2) + cn.$$

Finding a closed form for this recurrence is just slightly more complicated than it was for recursive binary search. Figure 11.8 illustrates the sequence of substitutions. Since there are two substitutions in each instance of the recurrence, the number of terms doubles in each level. But notice that the sum of the leftover values at each level is always equal to cn. As with the binary search recurrence, the size of the list (i.e., the argument of T) is divided in half with each level, so the number of levels is the same as it was with binary search, except that the base case with merge sort happens with $n = 1$ instead of $n = 0$. Looking back at Figure 11.2, we see that we reached $n = 1$ after $\log_2 n + 1$ levels. Therefore, the total number of steps must be

$$T(n) = (\log_2 n + 1)\, cn = cn \log_2 n + cn,$$

which is proportional to $n \log_2 n$ asymptotically.

How much faster is this than the quadratic-time selection and insertion sorts? Figure 11.9 illustrates the difference by comparing the merge sort and selection sort functions on small randomly shuffled lists. The merge sort algorithm is *much* faster. Recall that the algorithm behind the built-in `sort` method is based on merge sort, which explains why it was so much faster than our previous sorts. Exercise 11.4.1 asks you to compare the algorithms on much longer lists as well.

Exercises

11.4.1. To keep things simple, assume that the selection sort algorithm requires exactly n^2 steps and the merge sort algorithm requires exactly $n \log_2 n$ steps. About how many times slower is selection sort than merge sort when $n = 100$? $n = 1000$? $n = 1$ million?

11.4.2. Repeat Exercise 11.3.3 with the merge sort algorithm. How does the time required by the merge sort algorithm compare to that of the insertion sort algorithm and the built-in `sort` method?

11.4.3. Add merge sort to the running time plot in Exercise 11.3.4. How does its time compare to the other sorts?

11.4.4. Our `mergeSort` function is a stable sort, meaning that two items with the same value always appear in the sorted list in the same order as they appeared in the original list. However, if we changed the `<=` operator in line 18 of the `merge` function to a `<` operator, it would no longer be stable. Explain why.

11.4.5. We have seen that binary search is exponentially faster than linear search in the worst case. But is it always worthwhile to use binary search over linear search?

The answer, as is commonly the case in the "real world", is "it depends." In this exercise, you will investigate this question. Suppose we have an unordered list of n items that we wish to search.

(a) If we use the linear search algorithm, what is the time complexity of this search?

(b) If we use the binary search algorithm, what is the time complexity of this search?

(c) If we perform n (where n is also the length of the list) individual searches of the list, what is the time complexity of the n searches together if we use the linear search algorithm?

(d) If we perform n individual searches with the binary search algorithm, what is the time complexity of the n searches together?

(e) What can you conclude about when it is best to use binary search vs. linear search?

11.4.6. Suppose we have a list of n keys that we anticipate needing to search k times. We have two options: either we sort the keys once and then perform all of the searches using a binary search algorithm or we forgo the sort and simply perform all of the searches using a linear search algorithm. Suppose the sorting algorithm requires exactly $n^2/2$ steps, the binary search algorithm requires $\log_2 n$ steps, and the linear search requires n steps. Assume each step takes the same amount of time.

(a) If the length of the list is $n = 1024$ and we perform $k = 100$ searches, which alternative is better?

(b) If the length of the list is $n = 1024$ and we perform $k = 500$ searches, which alternative is better?

(c) If the length of the list is $n = 1024$ and we perform $k = 1000$ searches, which alternative is better?

11.4.7. Write a function that merges two sorted files into one sorted file. Your function should take the names of the three files as parameters. Assume that all three files contain one string value per line. Your function should not use any lists, instead reading only one item at a time from each input file and writing one item at a time to the output file. In other words, at any particular time, there should be at most one item from each file assigned to any variable in your function. You will know when you have reached the end of one of the input files when a read function returns the empty string. There are two files on the book web site that you can use to test your function.

*11.5 TRACTABLE AND INTRACTABLE ALGORITHMS

The merge sort algorithm is much faster than the quadratic-time selection and insertion sort algorithms, but all of these algorithms are considered to be *tractable*, meaning that they can generally be expected to finish in a "reasonable" amount of time. We consider any algorithm with time complexity that is a *polynomial function* of n (e.g., n, n^2, n^5) to be tractable, while algorithms with time complexities that are *exponential* functions of n are *intractable*. An intractable algorithm is essentially

Input size n	Logarithmic $\log_2 n$	Linear n	$n \log_2 n$	Quadratic n^2	Exponential 2^n
10	3×10^{-9} sec	10^{-8} sec	3×10^{-8} sec	10^{-7} sec	10^{-6} sec
20	4×10^{-9} sec	2×10^{-8} sec	9×10^{-8} sec	4×10^{-7} sec	0.001 sec
30	5×10^{-9} sec	3×10^{-8} sec	2×10^{-7} sec	9×10^{-7} sec	1.1 sec
50	6×10^{-9} sec	5×10^{-8} sec	3×10^{-7} sec	3×10^{-6} sec	13 days
100	7×10^{-9} sec	10^{-7} sec	7×10^{-7} sec	10^{-5} sec	41 trillion yrs
1,000	10^{-8} sec	10^{-6} sec	10^{-5} sec	0.001 sec	3.5×10^{284} yrs
10,000	1.3×10^{-8} sec	10^{-5} sec	0.0001 sec	0.1 sec	6.3×10^{2993} yrs
100,000	1.7×10^{-8} sec	10^{-4} sec	0.0016 sec	10 sec	3.2×10^{30086} yrs
10^6	2×10^{-8} sec	0.001 sec	0.02 sec	17 min	?
10^9	3×10^{-8} sec	1 sec	30 sec	31 yrs	?

Table 11.2 A comparison of the approximate times required by algorithms with various time complexities on a computer capable of executing 1 billion steps per second.

useless with all but the smallest inputs; even if it is correct and will eventually finish, the answer will come long after anyone who needs it is dead and gone.

To motivate this distinction, Table 11.2 shows the execution times of five algorithms with different time complexities, on a hypothetical computer capable of executing one billion operations per second. We can see that, for n up to 30, all five algorithms complete in about a second or less. However, when $n = 50$, we start to notice a dramatic difference: while the first four algorithms still execute very quickly, the exponential-time algorithm requires 13 days to complete. When $n = 100$, it requires *41 trillion years*, about 3,000 times the age of the universe, to complete. And the differences only get more pronounced for larger values of n; the first four, tractable algorithms finish in a "reasonable" amount of time, even with input sizes of 1 billion,[2] while the exponential-time algorithm requires an absurd amount of time. *Notice that the difference between tractable and intractable algorithms holds no matter how fast our computers get;* a computer that is one billion times faster than the one used in the table can only bring the exponential-time algorithm down to 10^{275} years from 10^{284} years when $n = 1,000$!

The dramatic time differences in Table 11.2 also illustrate just how important efficient algorithms are. In fact, advances in algorithm efficiency are often considerably more impactful than faster hardware. Consider the impact of improving an algorithm from quadratic to linear time complexity. On an input with n equal to 1 million, we would see execution time improve from 16.7 minutes to 1/1000 of a second, a factor of one million! According to *Moore's Law* (see Box 11.2), such an increase in hardware performance will not manifest itself for about 30 more years!

---

[2]Thirty-one years is admittedly a long time to wait, and no one would actually wait that long, but it is far shorter than 41 trillion years.

> **Box 11.2: Moore's Law**
>
> Moore's Law is named after Gordon Moore, a co-founder of the Intel Corporation. In 1965, he predicted that the number of fundamental elements (transistors) on computer chips would double every year, a prediction he revised to every 2 years in 1975. It was later noted that this would imply that the performance of computers would double every 18 months. This prediction has turned out to be remarkably accurate, although it has started to slow down a bit as we run into the limitations of current technology. We have witnessed similar exponential increases in storage and network capacity, and the density of pixels in computer displays and digital camera sensors.
>
> Consider the original Apple IIe computer, released in 1983. The Apple IIe used a popular microprocessor called the 6502, created by MOS Technology, which had 3,510 transistors with features as small as 8 μm and ran at a clock rate of 1 MHz. Current Apple Macintosh and Windows computers use Intel Core processors. The recent Intel Core i7 (Haswell) quad-core processor contains 1.4 billion transistors with features as small as 22 nm, running at a maximum clock rate of 3.5 GHz. In 30 years, the number of transistors has increased by a factor of almost 400,000, the transistors have become 350 times smaller, and the clock rate has increased by a factor of 3,500.

Hard problems

Unfortunately, there are many common, simply stated problems for which the only known algorithms have exponential time complexity. For example, suppose we have n tasks to complete, each with a time estimate, that we wish to delegate to two assistants as evenly as possible. In general, there is no known way to solve this problem that is any better than trying all possible ways to delegate the tasks and then choosing the best solution. This type of algorithm is known as a *brute force* or *exhaustive search* algorithm. For this problem, there are two possible ways to assign each task (to one of the two assistants), so there are

$$\underbrace{2 \cdot 2 \cdot 2 \cdot \ldots \cdot 2}_{n \text{ tasks}} = 2^n$$

possible ways to assign the n tasks. Therefore, the brute force algorithm has exponential time complexity. Referring back to Table 11.2, we see that even if we only have $n = 50$ tasks, the brute force algorithm would be of no use to us.

It turns out that there are thousands of such problems, known as the *NP-hard problems*, that, on the surface, do not seem as if they should be intractable, but, as far as we know, they are. Even more interesting, no one has actually *proven* that they are intractable. See Box 11.3 if you would like to know more about this fascinating problem.

When we cannot solve a problem exactly, one common approach is to instead use a *heuristic*. A heuristic is a type of algorithm that does not necessarily give a correct answer, but tends to work well in practice. For example, a heuristic for the task delegation problem might assign the tasks in order, always assigning the next

> **Box 11.3: Does P = NP?**
>
> The *P=NP problem* is the most important unresolved question in computer science. Formally, the question involves the tractability of *decision problems*, problems with a "yes" or "no" answer. For example, the decision version of the task delegation problem would ask, "Is there is a task allocation that is within m minutes of being even?" Problems like the original task delegation problem that seek the value of an optimal solution are called *optimization problems*.
>
> The "P" in "P=NP" represents the class of all tractable decision problems (for which algorithms with **P**olynomial time complexity are known). "NP" (short for "**N**ondeterministic **P**olynomial") represents the class of all decision problems for which a candidate solution can be *verified* to be correct in polynomial time. For example, in the task delegation problem, a verifier would check whether a task delegation is within m minutes of being even. (This is a much easier problem!) Because any problem that can be solved in polynomial time must have a polynomial time verifier, we know that a problem in the class P is also contained in the class NP.
>
> The *NP-complete* problems are the hardest problems in NP. (NP-hard problems are usually optimization versions of NP-complete problems.) "P=NP" refers to the question of whether all of the problems in NP are also in P or have polynomial time algorithms. Since so many brilliant people have worked on this problem for over 40 years, it is assumed that no polynomial-time algorithms exist for NP-hard problems.

task to the assistant with the least to do so far. Although this will not necessarily give the best solution, it may be "good enough" in practice.

11.6 SUMMARY

Sorting and searching are perhaps the most fundamental problems in computer science for good reason. We have seen how simply sorting a list can *exponentially* decrease the time it takes to search it, using the *binary search algorithm*. Since binary search is one of those algorithms that "naturally" exhibits self-similarity, we designed both iterative and recursive algorithms that implement the same idea. We also designed two basic sorting algorithms named *selection sort* and *insertion sort*. Each of these algorithms can sort a short list relatively quickly, but they are both very inefficient when it comes to larger lists. By comparison, the recursive *merge sort* algorithm is very fast. Merge sort has the added advantage of being an *external sorting algorithm*, meaning we can adapt it to sort very large data sets that cannot be brought into a computer's memory all at once.

Although the selection and insertion sort algorithms are quite inefficient compared to merge sort, they are still *tractable*, meaning that they will finish in a "reasonable" amount of time. In fact, all algorithms with time complexities that are polynomial functions of their input sizes are considered to be tractable. On the other hand, exponential-time algorithms are called *intractable* because even when their input sizes are relatively small, they require eons to finish.

11.7 FURTHER DISCOVERY

This chapter's epigraph is from an interview given by Marissa Mayer to the *Los Angeles Times* in 2008 [16].

The subjects of this chapter are fundamental topics in second-semester computer science courses, and there are many books available that cover them in more detail. A higher-level overview of some of the tricks used to make searching fast can be found in John MacCormick's *Nine Algorithms that Changed the Future* [30].

11.8 PROJECTS

Project 11.1 Creating a searchable database

Write a program that allows for the interactive search of a data set, besides the earthquake data from Section 11.2, downloaded from the web or the book web site. It may be data that we have worked with elsewhere in this book or it may be a new data set. Your program, at a minimum, should behave like the program that we developed in Section 11.2. In particular, it should:

1. Read the data into a table or parallel lists.

2. Sort the data by an appropriate key value.

3. Interactively allow someone to query the data (by a key value). When a key value is found, the program should print the satellite data associated with that key. Use binary search to search for the key and return the results.

Project 11.2 Binary search trees

In this project, you will write a program that allows for the interactive search of a data set using an alternative data structure called a *binary search tree (BST)*. Each key in a binary search tree resides in a *node*. Each node also contains references to a *left child* node and a *right child* node. The nodes are arranged so that the key of the left child of a node is less than or equal to the key of the node and the key of the right child is greater than the key of the node. For example, Figure 11.10 shows a binary search tree containing the numbers that we sorted in previous sections. The node at the top of the tree is called the tree's *root*.

To insert a new item into a binary search tree, we start at the root. If the new item is less than or equal to the root, we next look at the left child. On the other hand, if the new item is greater than the root, we next look at the right child. Then we repeat this step on the next node, and continue until we arrive at a position without a node. Figure 11.11 illustrates how the value 35 would be inserted into the binary search tree in Figure 11.10. Starting at the root, since 35 < 50, we move

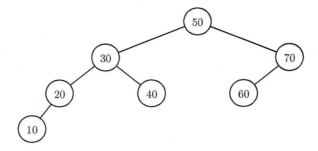

Figure 11.10 A binary search tree.

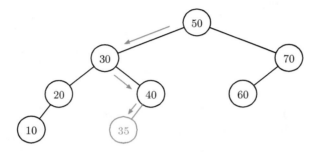

Figure 11.11 Insertion into a binary search tree.

to the left. Next, since 35 > 30, we next move to the right. Then, since 35 < 40, we move to the left. Since there is no node in this position, we create a new node with 35 and insert it there, as the left child of the node containing 40.

Question 11.2.1 *How would the values 5, 25, 65, and 75 be inserted into the binary search tree in Figure 11.10?*

Question 11.2.2 *Does the order in which items are inserted affect what the tree looks like? After the four values in the previous question are inserted into the binary search tree, insert the value 67. Would the binary search tree be different if 67 were inserted before 65?*

Searching a binary search tree follows the same process, except that we check whether the target value is equal to the key in each node that we visit. If it is, we return success. Otherwise, we move to the left or right, as we did above. If we eventually end up in a position without a node, we know that the target value was not found. For example, if we want to search for 20 in the binary search tree in Figure 11.10, we would start at the root and first move left because 20 < 50. Then we move left again because 20 < 30. Finally, we return success because we found our target. If we were searching for 25 instead, would have moved right when we arrived at node 20, but finding no node there, we would have returned failure.

Question 11.2.3 *What nodes would be visited in searches for 10, 25, 55, and 60 in the binary search tree in Figure 11.10?*

In Python, we can represent a node in a binary search tree with a three-item list. As illustrated below, the first item in the list is the key, the second item is a list representing the left child node, and the third item is a list representing the right child node.

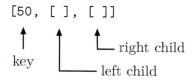

The list above represents a single node with no left or right child. Or, equivalently, we can think of the two empty lists as representing "empty" left and right children. To insert a child, we simply insert into one of the empty lists the items representing the desired node. For example, to make 70 the right child of the node above, we would insert a new node containing 70 into the second list:

[50, [ ], [70, [ ], [ ]]]

To insert 60 as the left child of 70, we would insert a new node containing 60 into the first list in 70:

[50, [ ], [70, [60, [ ], [ ]], [ ]]]

The list above now represents the root and the two nodes to the right of the root in Figure 11.10. Notice that an entire binary search tree can be represented by its root node. The complete binary search tree in Figure 11.10 looks like this:

```
bst = [50, [30, [20, [10, [], []], []], [40, [], []]],
       [70, [60, [], []], []]]
```

Question 11.2.4 *Parse the list above to understand how it represents the binary search tree in Figure 11.10.*

This representation quickly becomes difficult to read. But, luckily, we will rely on our functions to read them instead of us.

Let's now implement the insert and search algorithms we discussed earlier, using this list implementation. To make our code easier to read, we will define three constant values representing the indices of the key, left child, and right child in a node:

```
KEY = 0
LEFT = 1
RIGHT = 2
```

So if `node` is the name of a binary search tree node, then `node[KEY]` is the node's key value, `node[LEFT]` is the node's left child, and `node[RIGHT]` is the node's right child.

The following function inserts a new node into a binary search tree:

```
def insert(root, key):
    """Insert a new key into the BST with the given root.

    Parameters:
        root: the list representing the BST
        key: the key value to insert

    Return value: None
    """

    current = root
    while current != [ ]:
        if key <= current[KEY]:
            current = current[LEFT]
        else:
            current = current[RIGHT]
    current.extend([key, [ ], [ ]])
```

The variable named `current` keeps track of where we are in the tree during the insertion process. The `while` loop proceeds to "move" `current` left or right until `current` reaches an empty node. At that point, the loop ends, and the algorithm inserts a new node containing `key` by inserting `key` and two empty lists into the empty list assigned to `current`. (Recall that the `extend` method effectively appends each item in its list argument to the end of the list.) To use this function to insert the value 35 into our binary search tree named `bst` above, as in Figure 11.11, we would call `insert(bst, 35)`.

The function to search a binary search tree is very similar:

```
def search(root, key):
    """Search for a target key in the BST with the given root.

    Parameters:
        root: the list representing the BST
        key: the key value to search for

    Return value: a Boolean indicating whether key was found
    """

    current = root
    while current != [ ] and current[KEY] != key:
        if key < current[KEY]:
            current = current[LEFT]
        else:
            current = current[RIGHT]
    return current != [ ]
```

The only differences in the `search` function are (a) the loop now also ends if we find the desired `key` value in the node assigned to `current`, and (b) at the end of the loop, we return `False` (failure) if `current` ends at an empty node and `True` otherwise.

In this project, you will work with a data set of your choice, downloaded from the web or the book web site. It may be data that we have worked with elsewhere in this book or it may be a new data set. Your data must contain two or more attributes per entry, one of which will be an appropriate key value. The remaining attributes will constitute the satellite data associated with the entry.

Part 1: Extend the BST implementation

Extend the list representation of a node so that each node can store both a key and associated satellite data. Think about how your new design will affect the `insert` and `search` functions. Then modify these functions so that they work with your new representation.

With this extension, you are actually creating a new data structure that implements a dictionary abstract data type. The dictionary abstract data type in Python defines a way to insert (key, value) pairs and retrieve the value associated with a key using indexing. As explained in Box 8.2, Python dictionaries are usually implemented using a data structure called a hash table. In your extended binary search tree data structure, the `insert` function will insert a new (key, value) pair into the dictionary and the `search` function will return the value associated with `key`, if it is found, or `None` if it is not. We will revisit a dictionary implementation in Section 13.6.

Part 2: Read the data

Write a function that creates an empty binary search tree, reads your data from the web or a file, and then inserts each entry into the binary search tree.

Part 3: Allow queries

Write a function, like the `queryQuakes` function from Section 11.2, that allows someone to interactively query your data. The function should prompt for a key value, and then print the associated satellite data. To locate this data, search for the key in your binary search tree.

Question 11.2.5 *Is searching a binary search tree as efficient as using the binary search algorithm to search in a sorted list? In what situations might a binary search tree not be as efficient? Explain your answers.*

Write a `main` function that puts all of the pieces together to create a program that reads your data set and allows repeated queries of the data.

Part 4: Recursion

Every node in a binary search tree is the root of a `subtree`. In this way, binary search trees exhibit self-similarity. The subtrees rooted by a node's left and right children are called the node's *left subtree* and *right subtree*, respectively. Exploiting this self-similarity, we can think about inserting into (or searching) a binary search tree with root r as recursively solving one of two subproblems: inserting into the left subtree of r or inserting into the right subtree of r. Write recursive versions of the `insert` and `search` functions that use this self-similarity.

Part 5: Sorting

Once data is in a binary search tree, we have a lot of information about how it is ordered. We can use this structure to create a sorted list of the data. Notice that a sorted list of the keys in a binary search tree consists of a sorted list of the keys in the left subtree, followed by the root of the tree, followed by a sorted list of the keys in the right subtree. Using this insight, write a recursive function `bstSort(root)` that returns a sorted list of the keys in a binary search tree. Then use this function to add an option to your query function from Part 3 that prints the list of keys when requested.

Question 11.2.6 *How efficient do you think this sorting algorithm is? How do you think it compares to the sorting algorithms we discussed in this chapter? (Remember to take into account the time it takes to insert the keys into the binary search tree.)*

CHAPTER 12

Networks

> Fred Jones of Peoria, sitting in a sidewalk cafe in Tunis and needing a light for his cigarette, asks the man at the next table for a match. They fall into conversation; the stranger is an Englishman who, it turns out, spent several months in Detroit studying the operation of an interchangeable-bottlecap factory. "I know it's a foolish question," says Jones, "but did you ever by any chance run into a fellow named Ben Arkadian? He's an old friend of mine, manages a chain of supermarkets in Detroit..."
>
> "Arkadian, Arkadian," the Englishman mutters. "Why, upon my soul, I believe I do! Small chap, very energetic, raised merry hell with the factory over a shipment of defective bottlecaps."
>
> "No kidding!" Jones exclaims in amazement.
>
> "Good lord, it's a small world, isn't it?"
>
> Stanley Milgram
> *The Small-World Problem (1967)*

WHAT do Facebook, food webs, the banking system, and our brains all have in common? They are all *networks*: systems of interconnected units that exchange information over the links between them. There are networks all around us: social networks, road networks, protein interaction networks, electrical transmission networks, the Internet, networks of seismic faults, terrorist networks, networks of political influence, transportation networks, and semantic networks, to name a few.

The continuous and dynamic local interactions in large networks such as these make them extraordinarily complex and hard to predict. Learning more about networks can help us combat disease, terrorism, and power outages. Realizations that some networks are *emergent* systems that develop global behaviors based on local interactions have improved our understanding of insect colonies, urban planning, and even our brains. Too little understanding of networks has had unfortunate consequences, such as when invasive species have been introduced into poorly understood ecological networks.

As with the other types of "big data," we need computers and algorithms to understand large and complex networks. In this chapter, we will begin by discussing

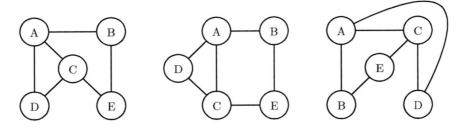

Figure 12.1   Three representations of the same graph.

how we can represent networks in algorithms so that we can analyze them. Then we will develop an algorithm to find the distance between any two nodes in a network. Recent discoveries have shown that many real networks exhibit a "small-world property," meaning that the average distance between nodes is relatively small. In later sections, we will computationally investigate the characteristics of small-world networks and their ramifications for solving real problems.

12.1   MODELING WITH GRAPHS

We model networks with a mathematical structure called a *graph*. A graph consists of a set of *nodes* (or *vertices*) and a set of *links* (or *edges*) that connect pairs of nodes. Nodes are usually drawn as circles (or another shape) and links are drawn as lines between them, as in Figure 12.1. If two nodes are connected by a link, we say they are *adjacent*. A graph is completely represented by its nodes and links. Although we often draw graphs to gain insight into their structure, the placement of the nodes and links on the page is arbitrary. In other words, there can be many different visual representations of the same graph. For example, all three of the graphs in Figure 12.1 are equivalent.

A social network (like Facebook or LinkedIn) can be represented by a graph in which the nodes are people and the links represent relationships (e.g., friends, connections, circles, followers). For example, in the social network in Figure 12.2, Caroline has three friends: Amelia, Lillian, and Nick. In a neural network, the nodes represent neurons and the links represent axons that transmit nerve impulses between the neurons. Figure 12.3 represents the interconnections between neurons in one of the simple neural networks that control digestion in the guts of arthropods. In a graph representing a power grid, like that in Figure 12.4, the nodes represent power stations and the links represent high-voltage transmission lines connecting them.

In an algorithm, a graph is usually represented in one of two ways. The first method is called an *adjacency matrix*. An adjacency matrix contains a row and a column for every node in the network. A one in a matrix entry represents a link between the nodes in the corresponding row and column. A zero means that there

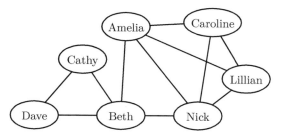

Figure 12.2 A small social network.

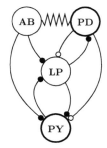

Figure 12.3 A model of the pyloric central pattern generator that controls stomach motion in lobsters.

Figure 12.4 The electrical transmission network in New Zealand. [58]

is no link. An adjacency matrix for the network in Figure 12.2 can be represented by the following table.

	Amelia	Beth	Caroline	Cathy	Dave	Lillian	Nick
Amelia	0	1	1	0	0	1	1
Beth	1	0	0	1	1	0	1
Caroline	1	0	0	0	0	1	1
Cathy	0	1	0	0	1	0	0
Dave	0	1	0	1	0	0	0
Lillian	1	0	1	0	0	0	1
Nick	1	1	1	0	0	1	0

For example, the first row in the adjacency matrix indicates that Amelia is only connected to Beth, Caroline, Lillian, and Nick. In Python, we would represent this matrix with the following nested list.

```
graph = [[0, 1, 1, 0, 0, 1, 1],
         [1, 0, 0, 1, 1, 0, 1],
         [1, 0, 0, 0, 0, 1, 1],
         [0, 1, 0, 0, 1, 0, 0],
         [0, 1, 0, 1, 0, 0, 0],
         [1, 0, 1, 0, 0, 0, 1],
         [1, 1, 1, 0, 0, 1, 0]]
```

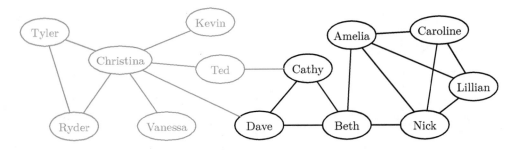

Figure 12.5 An expanded social network. The nodes and links in red are additions to the graph in Figure 12.2.

Although the nodes' labels are not stored in the adjacency matrix itself, they could be stored separately as strings in a list. The index of each string in the list should equal the row and column of the corresponding node in the adjacency matrix.

Reflection 12.1 *Create an adjacency matrix for the graph in Figure 12.1. (Remember that all three pictures depict the same graph.)*

The alternative representation, which we will use in this chapter, is an *adjacency list*. An adjacency list, which is actually a collection of lists, contains, for each node, a list of nodes to which it is connected. In Python, an adjacency list can be stored as a dictionary. For the network in Figure 12.2, this would look like the following:

```
graph = {'Amelia': ['Beth', 'Caroline', 'Lillian', 'Nick'],
         'Beth': ['Amelia', 'Cathy', 'Dave', 'Nick'],
         'Caroline': ['Amelia', 'Lillian', 'Nick'],
         'Cathy': ['Beth', 'Dave'],
         'Dave': ['Beth', 'Cathy'],
         'Lillian': ['Amelia', 'Caroline', 'Nick'],
         'Nick': ['Amelia', 'Beth', 'Caroline', 'Lillian']}
```

Each key in this dictionary, a string, represents a node, and each corresponding value is a list of strings representing the nodes to which the key node is connected. Notice that, if two nodes are connected, that information is stored in both nodes' lists. For example, there is a link connecting Amelia and Beth, so Beth is in Amelia's list and Amelia is in Beth's list.

Reflection 12.2 *Create an adjacency list for the graph in Figure 12.1.*

Making friends

Social networking sites often have an eerie ability to make good suggestions about who you should add to your list of "connections" or "friends." One way they do this is by examining the connections of your connections (or "friends-of-friends"). For example, consider the expanded social network graph in Figure 12.5. Dave currently has only three friends. But his friends have an additional seven friends that an algorithm could suggest to Dave.

Reflection 12.3 *Who are the seven friends-of-friends of Dave in Figure 12.5?*

In graph terminology, the connections of a node are called the node's *neighborhood*, and the size of a node's neighborhood is called its *degree*. In the graph in Figure 12.5, Dave's neighborhood contains Beth, Cathy and Christina, and therefore his degree is three.

Reflection 12.4 *How can you compute the degree of a node from the graph's adjacency matrix? What about from the graph's adjacency list?*

Once we have a network in an adjacency list, writing an algorithm to collect new friend suggestions is relatively easy. The function below iterates over the neighbors of the node for which we would like suggestions and then, for each of these neighbors, iterates over the neighbors' neighbors.

```
1  def friendsOfFriends(network, node):
2      """Find new neighbors-of-neighbors of a node in a network.
3
4      Parameters:
5          network: a graph represented by a dictionary
6          node: a node in the network
7
8      Return value: a list of new neighbors-of-neighbors of node
9      """
10
11     suggestions = [ ]
12     neighbors = network[node]
13     for neighbor in neighbors:                    # neighbors of node
14         for neighbor2 in network[neighbor]:       # neighbors of neighbors
15             if neighbor2 != node and \
16                neighbor2 not in neighbors and \
17                neighbor2 not in suggestions:
18                 suggestions.append(neighbor2)
19     return suggestions
```

On line 12, `network[node]` is the list of nodes to which `node` is connected in the adjacency list named `network`. We assign this list to `neighbors`, and then iterate over it on line 13. On line 14, in the inner `for` loop, we then iterate over the list of each neighbors' neighbors. In the `if` statement, we choose `suggestions` to be a list of unique neighbors-of-neighbors that are not the `node` itself or neighbors of the `node`.

Reflection 12.5 *Look carefully at the three-part* `if` *statement in the function above. How does each part contribute to the desired characteristics of* `suggestions` *listed above?*

Reflection 12.6 *Insert the additional nodes and links from Figure 12.5 (in red) into the dictionary on Page 590. Then call the* `friendsOfFriends` *function with this graph to find new friend suggestions for Dave.*

In the next section, we will design an algorithm to find paths to nodes that are farther away. The ability to compute the distance between nodes will also allow us to better characterize and understand large networks.

Exercises

12.1.1. Besides those presented in this section, describe three other examples of networks.

12.1.2. Draw the network represented by the following adjacency matrix.

	A	B	C	D	E
A	0	1	1	0	0
B	1	0	0	1	1
C	1	0	0	0	1
D	0	1	0	0	1
E	0	1	1	1	0

12.1.3. Draw the network represented by the following adjacency list.

```
graph = {'A': ['C', 'D', 'F'],
         'B': ['C', 'E'],
         'C': ['A', 'B', 'D'],
         'D': ['A', 'C'],
         'E': ['B', 'F'],
         'F': ['A', 'E']}
```

12.1.4. Show how the following network would be represented in Python as

(a) an adjacency matrix

(b) an adjacency list

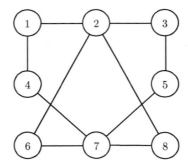

12.1.5. What is the neighborhood of each of the nodes in the network from Exercise 12.1.4?

12.1.6. What is the degree of each node in the network from Exercise 12.1.4? Which node(s) have the maximum degree?

12.1.7. Are the networks in each of the following pairs the same or different? Why?

(a)

(b)

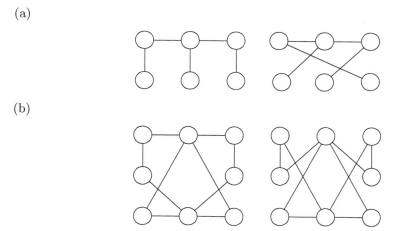

12.1.8. A graph can be represented in a file by listing one link per line, with each link represented by a pair of nodes. For example, the graph below is represented by the file on the right. Write a function that reads such a file and returns an adjacency list (as a dictionary) for the graph. Notice that, for each line A B in the file, your function will need to insert node B into the list of neighbors of A *and* insert node A into the list of neighbors of B.

graph.txt

A B
A C
A D
B E
C D
C E

12.1.9. In this chapter, we are generally assuming that all graphs are *undirected*, meaning that each link represents a mutual relationship between two nodes. For example, if there is a link between nodes A and B, then this means that A is friends with B *and* B is friends with A, or that one can travel from city A to city B *and* from city B to city A. However, the relationships between nodes in some networks are not mutual or do not exist in both directions. Such a network is more accurately represented by a *directed graph* (or *digraph*), in which links are directed from one node to another. In a picture, the directions are indicated arrows. For example, in the directed graph below, one can go directly from node A to node B, but not vice versa. However, one can go in both directions between nodes B and E.

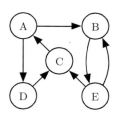

digraph.txt

A B
A D
B E
C A
D C
E B
E C

(a) Give three examples of networks that are better represented by a directed graph.

(b) How would an adjacency list representation of a directed graph differ from that of an undirected graph?

(c) Write a function that reads a file representing a directed graph (see the example above), and returns an adjacency list (as a dictionary) representing that directed graph.

12.1.10. Write a function that returns the maximum degree in a network represented by an adjacency list (dictionary).

12.1.11. Write a function that returns the average degree in a network represented by an adjacency list (dictionary).

12.2 SHORTEST PATHS

A sequence of links (or equivalently, linked nodes) between two nodes is called a *path*. A path with the minimum number of links is called a *shortest path*. For example, in Figure 12.5, a shortest path from Dave to Lillian starts with Dave, then visits Beth, then Nick, then Lillian. The *distance* between two nodes is the number of links on a shortest path between them. Because three links were crossed along the shortest path from Dave to Lillian, the distance between them is 3.

$$\text{Dave} \xrightarrow{1} \text{Beth} \xrightarrow{2} \text{Nick} \xrightarrow{3} \text{Lillian}$$

Computing the distance between two nodes is a fundamental problem in network analysis. In a transportation network, the distance between a source and destination gives the number of stops along the route. In an ecological network, the distance between two organisms may be a measure of how dependent one organism is upon the other. In a social network, the distance between two people is the number of introductions by friends that would be necessary for one person to meet the other.

Reflection 12.7 *Are shortest paths always unique? Is there another shortest path between Dave and Lillian?*

Yes, there is:

$$\text{Dave} \xrightarrow{1} \text{Beth} \xrightarrow{2} \text{Amelia} \xrightarrow{3} \text{Lillian}$$

There may be many shortest paths between two nodes in a network, but in most applications we are concerned with just finding one.

Shortest paths can be computed using an algorithm called *breadth-first search (BFS)*. A breadth-first search explores outward from a source node, first visiting all nodes with distance one from the source, then all nodes with distance two, etc., until it has visited every reachable node in the network. In other words, the BFS algorithm incrementally pushes its "frontier" of visited nodes outward from the source. When the algorithm finishes, it has computed the distances between the source node and every other node.

For example, suppose we wanted to discover the distance from Beth to every other person in the social network in Figure 12.5, reproduced below.

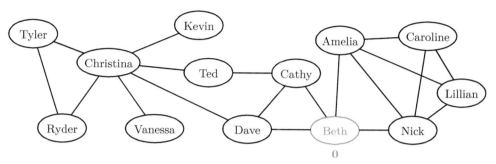

As indicated above, we begin by labeling Beth's node with distance 0, signifying that there are zero links between the node and itself. Then, in the first round of the algorithm, we explore all neighbors of Beth, colored red below.

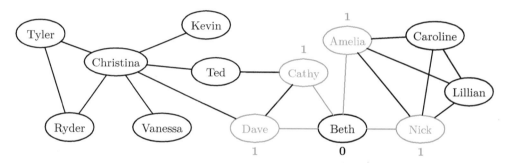

Since these nodes are one hop away from the source, we label them with distance 1. These nodes now comprise the "frontier" being explored by the algorithm. In the next round, we explore all unvisited neighbors of the nodes on this frontier, as shown below.

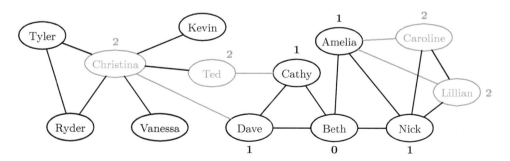

As indicated by the red links, Christina is visited from Dave, Ted is visited from Cathy, and both Caroline and Lillian are visited from Amelia. Notice that Caroline and Lillian could have been visited from Nick as well. The decision is arbitrary, depending, as we will see, on the order in which nodes are considered by the algorithm. Since all four of these nodes are neighbors of a node with distance 1, we label them with distance 2. Finally, in the third round, we visit all unvisited neighbors of the new frontier of nodes, as shown below.

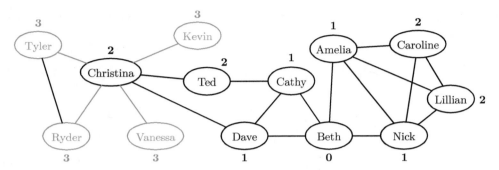

Since these newly visited nodes are all neighbors of a node labeled with distance 2, we label all of them with distance 3. At this point, all of the nodes have been visited, and the final label of each node gives its distance from the source.

Reflection 12.8 *If you also studied the depth-first search algorithm in Section 10.5, compare and contrast that approach with breadth-first search.*

In an algorithm, keeping track of the nodes on the current frontier could get complicated. The trick is to use a *queue*. A queue is a list in which items are always inserted at the end and deleted from the front. The insertion operation is called *enqueue* and the deletion operation is called *dequeue*.

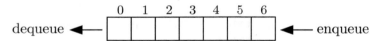

Reflection 12.9 *In Python, if we use a list named* `queue` *to implement a queue, how do we perform the enqueue and dequeue operations?*

An enqueue operation is simply an append:

```
queue.append(item)    # enqueue an item
```

And then a dequeue can be implemented by "popping" the front item from the list:

```
item = queue.pop(0)   # dequeue an item
```

In the breadth-first search algorithm, we use a queue to remember those nodes on the "frontier" that have been visited, but from which the algorithm has not yet visited new nodes. When we are ready to visit the unvisited neighbors of a node on the frontier, we dequeue that node, and then enqueue the newly visited neighbors so that we can remember to explore outward from them later.

Reflection 12.10 *Why can we not explore outward from these newly visited neighbors right away? Why do they need to be stored in the queue for later?*

We need to wait because there may be nodes further ahead in the queue that have smaller distances from the source. For the algorithm to work correctly, we have to explore outward from these nodes first.

The following Python function implements the breadth-first search algorithm.

```
1  def bfs(network, source):
2      """Perform a breadth-first search on network, starting from source.
3
4      Parameters:
5          network: a graph represented by a dictionary
6          source: the node in network from which to start the BFS
7
8      Return value: a dictionary with distances from source to all nodes
9      """
10
11     visited = { }
12     distance = { }
13     for node in network:
14         visited[node] = False
15         distance[node] = float('inf')
16     visited[source] = True
17     distance[source] = 0
18     queue = [source]
19     while queue != [ ]:
20         front = queue.pop(0)              # dequeue front node
21         for neighbor in network[front]:
22             if not visited[neighbor]:
23                 visited[neighbor] = True
24                 distance[neighbor] = distance[front] + 1
25                 queue.append(neighbor)    # enqueue visited node
26     return distance
```

The function maintains two dictionaries: `visited` keeps track of whether each node has been visited and `distance` keeps track of the distance from the `source` to each node. Lines 11–17 initialize the dictionaries. Every node, except the source, is marked as unvisited and assigned an initial distance of infinity (∞) because we do not yet know which nodes can be reached from the source. (The expression `float('inf')` creates a special value representing ∞ that is greater than every other floating point value.) The source is marked as visited and assigned distance zero. On line 18, the queue is initialized to contain just the source node. Then, while the queue is not empty, the algorithm repeatedly dequeues the front node (line 20), and explores all neighbors of this node (lines 21–25). If a neighbor has not yet been visited (line 22), it is marked as visited (line 23), assigned a distance that is one greater than the node from which it is being visited (line 24), and then enqueued (line 25). Once the queue is empty, we know that all reachable nodes have been visited, so we return the `distance` dictionary, which now contains the distance to each node.

Reflection 12.11 *Call the `bfs` function with the graph that you created earlier, to find the distances from Beth to all other nodes.*

Reflection 12.12 *What does it mean if the `bfs` function returns a distance of ∞ for a node?*

If the final distance is ∞, then the node must not have been visited by the algorithm, which means that there is no path to it from the source.

Reflection 12.13 *If you just want the distance between two particular nodes, named `source` and `dest`, how can you use the `bfs` function to find it?*

The `bfs` function finds the distance from a source node to every node, so we just need to call `bfs` and then pick out the particular distance we are interested in:

```
allDistances = bfs(graph, source)
distance = allDistances[dest]
```

Finding the actual paths

In some applications, just finding the distance between two nodes is not enough; we actually need a path between the nodes. For example, just knowing that you are only three hops away from that potential employer in your social network is not very helpful. You want to know who to ask to introduce you! And in a road network, we want to know the actual directions, not just the distance.

Fortunately, the breadth-first search algorithm is already finding the shortest paths; we just need to remember them. Consider how the distance from Beth to Tyler was computed in the previous example. As depicted below, from Beth, we visited Dave; from Dave, we visited Christina; and from Christina, we visited Tyler.

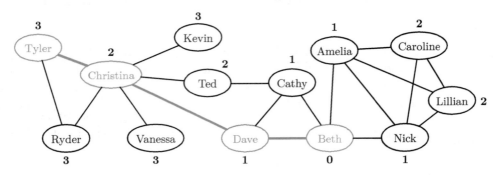

Since we incremented the distance to Tyler by one each time we crossed one of these links, this path must be a shortest path! Therefore, all we have to do is remember this order of nodes, as we visit them. This is accomplished by adding another dictionary, named `predecessor`, to the `bfs` function:

```
 1  def bfs(network, source):
 2      """ (docstring omitted) """
 3
 4      visited = { }
 5      distance = { }
 6      predecessor = { }
 7      for node in network:
 8          visited[node] = False
 9          distance[node] = float('inf')
10          predecessor[node] = None
11      visited[source] = True
12      distance[source] = 0
13      queue = [source]
14      while queue != [ ]:
15          front = queue.pop(0)          # dequeue front node
16          for neighbor in network[front]:
17              if not visited[neighbor]:
18                  visited[neighbor] = True
19                  distance[neighbor] = distance[front] + 1
20                  predecessor[neighbor] = front
21                  queue.append(neighbor)   # enqueue visited node
22      return distance, predecessor
```

The `predecessor` dictionary remembers the predecessor (the node that comes before) of each node on the shortest path to it from the source. This value is assigned on line 20: the predecessor of each newly visited node is assigned to be the node from which it is visited.

Reflection 12.14 *How can we use the final values in the `predecessor` dictionary to construct a shortest path between the source node and another node?*

To construct the path to any particular node, we need to follow the predecessors *backward* from the destination. As we follow them, we will insert each one into the *front* of a list so they are in the correct order when we are done. For example, in the previous example, to find the shortest path from Beth to Tyler, we start at Tyler.

> path = ['Tyler']

Tyler's predecessor was Christina (i.e., `predecessor['Tyler']` was assigned to be 'Christina' by the `bfs` algorithm), so we insert Christina into the front of the list:

> path = ['Christina', 'Tyler']

Christina's predecessor was Dave, so we next insert Dave into the front of the list:

> path = ['Dave', 'Christina', 'Tyler']

Finally, Dave's predecessor was Beth:

```
        path = ['Beth', 'Dave', 'Christina', 'Tyler']
```

Since Beth was the source, we stop. The following function implements this idea.

```
def path(network, source, dest):
    """Find a shortest path in network from source to dest.

    Parameters:
        network: a graph represented by a dictionary
        source: the source node in network
        dest: the destination node in network

    Return value: a list containing a path from source to dest
    """

    allDistances, allPredecessors = bfs(network, source)

    path = [ ]
    current = dest
    while current != source:
        path.insert(0, current)
        current = allPredecessors[current]
    path.insert(0, source)

    return path
```

With the destination node initially assigned to `current`, the `while` loop inserts each value assigned to `current` into the front of the list named `path` and then moves `current` one step closer to the source by reassigning it the predecessor of `current`. When `current` reaches the `source`, the loop ends and we insert the `source` as the first node in the `path`.

In the next section, we will use information about shortest paths to investigate a special kind of network called a *small-world* network.

Exercises

12.2.1. List the order in which nodes are visited by the `bfs` function when it is called to find the distance between Ted and every other node in the graph in Figure 12.5. (There is more than one correct answer.)

12.2.2. List the order in which nodes are visited by the `bfs` function when it is called to find the distance between Caroline and every other node in the graph in Figure 12.5. (There is more than one correct answer.)

12.2.3. By modifying one line of code, the `visited` dictionary can be completely removed from the `bfs` function. Show how.

12.2.4. Write a function that uses the `bfs` function to return the distance in a graph between two particular nodes. The function should take three parameters: the graph, the source node, and the destination node.

12.2.5. We say that a graph is *connected* if there is a path between any pair of nodes. The breadth-first search algorithm can be used to determine whether a graph is connected. Show how to modify the `bfs` algorithm so that it returns a Boolean value indicating whether the graph is connected.

12.2.6. A depth-first search algorithm (see Section 10.5) can also be used to determine whether a graph is connected. Recall that a depth-first search recursively searches as far from the source as it can, and then backtracks when it reaches a dead end. Writing a depth-first search algorithm for a graph is actually much easier than writing the one in Section 10.5 because there are fewer base cases to deal with.

   (a) Write a function

   ```
   dfs(network, source, visited)
   ```

   that performs a depth-first search on the given `network`, starting from the given `source` node. The third parameter, `visited`, is a list of nodes that have been visited by the depth-first search. The initial list argument passed in for `visited` should be empty, but when the function returns, `visited` should contain all of the visited nodes. In other words, you should call the function initially like this:

   ```
   visited = []
   dfs(network, source, visited)
   ```

   (b) Write another function

   ```
   connected(network)
   ```

   that calls your `dfs` function to determine whether `network` is connected. (The source node can be any node in the network.)

12.3 IT'S A SMALL WORLD...

In a now-famous 1967 experiment, sociologist Stanley Milgram asked several individuals in the midwest United States to forward a postcard to a particular person in Boston, Massachusetts. If they did not know this person on a first-name basis, they were asked instead to forward it to someone they thought might have a better chance of knowing the person. Each intermediate person was asked to follow the same instructions. Of the postcards that made it to the destination (many were simply not forwarded), the average number of hops was about six.

From this experiment later came the suggestion that there are only "six degrees of separation" between any two people on Earth, and that the human race must constitute a *small-world network*. In the late 1990's, using computers, researchers began to discover that networks representing a wide range of unrelated phemenona, from social networks to neural networks to the Internet, all exhibit the same *small-world* property: for most nodes in the network, there is a very short path connecting them. Put another way, in a small-world network, the average distance between any two nodes is small.

Intuitively, it seems as though a small-world network must have a lot of links to facilitate so many short paths. However, it has been shown that small-world networks can actually be quite *sparse*, meaning that the number of links is quite

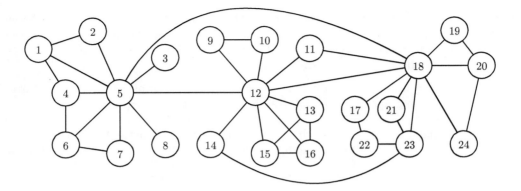

Figure 12.6 A small-world network.

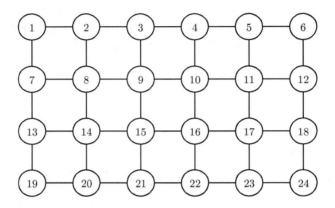

Figure 12.7 A grid network.

small relative to the number possible. The keys to a small-world network are a high degree of *clustering* and a few long-range shortcuts that facilitate short paths between clusters. A cluster is a set of nodes that are highly connected among themselves. In your social network, you probably participate in several clusters: family, friends at school, friends at home, co-workers, teammates, etc. Many of the members of each of these clusters are probably also connected to one another, but members of different clusters might be far apart if you did not act as a shortcut link between them.

Although it is too small to really be called a small-world network, the network in Figure 12.6 illustrates these ideas. The graph contains three clusters of nodes, centered around nodes 5, 12 and 18, that are connected by a few shortcut links (e.g., the links between nodes 5 and 18 and between nodes 14 and 23). These two characteristics together give an average distance between nodes of about 2.42. On the other hand, the highly structured grid or mesh network in Figure 12.7 has an average node distance of about 3.33. Both of these graphs have 24 nodes and 38 links, so they are both sparse relative to the $(24 \cdot 23)/2 = 276$ possible links that they could have.

Clustering coefficients

The extent to which the neighborhood of a node is clustered is measured by its *local clustering coefficient*. The local clustering coefficient of a node is the number of links between its neighbors, divided by the total number of possible links between neighbors. For example, consider the cluster on the left below surrounding the blue node in the center.

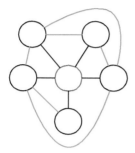 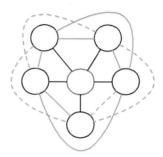

The blue node has five neighbors, with six links between them (in red). Notice that each of these links, together with two black links, forms a closed cycle, called a *triangle*. So we can also think about the local clustering coefficient as counting these triangles. As shown on the right, there are four dashed links between neighbors of the blue node (i.e., four additional triangles) that are not present on the left, for a total of ten possible links altogether. So the local clustering coefficient of the blue node is $6/10 = 0.6$. (The clustering coefficient will always be between 0 and 1.)

Reflection 12.15 *In general, if a node has k neighbors, how many possible links are there between pairs of these neighbors?*

Each of the k neighbors could be connected to $k-1$ other neighbors, for a total of $k(k-1)$ links. However, this counts each link twice, so the total number of possible links is actually $k(k-1)/2$. Therefore, the local clustering coefficient of a node with k neighbors is the number of neighbors that are connected, divided by $k(k-1)/2$. The clustering coefficient for an entire network is the average local clustering coefficient over all nodes in the network. The highly structured grid or mesh graph in Figure 12.7 does not have any triangles at all, so its clustering coefficient is 0. On the other hand, the graph in Figure 12.6 has a clustering coefficient of about 0.59.

Reflection 12.16 *If you had a small local clustering coefficient in your social network (i.e., if your friends are not friends with each other), what implications might this have?*

It has been suggested that situations like this breed instability. Imagine that, instead of a social network, we are talking about a network of nations and links represent the existence of diplomatic relations. A nation with diplomatic relations with many other nations that are enemies of each other is likely in a stressful situation. It might be helpful to detect such situations in advance to curtail potential conflicts.

To compute the local clustering coefficient for a node, we need to iterate over all of the node's neighbors and count for each one the number of links between it and the other neighbors of the node. Then we divide this number by the maximum possible number of links between the node's neighbors. This is accomplished by the following function.

```
def clusteringCoefficient(network, node):
    """Compute the local clustering coefficient for a node.

    Parameters:
        network: a graph represented by a dictionary
        node: a node in the network

    Return value: the local clustering coefficient of node
    """

    neighbors = network[node]
    numNeighbors = len(neighbors)
    if numNeighbors <= 1:
        return 0
    numLinks = 0
    for neighbor1 in neighbors:
        for neighbor2 in neighbors:
            if neighbor1 != neighbor2 and neighbor1 in network[neighbor2]:
                numLinks = numLinks + 1
    return numLinks / (numNeighbors * (numNeighbors - 1))
```

This function is relatively straightforward. The two `for` loops iterate over every possible pair of neighbors, and the `if` statement checks for a link between unique neighbors. However, this process effectively counts every link twice, so at the end we divide by `numNeighbors * (numNeighbors - 1)` (i.e., $k(k-1)$), which is twice what we discussed previously.

Reflection 12.17 *Do you see why the function counts every link twice? How can we fix this?*

The function effectively counts every link twice because it checks whether each neighbor is in every other neighbor's list of adjacent nodes. Therefore, for any two connected neighbors, call them A and B, we are counting the link once when we see A in the list of adjacent nodes of B and again when we see B in the list of adjacent nodes of A.

To count each link just once, we can use the following trick. In the list of `neighbors`, we first check whether the node at index 0 is connected to nodes at indices $1, 2, \ldots, k-1$. Then, to prevent counting a link twice, we never want to check whether any node is connected to node 0 again. So we next check whether node 1 is connected to nodes $2, 3, \ldots, k-1$. Now, to prevent double counting, we never want to check whether any node is connected to nodes 0 or 1. So we next check whether node 2 is connected to nodes $3, 4, \ldots, k-1$. Do you see the pattern? In general, we only want to check whether node i is connected to nodes $i+1, i+2, \ldots, k-1$. (This is the same trick you may have seen in Exercise 8.5.4.) This is implemented in the following improved version of the function.

```
def clusteringCoefficient(network, node):
    """ (docstring omitted) """

    neighbors = network[node]
    numNeighbors = len(neighbors)
    if numNeighbors <= 1:
        return 0
    numLinks = 0
    for index1 in range(len(neighbors) - 1):
        for index2 in range(index1 + 1, len(neighbors)):
            neighbor1 = neighbors[index1]
            neighbor2 = neighbors[index2]
            if neighbor1 != neighbor2 and neighbor1 in network[neighbor2]:
                numLinks = numLinks + 1
    return numLinks / (numNeighbors * (numNeighbors - 1) / 2)
```

Once we have this function, to compute the clustering coefficient for the network, we just have to call it for every node, and compute the average. We leave this, and writing a function to compute the average distance, as exercises.

Scale-free networks

In addition to having short paths and high clustering, researchers soon discovered that most small-world networks also contain a few highly connected (i.e., high degree) nodes called *hubs* that facilitate even shorter paths. In the network in Figure 12.6, nodes 5, 12, and 18 are hubs because their degrees are large relative to the other nodes in the network.

Reflection 12.18 *How do connected hubs facilitate short paths?*

The existence of hubs in a large network can be seen by plotting for each node degree in the network the fraction of the nodes that have that degree. This is called the *degree distribution* of the network. The degree distribution for a network with a few hubs will show the vast majority of nodes having relatively small degree and just a few nodes having very large degrees. For example, Figure 12.8 shows such a plot for a small portion of the web. Each node represents a web page and a directed link from one node to another represents a hyperlink from the first page to the

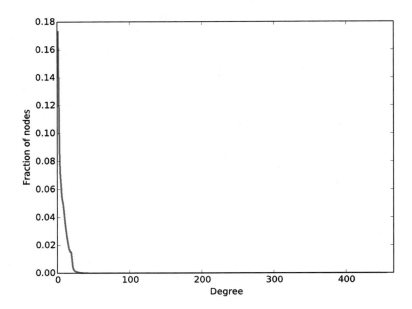

Figure 12.8 The degree distribution of 875,713 nodes in the web network.

second page.[1] In this network, 99% of the nodes have degree at most 25, while just a few have degrees that are much higher. (In fact, 98% of the nodes have degrees at most 20 and 90% have degrees at most 15.) These few hubs with high degree enable a small average distance and a clustering coefficient of about 0.37.

Reflection 12.19 *In the web network from Figure 12.8, the degree of a node is the number of hyperlinks from that page. How do you think the degree distribution might change if we instead counted the number of hyperlinks to each page?*

Networks with this characteristic shape to their degree distributions are called *scale-free networks*. The name comes from the observation that the fraction of nodes with degree d is roughly $(1/d)^a$, for some small value of a. Such functions are called "scale-free" because their plots have the same shape regardless of the scale at which you view them. A scale-free degree distribution is very different from the normal distribution that seems to describe most natural phenomena, which is why this discovery was so interesting.

Reflection 12.20 *How could recognizing that a network is scale-free and then identifying the hubs have practical importance?*

The presence of hubs in a network is a double-edged sword. On the one hand, hubs enable efficient communication and transportation. For this reason, the Internet

---

[1] Web network data obtained from http://snap.stanford.edu/data/web-Google.html

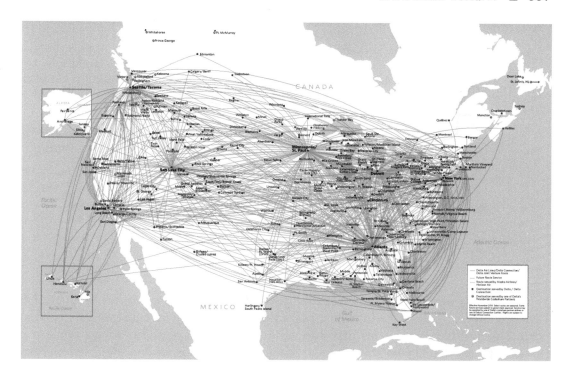

Figure 12.9 The North American routes of Delta Airlines. [9]

is structured in this way, as are airline networks (see Figure 12.9). Also, because so many of the nodes in a scale-free network are relatively unimportant, scale-free networks tend to be very robust when subjected to random attacks or damage. Some have speculated that, because some natural networks are scale-free, they may represent an evolutionary advantage. On the other hand, because the few hubs are so important, a directed attack on a hub can cause the network to fail. (Have you ever noticed the havoc that ensues when an airline hub is closed due to weather?) A directed attack on a hub can also be advantageous if we *want* the network to fail. For example, if we suspect that the network through which an epidemic is traveling is scale-free, we may have a better chance of stopping it if we vaccinate the hubs.

Exercises

12.3.1. Write a function that returns the average local clustering coefficient for a network. Test your function by calling it on some of the networks on the book web site. You will need the function assigned in Exercise 12.1.8 to read these files.

12.3.2. Write a function that returns the average distance between every pair of nodes in a network. If two nodes are not connected by a path, assign their distance to be equal to the number of nodes in the network (since this is longer than any possible path). Test your function by calling it on some of the networks on the book web site. You will need the function assigned in Exercise 12.1.8 to read these files.

12.3.3. The *closeness centrality* of a node is the total distance between it and all other nodes in the network. By this measure, the node with the smallest value is the most central (and perhaps most influential) node in the network. Write a function that computes the closeness centrality of a node. Your function should take two parameters: the network and a node. Test your function by calling it on some of the networks on the book web site. You will need the function assigned in Exercise 12.1.8 to read these files.

12.3.4. Using the function you wrote in the previous exercise, write another function that returns the most central node in a network (with the smallest closeness centrality).

12.3.5. Write a function that plots the degree distribution of a network, producing a plot like that in Figure 12.8. Test your function on a small network first. Then call your function on the large Facebook network (with 4,039 nodes and 88,234 links) that is available on the book web site. (You will need the function assigned in Exercise 12.1.8 to read these files.) Is the network scale-free?

12.4 RANDOM GRAPHS

Since small-world and scale-free networks seem to be so common, it is natural to ask whether such networks just happen randomly. To answer this question, we can compare the characteristics of these networks to a class of randomly generated graphs. In particular, we will look at the class of *uniform random graphs*, which are created by adding each possible edge with some probability p.

Creating a uniform random graph is straightforward. We first create an adjacency list with the desired number of nodes, then iterate over all possible pairs of nodes. For each pair of nodes, we link them with probability p, as shown below.

```
import random

def randomGraph(n, p):
    '''Return a uniform random graph with n vertices.

    Parameters:
        n: the number of nodes
        p: the probability that two nodes are connected

    Return value: a random graph
    '''

    graph = { }
    for node in range(n):           # label nodes 0, 1, ..., n-1
        graph[node] = [ ]           # graph has n nodes, 0 links

    for node1 in range(n - 1):
        for node2 in range(node1 + 1, n):
            if random.random() < p:
                graph[node1].append(node2)   # add edge between
                graph[node2].append(node1)   #   node1 and node2
    return graph
```

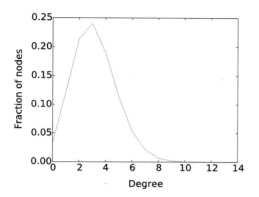
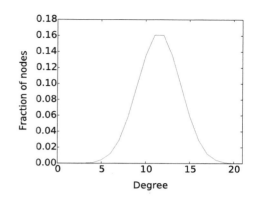

Figure 12.10 The degree distribution of random graphs with $n = 24$ and $p = 38/276$.

Figure 12.11 The degree distribution of random graphs with $n = 24$ and $p = 1/2$.

Because we will get a different random graph every time we call this function, any characteristics that we want to measure will have to be averages over several different random graphs with the same values of `n` and `p`. To illustrate, let's compute the average distance, clustering coefficient, and degree distribution for uniform random graphs with the same number of nodes and links as the graphs in Figures 12.6 and 12.7. Recall that those graphs had 24 nodes and 38 links.

Reflection 12.21 *What parameters should we use to create a uniform random graph with 24 nodes and 38 links?*

We cannot specify the number of links specifically in a uniform random graph, but we can set the probability so that we are likely to get a particular number, on average, over many trials. In particular, we want 38 out of a possible $(24 \cdot 23)/2$ links, so we set

$$p = \frac{38}{(24 \cdot 23)/2} = \frac{38}{276} \approx 0.14.$$

Averaging over 20,000 uniform random graphs, each generated by calling `randomGraph(24, 0.14)`, we find that the average distance between nodes is about 4.32 and the average clustering coefficient is about 0.12. The table below compares these results to what we computed previously for the other two graphs.

Graph	Average distance	Clustering coefficient
Figure 12.6 (clusters)	2.42	0.59
Figure 12.7 (grid)	3.33	0
Uniform random	4.32	0.12

The random graph with the same number of nodes and edges has a slightly longer average distance and a markedly smaller clustering coefficient than the graph in Figure 12.6 with the three clusters. Because these graphs are so small, these numbers alone, while suggestive, are not very strong evidence that random graphs do not have the small-world or scale-free properties. So let's also look at the average degree distribution of the random graphs, shown in Figure 12.10. The shape of the degree distribution is quite different from that of a scale-free network, and is much closer to a normal distribution. Because the probability of adding an edge was relatively low, the average degree was only about 3 and there were a number of nodes with 0 degree, causing the plot to "run into" the y-axis. If we perform the same experiment with $p = 0.5$, as shown in Figure 12.11, we get a much clearer bell curve. These distributions show that random graphs do not have hubs; instead, the nodes all tend to have about the same degree. So there is definitely something non-random happening to generate scale-free networks.

Reflection 12.22 *What kind of process do you think might create a scale-free network with a few high-degree nodes?*

The presumed process at play has been dubbed *preferential attachment* or, colloquially the "rich get richer" phenomenon. The idea is relatively intuitive: popular people, destinations, and web pages tend to get more popular over time as word of them moves through the network.

Exercises

12.4.1. Show how to create a uniform random graph with 30 nodes and 50 links, on average.

12.4.2. Exercise 12.3.1 asked you to write a function that returns the clustering coefficient for a graph. Use this function to write another function

   `avgCCRandom(n, p, trials)`

   that returns the average clustering coefficient, over the given number of trials, of random graphs with the given values of n and p.

12.4.3. Exercise 12.3.2 asked you to write a function that returns the average distance between any two nodes in a graph. Use this function to write another function

   `avgDistanceRandom(n, p, trials)`

   that returns the average of this value, over the given number of trials, for random graphs with the given values of n and p.

12.4.4. Exercise 12.3.5 asked you to write a function to plot the degree distribution of a graph and then call the function on the large Facebook network on the book web site. This network has 4,039 nodes and 88,234 links. To compare the degree distribution of this network to a random graph of the same size, write a function

   `degreeDistributionRandom(n, p, trials)`

   that plots the average degree distribution, over the given number of trials, of random graphs with the given values of n and p. Then use this function to plot the degree distribution of random graphs with 4,039 nodes and an average of 88,234 links. What do you notice?

12.4.5. We say that a graph is *connected* if there is a path between any pair of nodes. Random graphs that are generated with a low probability p are unlikely to be connected, while random graphs generated with a high probability p are very likely to be connected. But for what value of p does this transition between disconnected and connected graphs occur?

To determine whether a graph is connected, we can use either a breadth-first search (as in Exercise 12.2.5) or a depth-first search (as in Exercise 12.2.6). In either case, we start from any node in the network, and try to visit all of the other nodes. If the search is successful, then the graph must be connected. Otherwise, it must not be connected.

(a) Write a function

```
connectedRandom(n, minp, maxp, stepp, trials)
```

that plots the fraction of random graphs with n nodes that are connected for values of p ranging from `minp` to `maxp`, in increments of `stepp`. To compute the fraction that are connected for each value of p, generate `trials` random graphs and count how many of those are connected using your `connected` function from either Exercise 12.2.5 or Exercise 12.2.6.

(b) For $n = 24$, what do you find? For what value of p is there a 50% chance that the graph will be connected? Does the transition from disconnected graphs to connected graphs happen gradually or is change abrupt?

12.5 SUMMARY

In this chapter, we took a peek at one of the more exciting interdisciplinary areas in which computer scientists have become engaged. Networks are all around us, some obvious and some not so obvious. But they can all be described using the language of *graphs*. The shortest path and distance between any two nodes in a graph can be found with the *breadth-first search* algorithm. Graphs in which the distance between any two nodes is relatively short and the *clustering coefficient* is relatively high are called *small-world networks*. Networks that are also characterized by a few high-degree hubs are called *scale-free* networks. Scientists have discovered over the last two decades that virtually all large-scale natural and human-made networks are scale-free. Knowing this about a network can give a lot of information about how the network works and about its vulnerabilities.

12.6 FURTHER DISCOVERY

This chapter's epigraph is from a 1967 article by Stanley Milgram titled, *The Small-World Problem* [33]. Dr. Milgram was an influential American social psychologist. Besides his small world experiment, he is best known for experiments in which he demonstrated that ordinary people are capable of disregarding their consciences when instructed to do so by those in authority.

There are several excellent books for a general audience about the emerging field of network science. Three of these are *Six Degrees* by Duncan Watts [57], *Sync* by Steven Strogatz [54], and *Linked* by Albert-László Barabási [3]. *Think Complexity* by Allen Downey [10] introduces slightly more advanced material on implementing algorithms on graphs, small-world networks, and scale-free networks in Python.

12.7 PROJECTS

Project 12.1 Diffusion of ideas and influence

In this project, you will investigate how memes, ideas, beliefs, and information propagate through social networks, and who in a network has the greatest influence. We will simulate diffusion through a network with the *independent cascade model*. In this simplified model, an idea originates at one node (called the seed) and each of the neighbors of this node adopts it with some fixed probability (called the *propagation probability*). This probability measures how influential each node is. In reality, of course, people have different degrees of influence, but we will assume here that everyone is equally influential. Once the idea has propagated to one or more new nodes, each of these nodes gets an opportunity to propagate it to each of their neighbors in the next round. This process continues through successive rounds until all of the nodes that have adopted the idea have had a chance to propagate it to their neighbors.

For example, suppose node 1 in the network below is the seed of an idea.

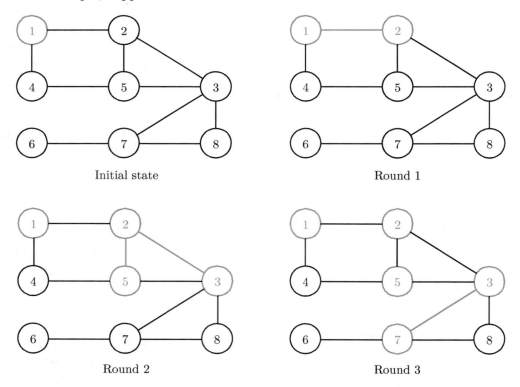

In round 1, node 1 is successful in spreading the idea to node 2, but not to node 4. In round 2, the idea spreads from node 2 to both node 3 and node 5. In round 3, the idea spreads from node 3 to node 7, but node 3 does not successfully influence node 8. Node 5 also attempts to influence node 4, but is unsuccessful. In round 4 (not shown), node 7 attempts to spread the idea to nodes 6 and 8, but is unsuccessful, and the process completes.

Part 1: Create the network

For this project, you will use a network that is an anonymized version of a very small section of the Facebook network with 333 nodes and 2,519 links. The file containing this network is available on the book web site. The format of the file is the same as the format described in Exercise 12.1.8: each node is represented by a number, and each line represents one link. If you have not already done so in Exercise 12.1.8, write a function

```
readNetwork(fileName)
```

that reads in a network file with this format, and returns an adjacency list representation of the network.

Part 2: Simulate diffusion with a single seed

Simulating the diffusion of an idea through the network, as described above, is very similar to a breadth-first search, except that nodes are visited probabilistically. Just as visited nodes are not revisited in a BFS, nodes that have already adopted the idea are not re-influenced. Using the `bfs` function from Section 12.2 as a starting point, write a function

```
diffusion(network, seed, p)
```

that simulates the independent cascade model starting from the given seed, using propagation probability `p`. The function should return the total number of nodes in the network who adopted the idea (i.e., were successfully influenced).

Part 3: Rate each node's influence

Because this is a random process, the `diffusion` function will return a different number every time it is run. To get a good estimate of how influential a node is, you will have to run the function many times and average the results. Therefore, write a function

```
rateSeed(network, seed, p, trials)
```

that calls `diffusion` `trials` times with the given values of `network`, `seed` and `p`, and returns the average number of nodes influenced over that many trials.

Part 4: Find the most influential node(s)

Finally, write a function

```
maxInfluence(network, p, trials)
```
that calls the `rateSeed` function for every node in the network, and returns the most influential node.

Combine these functions into a complete program that finds the most influential node(s) in the small Facebook network, using a propagation probability of 0.05. Then answer the following questions. You may write additional functions if you wish.

Question 12.1.1 *Which node(s) turned out to be the most influential in your experiments?*

Question 12.1.2 *Were the ten most influential nodes the same as the ten nodes with the most friends?*

Question 12.1.3 *Can you find any relationship between a person's number of friends and how influential the person is?*

Project 12.2 Slowing an epidemic

In Section 4.4, we simulated the spread of a flu-like virus through a population using the SIR model. In that simulation, we lumped all individuals into three groups: those who are susceptible to the infection, those who are currently infected, and those who recovered. In reality, of course, not every individual in the susceptible group is equally susceptible and not every infected person is equally likely to infect others. Some of the differences have to do with who comes into contact with whom, which can be modeled by a network. For example, consider the network in Figure 12.6 on Page 602. In that network, if node 3 becomes infected and there is a 10% chance of any individual infecting a contact, then the virus is very unlikely to spread at all. On the other hand, if node 12 becomes infected, there is a $9 \cdot 10\% = 90\%$ chance that the virus will spread to at least one of the node's nine neighbors.

If a vaccine is available for the virus, then knowledge about the network of potential contacts can be extremely useful. In this project, you will simulate and compare two different vaccination strategies: vaccinating randomly and vaccinating targeted nodes based on a criterion of your choosing.

Part 1: Create the network

For this project, you will use a network that is an anonymized version of a very small section of the Facebook network with 333 nodes and 2,519 links. The file containing this network is available on the book web site. The format of the file is the same as the format described in Exercise 12.1.8: each node is represented by a number, and each line represents one link. If you have not already done so in Exercise 12.1.8, write a function

```
readNetwork(fileName)
```

that reads in a network file with this format, and returns an adjacency list representation of the network.

Part 2: Simulate spread of the contagion through the network

Simulating the spread of the infection through the network is very similar to a breadth-first search, except that nodes are visited (infected) probabilistically and we will assume that each infection starts with a random source node. Once a node has been infected, it should be enqueued so that it has an opportunity to infect other nodes. When a node is dequeued, it should attempt to infect all of its uninfected neighbors, successfully doing so with some *infection probability p*. But once infected, a node should never be infected again or put back into the queue. Using the `bfs` function from Section 12.2 as a starting point, write a function

```
infection(network, p, vaccinated)
```

that simulates this infection spreading in `network`, starting from a randomly chosen source node, and using infection probability `p`. The fourth parameter, `vaccinated`, is a list of nodes that are immune to the infection. An immune node should never become infected, and therefore cannot infect others either; more on this in Part 3. The function should return when all infected nodes have had a chance to infect their neighbors. The function should return the total number of nodes who were infected.

Part 3: Simulate vaccinations

Now, you will investigate how the number of infected nodes is affected by vaccinations. Assume that the number of vaccine doses is limited, and compare two different strategies for choosing the sequence of nodes to vaccinate. For each strategy, first run your simulation with no vaccinations, then with only the first node vaccinated, then with the first two nodes vaccinated, etc. Because the infection is a random process, each simulation with a different number of vaccinations needs to be repeated many times to find an average number of infected nodes.

First, investigate what happens if you vaccinate a sequence of random nodes. Second, design a better strategy, and compare your results to the results of the random vaccinations. Use the ideas from this chapter to learn about the network, and choose a specific sequence of nodes that you think will do a better job of reducing the number of infections in the network.

Implement these simulations in a function

```
vaccinationSim(network, p, trials, numVaccs, randomSim)
```

If the last parameter `randomSim` is `True`, the function should perform the simulation with random vaccinations. Otherwise, it should perform the simulation with your improved strategy. The fourth parameter, `numVaccs`, is the number of individuals to vaccinate. For every number of vaccinations, the simulation should call the `infection` function `trials` times with the given `network` and infection probability

p. The function should return a list of `numVaccs + 1` values representing the average number of infections that resulted with $0, 1, 2, \ldots,$ `numVaccs` immune nodes.

Create a program that calls this function once for each strategy with the following arguments:

- infection probability 0.1
- 500 trials
- 50 vaccinations

Then your program should produce a plot showing for both strategies how the number of infections changes as the number of vaccinations increases.

Question 12.2.1 *What strategy did you choose?*

Question 12.2.2 *Why did you think that your strategy would be effective?*

Question 12.2.3 *How successful was your strategy in reducing infections, compared to the random strategy?*

Question 12.2.4 *Given an unlimited number of vaccinations, how many are required for the infection to not spread at all?*

Project 12.3 The Oracle of Bacon

In the 1990's, a group of college students invented a pop culture game based on the six degrees of separation idea called the "Six Degrees of Kevin Bacon." In the game, players take turns challenging each other to identify a connected chain of actors between Kevin Bacon and another actor, where two actors are considered to be connected if they appeared in the same movie. An actor's Bacon number is the distance of the actor from Kevin Bacon.

This game later spawned the "Oracle of Bacon" website, at `http://oracleofbacon.org`, which gives the shortest chain between Mr. Bacon and any other actor you enter. This works by constructing a network of actors from the Internet Movie Database (IMDb)[2], and then using breadth-first search to find the desired path. This network of movie actors provides fertile ground for investigating all of the concepts discussed in this chapter. Is the network really a small-world network, as suggested by the "Six Degrees of Kevin Bacon?" Is there really anything special about Kevin Bacon in this network? Or is there a short path between any two pair of actors? Is the network scale-free?

In this project, you will create an actor network, based on data from the IMDb, and then answer questions like these. You will also create your own "Oracle of Bacon."

---

[2] `http://www.imdb.com`

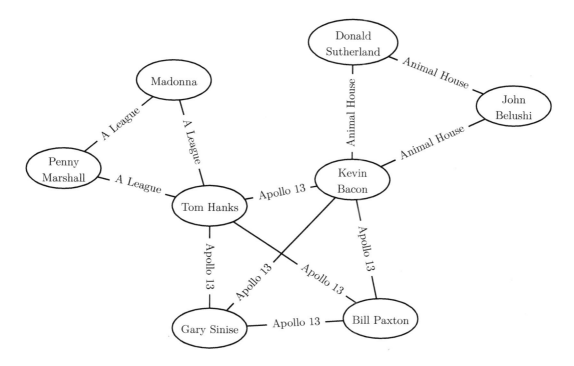

Figure 12.12 A simple actors network.

Part 1: Create the network

On the book web site, you will find several files, created from IMDb database files, that have the following format:

```
League of Their Own, A (1992) ▷Tom Hanks▷Madonna▷Penny Marshall▷ ···
Animal House (1978) ▷Kevin Bacon▷John Belushi▷Donald Sutherland▷ ···
Apollo 13 (1995) ▷Kevin Bacon▷Tom Hanks▷Bill Paxton▷Gary Sinise▷ ···
```

The files are tab-separated (the ▷ symbols represent tabs). The first entry on each line is the name of a movie. The movie is followed by a list of actors that appeared in that movie.

Write a function

```
createNetwork(filename)
```

that takes the name of one of these data files as a parameter, and returns an actor network (as an adjacency list) in which two actors are connected if they have appeared in the same movie. For example, for the very short file above, the network would look like that in Figure 12.12. Each link in this network is labeled with the name of a movie in which the two actors appeared. This is not necessary to determine a Bacon number, but it is necessary to display all the relationships, as required by the traditional "Six Degrees of Kevin Bacon" game (more on this later).

The following files were created from the IMDb data, and are available on the book web site.

File	Movies	Actors	Description
`movies2005.txt`	560	24,276	MPAA-rated movies from 2005
`movies2012.txt`	518	25,225	MPAA-rated movies from 2012
`movies2013.txt`	517	24,190	MPAA-rated movies from 2013
`movies2000p.txt`	8,396	362,798	MPAA-rated movies from 2000–2014
`movies2005p.txt`	5,590	251,881	MPAA-rated movies from 2005–2014
`movies2010p.txt`	2,547	117,908	MPAA-rated movies from 2010–2014
`movies_mpaa.txt`	12,404	514,780	All movies rated by MPAA
`moves_all.txt`	565,509	6,147,773	All movies

Start by testing your function with the smaller files, building up to working with the entire database. Keep in mind that, when working with large networks, this and your other functions below may take a few minutes to execute. For each file, we give the number of movies, the total number of actors in all casts (not unique actors), and a description.

Part 2: Is the network scale-free?

Because the actor network is so large, it will take too much time to compute the average distance between its nodes and its clustering coefficient. But we can plot the degree distribution and investigate whether the network is scale-free. To do so, write a function

```
degreeDistribution(network)
```

that plots the degree distribution of a network using `matplotlib`. Again, start with small files and work your way up.

Question 12.3.1 *What does your plot show? Is the network scale-free?*

Question 12.3.2 *Where does Kevin Bacon fall on your plot? Is he a hub? If so, is he unique in being a hub?*

Question 12.3.3 *Which actors have the ten largest degrees? (They might not be who you expect.)*

Part 3: Oracle of Bacon

You now know everything you need to create your own "Oracle of Bacon." Write a function

```
oracle(network)
```

that repeatedly prompts for the names of two actors, and then prints the distance between them in the network. Your function should call `bfs` to find the shortest distance.

For an extra challenge, modify the algorithm (and your adjacency list) so that it also prints how the two actors are related. For example, querying the oracle with Kevin Bacon and Zooey Deschanel might print

```
Zooey Deschanel and Kevin Bacon have distance 2.

1. Kevin Bacon was in "The Air I Breathe (2007)" with Clark Gregg.
2. Clark Gregg was in "(500) Days of Summer (2009)" with Zooey Deschanel.
```

Part 4: Frequencies of Bacon numbers

Finally, determine just how central Kevin Bacon is in the movie universe. To do so, create a chart with the frequency of Bacon numbers. Write a function

    `baconNumbers(network)`

that displays this chart for the given network. For example, given the file `movies2013.txt`, your function should print the following:

```
Bacon Number    Frequency
------------    ---------
     0              1
     1            152
     2           2816
     3          12243
     4           4230
     5            253
     6             47
     7              0
     8              0
infinity          819
```

An actor has infinite Bacon number if he or she cannot be reached from Kevin Bacon.

Once your function is working, call it with some of the larger files.

Question 12.3.4 *What does the chart tell you about Kevin Bacon's place in the movie universe?*

CHAPTER 13

Abstract data types

> Algorithms + Data Structures = Programs
>
> Niklaus Wirth
> *Textbook title (1976)*

THROUGHOUT this book, we have stressed the importance of both functional and data abstractions in problem solving. By decomposing algorithms into groups of functional abstractions, each implemented by a Python function, we have been able to more easily solve complex problems. Each functional abstraction provides a solution to a subproblem that we can use to solve our overall problem, without having to worry about how it works.

We have also made extensive use of a variety of data abstractions, called abstract data types (ADTs), such as turtles, strings, lists, and dictionaries. Each ADT is defined by the information it can store and a set of functions that operate on that information. An ADT is called *abstract* because its description is independent of its implementation. Barbara Liskov, the pioneering researcher and Turing Award winner who first developed the idea of an abstract data type, explained,

> What we desire from an abstraction is a mechanism which permits the expression of relevant details and the suppression of irrelevant details. In the case of programming, the use which may be made of an abstraction is relevant; the way in which the abstraction is implemented is irrelevant.

For example, our understanding of a list as a sequence of objects that may be appended, modified, and searched is independent of whether those objects are stored in sequential blocks of computer memory, archived on a hard drive or written on paper by a troop of monkeys. The hidden implementation of an ADT is called a *data structure*. Separating the ADT from the underlying data structure allows us to ignore the implementation details and use abstract data types at a high level to solve problems that would otherwise be out of reach.

In this chapter, we will discover how to implement new classes based on ADT descriptions. Programming languages like Python that allow us to create our own classes are called *object-oriented programming languages*. As we will see, object-oriented languages support the ability to create classes that behave the same way the built-in classes do. For instance, we will define new classes on which we can use both operators (e.g., +, in) and built-in functions (e.g., str, len). Implementing an ADT as a class also makes using the ADT more convenient and simultaneously protects the attributes of each object from being corrupted by mistake. Later in this chapter, we will design classes on which to build a bird-flocking simulation in which each bird, implemented as an object, has its own independent attributes and can interact with other birds in the flock.

13.1 DESIGNING CLASSES

Suppose you are on the planning commission for your local town, and are evaluating possible locations for a new high school. One consideration is how central the new school will be with respect to homes within the district. If we know the (longitude, latitude) location of each home, then we can compute the average distance to each proposed site or the most central location with respect to the homes. The most central point for a list of (x, y) points is commonly defined to the point whose x and y coordinates are the average of the x and y coordinates in the list. The result is called the *centroid*. (You may recall centroids from Section 8.7.) If we represent each point with a tuple, then the following function returns a tuple representing the centroid of a list of points.

```
def centroid(points):
    """Compute the centroid of a list of points.

    Parameter:
        points: a list of two-element tuples

    Return value: a tuple representing the centroid of points
    """
    n = len(points)
    if n == 0:
        return None
    sumX = 0
    sumY = 0
    for point in points:
        sumX = sumX + point[0]
        sumY = sumY + point[1]
    return (sumX / n, sumY / n)

def main():
    homes = [(0.5, 5), (3.5, 2), (4, 3.5), (5, 2), (7, 1)]
    central = centroid(homes)
    print(central)

main()
```

To keep things simple for now, instead of longitude and latitude, we will assume that x and y represent the east-west and north-south distances (in km), respectively, from some fixed point $(0,0)$. (In Project 13.1, we will look at how to compute distance between geographical points.)

For example, the points in the list named **homes** above are shown in black below and the computed centroid is shown in blue.

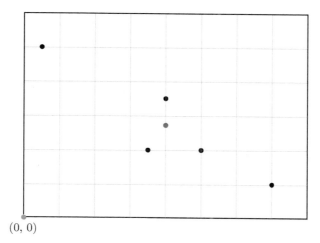

(0, 0)

We can simplify working with points by designing a new abstract data type for a point, and then implementing this ADT with a class. For example, we can define what it means to add two points and divide a point by a number. But, instead of designing an ADT that is specific to a geometric point, we will define a more general ADT for an *ordered pair* of numbers. An ordered pair is any pair of numbers in which we attribute different meanings to the values in the different positions. In other words, the ordered pair $(3, 14)$ is different from the ordered pair $(14, 3)$. We can, for example, use this ADT to represent a geographic location, the (x, y) position of a particle, a (row, column) position in a grid, or a vector.

Remember that an ADT serves as a blueprint for a category of data. The ADT defines both data, called *attributes* and the operations (or functions) that we are allowed to perform with those attributes. For a pair of numbers, the obvious attributes are the two numbers, which we will simply name a and b.

Pair ADT Attributes	
Name	Description
a	the pair's first value
b	the pair's second value

Reflection 13.1 *What kinds of operations or functions might we want to perform with a pair (a, b)?*

We might define the following operations for the **Pair** ADT.

Pair ADT Operations

Name	Arguments	Description
create	a and b values	create a new pair instance, assigning the given values to **a** and **b**
getFirst	—	return the first value of the pair
getSecond	—	return the second value of the pair
get	—	return a tuple (**a**, **b**) representing the pair
add	another pair	return a new pair representing the component-wise sum of pair and another pair
subtract	another pair	return a new pair representing the component-wise difference of pair and another pair
set	a and b values	set the **a** and **b** values of the pair
scale	a number	multiply the values of **a** and **b** in the pair by a number
destroy	—	destroy the pair instance

The operations of an abstract data type can be divided into four types:

1. A *constructor* creates a new instance of an ADT.

2. An *accessor* reads the attributes of an ADT instance and returns information derived from them, but does not modify the attributes' values.

3. A *mutator* modifies the values of the attributes of an ADT instance.

4. A *destructor* destroys an ADT instance and deallocates resources that it holds.

Reflection 13.2 *In which category does each of the nine operations above fall?*

Of the nine operations we identified above for the **Pair** ADT, the first is the constructor because it creates a new instance of a **Pair**. The next five operations are accessors because they give information that is derived from the attribute values of an ADT instance, without modifying it. The **set** and **scale** operations are mutators because they change the values of the attributes of an instance of the **Pair** ADT. Finally, the last operation is the destructor. In other words:

1. constructor: **create**

2. accessors: **getFirst**, **getSecond**, **get**, **add**, and **subtract**

3. mutators: **set** and **scale**

4. destructor: **destroy**

Implementing a class

In Python, we implement an abstract data type as a *class*. A class serves as a blueprint for a category of data. *Objects* are particular instances of a class. In Section 3.1, we described this difference by analogy to a species and the organisms that belong to that species. The species description is like a class; it describes a category of organisms but is not an organism itself. The individual organisms belonging to a species are like objects from the same class. For example, in

```
george = turtle.Turtle()
diego = turtle.Turtle()
```

distinct objects are assigned to the names `george` and `diego`, both belonging to the class `Turtle` (which is contained in the module named `turtle.py`).

When we implement an ADT as a class, the attributes defined by the ADT are assigned to a set of variables called *instance variables*. The operations defined by the ADT are implemented as *methods*. Instance variables remain hidden to a programmer using the class, and are only accessed or modified indirectly through methods. The `Turtle` class contains several hidden instance variables that store each `Turtle` object's position, color, heading, and whether its tail is up or down. The `Turtle` class also defines several familiar methods, such as `forward/backward`, `left/right`, `speed`, and `up/down` that we can call to indirectly interact with these instance variables.

Let's now implement the `Pair` abstract data type as a class. The definition of a new class begins with the keyword `class` followed by the name of the class and, of course, a colon. The class' methods are indented inside the class.

The constructor method

In Python, the constructor of a class is always named `__init__` (with two underscore characters at both the beginning and the end). The beginning of the `Pair` class, with its constructor, is shown below.

```
class Pair:
    """An ordered pair class."""

    def __init__(self):
        """Constructor initializes a Pair object to (0, 0).

        Parameter:
            self: a Pair object

        Return value: None
        """

        self._a = 0    # the pair's first value
        self._b = 0    # the pair's second value
```

The constructor is implicitly called when we call the function bearing the name of the class to create a new object. For example, when we created the two `Turtle` objects above, we invoked the `Turtle` constructor twice. To invoke the constructor of the `Pair` class to create a new `Pair` object, we call

```
pair = Pair()
```

The parameter of the `__init__` method, named `self`, is an *implicit* reference to the object on which the method is being called. In the assignment statement above, when `Pair()` implicitly calls the constructor, `self` is assigned to the new object being created. This same object is implicitly returned by the constructor and assigned to `pair`. We never *explicitly* pass anything in for `self`. We will see more examples of this soon.

Inside the constructor method, two *instance variables*, `self._a` and `self._b`, corresponding to the two attributes in the ADT, are assigned the value 0. These instance variables are also commented by the descriptions of their corresponding attributes in the ADT. An *instance* is another word for an object. Attributes are called instance variables because every instance of the class (i.e., an object) has its own copy of the variable. Every instance variable name is preceded by `self.` to signify that it belongs to the particular object assigned to `self`. Since `self` is assigned to the new object being created, the assignment statement above is, *in effect*, creating a new `Pair` object named `pair` and assigning

```
pair._a = 0
pair._b = 0
```

The underscore (_) character before each instance variable name is a Python convention that indicates that the instance variables should be *private*, meaning that they should never be accessed from outside the class. However, Python does not actually prevent someone from accessing private names (as other languages do). More specifically, any name in a class that is preceded by at least one underscore and not followed by two underscores like `__init__` is private by convention.

Reflection 13.3 *Why do you think it is important to keep instance variables private?*

The scope of an instance variable is the entire object. This means that we can access and change the values of `self._a` and `self._b` from within every method that we define for the class. In contrast, any variable name defined inside a method that is not preceded by `self` is considered to be a local variable within that method.

A constructor can also take additional parameters that are used to initialize the object's instance variables. In the case of the `Pair` class, it would be convenient to be able to pass the pair's *a* and *b* values into the constructor. This is accomplished by the following modified constructor:

```
def __init__(self, a = 0, b = 0):
    """Constructor initializes a Pair object to (a, b).

    Parameter:
        self: a Pair object

    Return value: None
    """

    self._a = a    # the pair's first value
    self._b = b    # the pair's second value
```

The = 0 following each of the a and b parameters is specifying a *default argument*. This allows us to call the constructor with either no arguments as before, in which case 0 will be assigned to a and b, or with explicit arguments for a and b that will override the default arguments. If we supply only one argument, then a will be assigned to it, and 0 will be assigned to b. In the constructor, the values assigned to the parameters a and b are assigned to the instance variables self._a and self._b by the two assignment statements in the method. For example:

```
pair1 = Pair()         # pair1 will represent (0, 0)
pair2 = Pair(3)        # pair2 will represent (3, 0)
pair3 = Pair(3, 14)    # pair3 will represent (3, 14)
```

Reflection 13.4 *How would you create a new* Pair *object named* myPair *with value* $(0, 18)$?

Earlier, we said that abstract data types also have destructors that destroy ADT instances when we are done with them. However, Python destroys objects automatically, so we do not need explicit destructors.

Accessor methods

Next, let's add three *accessor* methods to the Pair class, based on the ADT specification. (Notice that the descriptions of the methods in the docstrings are taken from the ADT specification.)

```
def getFirst(self):
    """Return the first value of self.

    Parameter:
        self: a Pair object

    Return value: the first value of self
    """

    return self._a
```

```
        def getSecond(self):
            """Return the second value of self.

            Parameter:
                self: a Pair object

            Return value: the second value of self
            """

            return self._b

        def get(self):
            """Return the (a, b) tuple representing self.

            Parameter:
                self: a Pair object

            Return value: the (a, b) tuple representing self
            """

            return (self._a, self._b)
```

The methods `getFirst` and `getSecond` return the values of the `self._a` and `self._b` instance variables, respectively. The `get` method returns a tuple containing both `self._a` and `self._b`. For example,

```
pair3 = Pair(3, 14)
a = pair3.getFirst()      # a is assigned 3
b = pair3.getSecond()     # b is assigned 14
pair = pair3.get()        # pair is assigned the tuple (3, 14)
```

Notice that, like the `__init__` method, the first (and only) parameter to these methods is `self`. Again, a reference to the object on which the method is called is assigned to `self`. For example, when we call `pair3.getFirst()`, the object `pair3` is implicitly passed in for the parameter `self`, even though it is not passed in the parentheses following the name of the method. This means that, inside the method, `self._a` really refers to `pair3._a`, the value of the instance variable `_a` that belongs to the object named `pair3`.

Reflection 13.5 *Take a moment, if you haven't already, to type in what we have so far for the* `Pair` *class. (You might omit the docstrings to save time.) Then write a* `main` *function that creates the new* `Pair` *object from the previous Reflection and prints its value using the* `get` *method as follows.*

```
def main():
    myPair = Pair(0, 18)
    print(myPair.get())
```

Reflection 13.6 *Create another* `Pair` *object in the* `main` *function and print its values as well.*

Arithmetic methods

As in the `centroid` function, we sometimes want to add (or subtract) the corresponding elements in a pair. So we can define addition of two pairs (a, b) and (c, d) as the pair $(a + c, b + d)$. For example, $(3, 8) + (4, 5) = (7, 13)$. If we represented pairs as two-element tuples, then an addition function would look like this:

```
def add(pair1, pair2):
    """Return a new tuple representing the component-wise sum
       of pair1 and pair2.

    Parameters:
        pair1: a tuple representing a pair
        pair2: a tuple representing a pair

    Return value: the sum of the two pairs
    """

    sum1 = pair1[0] + pair2[0]
    sum2 = pair1[1] + pair2[1]
    return (sum1, sum2)
```

To use this function to find the sum of (3, 8) and (4, 5), we would do the following:

```
duo1 = (3, 8)
duo2 = (4, 5)
sum = add(duo1, duo2)    # sum is assigned (7, 13)
print(sum)               # prints "(7, 13)"
```

To implement this functionality for `Pair` objects, we need to write a similar `add` method for the `Pair` class. Since we are writing a method, one of the points involved in the sum calculation will be assigned to `self` and the other will be assigned to a second parameter. The implementation of the new method is shown below.

```
def add(self, pair2):
    """Return a new Pair representing the component-wise sum
       of self and pair2.

    Parameters:
        self: a Pair object
        pair2: another Pair object

    Return value: a Pair object representing self + pair2
    """

    sumA = self._a + pair2._a
    sumB = self._b + pair2._b
    return Pair(sumA, sumB)
```

Now to find the sum of two `Pair` objects named `duo1` and `duo2`, we would call this method as follows:

```
duo1 = Pair(3, 8)
duo2 = Pair(4, 5)
sum = duo1.add(duo2)     # sum is assigned Pair(7, 13)
print(sum.get())         # prints "(7, 13)"
```

In this case, `duo1` is assigned to `self` and `duo2` is assigned to the second parameter `pair2`. The method creates and returns a *new* `Pair` object, as emphasized in red.

Reflection 13.7 *Add the* `add` *method to your* `Pair` *class. Then, define two new* `Pair` *objects in your* `main` *function and compute their sum.*

Reflection 13.8 *Using the* `add` *method as a template, write a* `subtract` *method that subtracts another* `Pair` *object from* `self`.

Mutator methods

Our `Pair` ADT also includes two mutator operations that change the values of the attributes: `set` and `scale`. Implementing methods for these operations is no harder than implementing the previous methods.

```
def set(self, a, b):
    """Set the two values in self.

    Parameters:
        self: a Pair object
        a: a number representing a new first value for self
        b: a number representing a new second value for self

    Return value: None
    """

    self._a = a
    self._b = b

def scale(self, scalar):
    """Multiply the values in self by a scalar value.

    Parameters:
        self: a Pair object
        scalar: a number by which to scale the values in self

    Return value: None
    """

    self.set(self._a * scalar, self._b * scalar)
```

The set method, which is almost identical to the constructor, sets the first and second values of a pair object to the values passed in for the parameters a and b, respectively. For example,

```
pair1 = Pair()          # pair1 is assigned Pair(0, 0)
pair1.set(3, 14)        # pair1 is now Pair(3, 14)
print(pair1.get())      # prints "(3, 14)"
```

changes the self._a instance variable of pair1 from 0 to 3, and the self._b instance variable of pair1 from 0 to 14. Then, when we call pair1.get(), we get the tuple (3, 14).

The scale method takes as a parameter a numerical value (which is called a *scalar* in mathematics) and uses the set method to multiply the values of self._a and self._b by scalar. Notice that, when we call a method of the class from within another method, we need to preface the name of the called method with self. just as we do with instance variables.

Reflection 13.9 *How can we write the* scale *method without calling* self.set*?*

Reflection 13.10 *Add the* set *and* scale *methods to your class. Then, in your main function, use the* set *method to change the value of* myPair *to* (43, 23).

Let's now revisit the centroid function, but pass in a list of Pair objects instead of a list of tuples.

```
def centroid(points):
    """Compute the centroid of a list of Pair objects.

    Parameter:
        points: a list of Pair objects

    Return value: a Pair object representing the centroid of the points
    """

    n = len(points)
    if n == 0:
        return None
    sum = Pair()                    # sum is the Pair (0, 0)
    for point in points:
        sum = sum.add(point)        # sum = sum + point
    sum.scale(1 / n)                # divide sum by n
    return sum

def main():
    homes = [Pair(0.5, 5), Pair(3.5, 2), Pair(4, 3.5),
             Pair(5, 2), Pair(7, 1)]
    central = centroid(homes)
    print(central.get())

main()
```

Notice that we have replaced the two `sumX` and `sumY` variables with a single `sum` variable, initialized to the pair (0,0). Inside the `for` loop, each value of `point`, which is now a `Pair` object, is added to `sum` using the `add` method. After the loop, we use the `scale` method to multiply the point by `1 / n` which, of course, is the same as dividing by `n`. In the `main` function, we assign `homes` to be a list of `Pair` objects instead of tuples. Printing the value of the centroid at the end is slightly more cumbersome because we have to convert it to a tuple first, but we will fix that in the next section.

Reflection 13.11 *Add the code above after your `Pair` class. What is the value of the centroid?*

Documenting a class

Just as we have consistently documented programs and functions, we have also documented the `Pair` class and its methods. Notice that, in the `Pair` class, there are docstrings for the class itself immediately after the `class` line and for each method, just as we have done for functions. These docstrings provide documentation on your class to anyone using it. Once the class is defined, calling

```
help(Pair)
```

will produce a summary of your class constructed from your docstrings.

```
class Pair(builtins.object)
 |  An ordered pair class.
 |
 |  Methods defined here:
 |
 |  __init__(self, a=0, b=0)
 |      Constructor initializes a Pair object to (a, b).
 |
 |      Parameter:
 |          self: a Pair object
 |
 |      Return value: None
 |
 |  add(self, pair2)
 |      Return a new Pair representing the component-wise sum
 |          of self and pair2.
 |
 |      Parameters:
 |          self: a Pair object
 |          pair2: another Pair object
 |
 |      Return value: a Pair object representing self + pair2
 |
 ⋮
```

Reflection 13.12 *In your program, call `help(Pair)` to view the complete documentation for the `Pair` class.*

Exercises

13.1.1. Name two accessor methods and two mutator methods in the `Turtle` class.

13.1.2. Name two accessor methods and two mutator methods in the `list` class.

13.1.3. Add a method to the `Pair` class named `round` that rounds the two values to the nearest integers.

13.1.4. Suppose you are tallying the votes in an election between two candidates. Write a program that repeatedly prompts for additional votes for both candidates, stores these votes in a `Pair` object, and then adds this `Pair` object to a running sum of votes, also stored in `Pair` object. For example, your program output may look like this:

```
Enter votes (q to quit): 1 2
Enter votes (q to quit): 2 4
Enter votes (q to quit): q

Candidate 1: 3 votes
Candidate 2: 6 votes
```

13.1.5. Suppose you are writing code for a runner's watch that keeps track of a list of split times and total elapsed times. While the timer is running, and the split button is pressed, the time elapsed since the last split is recorded in a `Pair` object along with the total elapsed time so far. For example, if the split button were pressed at 65, 67, and 62 second intervals, the list of (split, elapsed) pairs would be `[(65, 65), (67, 132), (62, 194)]` (where a tuple represents a `Pair` object). Write a function that is to be called when the split button is pressed that updates this list of `Pair` objects. Your function should take two parameters: the list of `Pair` objects and the current split time.

13.1.6. A data logging program for a jetliner periodically records the time along with the current altitude in a `Pair` object. Write a function that takes such a list of `Pair` objects as a parameter and plots the data using `matplotlib`.

13.1.7. Write a function that returns the distance between two two-dimensional points, each represented as a `Pair` object.

13.1.8. Write a function that returns the average distance between a list of points, each represented by a `Pair` object and a given site, also represented as a `Pair` object.

13.1.9. The file `africa.txt`, available on the book web site, contains (longitude, latitude) locations for cities on the African continent. The following program should read this file into a list of `Pair` objects, find the closest and farthest pairs of points in the list, and then plot all of the points using turtle graphics, coloring the closest pair blue and farthest pair red. To finish this program, add a method `draw(self, tortoise, color)` to the `Pair` class that plots a `Pair` object as an (x, y) point, and write the functions named `closestPairs` and `farthestPairs`.

```
import turtle

class Pair:
    # fill in here from the text

    def draw(self, tortoise, color):
        pass

def closestPairs(points):
    pass

def farthestPairs(points):
    pass

def main():
    points = []
    inputFile = open('africa.txt', 'r', encoding = 'utf-8')
    for line in inputFile:
        values = line.split()
        longitude = float(values[0])
        latitude = float(values[1])
        p = Pair(longitude, latitude)
        points.append(p)

    cpoint1, cpoint2 = closestPairs(points)
    fpoint1, fpoint2 = farthestPairs(points)

    george = turtle.Turtle()
    screen = george.getscreen()
    screen.setworldcoordinates(-37, -23, 37, 58)
    george.hideturtle()
    george.speed(0)
    screen.tracer(10)
    for point in points:
        point.draw(george, 'black')
    cpoint1.draw(george, 'blue')
    cpoint2.draw(george, 'blue')
    fpoint1.draw(george, 'red')
    fpoint2.draw(george, 'red')
    screen.update()
    screen.exitonclick()

main()
```

13.1.10. In this chapter, we implemented a Pair ADT with a class in which the two values were stored in two variables. These two variables comprised the extremely simple data structure that we used to implement the ADT. Now rewrite the Pair class so that it stores its two values in a two-element list instead. The way in which the class' methods are called should remain exactly the same. In other words, the way someone uses the class (the ADT specification) must remain the same even though the implementation (the data structure) changes.

13.1.11. Write a `BankAccount` class that has a single instance variable (the available balance), a constructor that takes the initial balance as a parameter, and methods `getBalance` (which should return the amount left in the account), `deposit` (which should deposit a given amount into the account), and `withdraw` (which should remove a given amount from the account).

13.1.12. Using your `BankAccount` class from the previous exercise, write a program that prompts for an initial balance, creates a `BankAccount` object with this balance, and then repeatedly prompts for deposits or withdrawals. After each transaction, it should update the `BankAccount` object and print the current balance. For example:

```
Initial balance? 100
(D)eposit, (W)ithdraw, or (Q)uit? d
Amount = 50
Your balance is now $150.00
(D)eposit, (W)ithdraw, or (Q)uit? w
Amount = 25
Your balance is now $125.00
(D)eposit, (W)ithdraw, or (Q)uit? q
```

13.1.13. Write a class that represents a U.S. president. The class should include instance variables for the president's name, party, home state, religion, and age when he or she took office. The constructor should initialize the president's name to a parameter value, but initialize all other instance variables to default values (empty strings or zero). Write accessor and mutator methods for all five instance variables.

13.1.14. On the book web site is a tab-separated file containing a list of all U.S. presidents with the five instance variables mentioned in the previous exercise. Write a function that reads this information and returns a list of president objects (using the class you wrote in the previous exercise) representing all of the presidents in the file. Also, write a function that, given a list of president objects and an age, prints a table with all presidents who where at least that old when they took office, along with their ages when they took office.

13.1.15. Write a `Movie` class that has as instance variables the movie title, the movie year, and a list of actors (all of which are initialized in the constructor). Write accessor and modifier functions for all the instance variables and an `addActor` method that adds an actor to the list of actors in the movie. Finally, write a method that takes as a parameter *another* movie object and checks whether the two movies have any common actors.

There is a program on the book web site with which to test your class. The program reads actors from a movie file (like those used in Project 12.3), and then prompts for movie titles. For each movie, you can print the actors, add an actor, and check whether the movie has actors in common with another movie.

13.1.16. Write a class representing a U.S. senator. The `Senator` class should contain instance variables for the senator's name, political party, home state, and a list of committees on which they serve. The constructor should initialize all of the instance variables to parameter values, except for the list of committees, which should be initialized to an empty list. Add accessor methods for all four instance variables, plus a mutator method that adds a committee to a senator's list of committees.

13.1.17. On the book web site is a function that reads a list of senators from a file and returns a list of senator objects, using the `Senator` class that you wrote in the previous exercise. Write a program that uses this function to create a list of `Senator` objects, and then iterates over the list of `Senator` objects, printing each senator's name, party, and committees. Then your program should prompt repeatedly for the name of a committee, and print the names and parties of all senators who are on that committee. For example,

```
Alexander, Lamar (R)
    Committee on Appropriations
    Committee on Energy and Natural Resources
    Committee on Health, Education, Labor, and Pensions
    Committee on Rules and Administration
Ayotte, Kelly (R)
    Commission on Security and Cooperation in Europe
    Special Committee on Aging
⋮

Committee name (q to quit): Select Committee on Intelligence
Burr, Richard (R)
Chambliss, Saxby (R)
⋮

Committee name (q to quit): Committee on Finance
Bennet, Michael F. (D)
Brown, Sherrod (D)
⋮

Committee name (q to quit): q
```

13.1.18. Write a class named `Student` that has the following instance variables: student name, exam grades, quiz grades, lab grades, and paper grades. The constructor should only take the student name as a parameter, but initialize all the other instance variables (to empty lists). Write an accessor method for the name and methods to add grades to the lists of exam, quiz, paper, and lab grades. Next, write methods for returning the exam, quiz, paper, and lab averages. Finally, write a method to compute the final grade for the course, assuming the average exam grade is worth 50%, the average quiz grade is worth 10%, and the average lab and paper grades are worth 20% each.

13.1.19. Write a class that represents a set of numerical data from which simple descriptive statistics can be computed. The class should contain five methods, in addition to the constructor: add a new value to the data set, return the minimum and maximum values in the data set, return the average of the values in the data set, and return the size of the data set. Think carefully about the instance variables needed for this class. It is not actually necessary for the class to include a list of all of the values that have been added to it.

13.1.20. This exercise assumes you read Section 6.7. Write a `Sequence` class to represent a DNA, RNA, or amino acid sequence. The class should store the type of sequence, a sequence identifier (or accession number), and the sequence itself. Identify and implement at least three useful methods, in addition to the constructor.

13.2 OPERATORS AND SPECIAL METHODS

Using built-in functions like `str` and `len`, and operators like `+`, `in`, and `<` considerably simplifies how we work with standard Python objects like numbers, strings, and lists. But these functions and operators have different meanings for each class. Think about, for example, how the meaning of the `+` operator differs among numbers, strings, and lists. Similarly, the `str` function returns a different string representation for each class. This ability to define operators differently for different classes is called *operator overloading*. By overloading appropriate operators for our `Pair` class as well, we can make it as convenient to use as the built-in classes.

String representations

When we print a `Pair` object like

```
counts = Pair(3, 14)
print(counts)
```

we would ideally get a convenient representation that shows the values in the ordered pair. Instead, we get

```
<__main__.Pair object at 0x10215ab50>
```

This is a default string representation of an object; it tells us that `counts` is a `Pair` object in the `__main__` namespace, located in memory at address `10215ab50` (in hexadecimal notation).

To override this default behavior, we can define a *special method* named `__str__` in the `Pair` class. Like `__init__`, `__str__` is an implicitly called method. The `__str__` method is called implicitly whenever we call the `str` function on an object. In other words, calling

```
str(counts)
```

is identical to calling

```
counts.__str__()
```

Since the `print` function implicitly calls `str` on an object to get the string representation that it prints, defining the `__str__` method also dictates how `print` behaves.

The following `__str__` method for the `Pair` class returns a string representing the pair in parentheses (like a tuple).

```
        def __str__(self):
            """Return an '(a, b)' string representation of self.

            Parameter:
                self: a Pair object

            Return value: an '(a, b)' string representation of self
            """

            return '(' + str(self._a) + ', ' + str(self._b) + ')'
```

With this method added to the class, if we want to include a string representation of a `Pair` object in a larger string, we can now use the `str` function. For example,

```
print('The current votes are ' + str(counts) + '.')
```

will now print

```
The current votes are (3, 14).
```

Also, just calling

```
print(counts)
```

will print

```
(3, 14)
```

Reflection 13.13 *Add the* `__str__` *method to your* `Pair` *class. Then print some of the* `Pair` *objects in your* `main` *function.*

Arithmetic

Just as we can define the special `__str__` method in the `Pair` class to define how `str` and `print` behave with `Pair` objects, we can define other special methods to define how operators work with `Pair` objects. For example, the `__add__` method is implicitly called when the + operator is used. An assignment statement like

```
name = first + last
```

is identical to

```
name = first.__add__(last)
```

The ability to define this special method for each class is precisely what allows us to use the + operator in different ways on different objects (e.g., numbers, strings, lists). We can implement the + operator on `Pair` objects by simply changing the name of our `add` method to `__add__`:

```
def __add__(self, pair2):
    """ (docstring omitted) """

    sumA = self._a + pair2._a
    sumB = self._b + pair2._b
    return Pair(sumA, sumB)
```

With this special method defined, we can carry out our previous example as follows instead:

```
duo1 = Pair(3, 8)
duo2 = Pair(4, 5)
sum = duo1 + duo2     # sum is assigned Pair(7, 13)
print(sum)            # prints "(7, 13)"
```

Reflection 13.14 *Incorporate the* `__add__` *method into your* `Pair` *class and experiment with adding* `Pair` *objects.*

Reflection 13.15 *The behavior of the − operator is similarly defined by the* `__sub__` *method. Modify our previous* `subtract` *method so that it is called when the − operator is used with* `Pair` *objects.*

Similarly, we can define * and / operators to implement multiplication with `Pair` objects. The methods corresponding to these operators are named `__mul__` and `__truediv__`, respectively. (Recall that / is called true division in Python while // is called floor division. The // operator is defined by the `__floordiv__` method.) Similarly to the `scale` method, let's define multiplication of a `Pair` object by a scalar quantity.

```
def __mul__(self, scalar):
    """Return a new Pair representing self multiplied by scalar.

    Parameters:
        self: a Pair object
        scalar: a number

    Return value: a Pair object representing self * scalar
    """

    return Pair(self._a * scalar, self._b * scalar)
```

Like the `__add__` method, this method returns a new `Pair` object, but this time it is simply the values in `self` multiplied by a number.

Reflection 13.16 *How is the* `__mul__` *method different from* `scale`? *(What does each return?)*

With this new method, we can easily scale pairs of numbers in one statement:

```
bets = Pair(150, 100)
double = bets * 2     # double is assigned Pair(300, 200)
```

Reflection 13.17 *Why would* `double = 2 * bets` *give an error?*

Reflection 13.18 *Using the* `__mul__` *operator as a template, write the* `__truediv__` *method to divide a* `Pair` *object by a scalar value.*

Applying the new addition and (true) division operators and the `__str__` method to our `centroid` function from the previous section makes the code much more elegant and natural.

```
def centroid(points):
    """ (docstring omitted) """

    n = len(points)
    if n == 0:
        return None
    sum = Pair()
    for point in points:
        sum = sum + point
    return sum / n

def main():
    homes = [Pair(0.5, 5), Pair(3.5, 2), Pair(4, 3.5),
             Pair(5, 2), Pair(7, 1)]
    central = centroid(homes)
    print(central)

main()
```

Comparison

We can also overload the comparison operators ==, <, <=, etc. using the following special methods.

Operator	==	!=	<	<=	>	>=
Method	__eq__	__ne__	__lt__	__le__	__gt__	__ge__

The special method named `__eq__` defines how the == operator behaves with a class. It is natural to say that two pairs are equal if their corresponding values are equal, as the following method implements.

```
        def __eq__(self, pair2):
            """Return whether self and pair2 contain the same ordered pair.

            Parameters:
                self: a Pair object
                pair2: another Pair object

            Return value:
                True if the corresponding values of self and pair2 are equal;
                False otherwise
            """

            return (self._a == pair2._a) and (self._b == pair2._b)
```

The special method named `__lt__` defines how the < operator behaves with a class. If duo1 and duo2 are two `Pair` objects, then duo1 < duo2 should return `True` if duo1._a < duo2._a, or if duo1._a == duo2._a and duo1._b < duo2._b. Otherwise, it should return `False`.

```
        def __lt__(self, pair2):
            """Return whether self < pair2.

            Parameters:
                self: a Pair object
                pair2: another Pair object

            Return value: True if self < pair2; False otherwise
            """

            return (self._a < pair2._a) or \
                   ((self._a == pair2._a) and (self._b < pair2._b))
```

Suppose we store the number of wins and ties in a `Pair` object for each of three teams. If a team is ranked higher when it has more wins and the number of ties is used to rank teams with the same number of wins, then the comparison operators we defined can be used to decide rankings.

```
    wins1 = Pair(6, 2)       # 6 wins, 2 ties
    wins2 = Pair(6, 4)       # 6 wins, 4 ties
    wins3 = Pair(6, 2)       # 6 wins, 2 ties

    print(wins1 < wins2)     # prints "True"
    print(wins2 < wins3)     # prints "False"
    print(wins1 == wins3)    # prints "True"
```

With the `__eq__` and `__lt__` methods defined, Python will automatically deduce the outcomes of the other four comparison operators. However, we leave their implementations to you as a practice exercise.

Reflection 13.19 *Add these two new methods to your* `Pair` *class. Experiment with some comparisons, including those we did not implement, in your* `main` *function.*

Indexing

When an element in a string, list, tuple, or dictionary is accessed with indexing, a special method named `__getitem__` is implicitly called. For example, if `maxPrices` and `minPrices` are lists, then

```
range = maxPrices[0] - minPrices[0]
```

is equivalent to

```
range = maxPrices.__getitem__(0) - minPrices.__getitem__(0)
```

Similarly, when we use indexing to change the value of an element in a sequence, a method named `__setitem__` is implicitly called. For example,

```
temperatures[1] = 18.9
```

is equivalent to

```
temperatures.__setitem__(1, 18.9)
```

We can define indexing, as an alternative to the `getFirst` and `getSecond` methods, to access the values in a `Pair` object. The following implements the `__getitem__` method for the `Pair` class.

```
def __getitem__(self, index):
    """Return the first (index 0) or second (index 1) value
       in self.  For other index values, return None.

    Parameter:
        index: an integer (0 or 1)

    Return value: an element in a pair (or None)
    """

    if index == 0:
        return self._a
    if index == 1:
        return self._b
    return None
```

The `__getitem__` method returns the value of `self._a` or `self._b` if `index` is 0 or 1, respectively. If index is anything else, it returns `None`.

Reflection 13.20 *Is this behavior consistent with what happens when you use an erroneous index with a list?*

When we use an erroneous index with an object from one of the built-in classes, we get a `IndexError`. We will look at how to implement this alternative behavior in Sections 13.5 and 13.6.

As an example, suppose we have a `Pair` object defined as follows:

```
counts = Pair(12, 15)
```

With the new `__getitem__` method, we can retrieve the individual values in `counts` with

```
first = counts[0]
second = counts[1]
```

as these statements are equivalent to

```
first = counts.__getitem__(0)
second = counts.__getitem__(1)
```

Next, we can implement the `__setitem__` method as follows.

```
def __setitem__(self, index, value):
    """Set the first (index 0) or second (index 1) value in
        self to the given value.

    Parameters:
        index: an integer (0 or 1)
        value: a number to which to set a value in self

    Return value: None
    """

    if index == 0:
        self._a = value
    elif index == 1:
        self._b = value
```

The `__setitem__` method assigns `self._a` or `self._b` to the given `value` if `index` is 0 or 1, respectively.

Reflection 13.21 *What does the* `__setitem__` *method do if* `index` *is not 0 or 1?*

With the new `__setitem__` method, we can assign a new value to `counts` with

```
counts[0] = 14
counts[1] = 16
print(counts)     # prints "(14, 16)"
```

With these indexing methods defined, we can now use indexing within other methods, where it is convenient. For example, we can use indexing in the `__add__` method to get the individual values and in the `set` method to assign new values.

```python
    def __add__(self, pair2):
        """Return a new Pair representing the component-wise sum
            of self and pair2."""

        sumA = self[0] + pair2[0]
        sumB = self[1] + pair2[1]
        return Pair(sumA, sumB)

    def set(self, a, b):
        """Set the two values in self."""

        self[0] = a
        self[1] = b
```

Reflection 13.22 *Add the two indexing methods to your* `Pair` *class. Then modify the* `__lt__` *method so that it uses indexing to access values of* `self._a` *and* `self._b` *instead.*

You can find a summary of these and other special methods in Appendix B.9.

Exercises

13.2.1. Implement alternative `__mul__` and `__truediv__` methods for the `Pair` class that multiply two `Pair` objects. The product of two `Pair` objects `pair1` and `pair2` is a `Pair` object in which the first value is the product of the first values of `pair1` and `pair2`, and the second value is the product of the second values of `pair1` and `pair2`. Division is defined similarly.

13.2.2. Implement the remaining four comparison operators (`!=`, `<=`, `>`, `>=`) for the `Pair` class.

13.2.3. Rewrite your `linearRegression` function from Exercise 8.6.1 so that it takes a list of `Pair` objects as a parameter.

13.2.4. Add a `__str__` method to the president class that you wrote in Exercise 13.1.13. The method should return a string containing the president's name and political party, for example, `'Kennedy (D)'`. Also, write a function that, given a list of president objects and a state abbreviation, prints the presidents in this list (indirectly using the new `__str__` method) that are from that state.

13.2.5. Add a `__lt__` method to the president class that you wrote in Exercise 13.1.13. The method should base its results on a comparison of the presidents' ages.

13.2.6. Add a `__str__` method to the `Senator` class from Exercise 13.1.16 that prints the name of the senator followed by their party, for example, `'Franken, Al (D)'`.

13.2.7. Rewrite the distance function from Exercise 13.1.7 so that it uses indexing to get the first and second values from each pair.

13.2.8. Modify your program from Exercise 13.1.17 so that it uses the new `__str__` method that you wrote in the previous exercise.

13.2.9. Write a class that represents a rational number (i.e., a number that can be represented as a fraction). The constructor for your class should take as parameters a numerator and denominator. In addition, implement the following methods for your class:

- arithmetic: __add__, __sub__, __mul__, __truediv__
- comparison: __lt__, __eq__, __le__
- __str__

When you are done, you should be able to perform calculations like the following:

```
a = Rational(3, 2)    # 3/2
b = Rational(1, 3)    # 1/3
sum = a + b
print(sum)            # should print 11/6
print(a < b)          # should print False
```

13.3 MODULES

When we define a new class, we usually intend it to be broadly useful in a variety of programs. Therefore, we should save each class as a *module* in its own file. Then we can use the `import` statement to import the class definition into a program. This is precisely the way we have been using the `Turtle` class all along.

For example, if we save our `Pair` class in its own module named `pair.py`, then we can write a very simple program using the `Pair` class that looks like the following:

```
import pair

def main():
    pair1 = pair.Pair(3, 8)
    pair2 = pair.Pair(4, 5)
    sum = pair1 + pair2
    difference = pair1 - pair2
    print(sum, difference)

main()
```

As you may have seen in Section 7.3, the `import` statement both imports names from another file *and* executes the code in that file. Therefore, if a module like `pair.py` calls its own `main` function, the `main` function in `pair.py` will be executed when `pair` is imported into a program, *before* the `main` function in the program is executed. To prevent this from happening, we can place the module's call to `main()` inside the following conditional statement:

```
if __name__ == '__main__':
    main()
```

For example, suppose the `pair.py` module looks like this:

```
"""pair.py"""

class Pair:
    def __init__(self, x = 0, y = 0):
        ⋮

    # rest of the Pair class here
def main():
    myPair = Pair(0, 18)
    print(myPair)

if __name__ == '__main__':
    main()
```

Without the red `if` statement, when `pair` is imported into the program above, the `main` function in `pair.py` will be executed before the `main` function in the program. But the `main` function in `pair.py` will not be called if we include the red `if` statement. The name `__name__` is the name of the current module, which is assigned the value `'__main__'` if the module is run as the main program. But when `pair.py` is imported into another module instead, `__name__` is set to `'pair'`. So this statement executes `main()` only if the module is run as the main program.

Namespaces, redux

Recall from Section 3.6 that each function and module defines its own namespace. The same is true for classes and objects, as illustrated in Figure 13.1. When we import the `pair` module with `import pair`, a new namespace is created with the name `pair`. Inside the `pair` namespace is the name `Pair`, which refers to the `Pair` class. The `Pair` class also has a namespace containing the names of all of the methods of that class, as shown on the right side of Figure 13.1. And each of these method names refers to a function with its own namespace (not shown in the figure).

When a new object is created, its namespace contains the names of all of its instance variables. For example, in Figure 13.1, the namespace for each of the `Pair` objects named `duo1` and `duo2` contains the instance variable names `_a` and `_b`. In addition, every object namespace contains the special name `__class__`, which refers to the class to which the object belongs. In our example, the name `__class__` in the namespaces for `duo1` and `duo2` refers to the `Pair` class. The `__class__` name connects an object to its methods. When we call a method for an object such as `duo1.get()`, this is equivalent to calling `duo1.__class__.get(duo1)`. (Yuck!)

To avoid having to preface the name of the class with the name of the module, we could substitute

```
import pair
```

with

Modules 647

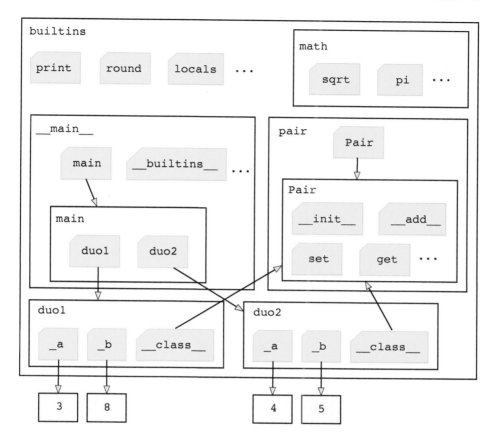

Figure 13.1 An illustration of namespaces in an object-oriented program.

```
from pair import Pair
```

or

```
from pair import *
```

The * character is a *wildcard* character that represents all of the names in the module. Using this alternative import style imports the names in the module into the current local namespace instead of creating a separate namespace. Since we always import modules at the top of our programs, the local namespace into which these names are imported is the global namespace.

Exercises

13.3.1. Create a new module for your `BankAccount` class from Exercise 13.1.11. Then rewrite your program from Exercise 13.1.12 so that it imports the `BankAccount` class from your new module.

13.3.2. Consider the following short program that uses the `BankAccount` class that you wrote in Exercise 13.1.11. The program assumes that the `BankAccount` class is in a module named `bankaccount.py`.

```
import bankaccount

def main():
    account = bankaccount.BankAccount(100)
    account.deposit(50)
    print('Your balance is $0:<4.2f'.format(account.getBalance()))

main()
```

Draw a namespace diagram like that in Figure 13.1 that includes namespaces for `__main__`, the `main` function, the `bankaccount` module, the `BankAccount` class, and the object named `account`.

13.3.3. Create a new module for the president class that you wrote in Exercise 13.1.13. Then rewrite your program from Exercise 13.1.14 so that it imports your president class from the new module.

13.3.4. Consider the following (silly) program that uses the president class that you wrote in Exercise 13.1.13. The program assumes that your president class is named `President` and is in a module named `president.py`.

```
import president

def main():
    taft = president.President('Taft')
    taft.setAge(51)
    washington = president.President('Washington')
    washington.setAge(57)

    print(washington.getName() + ' was older than '
            + taft.getName() + ' when he took office.')

main()
```

Draw a namespace diagram like that in Figure 13.1 that includes namespaces for `__main__`, the `main` function, the `president` module, the `President` class, and both `President` objects.

13.4 A FLOCKING SIMULATION

It was once assumed that the collective movements of flocks of birds, schools of fish, and herds of animals arise from many individuals following a designated leader. But we now know that this is not the case; instead each individual is independently carrying out an identical algorithm based on local interactions with its neighbors. The adaptive swarming and V-like patterns that result from these local interactions between individuals are known as *emergent* behaviors. This particular type of emergent behavior is, for obvious reasons, commonly known as *swarm intelligence*.

The *boids model*, originally conceived by Craig Reynolds, is a simulation of emergent flocking behavior. In this model, a "boid" is a representation of a bird, fish or other animal. Each boid independently follows three rules:

1. Avoid collisions with obstacles and nearby flockmates.

2. Attempt to match the velocity (heading plus speed) of nearby flockmates.

3. Attempt to move toward the center of the flock to avoid predators.

In this section, we will write an object-oriented *agent-based simulation* to visualize the flocking behavior that results from the boids model. An agent-based simulation consists of a set of independent individuals (the agents) that interact with each other in some way over time. Schelling's model of racial segregation from Project 9.1 is a simple example of an agent-based simulation. But agent-based simulations can also be much more involved, if the agents can roam more freely and/or be endowed with more attributes. For example, in an agent-based predator-prey simulation, each individual may be either a predator or a prey, endowed with characteristics describing its fitness, age, size, sex, and location. Each animal could then be allowed to move throughout its habitat and initiate events based on encounters with other animals. This differs substantially from the predator-prey model that you may have explored in Project 4.4, as the earlier model treated each population as a group of identical individuals, and was concerned only with the populations' sizes.

An agent-based simulation requires two main components: the agents (in our case, boids) and a "world" in which the agents live. Each of these components can be conceived as an abstract data type, and implemented as a class.

The World ADT

The simplest kind of world for agents to inhabit is a rectangular two-dimensional plane, like those we used in Chapter 9. Although we will limit our boids to dwell in two dimensions for now, we will think of the world as a continuous space rather than a grid. In other words, each boid may reside at any (x, y) position within its boundaries, not just those where x and y are integers. (In Project 13.4, you will have the opportunity to extend this to a three-dimensional world.) An abstract data type for this flat world will contain attributes for its dimensions and a list of agents inhabiting it, with their positions, as shown below.

World ADT Attributes	
Name	Description
agents	a list of agents, with their positions
width	the number of columns in the grid
height	the number of rows in the grid

The operations for a World ADT involve getting its dimensions, getting, setting and deleting the agent in a particular position, and querying the neighborhood of any position. We will represent a position with a (x, y) tuple, where x is a column number and y is a row number.

World ADT Operations

Name	Arguments	Description
create	width, height	create a new **World** instance with the given size
getWidth	—	return the width of the world
getHeight	—	return the height of the world
get	position	return the agent in the given position
set	position, agent	place an agent in the given position
delete	position	delete the agent in the given position
neighbors	position, distance	return a list of agents within some distance of a position
stepAll	—	advance all agents one step in the simulation

As suggested by the **stepAll** function in the table, the **World** ADT also drives the simulation. A program that implements an agent-based simulation using this ADT will repeatedly call **stepAll** to simulate the progression of time.

A class that implements the constructor, plus the first two accessor operations, is shown below. (We will use abbreviated docstrings to save space.)

```
class World:
    """A two-dimensional world class."""

    def __init__(self, width, height):
        """Construct a new flat world with the given dimensions."""

        self._width = width
        self._height = height
        self._agents = { }

    def getWidth(self):
        """Return the width of self."""

        return self._width

    def getHeight(self):
        """Return the height of self."""

        return self._height
```

As an alternative to the list-of-lists grid implementation from Chapter 9, we will store the agents in a dictionary with keys equal to the agents' positions (tuples). In this implementation, we start with an empty dictionary and add entries as agents are added to the world. This way, we only use as much space as we need to store the agents and assume that the rest of the world is empty. In contrast, the implementation that we used in Chapter 9 explicitly stores every cell, whether or not it is occupied. If most cells are always occupied, then this implementation is perfectly reasonable. However, in situations where most cells are not occupied at any particular time, as will be the case in our boids simulation, the dictionary implementation is more efficient. In these situations, we say that the space is *sparse* or a *sparse matrix*.

Two-dimensional indexing

To get or change the agent in a particular position in the world, we can define indexing for the World class using the __getitem__ and __setitem__ methods. In our two-dimensional world, an index needs to be a (x, y) pair. So we can define the __getitem__ and __setitem__ methods for our World class to interpret the index parameter as a two-element tuple that we directly use as a key in our dictionary.

```
def __getitem__(self, position):
    """Return the agent at the given position."""

    if position in self._agents:
        return self._agents[position]
    return None

def __setitem__(self, position, agent):
    """Set the given position to contain agent."""

    if (position not in self._agents) and \
       (position[0] >= 0) and (position[0] < self._width) and \
       (position[1] >= 0) and (position[1] < self._height):
        self._agents[position] = agent
```

With these methods, we can get and set particular positions in the world by simply indexing with tuples. As a simple example, we can place an integer value (in lieu of an agent for now) in the world at position $(2, 1)$ and then print the value at that position.

```
myWorld = World(10, 10)
myWorld[2, 1] = 5
print(myWorld[2, 1])
```

Notice that we did not need to put parentheses around the tuple values that we used in the square brackets; Python will automatically wrap the pair in parentheses and interpret it as a tuple. If we try to get an agent from a position that is empty, the __getitem__ method returns None. The __setitem__ method does nothing if we try to insert an agent into a position that is occupied or out of bounds.

Reflection 13.23 *How can a simulation determine whether a position in the world is occupied?*

When an agent moves, we will need to delete it from its current position in the world. To delete an item from a Python list or dictionary, we can use the `del` operator. For example,

```
del frequency[18.9]
```

deletes the (key, value) pair with key equal to `18.9` from the dictionary named `frequency`. We can mimic this behavior in a `World` object by defining the `__delitem__` method.

```
def __delitem__(self, position):
    """Delete the agent at the given position."""

    if position in self._agents:
        del self._agents[position]
```

With this method defined, we can delete an agent from a particular position like this:

```
del myWorld[2, 1]
```

Meeting the neighbors

In an agent-based simulation in general, and our boids simulation in particular, we commonly need to query the neighborhood of an agent. In our boids simulation, each boid will repeatedly adjust its velocity based on the positions and velocities of nearby boids. Since we are not maintaining an explicit grid structure in this world, as in the cellular automata from Chapter 9, we need to go about finding neighbors a bit differently. The following method returns a list of agents within some distance of a particular position by iterating over all positions in the dictionary and checking whether each is within range.

```
def neighbors(self, position, distance):
    """Return a list of agents within distance of
       position (a tuple)."""

    neighbors = []
    for otherPosition in self._agents:
        if (position != otherPosition) and \
           (_distance(position, otherPosition) <= distance):
            neighbors.append(self._agents[otherPosition])
    return neighbors
```

The `_distance` function (not shown) is a private function defined outside the class that returns the distance between two positions.

Reflection 13.24 *Why is the* `_distance` *function not a method of* `World` *instead?*

We decided to not make `_distance` a method because it does not need to access any attributes of the class. The leading underscore in its name prevents the function from being imported into other modules.

Simulating one step

Finally, the last method in the `World` class iterates over every agent in the world, and has that agent take one step forward in the simulation. We assume that an agent's actions are implemented in an agent method named `step`. In the case of our boid simulation, a boid will look at its neighbors in each step and adjust its velocity accordingly. We will tackle this next.

```
def stepAll(self):
    """All agents advance one step in the simulation."""

    agents = list(self._agents.values())
    for agent in agents:
        agent.step()
```

The complete `World` class is shown below:

```
""" world.py """

import math

def _distance(point1, point2):
    """Return the distance between two positions."""

    diffX = point1[0] - point2[0]
    diffY = point1[1] - point2[1]
    return math.sqrt(diffX ** 2 + diffY ** 2)

class World:
    """A two-dimensional world class."""

    def __init__(self, width, height):
        """Construct a new flat world with the given dimensions."""

        self._width = width
        self._height = height
        self._agents = { }
```

```python
def getWidth(self):
    """Return the width of self."""

    return self._width

def getHeight(self):
    """Return the height of self."""

    return self._height

def __getitem__(self, position):
    """Return the agent at the given position."""

    if position in self._agents:
        return self._agents[position]
    return None

def __setitem__(self, position, agent):
    """Set the given position to contain agent."""

    if (position not in self._agents) and \
        (position[0] >= 0) and (position[0] < self._width) and \
        (position[1] >= 0) and (position[1] < self._height):
        self._agents[position] = agent

def __delitem__(self, position):
    """Delete the agent at the given position."""

    if position in self._agents:
        del self._agents[position]

def neighbors(self, position, distance):
    """Return a list of agents within distance of
       position (a tuple)."""

    neighbors = []
    for otherPosition in self._agents:
        if (position != otherPosition) and \
            (_distance(position, otherPosition) <= distance):
            neighbors.append(self._agents[otherPosition])
    return neighbors

def stepAll(self):
    """All agents advance one step in the simulation."""

    agents = list(self._agents.values())
    for agent in agents:
        agent.step()
```

The Boid ADT

The design of an agent in an agent-based simulation depends quite a bit on the particular application, but the relationship between the agent and the world tends to share some common characteristics. First, because agents can only interact indirectly with other agents through their shared world, an agent must have access to the world in which it resides. Second, the agent must "know" where in the world it resides. In our boid simulation, each boid will also have a velocity, which combines both its speed and heading.

Boid ADT Attributes	
Name	Description
world	the world in which the boid resides
position	the boid's (x, y) position in its world
velocity	the boid's velocity (speed and heading)

In each step of an agent-based simulation, each agent carries out some application-specific tasks, such as querying its neighbors and moving to a new location. We encapsulate this activity in a function named **step**. In our boids simulation, in each step, a boid will look at its neighbors, adjust its velocity according to the three rules of the boids model, and then move to its new position. These intermediate steps will be handled by the **neighbors** and **move** operations below.

Boid ADT Operations		
Name	Arguments	Description
create	the world	create a new **Boid** instance with random position and velocity in the given world
neighbors	distance, angle	return a list of boids within some distance and viewing angle
move	—	move to a new position in the world based on current velocity
step	—	adjust the boid's velocity following the three rules of the model

Vectors

Velocity is represented by a *vector*. A vector is simply an ordered pair, like a geometric point, but it represents a quantity with both magnitude and direction (e.g., velocity, force, or displacement). A vector $\langle x, y \rangle$ is often represented by a directed line segment that extends from the origin $(0, 0)$ to the point (x, y), as shown below on the left.

The angle α that the vector makes with the horizontal axis is the direction of the vector and the length of the line segment is the vector's magnitude (or length). If a vector represents velocity, then α is the direction of movement and the magnitude is speed. We know from Pythagorean theorem that the magnitude of the vector is $\sqrt{x^2 + y^2}$. If you know some trigonometry, you also know that the angle $\alpha = \tan^{-1}(y/x)$. Also, if you only know the magnitude r and angle α, you can find the vector $\langle x, y \rangle$ with $x = r\cos\alpha$ and $y = r\sin\alpha$, as illustrated above on the right.

To facilitate some of the computations that will be necessary in our boids simulation, we have written a new **Vector** class, based on the **Pair** class, but with a few additional methods that apply specifically to vectors. You can download the **vector.py** module from the book web site. We will not dwell on its details here, but instead include an abbreviated description given by the **help** function.

```
class Vector(builtins.object)
 |  A two-dimensional vector class.
 |
 |  Methods defined here:
 |
 |  __add__(self, vector2)
 |      Return the Vector that is self + vector2.
 |
 |  __getitem__(self, index)
 |      Return the value of the index-th coordinate of self.
 |
 |  __init__(self, vector=(0, 0))
 |      Constructor initializes a Vector object to <x, y>.
 |
 |  __mul__(self, scalar)
 |      Return the Vector <x * scalar, y * scalar>.
 |
 |  __setitem__(self, index, value)
 |      Set the index-th coordinate of self to value.
 |
 |  __str__(self)
 |      Return an '<x, y>' string representation of self.
 |
 |  __sub__(self, vector2)
 |      Return the Vector that is self - vector2.
```

```
 |
 |    __truediv__(self, scalar)
 |        Return the Vector <x / scalar, y / scalar>.
 |
 |    angle(self)
 |        Return the angle made by self (in degrees).
 |
 |    diffAngle(self, vector2)
 |        Return the angle (in degrees) between self and vector2.
 |
 |    dotProduct(self, vector2)
 |        Return the dot product of self and vector2,
 |            which is the cosine of the angle between them.
 |
 |    get(self)
 |        Return the (x, y) tuple representing self.
 |
 |    magnitude(self)
 |        Return the magnitude (length) of self.
 |
 |    scale(self, scalar)
 |        Multiply the coordinates in self by a scalar value.
 |
 |    set(self, x, y)
 |        Set the two coordinates in self.
 |
 |    turn(self, angle)
 |        Rotate self by the given angle (in degrees).
 |
 |    unit(self)
 |        Return a unit vector in the same direction as self.
```

The Boid class

In the constructor of the `Boid` class below, a reference to the `World` object in which the boid resides is passed as the parameter `myWorld`, and stored in the instance variable named `self._world`. Each boid needs to have access to the `World` object so that the boid can change its position in the world and call the `World` object's `neighbors` method.

```
class Boid:
    """A boid in a agent-based flocking simulation."""

    def __init__(self, myWorld):
        """Construct a boid at a random position in the given world."""

        self._world = myWorld
        (x, y) = (random.randrange(self._world.getWidth()),
                  random.randrange(self._world.getHeight()))
        while self._world[x, y] != None:
            (x, y) = (random.randrange(self._world.getWidth()),
                      random.randrange(self._world.getHeight()))
        self._position = Pair(x, y)
        self._world[x, y] = self
        self._velocity = Vector((random.uniform(-1, 1),
                                 random.uniform(-1, 1))).unit()
        self._turtle = turtle.Turtle()
        self._turtle.speed(0)
        self._turtle.up()
        self._turtle.setheading(self._velocity.angle())
```

The constructor assigns the new `Boid` object a random, unoccupied position in `self._world`. Notice that this involves three steps: finding an unoccupied position using a `while` loop, assigning this position (a `Pair` object) to the instance variable `self._position`, and placing the new `Boid` object (`self`) in `self._world` at that position. Next, the instance variable `self._velocity` is assigned a `Vector` object with random value between $\langle -1, -1 \rangle$ and $\langle 1, 1 \rangle$ (which covers every angle between 0 and 360 degrees). The `unit` method scales the velocity vector so that it has magnitude (speed) 1. Finally, we add an instance variable for a `Turtle` object to visualize the boid, and set the turtle's initial heading to the angle of the velocity vector.

Reflection 13.25 *What angle does the vector $\langle 1, 1 \rangle$ represent? What about $\langle -1, -1 \rangle$ and $\langle 0, -1 \rangle$?*

Reflection 13.26 *How many total instance variables does each `Boid` object have?*

Moving a boid

In each step of the simulation, a boid will move to a new location based on its current velocity (which will change periodically based on its interaction with neighboring boids). The `move` method below moves the `Boid` object by adding the x and y coordinates of its velocity to the x and y coordinates of its current position, respectively.

```
    def move(self):
        """Move self to a new position in its world."""

        self._turtle.setheading(self._velocity.angle())

        width = self._world.getWidth()
        height = self._world.getHeight()

        newX = self._position[0] + self._velocity[0]
        newX = min(max(0, newX), width - 1)
        newY = self._position[1] + self._velocity[1]
        newY = min(max(0, newY), height - 1)

        if self._world[newX, newY] == None:
            self._world[newX, newY] = self           # place in new pos
            del self._world[self._position.get()]    # and del from old
            self._position = Pair(newX, newY)        # set new pos
            self._turtle.goto(newX, newY)            # move turtle

        if (self._velocity[0] < 0 and newX < 5) or \
           (self._velocity[0] > 0 and newX > width - 5) or \
           (self._velocity[1] < 0 and newY < 5) or \
           (self._velocity[1] > 0 and newY > height - 5):
            self._velocity.turn(TURN_ANGLE)
```

After the boid's new position is assigned to `newX` and `newY`, if this new position is not occupied in the boid's world, the `Boid` object moves to it. Finally, if the boid is approaching a boundary, we rotate its velocity by some angle, so that it turns in the next step. The constant value `TURN_ANGLE` along with some other named constants will be defined at the top of the boid module. To avoid unnaturally abrupt turns, we use a small angle like

```
TURN_ANGLE = 30
```

Reflection 13.27 *What is the purpose of the* `min(max(0, newX), width - 1)` *expression (and the analogous expression for* `newY`*)?*

Reflection 13.28 *In the last* `if` *statement, why do we check both the new position and the velocity?*

Implementing the boids' rules

We are finally ready to implement the three rules of the boids model. In each step of the simulation, we will compute a new velocity based on each of the rules, and then assign the boid a weighted sum of these and the current velocity, scaled to a magnitude of one so that all boids maintain the same speed. (A vector with magnitude one is called a *unit vector*.) Once the velocity is set, we call `move` to move the boid to its new position.

```python
def step(self):
    """Advance self one step in the flocking simulation."""

    newVelocity = (self._velocity * PREV_WEIGHT +
                   self._avoid() * AVOID_WEIGHT +    # rule 1
                   self._match() * MATCH_WEIGHT +    # rule 2
                   self._center() * CENTER_WEIGHT)   # rule 3

    self._velocity = newVelocity.unit()
    self.move()
```

The private `_match`, `_center`, and `_avoid` methods will compute each of the three individual velocities. The boids model suggests that the weights assigned to the rules decrease in order of rule number. So avoidance should have the highest weight, velocity matching the next highest weight, and centering the lowest weight. For example, the following values follow these guidelines.

```
PREV_WEIGHT = 0.5
AVOID_WEIGHT = 0.25
MATCH_WEIGHT = 0.15
CENTER_WEIGHT = 0.1
```

Note that because we always scale the resulting vector to a unit vector, these weights need not always sum to 1. Once we have the complete simulation, we can modify these weights to induce different behaviors.

In each of the `_match`, `_center` and `_avoid` methods, we will need to get a list of boids within some distance and viewing angle. This is accomplished by the `Boid` method named `neighbors`, shown below.

```python
def neighbors(self, distance, angle):
    """Return a list of boids within distance and viewing angle."""

    neighbors = self._world.neighbors(self._position.get(), distance)
    visibleNeighbors = []
    for boid in neighbors:
        neighborDir = Vector((boid._position - self._position).get())
        if self._velocity.diffAngle(neighborDir) < angle:
            visibleNeighbors.append(boid)
    return visibleNeighbors
```

The method begins by calling the `neighbors` method of the `World` class to get a list of boids within the given distance. Then we iterate over these neighbors, and check whether each one is visible within the given viewing angle. Doing this requires a little vector algebra. In a nutshell, `neighborDir` is the vector pointing in the direction of the neighbor named `boid`, from the point of view of this boid (i.e., `self`). The `diffAngle` method of the `Vector` class computes the angle between `neighborDir` and the velocity of this boid. If this angle is within the boid's viewing angle, we add the neighboring boid to the list of visible neighbors to return.

With this infrastructure in place, methods that follow the three rules are relatively straightforward. Let's review them before continuing:

1. Avoid collisions with obstacles and nearby flockmates.

2. Attempt to match the velocity (heading plus speed) of nearby flockmates.

3. Attempt to move toward the center of the flock to avoid predators.

Since rule 2 is slightly easier than the other two, let's implement that one first.

```
def _match(self):
    """Return the average velocity of neighboring boids."""

    neighbors = self.neighbors(MATCH_DISTANCE, MATCH_ANGLE)
    if len(neighbors) == 0:
        return Vector()
    sumVelocity = Vector()
    for boid in neighbors:
        sumVelocity = sumVelocity + boid._velocity
    return (sumVelocity / len(neighbors)).unit()
```

The method first gets a list of visible neighbors, according to a distance and viewing angle that we will define shortly. If there are no such neighbors, the method returns the zero vector $\langle 0, 0 \rangle$. Otherwise, we return the average velocity of the neighbors, normalized to a unit vector. Since we have overloaded the addition and division operators for `Vector` objects, as we did for the `Pair` class, finding the average of a list of vectors is no harder than finding an average of a list of numbers!

The method to implement the first rule is similar, but now we want to find the velocity vector that points *away* from the average position of close boids.

```
def _avoid(self):
    """Return a velocity away from close neighbors."""

    neighbors = self.neighbors(AVOID_DISTANCE, AVOID_ANGLE)
    if len(neighbors) == 0:
        return Vector()
    sumPosition = Pair()
    for boid in neighbors:
        sumPosition = sumPosition + boid._position
    avgPosition = sumPosition / len(neighbors)

    avoidVelocity = Vector((self._position - avgPosition).get())
    return avoidVelocity.unit()
```

The first part of the method finds the average position (rather than velocity) of neighboring boids. Then we find the vector that points away from this average position by subtracting the average position from the position of the boid. This is illustrated below.

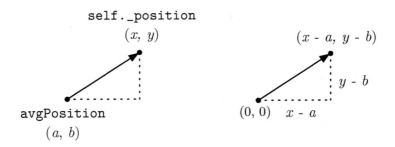

On the left, is an illustration of the vector we want, pointing from `avgPosition`, which we will call (a, b), to the boid's position (x, y). But vectors always start at $(0, 0)$, so the correct vector has horizontal distance $x - a$ and vertical distance $y - b$, as shown on the right. This is precisely the vector $\langle x - a, y - b \rangle$ that we get by subtracting `avgPosition` from `self._position`.

Finally, the method that implements rule 3 is almost identical to the `_avoid` method, except that we want a vector that points *toward* the average position of the flock, which we define to be a group of neighboring boids within a larger radius of the boid than those it is trying to avoid.

```
def _center(self):
    """Return a velocity toward center of neighboring flock."""

    neighbors = self.neighbors(CENTER_DISTANCE, CENTER_ANGLE)
    if len(neighbors) == 0:
        return Vector()
    sumPosition = Pair()
    for boid in neighbors:
        sumPosition = sumPosition + boid._position
    avgPosition = sumPosition / len(neighbors)

    centerVelocity = Vector((avgPosition - self._position).get())
    return centerVelocity.unit()
```

In the `_center` method, we perform the subtraction the other way around to produce a vector in the opposite direction. The distance and viewing angle will also be different from those in the `_avoid` method. We want `AVOID_DISTANCE` to be much smaller than `CENTER_DISTANCE` so that boids avoid only other boids that are very close to them. We will use the following values to start.

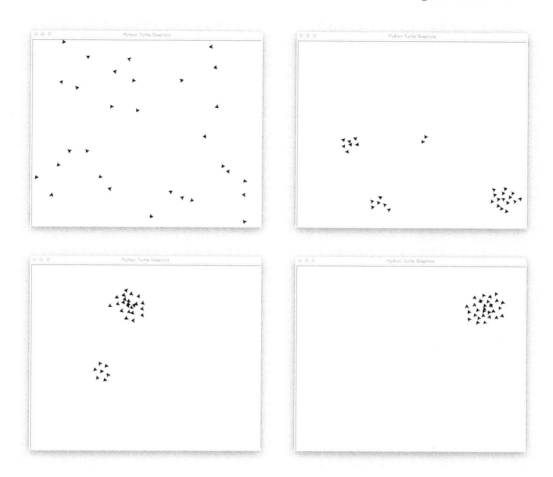

Figure 13.2 A progression of flocking boids.

```
AVOID_DISTANCE = 3      # avoid only close neighbors
AVOID_ANGLE = 300

MATCH_DISTANCE = 10     # match velocity of intermediate neighbors
MATCH_ANGLE = 240

CENTER_DISTANCE = 15    # move toward center of farther neighbors
CENTER_ANGLE = 180
```

The main simulation

We are finally ready to run the flocking simulation! The following program sets up a turtle graphics window, creates a world named `sky`, and then creates several `Boid` objects. The simulation is set in motion by the last `for` loop, which repeatedly calls the `step` method of the `World` class.

```
import turtle
from world import *
from boid import *

WIDTH = 100
HEIGHT = 100
NUM_BIRDS = 30
ITERATIONS = 2000

def main():
    worldTurtle = turtle.Turtle()
    screen = worldTurtle.getscreen()
    screen.setworldcoordinates(0, 0, WIDTH - 1, HEIGHT - 1)
    screen.tracer(NUM_BIRDS)
    worldTurtle.hideturtle()

    sky = World(WIDTH, HEIGHT)         # create the world
    for index in range(NUM_BIRDS):     # create the boids
        bird = Boid(sky)

    for step in range(ITERATIONS):     # run the simulation
        sky.step()

    screen.update()
    screen.exitonclick()

main()
```

The complete program is available on the book web site. Figure 13.2 shows a sequence of four screenshots of the simulation.

Reflection 13.29 *Run the program a few times to see what happens. Then try changing the following constant values. What is the effect in each case?*

(a) TURN_ANGLE = 90
(b) PREV_WEIGHT = 0
(c) AVOID_DISTANCE = 8
(d) CENTER_WEIGHT = 0.25

Exercises

13.4.1. Remove each of the three rules from the simulation, one at a time, by setting its corresponding weight to zero. What is the effect of removing each one? What can you conclude about the importance of each rule to successful flocking?

13.4.2. Implement the __eq__ method for the Vector class. Then modify the step method so that it slightly alters the boid's heading with some probability if the new velocity is the same as the old velocity (which will happen if it has no neighbors).

13.4.3. Modify the move method so that a boid randomly turns either left or right when it approaches a boundary. What is the effect?

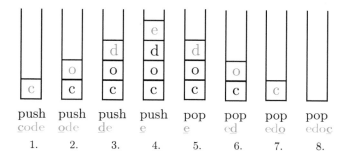

Figure 13.3 Reversing the string `'code'` with a stack.

13.5 A STACK ADT

An abstract data type that stores a group or sequence of data is known as a *collection*. We have used several of Python's collection abstract data types—strings, lists, tuples, and dictionaries—and built new types based on these ADTs (e.g., points, grids, graphs) to solve specific problems. In this section, we will demonstrate how to define and implement a new collection type by fleshing out the design and implementation of a *stack*.

A stack is a restricted list in which items are inserted and deleted only from one end, called the *top*. When an item is added to a stack, it *pushed* onto the top of the stack. When an item is deleted from a stack, it is *popped* from the same end. Therefore, when an item is popped, it is always the last item that was pushed. For this reason, a stack is known as a "last in, first out" (LIFO) structure.

Stacks have many uses, many of which have to do with reversing some process. (If you completed Project 10.1, you used a stack to draw figures from the strings produced by Lindenmayer systems.) As a simple example, suppose we want to reverse a string. (We used other approaches to do this in Exercises 6.3.11 and 10.2.7.) Using a stack, we can iterate over the characters of the string, pushing them onto the stack one at a time, as illustrated in steps 1–4 in Figure 13.3. Then, as illustrated in steps 5–8, we can pop the characters and concatenate them to the end of a new string as we do. Because the characters are popped in the reverse order in which they were pushed, the final string is the reverse of the original.

Based on our discussion above, the main attribute of the stack ADT is an ordered collection of items in the stack. We also need to know when the stack is empty, so we will also maintain the length of the stack as an attribute.

Stack ADT Attributes	
Name	Description
items	the items on the stack
length	the number of items on the stack

In addition to the push and pop functions, it is useful to have a function that allows us to peek at the item on the top of the stack without deleting it, and a function that tells us when the stack is empty.

Stack ADT Operations

Name	Arguments	Description
create	—	create a new empty Stack instance
top	—	return the item on the top of the stack
pop	—	return and delete the item on the top of the stack
push	an item	insert a new item on the top of the stack
isEmpty	—	return true if the stack is empty, false otherwise

Based on this specification, an implementation of a Stack class is fairly straightforward. (To save space, we are using short docstrings based on the ADT specification.)

```
class Stack:
    """A stack class."""

    def __init__(self):
        self._stack = []   # the items in the stack

    def top(self):
        """Return the item on the top of the stack."""

        if len(self._stack) > 0:
            return self._stack[-1]
        raise IndexError('stack is empty')

    def pop(self):
        """Return and delete the item on the top of the stack."""

        if len(self._stack) > 0:
            return self._stack.pop()
        raise IndexError('stack is empty')

    def push(self, item):
        """Insert a new item on the top of the stack."""

        self._stack.append(item)

    def isEmpty(self):
        """Return true if the stack is empty, false otherwise."""

        return len(self._stack) == 0
```

We implement the stack with a list named `self._stack`, and define the end of the list to be the top of the stack. We then use the **append** and **pop** list methods to push and pop items to and from the top of the stack, respectively. (This is the origin of the name for the `pop` method; without an argument, the `pop` method deletes the item from the end of the list.)

Reflection 13.30 *The stack ADT contained a length attribute. Why do we not need to include this in the class?*

We do not include the length attribute because the list that we use to store the stack items maintains its length for us; to determine the length of the stack internally, we can just refer to `len(self._stack)`. For example, in the `top` method, we need to make sure the stack is not empty before attempting to access the top element. If we tried to return `self._stack[-1]` when `self._stack` is empty, the following error will result:

```
IndexError: list index out of range
```

Likewise, in the `pop` method of `Stack`, if we tried to return `self._stack.pop()` when `self._stack` is empty, we would get this error:

```
IndexError: pop from empty list
```

These are not very helpful error messages, both because they refer to a list instead of the stack and because the first refers to an index, which is not a concept that is relevant to a stack. In the `top` and `pop` methods, we remedy this by using the `raise` statement. Familiar errors like `IndexError`, `ValueError`, and `SyntaxError` are called *exceptions*. The `raise` statement "raises" an exception which, by default, causes the program to end with an error message. The `raise` statement in red in the `top` and `pop` methods raises an `IndexError` and prints a more helpful message after it:

```
IndexError: stack is empty
```

Although we will not discuss exceptions in any more depth here, you may be interested to know that you can also create your own exceptions and respond to (or "catch") exceptions in ways that do not result in your program ending. (Look up `try` and `except` clauses.)

Reflection 13.31 *If you have not already done so, type the `Stack` class into a module named `stack.py`. Then run the following program.*

```
import stack

def main():
    myStack = stack.Stack()
    myStack.push('one')
    print(myStack.pop())    # prints "one"
    print(myStack.pop())    # exception

main()
```

Now let's explore a few ways we can use our new `Stack` class. First, the following function uses a stack to reverse a string, using the algorithm we described above.

```
import stack

def reverse(text):
    """ (docstring omitted) """

    characterStack = stack.Stack()
    for character in text:
        characterStack.push(character)

    reverseText = ''
    while not characterStack.isEmpty():
        character = characterStack.pop()
        reverseText = reverseText + character
    return reverseText
```

Reflection 13.32 *If the string* `'class'` *is passed in for* `text`, *what is assigned to* `characterStack` *after the* `for` *loop? What string is assigned to* `reverseText` *at the end of the function?*

For another example, let's write a function that converts an integer value to its corresponding representation in any other base. In Section 1.4 and Box 3.1, we saw that we can represent integers in other bases by simply using a number other than 10 as the base of each positional value in the number. For example, the numbers 234 in base 10, 11101010 in base 2, EA in base 16, and 1414 in base 5 all represent the same value, as shown below. To avoid ambiguity, we use a subscript with the base after any number not in base 10.

$$11101010_2 = 1 \times 2^7 + 1 \times 2^6 + 1 \times 2^5 + 0 \times 2^4 + 1 \times 2^3 + 0 \times 2^2 + 1 \times 2^1 + 0 \times 2^0$$
$$= 128 + 64 + 32 + 0 + 8 + 0 + 2 + 0$$
$$= 234.$$

$$EA_{16} = 14 \times 16^1 + 10 \times 16^0$$
$$= 224 + 10$$
$$= 234.$$

$$1414_5 = 1 \times 5^3 + 4 \times 5^2 + 1 \times 5^1 + 4 \times 5^0$$
$$= 125 + 100 + 5 + 4$$
$$= 234.$$

To represent an integer value in another base, we can repeatedly divide by the base and add the digit representing the remainder to the front of a growing string. For example, to convert 234 to its base 5 representation:

1. 234 ÷ 5 = 46 remainder 4
2. 46 ÷ 5 = 9 remainder 1
3. 9 ÷ 5 = 1 remainder 4
4. 1 ÷ 5 = 0 remainder 1

In each step, we divide the quotient obtained in the previous step (in red) by the base. At the end, the remainders (underlined), in reverse order, comprise the final representation. Since we obtain the digits in the opposite order that they appear in the number, we can push each one onto a stack as we get it. Then, after all of the digits are found, we can pop them off the stack and append them to the end of a string representing the number in the desired base.

```
import stack

def convert(number, base):
    """Represent a non-negative integer value in any base.

    Parameters:
        number: a non-negative integer
        base: the base in which to represent the value

    Return value: a string representing the number in the given base
    """

    digitStack = stack.Stack()
    while number > 0:
        digit = number % base
        digitStack.push(digit)
        number = number // base

    numberString = ''
    while not digitStack.isEmpty():
        digit = digitStack.pop()
        if digit < 10:
            numberString = numberString + str(digit)
        else:
            numberString = numberString + chr(ord('A') + digit - 10)
    return numberString
```

Exercises

13.5.1. Add a __str__ method to the Stack class. Your representation should indicate which end of the stack is the top, and whether the stack is empty.

13.5.2. Enhance the convert function so that it also works correctly for negative integers.

13.5.3. Write a function that uses a stack to determine whether a string is a palindrome.

13.5.4. Write a function that uses a stack to determine whether the parentheses in an arithmetic expression (represented as a string) are balanced. For example, the parentheses in the string '(1+(2+3)*(4 - 5))' are balanced but the parentheses in the strings '(1+2+3)*(4-5))' and '(1+(2+3)*(4-5)' are not. (Hint: push left parentheses onto a stack and pop when a right parenthesis is encountered.)

13.5.5. When a function is called recursively, a representation (which includes the values of its local variables) is pushed onto the top of a stack. The use of a stack can be seen in Figures 10.11, 10.12, and 10.16. When a function returns, it is popped from the top of the stack and execution continues with the function instance that is on the stack below it.

Recursive algorithms can be rewritten iteratively by replacing the recursive calls with an explicit stack. Rewrite the recursive depth-first function from Section 10.5 iteratively using a stack. Start by creating a stack and pushing the source position onto it. Then, inside a `while` loop that iterates while the stack is not empty, pop a position and check whether it can be visited. If it can, mark it as visited and push the four neighbors of the position onto the stack. The function should return `True` if the destination can be reached, or `False` otherwise.

13.5.6. Write a class that implements a *queue* abstract data type, as defined below.

Queue ADT Attributes	
Name	Description
items	the items in the queue
length	the number of items in the queue

Queue ADT Operations		
Name	Arguments	Description
front	—	return the item at the front of the queue
dequeue	—	return and delete the item at the front of the queue
enqueue	an item	insert a new item at the rear of the queue
isEmpty	—	return true if the queue is empty, false otherwise

13.5.7. Write a class named `Pointset` that contains a collection of points. Inside the class, the points should be stored in a list of `Pair` objects (from Section 13.2). Your class should implement the following methods:

- `insert(self, x, y)` adds a point (x, y) to the point set
- `length(self)` returns the number of points
- `centroid(self)` returns the centroid as a `Pair` object (see Section 13.2)
- `closestPairs(self)` returns the two closest points as `Pair` objects (see Exercise 13.1.9)
- `farthestPairs(self)` returns the two farthest points as `Pair` objects
- `diameter(self)` returns the distance between the two farthest points

- draw(self, tortoise, color) draws the points with the given turtle and color

Use your class to implement a program that is equivalent to that in Exercise 13.1.9. Your program should initialize a new PairSet object (instead of a list), insert each new point into the PairSet object, call your closestPairs, farthestPairs, and centroid methods, draw all of the points with the draw method of your PairSet class, and draw the closest pair, farthest pair and centroid in different colors using the draw method of the Pair class (as assigned in that exercise).

13.6 A DICTIONARY ADT

A dictionary abstract data type stores (key, value) pairs, and allows us to look up a value by specifying its key. In the Python implementation of a dictionary, we insert a (key, value) pair with the assignment statement dictionary[key] = value and retrieve a value with dictionary[key]. We have used dictionaries in a variety of applications: histograms in Section 8.3, Lindenmayer systems in Section 10.6, networks in Chapter 12, and the flocking simulation earlier in this chapter. Another simple application is shown below: storing the history of World Series champions by associating each year with the name of the champion team.

```
def main():
    worldSeries = { }
    worldSeries[1903] = 'Boston Americans'
    ⋮
    worldSeries[1979] = 'Pittsburgh Pirates'
    ⋮
    worldSeries[2014] = 'San Francisco Giants'

    print(worldSeries[1979])   # prints "Pittsburgh Pirates"

main()
```

In this section, we will implement our own **Dictionary** abstract data type to illustrate how more complex collection types can be implemented. Formally, a dictionary ADT simply contains a collection of (key, value) pairs and a length.

Dictionary ADT Attributes	
Name	Description
pairs	a collection of (key, value) pairs
length	the number of (key, value) pairs

The most fundamental operations supported by a **Dictionary** are **insert**, **delete**, and **lookup** (all of which are implemented using indexing in the built-in Python dictionary). These, and a few other useful operations, are defined below.

Dictionary ADT Operations

Name	Arguments	Description
create	—	create a new empty **Dictionary** instance
insert	key, value	insert a (key, value) pair
delete	key	delete the pair with a given key
lookup	key	return the value associated with a key
contains	key	return true if the key exists, false otherwise
items	—	return a list of all (key, value) pairs
keys	—	return a list of all keys
values	—	return a list of all values
length	—	return the number of (key, value) pairs
isEmpty	—	return true if the dictionary is empty, false otherwise

As we discussed in Box 8.2, Python dictionaries are implemented using a data structure called a *hash table*. A hash table consists of a fixed number of *slots* in which the key:value pairs are stored. A *hash function* is used to map a key value to the index of a slot. For example, in the illustration below from Box 8.2, the key:value pair `18.9:2` is mapped by the hash function to slot 3.

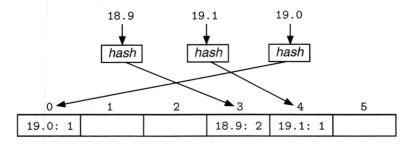

Hash tables

Although there are several other ways that a dictionary ADT could be implemented (such as with a binary search tree, as in Project 11.2), let's look at how to do so with our own hash table implementation. When implementing a hash table, there are two main issues that we need to consider. First, we need to decide what our hash function should be. Second, we need to decide how to handle *collisions*, which occur when more than one key is mapped to the same slot by the hash function. We will leave an answer to the second question as Project 13.2.

A hash function should ideally spread keys uniformly throughout the hash table to efficiently use the allocated space and to prevent collisions. Assuming that all key values are integers, the simplest hash functions have the form

$$\text{hash(key)} = \text{key mod } n,$$

where n is the number of slots in the hash table. Hash functions of this form are said to use the *division method*.

Reflection 13.33 *What range of possible slot indices is given by this hash function?*

To prevent patterns from emerging in the slot assignments, we typically want n to be a prime number.

Reflection 13.34 *If n were even, what undesirable pattern might emerge in the slot indices assigned to keys? (For example, what if all keys were odd?)*

Reflection 13.35 *Suppose $n = 11$. What is the value of* hash(key) *for key values 7, 11, 14, 25, and 100?*

There are many better hash functions that have been developed, but the topic is outside the scope of our discussion here. If you are interested, we give some pointers to additional reading at the end of the chapter. In the exercises, we discuss how you might define a hash function for keys that are strings.

Implementing a hash table

To demonstrate how hash tables work, we will start by implementing the dictionary ADT with a small hash table containing only $n = 11$ slots. We will also assume that no collisions occur, leaving the resolution of this issue to Project 13.2. The beginning of the class definition follows.

```
class Dictionary:
    """A dictionary class."""

    def __init__(self):
        """Construct a new Dictionary object."""

        self._length = 0      # number of (key, value) pairs
        self._size = 11       # number of slots in the hash table
        self._table = [None] * self._size   # hash table containing
                                            # (key, value) pairs
    def _hash(self, key):
        return key % self._size

    def insert(self, key, value):
        """Insert a key:value pair into self."""

        index = self._hash(key)
        self._table[index] = (key, value)
        self._length = self._length + 1
```

The constructor creates an empty hash table with 11 slots. We represent an empty slot with `None`. When a (key, value) pair is inserted into a slot, it will be represented with a `(key, value)` tuple. The `_hash` method implements our simple hash function. The leading underscore (`_`) character signifies that this method is *private* and should never be called from outside the class. (The leading underscore also tells the `help` function not to list this method.) We will discuss this further below. The `insert` method begins by using the hash function to map the given `key` value to the `index` of a slot in the hash table. Then the inserted `(key, value)` tuple is assigned to this slot (`self._table[index]`) and the number of entries in the dictionary is incremented.

Reflection 13.36 *With this implementation, how would you create a new dictionary and insert* (1903, Boston Americans), (1979, Pittsburgh Pirates), *and* (2014, San Francisco Giants) *into it (as in the* `main` *function at the beginning of this section)?*

With this implementation, these three insertions would be implemented as follows:

```
worldSeries = Dictionary()
worldSeries.insert(1903, 'Boston Americans')
worldSeries.insert(1979, 'Pittsburgh Pirates')
worldSeries.insert(2014, 'San Francisco Giants')
```

Reflection 13.37 *Using the* `insert` *method as a template, write the* `delete` *method.*

The `delete` method is very similar to the `insert` method *if* the key to delete exists in the hash table. In this case, the only differences are that we do not need to pass in a value, we assign `None` to the slot instead of a tuple, and we decrement the number of items.

Reflection 13.38 *How do we determine whether the key that we want to delete is contained in the hash table?*

If the slot to which the key is mapped by the hash function contains the value `None`, then the key must not exist. But even if the slot does not contain `None`, it may be that a different key resides there, so we still need to compare the value of the key in the slot with the value of the key we wish to delete. Since the key is the first value in the tuple assigned to `self._table[index]`, the key value is in `self._table[index][0]`. Therefore, the required test looks like this:

```
if self._table[index] != None and self._table[index][0] == key:
    # delete pair
else:
    # raise a KeyError exception
```

A dictionary ADT ■ 675

The `KeyError` exception is the exception raised when a key is not found in Python's built-in dictionary. The fleshed-out `delete` method is shown below.

```
_KEY = 0      # index of key in each pair
_VALUE = 1    # index of value in each pair

def delete(self, key):
    """Delete the pair with a given key."""

    index = self._hash(key)
    if self._table[index] != None and \
       self._table[index][self._KEY] == key:
        self._table[index] = None
        self._length = self._length - 1
    else:
        raise KeyError('key was not found')
```

To prevent the use of "magic numbers," we define two private constant values, `_KEY` and `_VALUE`, that correspond to the indices of the key and value in a tuple. Because they are defined inside the class, but not in any method, these are *class variables*. Unlike an instance variable, which can have a unique value in every object belonging to the class, a class variable is *shared* by every object in the class. In other words, if there were ten `Dictionary` objects in a program, there would be ten independent instance variables named `_size`, one in each of the ten objects. But there would be only one class variable named `_KEY`. A class variable can be referred to by prefacing it by either the name of the class or by `self`.

Reflection 13.39 *How do we delete* (2014, 'San Francisco Giants') *from the* `worldSeries` `Dictionary` *object that we created previously?*

To delete this pair, we simply call

   `worldSeries.delete(2014)`

Reflection 13.40 *Now write the* `lookup` *method for the* `Dictionary` *class.*

To retrieve a value corresponding to a key, we once again find the index corresponding to the key and check whether the key is present. If it is, we return the corresponding value. Otherwise, we raise an exception.

```
def lookup(self, key):
    """Return the value associated with a key."""

    index = self._hash(key)
    if self._table[index] != None and \
       self._table[index][self._KEY] == key:
        return self._table[index][self._VALUE]
    else:
        raise KeyError('key was not found')
```

Reflection 13.41 *How do we look up the 1979 World Series champion in the* `worldSeries` *Dictionary object that we created previously?*

To look up the winner of the 1979 World Series, we simply call the `lookup` method with the key 1979:

```
champion = worldSeries.lookup(1979)
print(champion)                          # prints "Pittsburgh Pirates"
```

We round out the class with a method that returns the number of (key, value) pairs in the dictionary. In the built-in collection types, the length is returned by the built-in `len` function. We can define the same behavior for our `Dictionary` class by defining the special `__len__` method.

```
def __len__(self):
    """Return the number of (key, value) pairs."""

    return self._length

def isEmpty(self):
    """Return true if the dictionary is empty, false otherwise."""

    return len(self) == 0
```

With the `__len__` method defined, we can find out how many entries are in the `worldSeries` dictionary with `len(worldSeries)`. Notice that in the `isEmpty` method above, we also call `len`. Since `self` refers to a `Dictionary` object, `len(self)` implicitly invokes the `__len__` method of `Dictionary` as well. The result is used to indicate whether the `Dictionary` object is empty.

The following `main` function combines the previous examples to illustrate the use of our class.

```
def main():
    worldSeries = Dictionary()
    worldSeries.insert(1903, 'Boston Americans')
    worldSeries.insert(1979, 'Pittsburgh Pirates')
    worldSeries.insert(2014, 'San Francisco Giants')
    print(worldSeries.lookup(1979))    # prints "Pittsburgh Pirates"
    worldSeries.delete(2014)
    print(len(worldSeries))            # prints 2
    print(worldSeries.lookup(2014))    # KeyError
```

Implementing indexing

As we already discussed, Python's built-in dictionary class implements insertion and retrieval of (key, value) pairs with indexing rather than explicit methods as we used above. We can also mimic this behavior by defining the `__getitem__` and `__setitem__` methods, as we did for the `Pair` class earlier. The `__getitem__` method takes a single index parameter and the `__setitem__` method takes an index and a value as parameters, just as our `lookup` and `insert` methods do. Therefore, to use indexing with our `Dictionary` class, we only have to rename our existing methods, as follows.

```python
def __setitem__(self, key, value):
    """Insert a (key, value) pair into self."""

    index = self._hash(key)
    self._table[index] = (key, value)
    self._length = self._length + 1

def __getitem__(self, key):
    """Return the value associated with a key."""

    index = self._hash(key)
    if self._table[index] != None and \
       self._table[index][self._KEY] == key:
        return self._table[index][self._VALUE]
    else:
        raise KeyError('key was not found')
```

The most general method for deleting an item from a Python collection is to use the `del` operator. For example,

```
del frequency[18.9]
```

deletes the (key, value) pair with key equal to `18.9` from the dictionary named `frequency`. We can mimic this behavior by renaming our `delete` method to `__delitem__`, as follows.

```python
def __delitem__(self, key):
    """Delete the pair with a given key."""

    index = self._hash(key)
    if self._table[index] != None and \
       self._table[index][self._KEY] == key:
        self._table[index] = None
        self._length = self._length - 1
    else:
        raise KeyError('key was not found')
```

With these three changes, the following `main` function is equivalent to the one above.

```
def main():
    worldSeries = Dictionary()
    worldSeries[1903] = 'Boston Americans'
    worldSeries[1979] = 'Pittsburgh Pirates'
    worldSeries[2014] = 'San Francisco Giants'
    print(worldSeries[1979])    # prints "Pittsburgh Pirates"
    del worldSeries[2014]
    print(len(worldSeries))     # prints 2
    print(worldSeries[2014])    # KeyError
```

ADTs vs. data structures

The difference between an abstract data type and a data structure is perhaps one of the most misunderstood concepts in computer science. As we have discussed, an ADT is an abstract description of a data type that is independent of any particular implementation. A data structure is a particular implementation of an abstract data type. In this section, we implemented a **Dictionary** ADT using a hash table data structure. But we could also have used a simple list of (key, value) pairs or something more complicated like a binary search tree from Project 11.2. *Our choice of data structure does not change the definition of the abstract data type.*

When someone uses a class that implements an abstract data type, the data structure should remain completely hidden. In other words, we must not make any public methods (those without leading underscores) dependent upon the underlying data structure. For example, we did not make the number of slots in the hash table a parameter to the constructor because this parameter is not part of the ADT and would not make sense if the ADT were implemented with a different data structure. Likewise, we indicated with a leading underscore that the `_hash` method should be private because this method is only necessary if the **Dictionary** is implemented with a hash table. Similarly, we indicated that instance and class variables in the **Dictionary** class should also remain private (with leading underscores) because they only make sense in the context of the chosen data structure.

Collectively, the non-private methods of the class constitute the only information that a programmer using the class should need to know. This is precisely what is displayed by the `help` function.

```
class Dictionary(builtins.object)
 |  A dictionary class.
 |
 |  Methods defined here:
 |
 |  __delitem__(self, key)
 |      Delete the pair with a given key.
 |
 |  __getitem__(self, key)
 |      Return the value associated with a key.
 |
 |  __init__(self)
 |      Construct a new Dictionary object.
 |
 |  __len__(self)
 |      Return the number of (key, value) pairs.
 |
 |  __setitem__(self, key, value)
 |      Insert a (key, value) pair into self.
 |
 |  isEmpty(self)
 |      Return true if the dictionary is empty, false otherwise.
```

Exercises

13.6.1. Add a private method named `_printTable` to the `Dictionary` class that prints the contents of the underlying hash table. For example, for the dictionary created in the `main` function above, the method should print something like this:

```
0: (1903, 'Boston Americans')
1: (2014, 'San Francisco Giants')
2: None
3: None
4: None
5: None
6: None
7: None
8: None
9: None
10: (1979, 'Pittsburgh Pirates')
```

13.6.2. Add a `__str__` method to the `Dictionary` class. The method should return a string similar to that printed for a built-in Python dictionary. It should not divulge any information about the underlying hash table implementation. For example, for the dictionary displayed in the previous exercise, the method should return a string like this:

```
{1903: 'Boston Americans', 2014: 'San Francisco Giants',
1979: 'Pittsburgh Pirates'}
```

13.6.3. Implement use of the `in` operator for the `Dictionary` class by adding a method named `__contains__`. The method should return `True` if a key is contained in the `Dictionary` object, or `False` otherwise.

13.6.4. Implement `Dictionary` methods named `items`, `keys`, and `values` that return the list of (key, value) tuples, keys, and values, respectively. In any sequence of calls to these methods, with no modifications to the `Dictionary` object between the calls, they must return lists in which the orders of the items correspond. In other words, the first value returned by `values` will correspond to the first key returned by `keys`, the second value returned by `values` will correspond to the second key returned by `keys`, etc. The order of the tuples in the list returned by `items` must be the same as the order of the lists returned by `keys` and `values`.

13.6.5. Show how to use the `keys` method that you wrote in the previous exercise to print a list of keys and values in a `Dictionary` object in alphabetical order by key.

13.6.6. Use the `Dictionary` class to solve Exercise 8.3.4.

13.6.7. Use the `Dictionary` class to implement the `removeDuplicates3` function on Page 401.

13.6.8. If keys are string values, then a new hash function is needed. One simple idea is to sum the Unicode values corresponding to the characters in the string and then return the sum modulo n. Implement this new hash function for the `Dictionary` class.

13.6.9. Design and implement a hash function for keys that are tuples.

13.6.10. Write a class named `Presidents` that maintains a list of all of the U.S. presidents. The constructor should take the number of presidents as a parameter and initialize a list of empty slots, each containing the value `None`. Then add `__setitem__` and `__getitem__` methods that insert and return, respectively, a `President` object (from Exercise 13.1.13) with the given chronological number (starting at 1). Be sure to check in each method whether the parameters are valid. Also add a `__str__` method that returns a complete list of the presidents in chronological order. If a president is missing, replace the name with question marks. For example, the following code

```
presidents = Presidents(44)
washington = President('George Washington')
kennedy = President('John F. Kennedy')
presidents[1] = washington
presidents[35] = kennedy
print(presidents[35])    # prints Kennedy
print(presidents)
```

should print

```
    John F. Kennedy
1.  George Washington
2.  ???
3.  ???
    ⋮
34. ???
35. John F. Kennedy
36. ???
    ⋮
44. ???
```

Also include a method that does the same thing as the function from Exercise 13.1.14. In other words, your method should, given an age, print a table with all presidents who where at least that old when they took office, along with their ages when they took office.

13.6.11. Write a class that stores all of the movies that have won the Academy Award for Best Picture. Inside the class, use a list of **Movie** objects from Exercise 13.1.15. Your class should include **__getitem__** and **__setitem__** methods that return and assign the movie winning the award in the given edition of the ceremony. (The 87th Academy Awards were held in 2015.) In addition, include a **__str__** method that returns a string containing the complete list of the winning titles in chronological order. If a movie is missing, replace the title with question marks. Finally, include a method that checks whether the winners in two given editions have actors in common (by calling the method from Exercise 13.1.15). You can find a complete list of winners at **http://www.oscars.org/oscars/awards-databases**

13.6.12. Write a class named **Roster** that stores information about all of the students enrolled in a course. In the **Roster** class, store the information about the students using a list of **Student** objects from Exercise 13.1.18. Each student will also have an associated ID number, which can be used to access the student through the class' **__getitem__** and **__setitem__** methods. Also include a **__len__** method that returns the number of students enrolled, a method that returns the average of all of the exam grades for all of the students, a method that returns the average overall grade (using the weights in Exercise 13.1.18), and a **__str__** method that returns a string representing the complete roster with current grades. Use of this class is illustrated with the following short segment of code:

```
roster = Roster()
alice = Student('Alice Miller')
bob = Student('Bob Smith')
roster[101] = alice
roster[102] = bob
roster[101].addExam(100)
roster[102].addExam(85)
print(roster.examAverage())    # prints 92.5
print(roster.averageGrade())   # prints 46.25
print(roster)                  # prints:
                               # 101   Alice Miller       50.00
                               # 102   Bob Smith          42.50
```

13.7 SUMMARY

As the quotation at the beginning of the chapter so concisely puts it, a program consists of two main parts: the algorithm and the data on which the algorithm works. When we design an algorithm using an object-oriented approach, we begin by identifying the main objects in our problem, and then define abstract data types for them. When we design a new ADT, we need to identify the data that the ADT will contain and the operations that will be allowed on that data. These operations are generally organized into four categories: *constructors*, a *destructor*, *accessors*, and *mutators*.

In an *object-oriented programming language* like Python, abstract data types are implemented as *classes*. A Python class contains a set of functions called *methods* and a set of *instance variables* whose names are preceded by `self` within the class. The name `self` always refers to the object on which a method is called. A class can also define the meaning of several *special methods* that dictate how operators and built-in functions behave on the class. The class implements a *data structure* that implements the specification given by the abstract data type. There may be many different data structures that one can use to implement a particular abstract data type. For example, the `Pair` ADT from the beginning of this chapter may be implemented with two individual variables, a list of length two, a two-element tuple, or a dictionary with two entries.

To illustrate how classes are used in larger programs, we designed an *agent-based simulation* that simulates flocking birds or schooling fish. This simulation consists of two main classes that interact with each other: an agent class and a class for the world that the agents inhabit. Agent-based simulations can be used in a variety of disciplines including sociology, biology, economics, and the physical sciences.

13.8 FURTHER DISCOVERY

This chapter's epigraph is actually the title of a book. *Algorithms + Data Structures = Programs* was written programming language pioneer Niklaus Wirth in 1976 [59].

The quote on Page 621 is from one of the first papers written by Barbara Liskov, in 1974 [28]. In 1968, Dr. Liskov was one of the first women to earn a Ph.D. in computer science in the United States. She has taught computer science at MIT since 1972, and was honored with the Turing Award in 2008.

The boids model was created by Craig Reynolds [45]. For more information on agent-based simulations, we suggest looking at *The Computational Beauty of Nature* by Gary Flake [14], *Think Complexity* by Allen Downey [10], *Agent-based Models* by Nigel Gilbert [15], and *Introduction to Computational Science* by Angela Shiflet and George Shiflet [52].

13.9 PROJECTS

Project 13.1 Tracking GPS coordinates

A GPS (short for Global Positioning System) receiver (like those in most mobile phones) is a small computing device that uses signals from four or more GPS satellites to compute its three-dimensional position (latitude, longitude, altitude) on Earth. By recording this position data over time, a GPS device is able to track moving objects. The use of such tracking data is now ubiquitous. When we go jogging, our mobile phones can track our movements to record our route, distance, and speed. Companies and government agencies that maintain vehicle fleets use GPS to track their locations to streamline operations. Biologists attach small GPS devices to animals to track their migration behavior.

In this project, you will write a class that stores a sequence of two-dimensional geographical points (latitude, longitude) with their timestamps. We will call such a sequence a *track*. We will use tracking data that the San Francisco Municipal Transportation Agency (SF MTA) maintains for each of the vehicles (cable cars, streetcars, coaches, buses, and light rail) in its Municipal Railway ("Muni") fleet. For example, the following table shows part of a track for the Powell/Hyde cable car in metropolitan San Francisco.

Time stamp	Longitude	Latitude
2014-12-01 11:03:03	−122.41144	37.79438
2014-12-01 11:04:33	−122.41035	37.79466
2014-12-01 11:06:03	−122.41011	37.7956
2014-12-01 11:07:33	−122.4115	37.79538
2014-12-01 11:09:03	−122.4115	37.79538

Negative longitude values represent longitudes west of the prime meridian and negative latitude values represent latitudes south of the equator. After you implement your track class, write a program that uses it with this data to determine the Muni route that is closest to any particular location in San Francisco.

An abstract data type for a track will need the following pair of attributes.

Track ADT Attributes	
Name	Description
points	a list of two-dimensional geographical points with associated times
name	an identifier for the track

In addition, the ADT needs operations that allow us to add new data to the track, draw a picture of the track, and compute various characteristics of the track.

Track ADT Operations

Name	Arguments	Description
append	a point, time	add a point and time to the end of the track
length	—	return the number of points on the track
averageSpeed	—	return the average speed over the track
totalDistance	—	return the total distance traversed on the track
diameter	—	return the distance between the two points that are farthest apart on the track
closestDistance	a point, error	find the closest distance a point on the track comes to the given point; return this distance and the time(s) when the track comes within error of this distance
draw	conversion function	draw the track, using the given function to convert each geographical point to an equivalent pixel location in the graphics window

Part 1: Write a Time class

Before you implement a class for the `Track` ADT, implement a `Time` class to store the timestamp for each point. On the book web site is a skeleton of a `time.py` module that you can use to guide you. The constructor of your `Time` class should accept two strings, one containing the date in YYYY-MM-DD format and one containing the time in HH:MM:SS format. Store each of these six components in the constructed object. Your class should also include a `duration` method that returns the number of seconds that have elapsed between a `Time` object and another `Time` object that is passed in as a parameter, a `__str__` method that returns the time in YYYY-MM-DD HH:MM:SS format, a `date` method that returns a string representing just the date in YYYY-MM-DD format, and a `time` method that returns a string representing just the time in HH:MM:SS format. Write a short program that thoroughly tests your new class before continuing.

Part 2: Add a timestamp to the `Pair` class

Next, modify the `Pair` class from earlier in the chapter so that it also includes an instance variable that can be assigned a `Time` object representing the timestamp of the point. Also, add a new method named `time` that returns the `Time` object representing the timestamp of the point. Write another short program that thoroughly tests your modified `Pair` class before continuing.

Part 3: Write a Track class

Now implement a `Track` class, following the ADT description above. The points should be stored in a list of `Pair` objects. On the book web site is a skeleton of a `track.py` module with some utility methods already written that you should use as a starting point.

Question 13.1.1 *What does the `_distance` method do? Why is its name preceded with an underscore?*

Question 13.1.2 *What is the purpose of the `degToPix` function that is passed as a parameter to the `draw` method?*

After you write each method, be sure to thoroughly test it before moving on to the next one. For this purpose, design a short track consisting of four to six points and times, and write a program that tests each method on this track.

Part 4: Mysteries on the Muni

You are developing a forensic investigation tool that shows, for any geographical point in metropolitan San Francisco, the closest Muni route and the times at which the vehicle on that route passed by the given point. On the book web site is almost-complete program that implements this tool, named `muni.py`. The program should read one day's worth of tracking data from the San Francisco Municipal Railway into a list or dictionary of `Track` objects, set up a turtle graphics window with a map of the railway system, and then wait for a mouse click on the map. The mouse click triggers a call to a function named `clickMap` that draws a red dot where the click occurred, and then uses the `closestDistance` and `draw` methods of the `Track` class to find the closest Muni route to that click and then draws it on the map with the times that the `Track` passed the clicked-upon position. The output from the finished tool is visualized below.

686 ■ Abstract data types

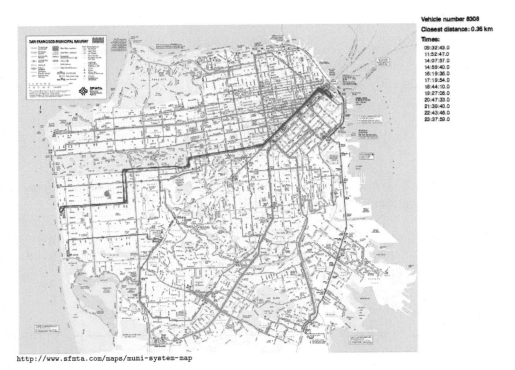

http://www.sfmta.com/maps/muni-system-map

In the `main` function of the program, the second-to-last function call

```
screen.onclick(clickMap)
```

registers the function named `clickMap` as the function to be called when a mouse click happens. Then the last function call

```
screen.mainloop()
```

initiates an *event loop* in the background that calls the registered `clickMap` function on each mouse click. The x and y coordinates of the mouse click are passed as parameters into `clickMap` when it is called.

The tracking data is contained in a comma-separated (CSV) file on the book web site named `muni_tracking.csv`. The file contains over a million individual data points, tracking the movements of 965 vehicles over a 24-hour period. The only function in `muni.py` that is left to be written is `readTracks`, which should read this data and return a dictionary of `Track` objects, one per vehicle. Each vehicle is labeled in the file with a unique vehicle tag, which you should use for the `Track` objects' names and as the keys in the dictionary of `Track` objects.

Question 13.1.3 *Why is* `tracks` *a global variable? (There had better be a very good reason!)*

Question 13.1.4 *What are the purposes of the seven constant named values in all caps at the top of the program?*

Question 13.1.5 *What do the functions* `degToPix` *and* `pixToDeg` *do?*

Question 13.1.6 *Study the* `clickMap` *function carefully. What does it do?*

Once you have written the `readTracks` function, test the program. The program also uses the `closestDistance` method that you wrote for the `Track` class, so you may have to debug that method to get the program working correctly.

Project 13.2 Economic mobility

In Section 13.6, we designed a `Dictionary` class that assumes that no collisions occur. In this project, you will complete the design of the class so that it properly (and transparently) handles collisions. Then you will use your finished class to write a program that allows one to query data on upward income mobility in the United States.

To deal with collisions, we will use a technique called *chaining* in which each slot consists of a *list* of (key, value) pairs instead of a single pair. In this way, we can place as many items in a slot as we need.

Question 13.2.1 *How do the implementations of the* insert, delete, *and* lookup *functions need to change to implement chaining?*

Question 13.2.2 *With your answer to the previous question in mind, what is the worst case time complexity of each of these operations if there are n items in the hash table?*

Part 1: Implement chaining

First, modify the `Dictionary` class from Section 13.6 so that the underlying hash table uses chaining. The constructor should initialize the hash table to be a list of empty lists. Each `__getitem__`, `__setitem__`, and `__delitem__` method will need to be modified. Be sure to raise an appropriate exception when warranted. You should notice that these three methods share some common code that you might want to place in a private method that the three methods can call.

In addition, implement the methods named `_printTable`, `__str__`, `__contains__`, `items`, `keys`, and `values` described in Exercises 13.6.1–13.6.4.

Test your implementation by writing a short program that inserts, deletes, and looks up several entries with integer keys. Also test your class with different values of `self._size`.

Part 2: Hash functions

In the next part of the project, you will implement a searchable database of income mobility data for each of 741 commuting zones that cover the United States. A commuting zone is an area in which the residents tend to commute to the same city for work, and is named for the largest city in the zone. This city name will be the key for your database, so you will need a hash function that maps strings to hash table indices. Exercise 13.6.8 suggested one simple way to do this. Do some independent research to discover at least one additional hash function that is effective for general strings. Implement this new hash function.

Question 13.2.3 *According to your research, why is the hash function you discovered better than the one from Exercise 13.6.8?*

Part 3: A searchable database

On the book web site is a tab-separated data file named `mobility_by_cz.txt` that contains information about the expected upward income mobility of children in each of the 741 commuting zones. This file is based on data from The Equality of Opportunity Project (`http://www.equality-of-opportunity.org`), based at Harvard and the University of California, Berkeley. The researchers measured potential income mobility in several ways, but the one we will use is the probability that a child raised by parents in the bottom 20% (or "bottom quintile") of income level will rise to the top 20% (or "top quintile") as an adult. This value is contained in the seventh column of the data file (labeled `"P(Child in Q5 | Parent in Q1), 80-85 Cohort"`).

Write a program that reads this data file and returns a `Dictionary` object in which the keys are names of commuting zones and the values are the probabilities described above. Because some of the commuting zone names are identical, you will need to concatenate the commuting zone name and state abbreviation to make unique keys. For example, there are five commuting zones named "Columbus," but your keys should designate Columbus, GA, Columbus, OH, etc. Once the data is in a `Dictionary` object, your program should repeatedly prompt for the name of a commuting zone and print the associated probability. For example, your output might look like this:

```
Enter the name of a commuting zone to find the chance that the
income of a child raised in that commuting zone will rise to
the top quintile if his or her parents are in the bottom quintile.
Commuting zone names have the form "Columbus, OH".

Commuting zone (or q to quit): Columbus, OH
Percentage is 4.9%.
Commuting zone (or q to quit): Columbus
Commuting zone was not found.
Commuting zone (or q to quit): Los Angeles, CA
Percentage is 9.6%.
Commuting zone (or q to quit): q
```

Part 4: *State analyses*

Finally, enhance your program so that it produces the following output, organized by state. You should create additional **Dectionary** objects to produce these results. (Do not use any built-in Python dictionary objects!)

1. Print a table like the following of all commuting zone data, alphabetized by state then by commuting zone name. (Hints: (a) create another **Dictionary** object as you read the data file; (b) the **sort** method sorts a list of tuples by the first element in the tuple.)

   ```
   AK
       Anchorage: 13.4%
       Barrow: 10.0%
       Bethel: 5.2%
       Dillingham: 11.8%
       Fairbanks: 16.0%
       Juneau: 12.6%
       Ketchikan: 12.0%
       Kodiak: 14.7%
       Kotzebue: 6.5%
       Nome: 4.7%
       Sitka: 7.1%
       Unalaska: 13.0%
       Valdez: 15.4%
   AL
       Atmore: 4.8%
       Auburn: 3.5%
       ⋮
   ```

2. Print a table, like the following, alphabetized by state, of the average probability for each state. (Hint: use another **Dictionary** object.)

   ```
   State   Percent
   -----   -------
   AK      11.0%
   AL       5.4%
   AR       7.2%
   ⋮
   ```

3. Print a table, formatted like that above, of the states with the five lowest and five highest average probabilities. To do this, it may be helpful to know about the following trick with the built-in **sort** method. When the **sort** method sorts a list of tuples or lists, it compares the first elements in the tuples or lists. For example, if **values = [(0, 2), (2, 1), (1, 0)]**, then **values.sort()** will reorder the list to be **[((0, 2), (1, 0), (2, 1)]**. To have the **sort** method use another element as the key on which to sort, you can define a simple function like this:

```
def getSecond(item):
    return item[1]

values.sort(key = getSecond)
```

When the list named `values` is sorted above, the function named `getSecond` is called for each item in the list and the return value is used as the key to use when sorting the item. For example, suppose `values = [(0, 2), (2, 1), (1, 0)]`. Then the keys used to sort the three items will be 2, 1, and 0, respectively, and the final sorted list will be `[(1, 0), (2, 1), (0, 2)]`.

Project 13.3 Slime mold aggregation

In this project, you will write an *agent-based simulation* that graphically depicts the emergent "intelligence" of a fascinating organism known as slime mold (*Dictyostelium discoideum*). When food is plentiful, the slime mold exists in a unicellular amoeboid form. But when food becomes scarce, it emits a chemical known as cyclic AMP (or cAMP) that attracts other amoeboids to it. The congregated cells form a *pseudoplasmodium* which then scavenges for food as a single multicellular organism. We will investigate how the pseudoplasmodium forms. A movie linked from the book web site shows this phenomenon in action.

The following sequence of images illustrates what your simulation may look like. The red triangles represent slime mold amoeboids and the varying shades of green represent varying levels of cAMP on the surface. (Darker green represents higher levels.)

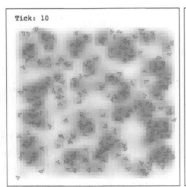
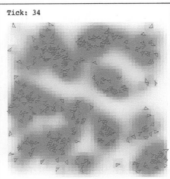
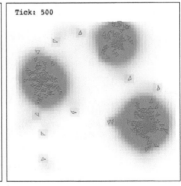

Slime world

In our simulation, the slime mold's world will consist of a grid of square patches, each of which contains some non-negative level of the chemical cAMP. The cAMP will be deposited by the slime mold (explained next). In each time step, the chemical in each patch should:

1. Diffuse to the eight neighboring patches. In other words, after the chemical in a patch diffuses, 1/8 of it will be added to the chemical in each neighboring patch. (Note: this needs to be done carefully; the resulting levels should be as if all patches diffused simultaneously.)

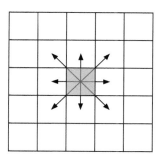

2. Partially evaporate. (Reduce the level in each patch to a constant fraction, say 0.9, of its previous level.)

Slime world will be modeled as an instance of a class (that you will create) called `World`. Each patch in slime world will be modeled as an instance of a class called `Patch` (that you will also create). The `World` class should contain a grid of `Patch` objects. You will need to design the variables and methods needed in these new classes.

There is code on the book web site to visualize the level of chemical in each patch. Higher levels are represented with darker shades of green on the turtle's canvas. Although it is possible to recolor each patch with a Turtle during each time step, it is far too slow. The supplied code modifies the underlying canvas used in the implementation of the `turtle` module.

Amoeboid behavior

At the outset of the simulation, the world will be populated with some number of slime mold amoeboids at random locations on the grid. At each time step in the simulation, a slime mold amoeboid will:

1. "Sniff" the level of the chemical cAMP at its current position. If that level is above some threshold, it will next sniff for chemical `SNIFF_DISTANCE` units ahead and `SNIFF_DISTANCE` units out at `SNIFF_ANGLE` degrees to the left and right of its current position. `SNIFF_ANGLE` and `SNIFF_DISTANCE` are parameters that can be set in the simulation. In the graphic below, the slime mold is represented by a red triangle pointing at its current heading and `SNIFF_ANGLE` is 45 degrees. The `X`'s represent the positions to sniff.

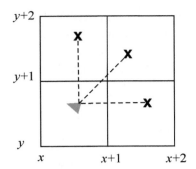

Notice that neither the current coordinates of the slime mold cell nor the coordinates to sniff may be integers. You will want to write a function that will round coordinates to find the patch in which they reside. Once it ascertains the levels in each of these three patches, it will turn toward the highest level.

2. Randomly wiggle its heading to the left or right a maximum of `WIGGLE_ANGLE` degrees.

3. Move forward one unit on the current heading.

4. Drop `CHEMICAL_ADD` units of cAMP at its current position.

A slime mold amoeboid should, of course, also be modeled as a class. At the very least, the class should contain a `Turtle` object that will graphically represent the cell. Set the speed of the `Turtle` object to 0 and minimize the delay between updates by calling `screen.tracer(200, 0)`. The remaining design of this class is up to you.

The simulation

The main loop of the simulation will involve iterating over some number of time steps. In each time step, every slime mold amoeboid and every patch must execute the steps outlined above.

Download a bare-bones skeleton of the classes needed in the project from the book web site. These files contain *only the minimum amount of code necessary* to accomplish the drawing of cAMP levels in the background (as discussed earlier).

Before you write any Python code, think carefully about how you want to design your project. Draw pictures and map out what each class should contain. Also map out the main event loop of your simulation. (This will be an additional file.)

Project 13.4 Boids in space

In this project, you will generalize the two-dimensional flocking simulation that we developed in Section 13.4 to three dimensions, and visualize it using three-dimensional graphics software called VPython. For instructions on how to install VPython on your computer, visit `http://vpython.org` and click on one of the download links on the left side of the page.

As of this writing, the latest version of VPython only works with Python 2.7, so installing this version of Python will be part of the process. To force your code to behave in most respects like Python 3, even while using Python 2.7, add the following as the first line of code in your program files, above any other `import` statements:

```
from __future__ import division, print_function
```

The three-dimensional coordinate system in VPython looks like this, with the positive z-axis coming out of the screen toward you.

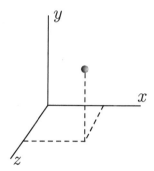

The center of the screen is at the origin $(0, 0, 0)$.

Part 1: Generalize the `Vector` class

The `Pair` and `Vector` classes that we used to represent positions and velocities, respectively, are limited to representing two-dimensional quantities. For this project, you will generalize the `Vector` class that we introduced in Section 13.4 so that it can store vectors of any length. Then use this new `Vector` class in place of both `Pair` and the old `Vector` class.

The implementation of every method will need to change, except as noted below.

- The constructor should require a list or tuple parameter to initialize the vector. The length of the parameter will dictate the length of the `Vector` object. For example,

    ```
    velocity = Vector((1, 0, 0))
    ```

    will assign a three-dimensional `Vector` object representing the vector $\langle 1, 0, 0 \rangle$.

- Add a `__len__` method that returns the length of the `Vector` object.

- The division operator in Python 2.7 is named `__div__` instead of `__truediv__`. To make your class compatible with both versions of Python, add a `__div__` method that simply calls your modified `__truediv__` method.

- The `unit` and `diffAngle` methods can remain unchanged.

- You can delete the `angle` and `turn` methods, as you will no longer need them.

- The dot product of two vectors is the sum of the products of corresponding elements. For example, the dot product of $\langle 1, 2, 3 \rangle$ and $\langle 4, 5, 6 \rangle$ is $1 \cdot 4 + 2 \cdot 5 + 3 \cdot 6 = 4 + 10 + 18 = 32$. The `dotproduct` method needs to be generalized to compute this quantity for vectors of any length.

Part 2: Make the `World` three-dimensional

Because we were careful in Section 13.4 to make the `World` class very general, there is little to be done to extend it to three dimensions. The main difference will be that instead of enforcing boundaries on the space, you will incorporate an object like a light to which the boids are attracted. Therefore, the swarming behavior will be more similar to moths around a light than migrating birds.

- Generalize the constructor to accept a `depth` in addition to `width` and `height`. These attributes will only be used to size the VPython display and choose initial positions for the boids.

- Once you have VPython installed, you can access the module with

    ```
    import visual
    ```

    To create a new window, which is an object belonging to the class `display`, do the following:

    ```
    self._scene = visual.display(title = 'Boids',
                                 width = 800, height = 800,
                                 range = (width, height, depth),
                                 background = (0.41, 0.46, 0.91))
    ```

    In the `display` constructor, the `width` and `height` give the dimensions of the window and the `range` argument gives the visible range of points on either side of the origin. The `background` color is a deep blue; feel free to change it. Then, to place a yellow sphere in the center to represent the light, create a new `sphere` object like this:

    ```
    self._light = visual.sphere(scene = self._scene,
                                pos = (0, 0, 0),
                                color = visual.color.yellow)
    ```

- In the `stepAll` method, make the light follow the position of the mouse with

    ```
    self._light.pos = self._scene.mouse.pos
    ```

    You can change the position of any `visual` object by changing the value of its `pos` instance variable. In the display object (`self._scene`), `mouse` refers to the mouse pointer within the VPython window.

- Finally, generalize the `_distance` function, and remove all code from the class that limits the position of an agent.

Part 3: Make a `Boid` *three-dimensional*

- In the constructor, initialize the position and velocity to three-dimensional `Vector` objects. You can represent each boid with a cone, pointing in the direction of the current velocity with:

  ```
  self._turtle = visual.cone(pos = (x, y, z),
                             axis = (self._velocity * 3).get(),
                             color = visual.color.white,
                             scene = self._world._scene)
  ```

- The `move` method will need to be generalized to three dimensions, but you can remove all of the code that keeps the boids within a boundary.

- The `_avoid`, `_center`, and `_match` methods can remain mostly the same, except that you will need to replace instances of `Pair` with `Vector`. Also, have the `_avoid` method avoid the light in addition to avoiding other boids.

- Write a new method named `_light` that returns a unit vector pointing toward the current position of the light. Incorporate this vector into your `step` method with another weight

  ```
  LIGHT_WEIGHT = 0.3
  ```

The main simulation

Once you have completed the steps above, you can remove the turtle graphics setup, and simplify the main program to the following.

```
from __future__ import division

from world import *
from boid import *
import visual

WIDTH = 60
HEIGHT = 60
DEPTH = 60
NUM_MOTHS = 20

def main():
    sky = World(WIDTH, HEIGHT, DEPTH)
    for index in range(NUM_MOTHS):
        moth = Boid(sky)

    while True:
        visual.rate(25)      # 1 / 25 sec elapse between computations
        sky.stepAll()

main()
```

APPENDIX A

Installing Python

THE Python programs in this book require Python version 3.4 or later, and two additional modules: `numpy` and `matplotlib`. Generally speaking, you have two options for installing this software on your computer.

A.1 AN INTEGRATED DISTRIBUTION

The easiest approach is to install an integrated distribution that includes Python and the additional modules that we use. At the time of this writing, we know of only one such distribution that includes Python 3. The **Anaconda** distribution from Continuum Analytics is freely available from

> https://store.continuum.io/cshop/anaconda/

Be sure to install the version that includes Python 3.4 or later (not Python 2.7).

A.2 MANUAL INSTALLATION

Your second option is to download and install Python, and the two additional modules, manually. The instructions for doing this are different for Macintosh and Windows computers.

Macintosh

1. Download Python 3.4 (or later) from

   > http://www.python.org/downloads/

   and follow the directions to install it. The Python installation and **IDLE** can then be found in the **Python 3.4** folder in your **Applications** folder.

2. To install `numpy` and `matplotlib`, open the **Terminal** application and type

   ```
   pip3 install matplotlib
   ```

in the window. This command will download and install `matplotlib` in addition to other modules on which `matplotlib` depends (including `numpy`). When you are done, you can quit **Terminal**.

3. (Optional) We recommend that Macintosh users use **TextWrangler** from Bare Bones Software, in place of IDLE. You can download this free program from

    http://www.barebones.com/products/textwrangler/

    To run a Python program in **TextWrangler**, save the file with a `.py` extension, and then select **Run** from the **#!** (pronounced "shebang") menu. To ensure that **TextWrangler** uses the correct version of Python, you may need to include the following as the first line of every program:

    `#!/usr/bin/env python3`

Windows

1. Download Python 3.4 (or later) from

    http://www.python.org/downloads/

    and follow the directions to install it. After installation, you should find **IDLE** in the **Start** menu, or you can use Windows' search capability to find it. The Python distribution is installed in the `C:\Python34` folder.

2. To install `numpy`, go to

    http://sourceforge.net/projects/numpy/files/NumPy/

    and then click on the highest version number next to a folder icon (`1.9.1` at the time of this writing). Next download the file that looks like `numpy-...-python3.4.exe` (e.g., `numpy-1.9.1-win32-superpack-python3.4.exe`). Save the file and double-click to install it.

3. To install `matplotlib`, go to

    http://matplotlib.org/downloads.html .

    In the section at the top of the page, labeled "Latest stable version" (version `1.4.3` at the time of this writing), find the file that looks like `matplotlib-...-py3.4.exe` (e.g., `matplotlib-1.4.3.win32-py3.4.exe`) and click to download it. Save the file and double-click to install it.

4. Open the **Command Prompt** application (in the **Accessories** folder) and type the following commands, one at a time, in the window.

```
cd \python34\scripts
pip3 install six
pip3 install python-dateutil
pip3 install pyparsing
```

The first command changes to the folder containing the `pip3` program, which is used to install additional modules in Python. The next three commands use `pip3` to install three modules that are needed by `matplotlib`. When you are done, you can quit **Command Prompt**.

APPENDIX B

Python library reference

THE following tables provide a convenient reference for the most common Python functions and methods used in this book. You can find a complete reference for the Python standard library at https://docs.python.org/3/library/.

B.1 MATH MODULE

The following table lists commonly used functions and constants in the `math` module. The variable name `x` represents a generic numerical argument. Arguments in square brackets are optional.

`acos(x)`	returns the arccosine of `x` ($\cos^{-1} x$)
`asin(x)`	returns the arcsine of `x` ($\sin^{-1} x$)
`atan(x)`	returns the arctangent of `x` ($\tan^{-1} x$)
`atan2(y, x)`	returns the arctangent of `y/x` ($\tan^{-1}(y/x)$)
`cos(x)`	returns the cosine of `x` radians ($\cos x$)
`degrees(x)`	returns the number of degrees in `x` radians
`exp(x)`	returns e^x
`log(x, [b])`	returns the logarithm base `b` of `x` ($\log_b x$); if `b` is omitted, returns the natural logarithm of `x` ($\ln x$)
`radians(x)`	returns the number of radians in `x` degrees
`sin(x)`	returns the sine of `x` radians ($\sin x$)
`sqrt(x)`	returns the square root of `x` ($\sqrt{x}$)
`tan(x)`	returns the tangent of `x` radians ($\tan x$)
`e`	the value of e (Euler's number), the base of the natural logarithm
`pi`	the value of π

B.2 TURTLE METHODS

The following table lists commonly used methods of the `Turtle` class (in the `turtle` module). The following descriptions assume default settings: angles are in degrees and right turns are clockwise. Arguments in square brackets are optional.

`backward(distance)`	moves turtle `distance` opposite to its current direction
`begin_fill()`	marks the beginning of a shape to be filled
`circle(radius, [extent, steps])`	draws a circle with given `radius`; if `extent` is given, draws an arc of `extent` degrees; if `steps` is given, draw a regular polygon with `steps` sides
`dot([size, color])`	draws a dot with diameter `size` in given `color` (defaults 1, 'black')
`down()`	puts the turtle's tail down, enabling drawing
`end_fill()`	fills the shape drawn since the last call to `begin_fill()`
`fillcolor(color)`	sets the turtle's fill color to `color` (see Box 3.2)
`forward(distance)`	moves turtle `distance` forward in its current direction
`getscreen()`	returns the `Screen` object in which the turtle is drawing
`goto(x, y)`	moves turtle to position (`x`, `y`) without changing heading
`heading()`	returns the turtle's heading
`hideturtle()`	hides the turtle while drawing
`home()`	moves turtle to the origin and resets to original heading
`left(angle)`	turns turtle `angle` degrees counterclockwise
`pencolor(color)`	sets the turtle's pen color to `color` (see Box 3.2)
`pensize(width)`	sets the pen to the given width
`position()`	returns the turtle's current position as a tuple
`right(angle)`	turns turtle `angle` degrees clockwise
`setheading(angle)`	sets turtle's heading to `angle` degrees
`speed(s)`	sets turtle's speed to `s`, a number 0 to 10; 1 is slowest, 10 is fast, 0 is fastest
`up()`	puts the turtle's tail up, disabling drawing
`write(message)`	writes `message` at the current turtle position
`xcor()`	returns the turtle's x coordinate
`ycor()`	returns the turtle's y coordinate

B.3 SCREEN METHODS

The following table lists commonly used methods of the `Screen` class (in the `turtle` module). Arguments in square brackets are optional.

`bgcolor(color)`	sets the color of the background to `color`
`bgpic(filename)`	sets the background to contain the named GIF image
`colormode(mode)`	if `mode` is 1.0, RGB colors are specified by numbers between 0 and 1.0; if `mode` is 255, they are specified by numbers between 0 and 255
`exitonclick()`	causes the drawing window to close when clicked
`mainloop()`	must be called at the end of any program handling mouse clicks or other events
`onclick(function)`	call `function` when there is a mouse click; `function` must take `x` and `y`, the location of the click, as parameters
`setup(width, height, startx, starty)`	sets the size and location of the drawing window
`setworldcoordinates (x1, y1, x2, y2)`	sets the coordinates of the window with (`x1`, `y1`) at the lower left and (`x2`, `y2`) at the upper right
`tracer(n)`	only perform every `n`th screen update
`update()`	updates the screen to reflect all drawing so far
`window_height()`	returns the height of the drawing window
`window_width()`	returns the width of the drawing window

B.4 MATPLOTLIB.PYPLOT MODULE

The following table lists commonly used functions in the `matplotlib.pyplot` module. The parameters in square brackets are optional. For a complete reference, see http://matplotlib.org/api/pyplot_summary.html.

`bar(x, y)`	creates a bar graph with the given `x` and `y` values
`hist(values, [bins])`	creates a histogram of `values` using the given number of bins (default is 10)
`legend()`	creates a legend using labels from the plotting calls
`plot(x, y, [options])`	creates a line graph with the given lists of `x` and `y` values; common optional keyword arguments are `color = 'blue'` or another color string `linewidth = 2` or another width `linestyle = 'dashed'` or `'solid'` or `'dotted'` `label = 'mylabel'` (used by `legend()`)
`scatter(x, y, [options])`	creates a scatter plot with the given lists of `x` and `y` values; common options are `color` and `label`
`title(titlestring)`	sets the title of the graph to be `titlestring`
`xlabel(xstring)`	labels the x axis of the current graph with `xstring`
`xticks(range, [labels], [options])`	set the locations (and optionally, labels) of the ticks on the x-axis
`ylabel(ystring)`	labels the y axis of the current graph with `ystring`
`yticks(range, [labels], [options])`	set the locations (and optionally, labels) of the ticks on the y-axis

B.5 RANDOM MODULE

The following table lists commonly used functions in the `random` module.

`gauss(mean, stdDev)`	returns a value according to the Gaussian (i.e., normal) distribution with the given mean and standard deviation
`random()`	returns a pseudorandom number in $[0, 1)$
`randrange(start, stop, step)`	returns a randomly selected integer value from `range(start, stop, step)`
`seed(s)`	sets the seed for the PRNG; default is the current time
`uniform(a, b)`	returns a pseudorandom number in $[a, b]$

B.6 STRING METHODS

The following table lists commonly used methods of the `str` class. Optional parameters are denoted in square brackets.

`count(substring)`	returns number of times `substring` appears in the string
`find(substring)`	returns the index of the first instance of `substring` in the string, or -1 if `substring` is not found
`lower()`	returns a copy of the string with all letters in lower case
`lstrip([chars])`	returns a copy of the string with all instances of the characters in the string `chars` removed from its beginning; if `chars` is omitted, whitespace characters are removed
`replace(old, new)`	returns a copy of the the string with all instances of the string `old` replaced with the string `new`
`rstrip([chars])`	returns a copy of the string with all instances of the characters in the string `chars` removed from its end; if `chars` is omitted, whitespace characters are removed
`split([sep])`	returns a list of "words" in the string that are separated by the delimiter string `sep`; if `sep` is omitted, the string is split at runs of whitespace characters
`strip([chars])`	returns a copy of the string with all leading and trailing instances of the characters in the string `chars` removed; if `chars` is omitted, whitespace characters are removed
`upper()`	returns a copy of the string with all letters in upper case

B.7 LIST METHODS

The following table lists commonly used methods of the `list` class and three list functions from the `random` module. Optional parameters are in square brackets.

`append(item)`	appends `item` to the end of the list; returns `None`
`clear()`	clears the contents of the list
`copy()`	returns a shallow copy of the list
`count(item)`	returns number of times `item` appears in the list
`extend(items)`	appends all of the values in the list named `items` to the end of the list; returns `None`
`index(item)`	returns the index of the first occurrence of `item` in the list; raises a `ValueError` if `item` is not found
`insert(index, item)`	inserts `item` in the list at `index`; returns `None`
`pop([index])`	deletes the item in position `index` from the list and returns it; if `index` is omitted, deletes and returns the last item in the list
`remove(item)`	removes the first instance of `item` from the list; returns `None`; raises `ValueError` if `item` is not found
`reverse()`	reverses the items in the list in place; returns `None`
`sort([key, reverse])`	sorts the list in place using a stable sort; if provided, `key` is a function that returns a key to be used for a list item in the sort; if `reverse` is `True`, the list is sorted in reverse order; returns `None`
`random.choice(data)`	returns a random element from the list `data`
`random.sample(data, k)`	returns a list of `k` unique elements from list `data`
`random.shuffle(data)`	shuffles the list `data` in place; returns `None`

B.8 IMAGE MODULE

The module `image.py` which contains the `Image` class is available on the book web site. The first table lists the functions in the `image` module.

`Image(width, height, [title = 'Title'])`	returns a new empty `Image` object with the given `width` and `height`; optionally sets the `title` of the image window displayed by `show`
`Image(file = 'file.gif', [title = 'Title'])`	returns a new `Image` object containing the image in the given GIF `file`; optionally sets the `title` of the image window displayed by `show`
`mainloop()`	waits until all image windows have been closed, then quits the program

The second table lists the methods of the `Image` class.

`get(x, y)`	returns a tuple representing the RGB color of the pixel at coordinates (x,y) of the image
`height()`	returns the height of the image
`set(x, y, color)`	sets the color of the pixel at coordinates (x,y) of the image to `color` (a RGB tuple)
`show()`	displays the image in its own window
`width()`	returns the width of the image
`update()`	updates the image in its existing window

B.9 SPECIAL METHODS

The following table lists commonly used special methods that may be overridden in new classes.

Method	Called by	Comments
`__init__(self)`		class constructor
`__str__(self)`	`str(self)`	string representing `self`
`__lt__(self, other)`	`self < other`	
`__le__(self, other)`	`self <= other`	
`__gt__(self, other)`	`self > other`	
`__ge__(self, other)`	`self >= other`	
`__eq__(self, other)`	`self == other`	
`__ne__(self, other)`	`self != other`	
`__len__(self)`	`len(self)`	length of `self`
`__getitem__(self, index)`	`self[index]`	returns the item in `self` at `index`
`__setitem__(self, index, value)`	`self[index] = value`	assigns `value` to the item in `self` at `index`
`__delitem__(self, index)`	`del self[index]`	deletes the item in `self` at `index`
`__contains__(self, item)`	`item in self`	returns whether `item` is in `self`
`__add__(self, other)`	`self + other`	
`__sub__(self, other)`	`self - other`	
`__mul__(self, other)`	`self * other`	
`__truediv__(self, other)`	`self / other`	true division
`__floordiv__(self, other)`	`self // other`	floor division

Bibliography

[1] Edwin A. Abbott. Flatland: A Romance of Many Dimensions. Dover Publications, 1992.

[2] Kevin A. Agatstein. Oscillating Systems II: Sustained Oscillation. *MIT System Dynamics in Education Project*, http://ocw.mit.edu/courses/sloan-school-of-management/15-988-system-dynamics-self-study-fall-1998-spring-1999/readings/oscillating2.pdf, 2001.

[3] Albert-László Barabási. Linked: How Everything is Connected to Everything Else and What It Means for Business, Science, and Everyday Life. Plume, 2003.

[4] Frank Bass. A new product growth model for consumer durables. *Management Science* 15(5):215–227, 1969.

[5] Peter J. Bentley. Digitized: The Science of Computers and How It Shapes Our World. Oxford University Press, 2012.

[6] Yogi Berra with Dave Kaplan. When you Come to a Fork in the Road, Take it! Inspiration and Wisdom from One of Baseball's Greatest Heroes. Hyperion, 2001.

[7] Carey Caginalp. Analytical and Numerical Results on Escape of Brownian Particles. B. Phil. thesis, University of Pittsburgh, 2011.

[8] William J. Cook. In Pursuit of the Traveling Salesman: Mathematics at the Limits of Computation. Princeton University Press, 2012.

[9] Delta Airlines North American route map. https://www.delta.com/content/dam/delta-www/pdfs/route-maps/us-route-map.pdf

[10] Allen B. Downey. Think Complexity. O'Reilly Media, 2012.

[11] Sir Arthur Conan Doyle. The Adventures of Sherlock Holmes. Harper, 1892.

[12] Charles Duhigg. How Companies Learn Your Secrets. *The New York Times*, http://www.nytimes.com/2012/02/19/magazine/shopping-habits.html, published February 16, 2012.

[13] Michael R. Fellows and Ian Parberry. SIGACT trying to get children excited about CS. *Computing Research News*, p. 7, January 1993.

[14] Gary W. Flake. The Computational Beauty of Nature. MIT Press, 2000.

[15] Nigel Gilbert. Agent-Based Models. SAGE Publications, 2008.

[16] Jessica Guynn. Marissa Mayer talks about Google at 10 — and 20. *Los Angeles Times*, http://latimesblogs.latimes.com/technology/2008/09/marissa-mayer-t.html, September 7, 2008.

[17] David Harel. Algorithmics: The Spirit of Computing, third edition. Addison-Wesley, 2004.

[18] W. Daniel Hillis. The Pattern On The Stone: The Simple Ideas That Make Computers Work. Basic Books, 1998.

[19] Eric Hobsbawm. The Age of Extremes: A History of the World, 1914-1991. Vintage Books, 1994.

[20] Andrew Hodges. Alan Turing: The Enigma. Princeton University Press, 1983.

[21] Deborah G. Johnson with Keith W. Miller. Computer Ethics: Analyzing Information Technology, fourth edition. Prentice Hall, 2009.

[22] Steven Johnson. Emergence: The Connected Lives of Ants, Brains, Cities, and Software. Scribner, 2001.

[23] William O. Kermack and Anderson G. McKendrick. A Contribution to the Mathematical Theory of Epidemics. *Proceedings of the Royal Society A* 115: 700–721, 1927.

[24] Donald E. Knuth. The Art of Computer Programming, volumes 1–4A. Addison-Wesley, 1968–2011.

[25] Donald E. Knuth. Computer Programming as an Art. *Communications of the ACM* 17(12):667–673, 1974.

[26] Donald E. Knuth. Computer Science and Mathematics. *American Scientist* 61(6), 1973.

[27] Ehud Lamm and Ron Unger. Biological Computation. Chapman & Hall/CRC Press, 2011.

[28] Barbara Liskov and Stephen Zilles. Programming with Abstract Data Types. In *Proceedings of the ACM Conference on Very High Level Languages*, SIGPLAN Notices 9(4):50–59, 1974.

[29] Mark Lutz. Programming Python. O'Reilly Media, 1996.

[30] John MacCormick. Nine Algorithms that Changed the Future. Princeton University Press, 2013.

[31] Benoit B. Mandelbrot. The Fractal Geometry of Nature. Macmillan, 1983.

[32] Steve McConnell. Software Project Survival Guide: How to Be Sure Your First Important Project Isn't Your Last. Microsoft Press, 1998.

[33] Stanley Milgram. The Small-World Problem. *Psychology Today* 1(1):61–67, May 1967.

[34] Leonard Mlodinow. The Drunkard's Walk: How Randomness Rules Our Lives. Vintage Books, 2009.

[35] Philip Morrison and Phylis Morrison. 100 or so Books that shaped a Century of Science. *American Scientist* 87(6), November-December 1999.

[36] Alexander J. Nicholson and Victor A. Bailey. The Balance of Animal Populations—Part I. *Proceedings of the Zoological Society of London* 105:551-598, 1935.

[37] Stephen K. Park and Keith W. Miller. Random Number Generators: Good Ones Are Hard to Find. *Communications of the ACM* 31(10):1192–1201, 1988.

[38] William R. Pearson and David J. Lipman. Improved Tools for Biological Sequence Comparison. *Proceedings of the National Academy of Science* 85(8):2444–2448, 1988.

[39] Jean R. Petit, *et al.* Climate and Atmospheric History of the Past 420,000 years from the Vostok Ice Core, Antarctica. *Nature* 399:429–436.

[40] Charles Petzold. CODE: The Hidden Language of Computer Hardware and Software. Microsoft Press, 2000.

[41] George Polya. How to Solve It: A New Aspect of Mathematical Method. Princeton University Press, 1945.

[42] Jean-Yves1 Potvin. Genetic algorithms for the traveling salesman problem. *Annals of Operations Research* 63(3):337–370, 1996.

[43] Przemyslaw Prusinkiewicz and Aristid Lindenmayer. The Algorithmic Beauty of Plants. http://algorithmicbotany.org/papers/abop/abop.pdf, 1990.

[44] Mitchell Resnick. Turtles, Termites, and Traffic Jams: Explorations in Massively Parallel Microworlds. MIT Press, 1994.

[45] Craig W. Reynolds. Flocks, Herds, and Schools: A Distributed Behavioral Model. *Computer Graphics, SIGGRAPH '87* 21(4):25–34, 1987.

[46] Michael C. Schatz and Ben Langmead. The DNA Data Deluge. *IEEE Spectrum* 50(7):28–33, 2013.

[47] Thomas C. Schelling. Dynamic Models of Segregation. *Journal of Mathematical Sociology* 1:143–186, 1971.

[48] Thomas C. Schelling. Micromotives and Macrobehavior. W. W. Norton & Company, 1978.

[49] Rachel Schutt and Cathy O'Neil. Doing Data Science. O'Reilly Media, 2014.

[50] Dr. Seuss. The Sneetches and Other Stories. Random House, 1961.

[51] Dennis Shasha and Cathy Lazere. Natural Computing: DNA, Quantum Bits, and the Future of Smart Machines. W. W. Norton & Company, 2010.

[52] Angela B. Shiflet and George W. Shiflet. Introduction to Computational Science: Modeling and Simulation for the Sciences, second edition. Princeton University Press, 2014.

[53] Ian Stewart. The Mathematics of Life. Basic Books, 2011.

[54] Steven Strogatz. Sync: How Order Emerges from Chaos in the Universe, Nature, and Daily Life. Hyperion, 2003.

[55] Alexander L. Taylor III, Peter Stoler, and Michael Moritz. The Wizard inside the Machine *Time* 123(6):64–73, April 1984.

[56] Christophe Van den Bulte and Yogesh V. Joshi. New Product Diffusion with Influentials and Imitators. *Marketing Science* 26(3):400–421, 2007.

[57] Duncan J. Watts. Six Degrees: The Science of a Connected Age. W. W. Norton & Company, 2004.

[58] Wikipedia. Electricity sector in New Zealand. http://en.wikipedia.org/w/index.php?title=Electricity_sector_in_New_Zealand, 2014.

[59] Niklaus Wirth. Algorithms + Data Structures = Programs. Prentice-Hall, 1976.

[60] James Zachos, *et al.* Trends, Rhythms, and Aberrations in Global Climate 65 Ma to Present. *Science* 292: 686–692, 2001.

[61] http://www.pd4pic.com/nautilus-cephalopods-sea-holiday-memory-housing-2.html, 2015.

[62] http://photojournal.jpl.nasa.gov/jpeg/PIA03424.jpg, 2015.

[63] http://www.pd4pic.com/leaf-green-veins-radiating-patterned.html, 2015.

[64] http://www.pd4pic.com/strait-of-malacca-sky-clouds-lightning-storm.html, 2015.

Index

__future__ module, 476, 693

Abbott, Edwin, 443, 471
abs(), 45
abstract data type, 66, 111, 242, 351, 621–682
    accessor, 624
    attribute, 66, 623, 649, 655, 665, 670, 671, 683
    constructor, 624
    destructor, 624, 627
    mutator, 624
    operation, 66, 623, 650, 655, 666, 670, 672, 684
abstraction, 3–4, 10, 21, 29, 66, 113, 250, 621
    data, *see* abstract data type
    functional, *see* functional abstraction
accumulator, 116–136, 353
    list, *see* list accumulator
    string, *see* string accumulator
actual parameters, *see* arguments
adder, 57
adjacency list, 590
adjacency matrix, 588
admissions, 428
ADT, *see* abstract data type
Africa, 388, 439, 633
agent-based simulation, 471, 649, 690
algorithm, 4, 7, 21, 390
    binary search, *see* binary search algorithm
    breadth-first search, *see* breadth-first search
    depth-first search, *see* depth-first search
    divide and conquer, *see* divide and conquer algorithm
    Euclid's GCD, *see* Euclid's GCD algorithm
    flood fill, *see* flood fill algorithm
    insertion sort, *see* insertion sort algorithm
    linear search, *see* linear search algorithm
    merge sort, *see* merge sort algorithm
    Metropolis, *see* Metropolis algorithm
    recursive, *see* recursion
    removing duplicates, *see* duplicates, removing
    selection sort, *see* selection sort algorithm
amino acid, 299, 637
amoeba, 144
amoeboid, 690
Analytical Engine, 112
app, *see* program
Apple, Inc., 579
application, *see* program
area, 35, 197, 484, 532
argument, 45
    default, *see* default argument
    keyword, *see* keyword argument
arguments, 45
array, *see* NumPy
ASCII, 263
assert statement, 335
assignment operator, 38
asymptote, 275
asymptotic time complexity, 275
atmosphere, 101, 148, 422
attribute, *see* abstract data type, attribute
average, 7, 36, 134, 186, 192, 204, 234, 237, 323, 372, 464, 532, 538, 588, 594, 603, 622, 633, 636, 661, 681, 689, *see also* mean

Babbage, Charles, 112
Babylonian method, 164
backtracking, 512, 515, 538
Bacon, Kevin, 616
bacteria, 135, 157, 299
Barabási, Albert-László, 611
base case, 479, 481, 490
base pair, 298
Bass diffusion model, 177
Bass, Frank, 177
Bentley, Peter, 30

Berra, Yogi, 185, 234
bicycling, 416
binary notation, 23
binary number system, 24, 54
binary search algorithm, 542–552, 576
binary search tree, 581, 672, 678
biology, 171, 180, 186, 237, 297, 316, 518,
      526, 648, 690, 692
bird, 648
birthday problem, 373
bit, 22
bitmap, 23, 461, 463
BMP image format, 463
Boid ADT, 655
boids model, 648–664
Boolean algebra, *see* Boolean logic
Boolean expression, 57, 59, 140, 187
Boolean logic, 25, 209, 233
Boolean operators, 25, 210
Borda count, 436
Borda, Jean-Charles de, 436
botany, 186, 518, 526
Botswana, 390
boundary case, 344
breadth-first search, 514, 594
Brown, Robert, 186, 234
Brownian motion, 186
Brownian tree, 474
brute force algorithm, 579
bus, 21
byte, 23
`bytes` object, 255

Caginalp, Carey, 240
calculus, 151, 162, 275, 404, 485
camelCase, 95
cancer, 419
carbon, 148, 158, 422
carrying capacity, 172, 183
cell
    biological, 44, 297, 519, 690
    in a grid, 449, 471, 512, 651
    memory, *see* memory cell
    Turing machine, 27
cellular automaton, 449
central limit theorem, 207
centroid, 410, 419, 532, 622, 631
chaining, 687
checksum, 269, 359
chi-squared distribution, 209

Chomsky hierarchy, 519
Chomsky, Noam, 519
`chr()`, 264
chromosome, 44, 440
Church-Turing thesis, 28
circuit, 59
class, 68, 625
class variable, 675
climate, 7, 422
clock rate, 22, 579
`close()`, 251
closed form, 503
clustering coefficient, 603
clustering, data, 410
`cmath` module, 52
CMYK, 462
code, *see* source code
coding region, 298
codon, 299
coffee, 110, 427
collection, 665
college, 155, 173, 403, 425, 616
collision, 186, 410, 649
    hash table, 376, 672, 687
color models, 74, 462
comma-separated values (CSV), 384, 389,
      390, 408, 419, 423, 425, 429, 444,
      448, 561, 686
comments, 91
compactness, 532
comparison operators, 188, 267, 640
compiler, 20, 350
complement, 298, 302
complex numbers, 37, 52
compression, 463
    image, 463
    text, 266
computation, 1
concordance, 284
conditional statement, 187–233
Condorcet voting, 437
consensus sequence, 383
constant-time, 276, 313, 376, 397, 401
constructor, 625
contagion, 615
    social, 177
continuous, 53, 146, 153, 462, 649
control characters
    newline character, *see* newline character

tab character, *see* tab character
converting between bases, 668
Conway, John, 449
coordinates, 69, 84, 128, 187, 204, 237, 290, 352, 405, 452, 464, 517, 622, 656
`copy` module, 458
    `deepcopy()`, 458
core, *see* processor
corner case, 347
CPU, *see* processor
CSV format, *see* comma-separated values (CSV)

Darcy's law, 101
data abstraction, *see* abstract data type
data mining, 351, 403, 410, 421
data science, 351, 421
data structure, 541, 621, 634
database, 302, 306, 543, 545, 574, 581, 616, 687
De Morgan's laws, 28, 227
`def`, 77
default argument, 627
degree distribution, 605
degrees, 52
`del` operator, 652, 677
dendrite, 474
depth-first search, 512, 534, 538, 601, 611, 670
design by contract, 331–340, 350
deterministic, 185
Dickens, Charles, 259
`dict` class, 105, 374–383, 401, 427, 436, 441, 447, 461, 522, 590, 650
    `in` operator, 376
    `keys()`, 377
    `values()`, 377
Dictionary ADT, 671
dictionary ADT, 374, 585, 671–681, *see also* `dict` class
Difference Engine, 112
difference equation, 145, 201, 503, *see also* recurrence relation
    coupled, 153
differential equation, 151
diffusion, 156, 177, 612
diffusion-limited aggregation, 474
Dijkstra, Edsger, 30
disease, 153, 300, 537, 587
distance, 53, 101, 191, 404, 410, 419, 439, 534, 594, 601, 622, 633, 644, 653, 684
distribution
    degree, *see* degree distribution
    probability, *see* probability distribution
divide and conquer algorithms, 508, 570
divide by zero error, 34
division method, 673
division operators, 34
DNA, 19, 44, 297, 371, 373, 382, 440, 637
docstring, 91, 332, 632
documentation, 91, 632
Downey, Allen, 612, 682
Doyle, Sir Arthur Conan, 421
duplicates, removing, 391

earthquake, 384, 448, 543, 558, 561
economics, 110, 173, 427, 471, 687
education, 425
efficiency, 16, 21, 170, 220, 254, 261, 330, 391, 505, 546, 550, 557, 566, 574, 578, *see also* time complexity
Einstein, Albert, 94
election, 435, 531, 633
Electronic Frontier Foundation, 422
elementary step, 16, 270
embryo, 312
emergence, 449, 472, 474, 587, 648, 690
employment, 425
empty string, 243
encoding, 23, 252, 263, 268
ENIAC, 201
epidemic, 153, 607, 614
epidemiology, 614
epoch, 402
equilibrium, 473
error, 13, 22, 33, 118, 143, 151, 234, 236, 249, 321, 409, *see also* exception
    divide by zero, *see* divide by zero error
    index, *see* index error
    name, *see* name error
    overflow, *see* overflow error
    propagation of, 152
    syntax, *see* syntax error
    type, *see* type error
    value, *see* value error
Escherichia coli, 310, 317
Euclid's GCD algorithm, 499
eukaryote, 299
Euler's method, 151

Euler's number, 53, 165, 171, 473, 701
event loop, 686
evolution, 4, 297, 304, 439, 537, 607
exception, 334, 667
    `AssertionError`, 335
    `IndexError`, 260, 643
    `KeyError`, 675
    `NameError`, 43, 104
    `SyntaxError`, 40, 72, 210
    `TypeError`, 50, 261, 332, 366
    `ValueError`, 52
    `ZeroDivisionError`, 34
exhaustive search, *see* brute force algorithm
exponential growth, 170, 183, 434
exponential-time algorithm, 276, 578, 579
exponentiation, 33
external sorting algorithm, 574

Facebook, 587, 588, 608, 610, 613, 614
factorial, 134, 439, 498
FASTA, 306
Fellows, Michael R., 1, 30
ferromagnetism, 473
fetch and execute cycle, 22
Fibonacci numbers, 157, 516
file object, 250
file pointer, 256
file system, 250
files, 250
    closing, 251, 254
    opening, 250
    reading, 251
    writing, 253
finite precision, *see* precision
finite state machine, 26
fish, 114, 136, 140, 648
fitness, 440, 649
flag variable, 192, 238
Flake, Gary, 682
Flatland, 443, 471
`float()`, 46, 386
floating point numbers, 33
    testing equality of, 348
flocking, 648
flood fill algorithm, 534
flooding, 430
flower, 70
fluid, 186, 474
`for` loop, 72, 245
forensics, 683

formal grammar, 518
formal parameters, *see* parameters
format string, 119
`format()`, 119, 130
fractal, 479
    Brownian tree, 474
    dragon curve, 521
    Hilbert curve, 486
    Koch curve, 481
    Lindenmayer system, 521
    quadratic Koch curve, 485
    Sierpinski carpet, 487
    Sierpinski triangle, 486
    tree, 479
function, 5, 45, 97, 324
function call, 45
function invocation, *see* function call
functional abstraction, 3–4, 45, 62, 76, 79, 83, 111, 324, 329, 621

game, 11, 20, 185, 224, 500, 552, 616
Game of Life, 449
garbage collection, 39
gas, 474
    greenhouse, 422
    ideal, *see* ideal gas
    natural, 537
Gaussian distribution, *see* normal distribution
gene, 206, 298, 316
    Hox, *see Hox* gene
genetic algorithm, 440
genetic code, 299
genome, 300, 439
genomics, 297
geology, 382, 384, 422, 430, 474, 537
geometric mean, 134
gerrymandering, 531
GIF image format, 463, 706
gigabyte (GB), 23, 102
gigahertz (GHz), 22, 102
Gilbert, Nigel, 471, 682
`global`, 91
Global Positioning System, 2, 683
global variable, 80, 90, 91
`globals()`, 107
Google, 112, 241, 286
googol, 33
GPA, 103, 219, 403, 428
GPS, *see* Global Positioning System

grades, 103, 200, 219, 221, 269, 636
grammar, 40
    formal, *see* formal grammar
graph, 588
    connected, 601, 611
    random, *see* random graph
grayscale, 461, 465, 469
greatest common divisor (GCD), 499
grid, 186, 449, 461, 471, 473, 474, 512, 536, 602, 649, 690
growth rate, 114, 132, 135, 144, 145, 157, 389

hailstone numbers, 231
half-life, 157
Hall, Monty, 197
Hamming distance, 295
hardware, 5
Harel, David, 30
harmonic series, 162
hash function, 376, 672, 674, 680, 688
hash table, 376, 543, 585, 672–676
Hawking, Stephen, 113, 171
Hello, world!, 32, 47, 63
help(), 93, 674, 678
hertz, 22
heuristic, 410, 439, 579
    genetic algorithm, *see* genetic algorithm
    k-means clustering, *see* k-means clustering
hexadecimal, 69, 637
high performance computing, 22
Hilbert curve, 486
Hillis, Danny, 30
histogram, 195, 200, 203, 206, 378, 380, 441, 671
Hobsbawm, Eric, 294
Hodges, Andrew, 30
Holmes, Sherlock, 351, 421
Hopper, Grace, 321, 350
House of Representatives, 313, 531
Hox gene, 312
HTML, *see* Hypertext Markup Language
hub, 605, 610, 618
Huffman coding, 266
Hypertext Markup Language, 255

ideal gas, 101, 205
IDLE, 32, 63
Idle, Eric, 63
if, 187

Image class, 465, 707
    get(), 466, 707
    height(), 465, 707
    set(), 466, 707
    show(), 466, 707
    update(), 707
    width(), 465, 707
image file formats, 463
    BMP, *see* BMP image format
    GIF, *see* GIF image format
    JPEG, *see* JPEG image format
image filter, 463–470
image module, 465
    mainloop(), 707
images, digital, 461–470
immutable, 261, 365, 452
import, 51, 89, 136, 646
    from, 342
in operator, 73, 115, 242, 245, 376, 392, 680
income, 125, 210, 687
increment, 42
independent cascade model, 612
index error, 260
index variable, 73, 115
indexing, 260, 354
    negative, *see* negative indexing
indirect competition, 159
infection, 153, 614
infinite loop, 142, 226
infinity, 597
initial condition, 146
input(), 49, 225
insertion sort algorithm, 563–569
instance, 626
instance variable, 68, 625, 675
    private, 626
int(), 46, 225
Intel Corporation, 23, 579
intelligence, 7, 20, 648, 690
interest rate, 53, 135, 140, 143, 173, 221, 231
internal sorting algorithm, 574
Internet, 2, 21, 185, 255, 359, 587, 601
Internet Movie Database (IMDb), 616
interpreter, 20, 33, 40, 79, 89, 92, 109
interval notation, 186
intractable, 577
investment, 135, 140, 174, 221
Ising model, 473
isinstance(), 333

isotope, 148, 422
iteration, 72, 488
    over a list, 352
    over a string, 245
    over indices, 280, 355

Johnson, Deborah, 422
Johnson, Steven, 471
Johnston, Leslie, 241, 313
JPEG image format, 463

k-means clustering, 410–421
Kasparov, Garry, 20
Keeling curve, 425
Keeling, Charles David, 425
Kermack, William, 153
key, 374, 447, 522, 543, 545, 581, 590, 671, 687
keyword argument, 139, 387
Keyword in Context (KWIC), 313
keywords, 39
kilobyte (KB), 23
Knuth, Donald E., 31, 62, 65, 112
Koch curve, 481, 521

Lamm, Ehud, 471
latitude, 384, 389, 390, 410, 416, 448, 558, 622, 633, 683
Lazere, Cathy, 313
Lehmer peudorandom number generator, 201
Lehmer, Derrick, 201
Leibniz series, 162
Leibniz, Gottfried, 162
Lempel-Ziv-Welch algorithm, 266, 463
`len()`, 352
Let's Make a Deal, 197
Library of Congress, 37, 241
LIFO, 665
Lindenmayer system, 518–531
    dragon curve, 521
    Koch curve, 521
    tree, 527
Lindenmayer, Aristid, 526
line continuation character, 214
linear growth, 168
linear regression, 404, 644
linear search algorithm, 284, 392, 503, 516, 542, 576
linear-time algorithm, 275, 284, 313, 396, 398, 401, 507, 567, 578

LinkedIn, 588
liquid, 101, 474, 537
Liskov, Barbara, 621, 682
list, 7, 14, 115, 290, 488, 541, 542, 553, 563, 570, 665, see also `list` class
    accumulator, see list accumulator
    comprehension, see list comprehension
    of lists, 445–449, 589
list accumulator, 138, 147, 202, 360–361, 368, 395
`list` class, 123, 136, 351–373, 706
    `append()`, 137, 361, 445, 596, 666, 706
    `clear()`, 573, 706
    concatenation operator +, 366
    `copy()`, 365, 706
    `count()`, 706
    `extend()`, 397, 584, 706
    `index()`, 560, 706
    `insert()`, 367, 565, 706
    `pop()`, 367, 526, 565, 596, 666, 706
    `remove()`, 367, 706
    repetition operator *, 366
    `reverse()`, 706
    `sort()`, 367, 548, 689, 706
list comprehension, 368–370
`list()`, 123
loan, 135, 145, 173
local variable, 80, 104
`locals()`, 105
logarithm, 53, 162, 165, 171, 433, 473, 701
logarithmic-time algorithm, 547, 550, 552
logic gates, 59
logical operators, see Boolean operators
Lonesome George, 68, 112
longitude, 384, 389, 390, 410, 416, 448, 558, 622, 633, 683
loop, 8, 72
Lotka-Volterra model, 180
Lovelace, Ada, 65, 112
Luhn algorithm, 360
Lutz, Mark, 63

Mac OS X, 250
MacCormick, John, 581
machine cycle, see fetch and execute cycle
machine language, 20
Madhava of Sangamagrama, 166
magic numbers, 95, 675
`main` function, 90
    conditionally calling, 343, 645

Mandelbrot, Benoît, 477, 526
mantissa, 33
market, 110, 124, 177, 185
`math` module, 51, 701
    `acos()`, 701
    `asin()`, 701
    `atan()`, 238, 701
    `atan2()`, 701
    `cos()`, 52, 701
    `degrees()`, 238, 701
    `e`, 51, 701
    `exp()`, 53, 162, 701
    `log()`, 172, 701
    `pi`, 51, 98, 701
    `radians()`, 52, 701
    `sin()`, 132, 701
    `sqrt()`, 51, 165, 192, 701
    `tan()`, 701
MATLAB, 136
`matplotlib.pyplot` module, 136, 697, 704
    `bar()`, 380, 704
    `hist()`, 195, 704
    `legend()`, 139, 704
    `plot()`, 138, 704
    `scatter()`, 290, 704
    `title()`, 704
    `xlabel()`, 139, 704
    `xticks()`, 380, 389, 390, 426, 704
    `ylabel()`, 139, 704
    `yticks()`, 704
`max()`, 103
Mayer, Marissa, 541, 581
McConnell, Steve, 321, 350
McKendrick, Anderson, 153
mean, 7, 13, 205, 352, 372, 410, 423, 427, 535, *see also* average
    geometric, *see* geometric mean
megabtye (MB), 23
Melville, Herman, 273
meme, 156, 612
memory, 20, 105
memory address, 21, 69, 637
memory cell, 21, 23, 38, 105, 259
merge sort algorithm, 570–577
Mersenne twister, 204
meteorite, 389
method, 68, 625
Metropolis algorithm, 473
microsatellite, 310

Milgram, Stanley, 611
Miller, Keith, 234, 422
Mlodinow, Leonard, 234
Moby Dick, 250, 273, 284, 329
mode, 377
model, 27, 109, 113, 145, 149, 153, 171, 177, 180, 186, 648
module, 51, 68, 186, 645
modulo operator, 35
Mohs hardness scale, 382
mole, 101
molecule, 59, 149, 186, 205, 297, 440, 474
Monte Carlo simulation, 185, 192–197, 203, 208, 230, 232, 237, 373, 538
Monty Python's Flying Circus, 31
Moore's Law, 578
Moore, Gordon, 579
movie, 616, 635, 681
mutable, 361

name error, 43
namespace, 105, 342, 363, 637, 646
narrow escape problem, 237
natural language processing, 286
Natural Language Toolkit (NLTK), 286
negative indexing, 260
neighborhood, 449, 454, 459, 591, 603, 652
nested loop, 258, 295, 313, 402, 452, 455, 458, 464, 556, 562, 565
network, 12, 587
    actors, 616
    airline, 607
    data, 22
    electrical, 588
    epidemic, 614
    neural, *see* neural network
    simulation on a, 612, 614
    social, *see* social network
neural network, 471, 588, 601
newline character, 243
Newton's method, *see* Babylonian method
Nicholson-Bailey model, 171
Nilakantha series, 167
`None`, 47, 93, 100, 193, 331, 353, 367, 674
normal distribution, 205, 208, 237
`not in` operator, 242
NP-complete, 580
NP-hard, 579
nucleotide, 297
numerical analysis, 159

NumPy, 369, 453, 697

O'Neil, Cathy, 421
object, 68, 136, 242, 351, 365, 625
object-oriented programming language, 622
ocean, 7, 30, 373, 422
oil, 537
Olson, Brian, 531
open reading frame (ORF), 316
`open()`, 250
operating system, 4
operation, *see* abstract data type, operation
operator overloading, 637
operator precedence, 36, 215
`ord()`, 264
ordered pair, 623
ORF, *see* open reading frame (ORF)
`os` module
    `access()`, 340
`os.path` module
    `isfile()`, 340
overflow error, 34

P=NP problem, 580
Pair ADT, 623
Palaeocene-Eocene Thermal Maximum (PETM), 424
palindrome, 310, 492, 670
parallel lists, 385, 559
parallel programming, 22
parameters, 78, 97
parasitoid, 171
Parberry, Ian, 1, 30
parity checking, 249
Park, Stephen, 234
Park-Miller PRNG, 202
particle, 88, 185, 237, 473, 474
`pass`, 327
password, 220, 381, 495
percolation, 537
Petzold, Charles, 30
phase transition, 472, 473, 539
phone tree, 11
physics, 473
pixel, 461, 532, 579, 684
plagiarism, 294
plurality voting, 436
political science, 234, 313, 435, 531, 636
polling, 234
Polya, George, 322

polynomial-time algorithm, 577, 580
population, 34, 113, 171, 177, 180, 185, 234, 389, 425, 440, 534, 614, 649
population dynamics simulation, 114–159
postcondition, 332, 350
power set, 499
precision
    finite, 34, 55, 152, 348, 462
    unlimited, 33
precondition, 331, 350
predator-prey model, 180, 649
prefix code, 266
prime number, 201, 562, 673
`print()`, 47, 119
privacy, 411
private method, 674
PRNG, *see* pseudorandom number generator
probability distribution, 205
    chi-squared, *see* chi-squared distribution
    Gaussian, *see* normal distribution
    normal, *see* normal distribution
    uniform, *see* uniform distribution
probability mass function, 379
problem, 1, 2, 113, 323, 477, 621
processor, 20
profit, 95, 125, 139, 144, 403, 427, 508, 516
program, 5
programming languages, 5
Project Gutenberg, 250, 258
prokaryote, 299, 316
promoter, 299
prompt, 32
protein, 298, 305, 312, 587
pseudocode, 326, 520
pseudoplasmodium, 690
pseudorandom number generator, 186, 200
Pythagorean theorem, 656
Python shell, 32

quadratic formula, 53
quadratic growth, 169
quadratic Koch curve, 485
quadratic-time algorithm, 276, 313, 398, 400, 558, 567, 569, 576, 578
query, 49, 307, 543, 558, 581
queue, 596, 670
Queue ADT, 670

radians, 52
radiocarbon dating, 148, 158

raise, 667
RAM, *see* memory
random graph, 608
random module, 84, 186, 205, 704
    choice(), 706
    gauss(), 704
    random(), 187, 205, 704
    randrange(), 704
    sample(), 706
    seed(), 704
    shuffle(), 706
    uniform(), 205, 209, 658, 704
random number generator, 186
    true, 186, 234
random walk, 88, 186, 208, 214, 237, 475
    biased, 196
randomness, 185, 203
range(), 73, 122, 130
rank, 427, 432, 435, 641
rate
    birth, 158, 181
    clock, *see* clock rate
    death, 158, 181
    decay, 148
    growth, *see* growth rate
    infection, 154, 158
    interest, *see* interest rate
    product adoption, 177
    recovery, 153
    sampling, 102
    streamflow, 432
    unemployment, *see* unemployment rate
read(), 251
readline(), 307
recurrence relation, 145, 503, 505, 550, *see also* difference equation
    closed form, 507
recursion, 477–525, 549–552, 570–577
register, 21
regression analysis, 403
regression testing, 343, 350
relational database, 543
relational operators, *see* comparison operators
remainder, 35, 270, 359, 499, 668
repetition, *see* iteration
Resnick, Mitchell, 471
return, 97, 99, 218
return value, 46, 97

reversing a string, 668
Reynolds, Craig, 648, 682
RGB color model, 74, 461
RNA, 298
RNG, *see* random number generator
robot, 439, 512
rock-paper-scissors-lizard-Spock, 231
round(), 46

salary, 36, 381
sampling, 185, 234, 423
satellite data, 558, 581, 585
scalability, 274
scale-free networks, 606
Schelling model, 471, 649
Schelling, Thomas, 471
Schutt, Rachel, 421
scope, 80, 104
scoping rule, 80
Screen class, 72, 703
    bgcolor(), 703
    bgpic(), 703
    colormode(), 703
    exitonclick(), 72, 703
    mainloop(), 85, 703
    numinput(), 703
    onclick(), 84, 703
    setup(), 703
    setworldcoordinates(), 128, 132, 204, 439, 664, 703
    textinput(), 703
    tracer(), 204, 664, 703
    update(), 204, 703
search, 280, 541
    binary, *see* binary search algorithm
    binary search tree, 582
    breadth-first, *see* breadth-first search
    database, 543
    depth-first, *see* depth-first search
    hash table, 675, 687
    linear, *see* linear search algorithm
    open reading frames, 316
    table, 446
    web, 2, 541
secondary storage, 21
segregation, 471
selection sort algorithm, 553–562
self-documenting code, 95
self-similarity, 477, 488, 549, 586
Senate, 635, 644

sequential search, *see* linear search algorithm
Shakespeare, William, 477, 526
Shasha, Dennis, 313
shell, *see* Python shell
Shiflet, Angela, 171, 682
Shiflet, George, 171, 682
short circuit evaluation, 212, 493
side effect, 90, 332
Sierpinski carpet, 487
Sierpinski triangle, 486
Sieve of Eratosthenes, 562
sign and magnitude, 56
simulation, 113
    agent-based, *see* agent-based simulation
    cellular automaton, *see* cellular automaton
    Monte Carlo, *see* Monte Carlo simulation
    network, *see* network, simulation on a
    population dynamics, *see* population dynamics simulation
SIR model, 153, 614
SIS model, 158
slicing, 261, 354
small-world networks, 601
smoothing data, 13, 372, 423
snake_case, 95
Sneetches, 471
social network, 121, 186, 241, 588, 590, 612
sociology, 472, 612, 688
software, 5
software engineering, 322
sorting
    insertion sort, *see* insertion sort algorithm
    merge sort, *see* merge sort algorithm
    selection, *see* selection sort algorithm
    Timsort, *see* Timsort
source code, 5
sparse, 601, 651
special methods, 637–645, 707
    __add__(), 638, 707
    __contains__(), 680, 707
    __delitem__(), 652, 677, 707
    __eq__(), 640, 707
    __floordiv__(), 639, 707
    __ge__(), 640, 707
    __getitem__(), 642, 651, 677, 707
    __gt__(), 640, 707
    __init__(), 625, 626, 707
    __le__(), 640, 707
    __len__(), 676, 707
    __lt__(), 640, 641, 707
    __mul__(), 639, 707
    __ne__(), 640, 707
    __setitem__(), 642, 651, 677, 707
    __str__(), 637, 707
    __sub__(), 639, 707
    __truediv__(), 639, 707
sphere, 4, 38, 98
SQL, 543
stack, 526, 665–670
Stack ADT, 665
standard deviation, 205, 209, 237, 427, 532
start codon, 299, 316
statement, 5
statistics, descriptive, 352–359, 636
Stewart, Ian, 313
stop codon, 299, 316
str class, 242–313, 705
    concatenation operator +, 48, 242
    count(), 243, 705
    find(), 260, 705
    lower(), 290, 705
    repetition operator *, 48, 242
    replace(), 243, 705
    split(), 385, 705
    upper(), 242, 705
str(), 50
strategy, 230, 232, 615
streamflow, 431
string, 32, 47, 74, *see also* str class
string accumulator, 263, 302, 312
Strogatz, Steven, 611
subproblem, 324, 489, 491, 492, 501, 504, 508, 512, 549, 570, 586, 621
substring, 242
Super Diego, 68, 112
supercomputer, 22
swap, 11, 44, 102, 371, 440, 553, 559, 560, 569
swarm intelligence, 648
syntax, 40, 115, 120
syntax error, 40, 329
sys module
    setrecursionlimit(), 536

tab character, 243

tab-separated values, 389, 390, 416, 424, 425, 617, 635, 688
technology, 3, 152, 241, 579
tectonic plates, 384, 388
temperature, 7, 13, 30, 41, 101, 103, 220, 276, 373, 389, 422, 444, 473
test case, 344
tetrahedron, 51
threads, 21
three-dimensional, 298, 443, 475, 534, 683, 692
threshold, 182, 472, 535, 539, 691
time complexity, 17, 270
    asymptotic, *see* asymptotic time complexity
    best-case, 271, 567
    exponential, *see* exponential-time algorithm
    linear, *see* linear-time algorithm
    logarithmic, *see* logarithmic-time algorithm
    quadratic, *see* quadratic-time algorithm
    worst-case, 271, 567
Timsort, 570
tipping point, *see* phase transition
`tkinter` module, 465
token, 40
top-down design, 324
tortoise, 68
tour, 438
Tower of Hanoi, 500
Track ADT, 683
tracking, 7, 390, 683
tractable, 577
tradeoff, 33, 150, 195
transcription, 298
transistor, 537, 579
translation, 299
transportation, 683
traveling salesperson problem (TSP), 438
tree, 477
    binary search, *see* binary search tree
    Brownian, *see* Brownian tree
    file system, 250
    fractal, *see* fractal, tree
    Lindenmayer system, *see* Lindenmayer system, tree
    phone, *see* phone tree
    phylogenetic, 300

triangular numbers, 169
tribbles, 144
trigonometry, 52, 237, 475, 656, 701
truth table, 25
tuple, 365, 382, 403, 410, 439, 452, 461, 486, 534, 559, 622, 624, 629, 637, 650, 674, 680
    as dictionary key, 461, 650
Turing Award, 63, 112, 621, 682
Turing complete, 28
Turing machine, 27, 519
Turing, Alan, 27, 30, 63
Turtle ADT, 66
`Turtle` class, 68, 702
    `backward()`, 702
    `begin_fill()`, 74, 702
    `circle()`, 702
    `dot()`, 204, 702
    `down()`, 81, 702
    `end_fill()`, 74, 702
    `fillcolor()`, 74, 702
    `forward()`, 70, 702
    `getscreen()`, 72, 702
    `goto()`, 127, 702
    `heading()`, 71, 702
    `hideturtle()`, 702
    `home()`, 702
    `left()`, 70, 702
    `pencolor()`, 73, 702
    `pensize()`, 81, 702
    `position()`, 70, 702
    `right()`, 81, 702
    `setheading()`, 84, 658, 702
    `speed()`, 72, 702
    `up()`, 81, 702
    `xcor()`, 702
    `ycor()`, 702
turtle graphics, 66
`turtle` module, 68
two's complement, 56
type error, 261

unemployment rate, 361, 426
Unger, Ron, 471
Unicode, 264, 680
uniform distribution, 205
uniform resource locator, 254
unit testing, 340, 350
    `doctest` module, 344
    `unittest` module, 344

universal, *see* Turing complete
URL, *see* uniform resource locator
UTF-8, 252, 264

vaccination, 158, 614
value error, 52
vampire, 144
van Rossum, Guido, 31, 63
variable, 38
variance, 354, 357, 409, 532
vector, 369, 623, 655, 693
    unit, 659
velocity, 649, 655, 693
Viète's product, 167
virus, 153, 311, 614
voting, 313, 435
VPython, 475, 692

Wallis product, 166
`while` loop, 140–144, 225–233, 238, 256, 277, 310, 333, 490, 530, 546, 564, 600, 658, 670
whitespace characters, 243, 259, 263, 385
wildcard, 647
wind chill, 41, 46
Wirth, Niklaus, 621, 682
World ADT, 649
`write()`, 253

XML, 256

zebra, Burchell's, 390
zodiac, Chinese, 222
zombie, 144